**The Gardener's
Guide to
Prairie Plants**

The Gardener's Guide to Prairie Plants

**Neil Diboll
and Hilary Cox**

The University of Chicago Press
Chicago and London

The University of Chicago Press, Chicago 60637
The University of Chicago Press, Ltd., London
Published 2023
Printed in China

33 32 31 30 29 28 27 26 25 24   2 3 4 5 6

ISBN-13: 978-0-226-80593-1 (paper)
ISBN-13: 978-0-226-80609-9 (e-book)
DOI: https://doi.org/10.7208/chicago/9780226806099.001.0001

Library of Congress Cataloging-in-Publication Data

Names: Diboll, Neil, author. | Cox, Hilary, author.
Title: The gardener's guide to prairie plants / Neil Diboll and Hilary
    Cox.
Description: Chicago ; London : The University of Chicago Press,
    2023. | Includes bibliographical references and index.
Identifiers: LCCN 2022007567 | ISBN 9780226805931
    (paperback) | ISBN 9780226806099 (ebook)
Subjects: LCSH: Prairie plants. | Prairie gardening. | Prairie
    planting. | Prairie ecology.
Classification: LCC SB434.3 .D54 2023 | DDC 639.9/9 — dc23/
    eng/20220223
LC record available at https://lccn.loc.gov/2022007567

♾ This paper meets the requirements of ANSI/NISO Z39.48-1992
(Permanence of Paper).

This book is dedicated to our respective parents,
Mary and Wally Diboll
and
Marjorie and Paul Lee,
who inspired in us a love of the natural world in all
its wonder and supported us in following our dreams.

Lost on the Prairie

BY MICHAEL YANNY

Last week, I went to a prairie
A real live one,
     With towering grasses,
          taller than me
And oodles of different flowers,
Attracting mysterious insects,
Ones I'd never seen before.

This was my native landscape.
That's what I'd been told.

But I am native — I thought.
I was born right around here.
I don't know any of these plants.
I'm lost.

I kept walking and walking,
     observing the amazing diversity.
There wasn't any ragweed or dandelion.
Not even quackgrass.
I was flabbergasted.

I didn't know a single plant.
Except one.
     The Showy Goldenrod.
Now I was really lost.
I thought that was a weed.
That's what I'd been told.

Nothing familiar was in sight for
     as far as I could see.
Except for an ocean.
An ocean of Goldenrod.
It was magnificent!
Not a weed.

As the sun was setting in the western sky,
    the prairie seemed to glow.
The yellows competed with the sun.
The reds were on fire.
The tall grasses blushed in the shimmering breeze.
And I had an appointment at 7:30.
I was lost

A moment grabbed me.
I had a feeling of heritage.
The same feeling I have when I see
    a Red-Tailed Hawk gliding above
        or a monstrous 300 year old Bur Oak.
But this place and these plants
    I didn't know.
I reverted to feeling lost.

The darkness set down.
Couldn't see a thing
    except for the silhouettes of the nearest flowers.
The sky was huge.
The stars — better than any fireworks.
The insects were singing.
    I'd never heard anything like it.
I had to get going.
Had to get to my appointment.

I looked to my left.
Then to my right.
I had no idea of where to go.
I just didn't know.

I JUST DID NOT KNOW!

So I stayed.

Impressions of the Greene Prairie at the
University of Wisconsin–Madison Arboretum
July 14, 1991

Contents

Acknowledgments

The authors would like to collectively thank Michael Yanny of JN Plant Selections for the use of his poem "Lost on the Prairie." Michael is a renowned woody plant propagator who writes plant poetry in his spare moments. He can be reached at jnplant selections@gmail.com.

We would like to extend our special gratitude to Dr. John Kartesz of the Biota of North America Project (BONAP) for the use of his outstanding plant range maps, which are included with each plant profile in chapter 5. These maps are the result of his painstaking, decades-long work to document confirmed records of plant occurrences in the United States. His website at www.bonap.org is a phenomenal resource and deserves the support of all native plant enthusiasts so he can continue his essential work, from which we all benefit.

Thanks are due to Dr. Daryl Smith of the Tallgrass Prairie Center at the University of Northern Iowa for use of the detailed North American tallgrass prairie map in chapter 2.

Filmmaker and author Catherine Zimmerman of the Meadow Project kindly granted permission to use the prairie versus lawn costs in tables 12.29 and 12.30 that she and Neil developed collaboratively for her previously published book *Urban and Suburban Meadows: Bringing Meadowscaping to Big and Small Spaces*. Neil updated the cost figures to reflect 2022 pricing. Catherine can be contacted at themeadowproject414 @gmail.com.

Thanks are due to Dr. David Atwell for his ingenious four-step mowing program for controlling Canada Goldenrod infestations in prairies without the use of herbicides, and to Gary Eldred, founder of the Prairie Enthusiasts, who kindly provided his insights on controlling taller invasive weeds such as Canada Goldenrod and Canada Thistle through the carefully targeted application of broadleaf herbicides on bundled groups of plant stems in late summer.

Our deep gratitude goes to Dr. Doug Tallamy for his practical advice and unwavering support.

NEIL DIBOLL'S ACKNOWLEDGMENTS

First and foremost, Neil would like to thank his former professor Dr. Keith White, whose class "Vegetation Management" in 1977 inspired Neil to investigate the young prairie seedings on the fledgling Cofrin Arboretum at the University of Wisconsin–Green Bay. This opened Neil's eyes to the incredibly complex plant community known as the North American prairie. He was hooked, and the restoration of prairies and related ecosystems became his passion and life's work. Without the inspiration and influence of Dr. White, this book would never have come into existence.

Neil is also deeply indebted to J. R. "Bob" Smith, who founded Prairie Nursery in 1972 when he sold his first prairie plants, which he had been cultivating as a hobby in his backyard garden. Bob was a leader in bringing native prairie plants to the marketplace back when they were widely considered to be weeds. Bob helped Neil learn the ropes as he took over the reins at Prairie Nursery in 1982 following Bob's retirement.

Dale Hendricks of Green Light Plants in Landenberg, PA, generously shared his experience with and insights into propagating prairie plants by cuttings. Dale can be reached at greenlightplants.com.

Neil's brilliant and talented niece Kate Castronovo donated many hours of her time to Photoshop many of the images in the book despite her busy schedule. Julia Roberts similarly contributed her Photoshop skills to adjust and improve many of our images by removing distracting background material and other extraneous imagery.

HILARY COX'S ACKNOWLEDGMENTS

Our shared gratitude goes to so many friends, colleagues, and especially family members for their encouragement and support through the 22 years of gestation it took to bring this book into the world. My personal thanks go to:

- My long-suffering children, Martin and Hazel, for their patience and technical expertise.
- My sister-in-law Judy, who provided Neil with lodgings in the foothills of Tucson for weeks at a time every winter, sometimes twice, as we edited pictures and text — and who became his "partner-in-crime" as they cruised the Tucson Gem Show and the Fourth Avenue Street Fair.
- Neil's partner, Reen, who put me up and took care of me in spring and summer visits, even doing my laundry and providing me with my vodka martini (or "Reenie Tini" as I called it) at the end of each marathon day.
- Dee Ann Peine, my botanizing friend, who drove her (pink) jeep on our botanizing jaunts on her farm and was able to identify what I saw as dead twigs as various terrestrial orchids while driving. Our adventures at a barrens in Kentucky have been documented (see the INPS [INPAWS] Journal). Our seed-collecting for the Millennium Seed Bank at the farm and the barrens was an experience beyond words.
- Rich Peine, who gave me the mandate in 1994 to "attract every form of wildlife possible" using prairie plants at his heating and air-conditioning company, Peine Engineering, which was situated by the Indiana Central Canal in downtown Indianapolis. The site had been a coal yard previously and became my first bioremediation of a brownfield.
- C. Colston (Cole) Burrell, for his constant friendship and for being my fount of wisdom.
- Anni and Prof. William (Bill) C. Clark, who have rescued me more than once and in many ways since we met at the International Institute for Applied Systems Analysis in Austria in the early 1970s.
- Wayne and Mary in Kentucky, who allowed Dee Ann and myself free range to explore their farm, and whose management of the property had ensured the survival of a barrens plant community that included rare and endangered species.
- Sarah and Nick Gray, who have put me up for months at a time at their beautiful house with stunning gardens on my return trips to Indiana.
- Our friends and colleagues in our various fields, who have aided and abetted us along the way and are now doing a lovely job of promoting our book.

1

How to Use
This Book

As with many things in life, the inspiration for this book came of necessity. Hilary's clients were having a problem identifying their native plants as they emerged in spring, mistaking them for weeds and often pulling them out of the ground. To preserve the integrity of her garden designs, she determined to produce a photographic field guide for identifying native herbaceous plants as they appeared in spring. Thus, out of frustration, this book was born.

Hilary then contacted Neil and asked him to collaborate on the book, probably because she thought him an easy mark. (Actually, she realized he had all the plants at his nursery). Neil's first response was "You want to do what? Now I know you've gone completely bonkers!"

So naturally, Neil agreed to the book.

The next spring we were pursuing the elusive prairie flowers and grasses, crawling around on cold, wet ground to get close-up shots. It was not long before we realized that gardeners also need to identify plants by their leaves, prior to bloom. One thing led to another, and soon we knew we had to include photos of the flowers, as well as of the complete plant. And gardeners really need to know what the seedheads look like, especially for those who wish to harvest their own seed. By the way, you know one of the questions I get all the time? "What do the seedlings look like?" That settled it. We were going to create a book that documented each species from birth to bloom to seed.

Chapters on prairie history and ecology, and establishing and managing prairie gardens and meadows were added. We added tables in the back of the book which provide detailed information on each plant's characteristics, growing conditions, and the wildlife it attracts. In this book we have endeavored to include as much pertinent information about prairies and prairie gardening as possible, short of turning it into a doorstop. We hope you find it helpful in solving the ever-present question: "What the heck is that plant?"

This book is a field guide to plant characteristics of 148 selected grasses, sedges, and flowers of the North American tallgrass prairie commonly used in gardens and prairie restorations. Hundreds of species occur in tallgrass prairies, not to mention others found in eastern meadows and shortgrass prairies of the Great Plains to the west. It is impossible to include them all. We have focused on commonly used plants, plus a few lesser-known ones.

Chapter 2 documents the history and ecology of the prairie. Chapters 3 and 4 cover soils and designing, planting, and maintaining prairie gardens using transplants. Chapter 5 is a field guide and contains descriptions and photos of plants at each life stage, from seedling to seedhead. Detailed cultural information and garden uses for each species are provided. Chapters 6 and 7 provide detailed information on prairie restoration and management on larger acreages using seeds instead of transplants. Chapters 8 and 9 share tried-and-true techniques for propagating native prairie plants, both from seed and by vegetative means. Chapter 10 focuses on the prairie food web and the wildlife that commonly inhabits and visits prairie gardens and meadows. Chapter 12 is a compendium of tables containing information about plant characteristics, growing requirements, plant uses, attracting wildlife, and so on.

Throughout this book, plants are grouped by family. The families are grouped by botanical classification of monocots, or grasses, sedges, and grasslike flowers; and dicots, or broadleaf flowers. Within families, species are listed in alphabetical order by genus in their scientific name. Table 1.1 lists the species in order by total number of species per family, from most to least.

Plant Nomenclature

The binomial plant classification system is based on the 1753 publication of *Species plantarum* by Swedish botanist Carl von Linné (1707–1778), commonly referred to as Linnaeus. He based his standardized naming system on Latin and Greek, so that people around the world could then apply a

TABLE 1.1.   PRAIRIE PLANTS

	Family	Genus	Number of species
Monocots	Grasses	*Poaceae*	18
	Sedge	*Cyperaceae*	1
Grasslike flowers	Lily	*Liliaceae*	4
	Iris	*Iridaceae*	4
	Spiderwort	*Commelinaceae*	3
Dicots (broad-leaved flowers)	Aster	*Asteraceae*	45
	Legume	*Fabaceae*	14
	Figwort	*Scrophulariaceae*	8
	Mint	*Lamiaceae*	7
	Rose	*Rosaceae*	6
	Buttercup	*Ranunculaceae*	5
	Milkweed	*Asclepiadaceae*	4
	Borage	*Boraginaceae*	4
	Mallow	*Malvaceae*	3
	Carrot	*Apiaceae*	3
	Gentian	*Gentianaceae*	2
	Vervain	*Verbenaceae*	2
	Bellflower	*Campanulaceae*	2
	Pink	*Caryophyllaceae*	1
	Spurge	*Euphorbiaceae*	1
	St. Johnswort	*Clusiaceae*	1
	Buckthorn	*Rhamnaceae*	1
	Violet	*Violaceae*	1
	Evening Primrose	*Onagraceae*	1
	Primrose	*Primulaceae*	1
	Dogbane	*Apocynaceae*	1
	Phlox	*Polemoniaceae*	1
	Acanthus	*Acanthaceae*	1

single, reliable scientific name to every plant, preventing confusion among various common names in use in every language.

Today's botanists focus on evolutionary relationships between plants and utilize relatively new tools, such as DNA, to determine species classifications and names. This has led to significant turmoil in what was previously a fairly stable system. Many plant species have been reclassified as new information about their genetic constitution has become available. People who aren't botanists often struggle with scientific names, and recent changes have served only to exacerbate the situation. For example, most members of the genus *Aster* have been reclassified as *Symphyotrichum*, *Oligoneuron*, *Eurybia*, and *Doellingeria*. A common response among native-plant enthusiasts and professionals alike is "I'll use the common name, because they'll just change the scientific name again anyway."

Because this book is intended for use by both gardeners and professionals, scientific names are those currently in use at time of publication. One most recent scientific name is listed in parentheses. In chapter 5 up to three scientific and common names for each species are listed. All current and previous names are listed in the index.

Biota of North America Project (BONAP) Plant Range Map Color Key for Chapter 5

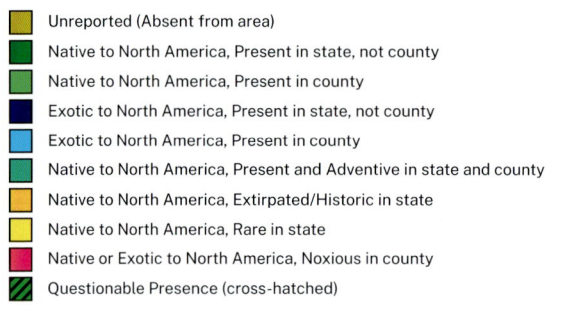

Unreported (Absent from area)

Native to North America, Present in state, not county

Native to North America, Present in county

Exotic to North America, Present in state, not county

Exotic to North America, Present in county

Native to North America, Present and Adventive in state and county

Native to North America, Extirpated/Historic in state

Native to North America, Rare in state

Native or Exotic to North America, Noxious in county

Questionable Presence (cross-hatched)

The Biota of North America Project (BONAP) has created plant range maps that indicate nativity at both state and county levels for each plant species. These maps are present for each plant species in chapter 5. If a county is dark green, the plant is known to have been native to the respective US state or Canadian province at the time of European settlement. If a county within a state is bright green (against a dark green state color), the plant is known to have been originally native to that particular county. If a county is teal, the plant is adventive, meaning that it was not recorded as originally native but has since either migrated to or been planted in the county or has spread from planted locations. Each county will have a single color designation in a state where the species occurs:

- Bright green — Native at the time of European settlement
- Teal — Adventive, not known to be native at the time of European settlement
- Yellow — Native, but only occurs rarely
- Orange — Historically native, but has since been extirpated in the state
- Pink — Considered a noxious weed by state authorities

These maps have been reproduced with grateful acknowledgement to Dr. John Kartesz of the Biota of North America Project (BONAP). Unless otherwise noted, credit for all maps in this book belongs to J. T. Kartesz, The Biota of North America Program (BONAP), *North American Plant Atlas* (http://bonap.net/napa) (Chapel Hill, NC, 2015) (maps generated from J. T. Kartesz, Floristic Synthesis of North America, Version 1.0, Biota of North America Program [BONAP], 2015 [in press]).

2

History and Ecology of the Prairie

Origin of the Prairie

Like the steppes of Asia and savannas of Africa, the North American prairie is the child of a continental climate. Great regions in centers of continents experience temperature and rainfall extremes. Because of their distance from major water bodies, they do not benefit from the thermal buffering that seas and large lakes provide and are typified by hot summers and frequent, extended droughts. The Midwest today is still subject to extended periods of drought and blistering summer heat.

The Rocky Mountains arose approximately 55 million to 80 million years ago. Air masses that flowed across the continent from west to east lost much of their moisture as they rose and cooled upon encountering the mountains. As air descended from the eastern Rockies onto the Great Plains it decompressed, further reducing relative humidity. As a result, a "rain shadow" developed to approximately three hundred miles directly east of the Rockies. This area, now known as the Great Plains, received limited precipitation and later became home to the shortgrass prairie.

As the air mass moved eastward into what is today the central Midwest, it picked up moist air moving northward from the Gulf of Mexico, resulting in a higher average rainfall than in the rain shadow. These conditions favored the growth of herbaceous grasses and flowers over trees and shrubs, which are generally more sensitive to heat and drought, supporting the lush plant growth of the tallgrass prairie; and the North American grasslands reached their floriferous zenith in this region. Fires were common in fall and early spring when grasses were flammable. Woody plants that might otherwise have survived in this climate were often consumed by raging fires, set by lightning or people, which swept across the prairie.

It is believed that the North American prairie and other grasslands across the planet expanded greatly during the Miocene epoch (5.3 million to 23 million years ago) of the late Cenozoic era. This period was characterized by a cooler, drier climate due to less moisture being driven to northern latitudes, which favored more drought-tolerant grasslands over what had once been tropical forests.

Mammals began to appear beginning in the Eocene epoch (34 million to 56 million years ago). Grasslands had expanded rapidly during the Miocene epoch, and cloven-hoofed grazers (ungulates) appeared there around 20 million years ago. Present-day grazers of the North American prairie, including bison, elk, and antelope, are believed to have arrived from Asia over the Bering land bridge during the Pleistocene Ice Age, some 15,000 to 35,000 years ago. These herds of grazing animals had a significant impact on plants of the North American prairie. The survival strategy of most prairie grasses was to maintain their growing points (where new leaf growth is initiated) at or just below soil surface, where they would not be eaten. This allowed them to recover rapidly following periods of intensive grazing. Most grazers prefer grasses over flowers, which resulted in the massive flower displays many early European explorers noted. Today's prairie remnants consist primarily of grasses, likely due to a dearth of grazers.

Prairie versus Meadow

When French explorers first encountered Midwestern prairies in the seventeenth century, they named them for the only similar landscape they knew. The word *prairie*, French for "meadow," has since become synonymous with the grasslands of the central and western United States and Canada. Over time, the term *meadow* has come to denote grasslands of the eastern United States that are composed primarily of lower-growing cool-season grasses. The term *prairie* now applies to the large expanses of primarily warm-season tallgrass prairies of the Midwest and the shortgrass prairies of the Great Plains. The North American prairie is split up into three distinct biological communities or biomes.

Midwestern Tallgrass Prairie

The Midwestern tallgrass prairie receives 25–40 inches of precipitation per year. Tallgrass prairies once covered millions of acres, including most of Iowa, Illinois, and northern Missouri, as well as large sections of other

Midwestern states. Once its fertility was recognized, and subsequently its agricultural value, it was converted to cropland as fast as settlers could plow it under. Today, less than 0.001% of native tallgrass prairie remains. It is recognized as one of the rarest plant communities on the planet.

Great Plains Shortgrass Prairie

The Great Plains shortgrass prairie receives 10–15 inches of precipitation per year. The shortgrass prairie occurred in eastern Colorado and Wyoming, western Kansas, Nebraska, Oklahoma, Texas, and the Dakotas — it exists today but in such a reduced state that only remnants survive. The rain shadow of the Rocky Mountains, combined with low relative humidity and strong, drying winds, created a harsh growing environment. This plant community is dominated by low-growing flowers and grasses, averaging from a few inches to 3 feet in height. Taller plants are more subject to desiccation and are uncommon in this landscape.

Mixed-Grass Transitional Prairie Zone

The mixed-grass transitional prairie zone receives 15–25 inches of precipitation per year. A transition zone between tallgrass and shortgrass prairies occurs in central Kansas, central Oklahoma, central Nebraska, and the eastern Dakotas. Members of the tallgrass prairie community tend to occur in lower, moister areas in this region, whereas shortgrass prairie species occupy drier uplands.

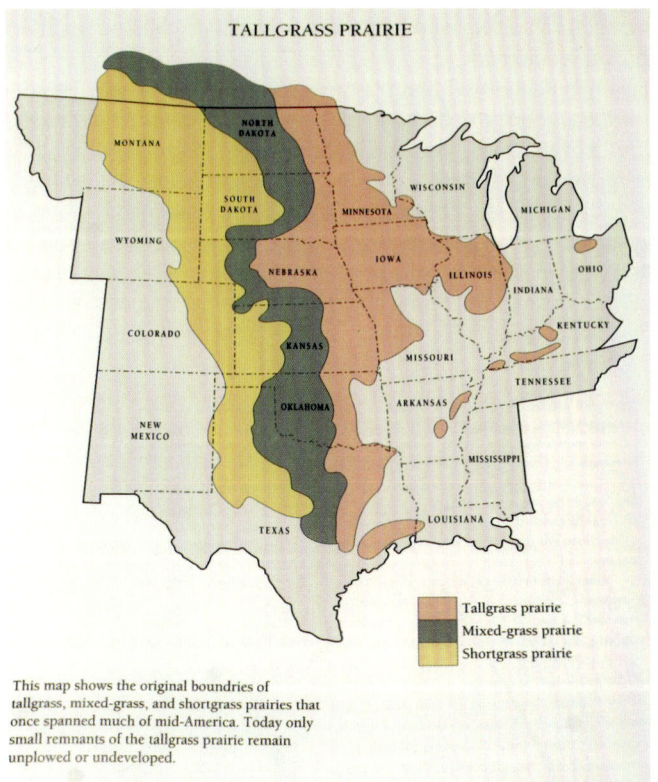

Range of presettlement North American prairie

This map shows the original boundries of tallgrass, mixed-grass, and shortgrass prairies that once spanned much of mid-America. Today only small remnants of the tallgrass prairie remain unplowed or undeveloped.

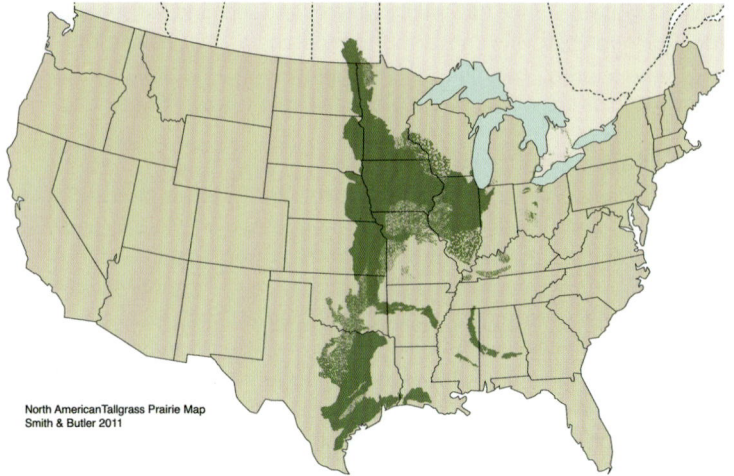

Detailed range of presettlement eastern tallgrass prairie. Photo courtesy of Daryl Smith.

Early explorers reported that prairie lands were unfit for cultivation of crops, since it was "too poor to support the growth of trees." Until then, agricultural activities of Europeans and Americans in the New World had been limited to forested lands of the east. These prairies were strange and new to white settlers of the early nineteenth century. Those few who were foolhardy enough to attempt to break the dense prairie sod with their wooden and cast-iron plows were rewarded only with exhausted teams of oxen and broken equipment.

Eventually people learned that prairie soils, once broken up, yielded a bounty of crops without the addition of fertilizer. Yet the prairie resisted the best efforts of these early sod busters for want of a proper plow, until John Deere arrived on the scene. In 1836, Deere set to work in his Illinois blacksmith shop to devise a plow that could break prairie sod. It had to shed sticky Illinois clay that fouled other plows. In 1837 he developed his first polished steel moldboard plow that soon gained fame across the prairie. By 1855 he was manufacturing 10,000 plows a year. Millions of acres of prairie, which had taken millennia to develop, were plowed up and converted to cropland within a single generation.

This book concerns itself with the plants of the tallgrass prairie which, prior to settlement, covered 170 million acres. Rich prairie soils that had accumulated over thousands of years were infused with the organic matter of generations of prairie grasses and flowers. Approximately 70% of the average prairie plant's living matter resides underground in its roots, and only about 30% is evident aboveground. Prairie grasses shed roughly 30% of their fine feeder roots at the end of every growing season, adding to the soil's organic storehouse. It is not unusual for virgin prairie soils to contain organic matter in excess of 10%, with a black topsoil layer extending 3 feet deep or more.

Many of the showiest flowers and grasses that occur in the tallgrass prairie make excellent garden plants. They require no fertilizers; few, if any, pesticides; and no irrigation once established. The great majority are adapted to growing in climates that receive between 20 and 40 inches of precipitation per year, 60% of which occurs as rain during the growing season. Within this climate zone is a wide variety of soil types: dry prairies on sandy, gravelly and rocky soils, medium or "mesic" prairies on well-drained loam and clay-based soils, and wet prairies in low areas close to the water table.

Modern Post-Glacial History

Glacial ice sheets covered most of Canada and the northern half of the United States during the Pleistocene, between 10,000 and 1.8 million years ago. The last ice sheets retreated from the Upper Midwest approximately 12,000 years ago. This resulted in massive climate change and a completely rearranged landscape. The resulting ecological disruption created a vacuum that was filled by many new plants and animals.

During this period plants were pushed southward into non-glaciated areas. When the glaciers receded, many plants moved northward to colonize the fresh soil created by the grinding of rock into sand, silt, and clay. The rich soils of the American Midwest are a legacy of the glaciers' work in conjunction with the organic matter created by prairie plants over thousands of years.

During the Xerothermic period — 3,500 to 7,000 years ago — prairies apparently expanded all the way to the eastern coast of North America.

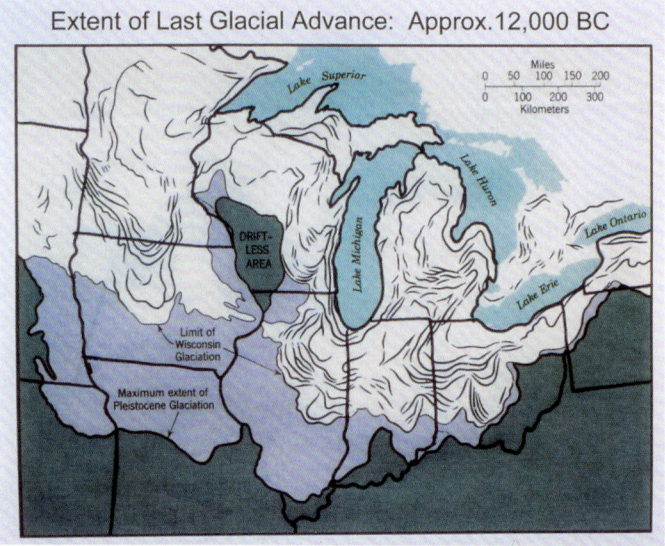

Extent of Last Glacial Advance: Approx. 12,000 BC

Prairielike grassland remnants in Pennsylvania, Virginia, New Jersey, Connecticut, and even Long Island indicate that the reach of the prairie biome was once far greater than it is today. Pockets of southern prairie with strong floristic relationships to tallgrass prairies of the Midwest also occurred in Texas, Louisiana, Arkansas, Kentucky, and Alabama. Many prairie plant species also occur in forests and woodlands. Most prairie grasses are believed to have evolved in eastern forest openings, southwestern deserts, and mountain meadows. Prairies contain relatively few endemic plant species. There are also few endemic prairie insect and bird species. Only 5.3% of North American bird species apparently evolved on prairies. Endemic species need longer to evolve, suggesting prairies are a relatively young ecosystem.

Climate + Fire = Prairie

East of the hundredth meridian in central Kansas and Nebraska there is sufficient precipitation to favor forests over prairies. This begs the question: Why did vast expanses of prairie extend across the Midwest and eastward?

Many theories have been suggested as to the origin of the prairie: soil type, climate, grazing, and fire. Studies have shown a strong correlation between the continental climate and distribution of prairies. Dry sandy and rocky soils tend to favor herbaceous plants over woody plants. Summer heat and drought often cause tree mortality and favor grasslands. Winter browsing of woody plants by ungulates may have kept woody plants at bay. All of these may be contributing factors; but no individual theory sufficiently explains the prairie's advance so far east into the realm of what should have been deciduous forests or why prairies came to dominate much of the Midwest.

Humans who lived on the prairie were supported by large grazing animals that provided hundreds of pounds of meat per kill. Forests, frequented by small game, yielded far less food per animal. Even deer, which were much less common in presettlement times than they are today, are an "edge species" that thrived at the transition between grassland and woodland. It is now known that Native Americans regularly burned prairies as an economic land management tool. Prairies burned in fall collected less snow in dead vegetation, making winter travel easier. These areas greened up earlier in spring, attracting bison and other herbivores, thereby improving post-winter hunting success. To a large degree, human activities assisted the prairie in expanding and maintaining its range in North America.

When white European settlers arrived, one of their first priorities was to halt the prairie fires they feared so much. In many prairies, suppressed tree "grubs" of oak, hickory, sumac, and other savanna species were no longer burned back to the ground. Forest soon asserted itself, and within a single generation young woodlands appeared where prairies had recently ruled.

Ecological Structure of Prairies

Prairies are grassland ecosystems dominated by a few species of highly adaptable grasses, as well as a wide variety of flowers and some less common grasses. Eastern tallgrass prairies support more than 300 species of herbaceous plants. Although grasses account for the highest percentage of total plant cover in prairies, there are many more flower species than grasses.

Peak flowering in dry prairies occurs in spring and fall, thereby avoiding higher moisture loss during hot summer months. Mesic prairies peak in midsummer, and wet prairies are most floriferous in late summer after their damp soils warm up. White flowers are common in spring and early summer, while yellows dominate in late summer and early fall. Purples, blues and pinks are also common; red and orange are rare.

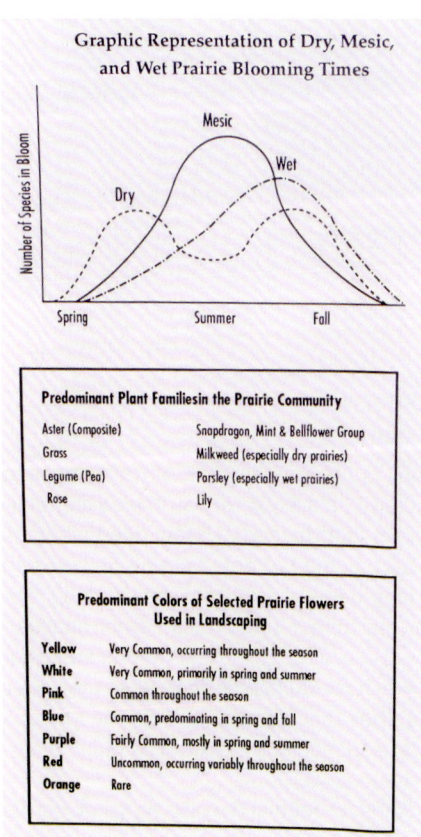

Graphic Representation of Dry, Mesic, and Wet Prairie Blooming Times

Number of Species in Bloom

Mesic

Wet

Dry

Spring   Summer   Fall

Predominant Plant Families in the Prairie Community

Aster (Composite)	Snapdragon, Mint & Bellflower Group
Grass	Milkweed (especially dry prairies)
Legume (Pea)	Parsley (especially wet prairies)
Rose	Lily

Predominant Colors of Selected Prairie Flowers Used in Landscaping

Yellow	Very Common, occurring throughout the season
White	Very Common, primarily in spring and summer
Pink	Common throughout the season
Blue	Common, predominating in spring and fall
Purple	Fairly Common, mostly in spring and summer
Red	Uncommon, occurring variably throughout the season
Orange	Rare

Ecological Adaptations of Prairie Plants

To withstand an often brutal climate, as well as grazing and trampling by herds of herbivores, prairie plants have devised a number of adaptations over millennia to ensure survival.

Root Systems

| A. Prairie Dropseed | C. Big Bluestem | E. Little Bluestem | G. Indiangrass | I. White False Indigo |
| B. Prairie Dock | D. Purple Coneflower | F. Black-eyed Susan | H. Showy Sunflower | J. Prairie Cordgrass |

Root systems of selected prairie flowers and grasses

The typical prairie plant dedicates approximately 70% of its biomass to root systems. Most woody plants have an average of half their living matter below ground. Short prairie-grass roots extend 2–5 feet deep, whereas those of tall grasses reach 5–8 feet. Some taprooted prairie flowers can extend 15 feet or deeper. It has long been believed that these deep root systems provide access to subsoil moisture during drought, when shallow-rooted plants may suffer. However, more recent research indicates that the majority of soil moisture uptake in prairie grasses occurs in the upper 2 feet of soil, where the majority of their root biomass resides. The function of the deeper roots is now being questioned and explored by plant ecologists studying the topic.

Roots also serve as storage organs for food and moisture. Large corms (bulblike roots) of Blazing Star (*Liatris* spp.) allow them to survive extended dry periods. The massive taproots of members of the genera *Baptisia, Ceanothus,* and *Silphium* can reach diameters of 3 inches or more. Rhizomes, such as the roots of Sunflowers (*Helianthus*), make admirable repositories for food and water, and also serve as a means of expanding territory.

Leaves and Shoots

Leaves have a number of adaptations that help reduce internal leaf temperature and moisture loss. Some prairie plants, such as Rattlesnake Master (*Eryngium yuccifolium*), have a thick, waxy cuticle layer on their leaf surfaces that prevents water loss. Leaves of other species are covered with dense white hairs that reflect sunlight and help ameliorate leaf temperatures on hot, sunny days. Hairs also create a "boundary layer" around the leaf, reducing wind speed and thus water loss, especially on windy days with low relative humidity.

The Compassplant (*Silphium laciniatum*) orients its leaves in a north-south direction to minimize the sunlight that strikes them during midday, when the sun is usually brightest.

Silvery leaves of Pussytoes reflect sunlight

Hairy leaves of Downy Sunflower create a protective "boundary layer"

Leaves of Compassplant orient north-south to minimize midday sun exposure

During drought, leaves of many prairie grasses turn from deep green to powder blue and curl inward to minimize exposure to sun and wind. During severe drought, leaves of Little Bluestem (*Schizachyrium scoparium*) and Sideoats Grama (*Bouteloua curtipendula*) become so dry that they literally crunch underfoot. They may appear to be dying, but are merely going temporarily dormant. With the first significant rainfall they snap right back, regain their green color literally overnight, unfold their leaves, and return to business as usual.

The prairie ecologist John E. Weaver, of the University of Nebraska, studied the response of prairie plants in Nebraska and surrounding states during the great drought of the dust bowl years in the 1930s. He documented plant occurrences on his study plots annually, noting those species that disappeared during the drought. Amazingly, when rains returned in the late 1930s he found Big Bluestem (*Andropogon gerardii*) and Purple Prairie Clover (*Dalea purpurea*) resprouting from roots that had been dormant for seven years.

Some plants protect themselves from grazing with furry leaves that are unpalatable to most herbivores. Members of the Mint family such as Lavender Hyssop (*Agastache foeniculum*), Bergamot (*Monarda fistulosa*), and Dotted Mint (*Monarda punctata*) produce powerfully scented oils in their leaves and stems that deter grazers. Many of the Milkweed family (*Asclepias* spp.) and Flowering Spurge (*Euphorbia corollata*) produce milky sap that few animals find appealing. Most prairie grasses protect themselves by keeping their growing points at or just below soil surface, enabling them to regrow quickly after periods of intense grazing.

Plant Size, Seasonal Activity, and Bloom Times

The average height of prairie plants that grow in dry soils is less than the average height of those of medium (mesic) or moist soils. A shorter stature has two primary advantages: reduction in total leaf material exposed to the sun reduces moisture loss, and desiccation from wind is less because of the lower wind speeds closer to the ground.

Summer soil temperatures are typically higher on exposed sandy soils than on medium or wet soils. The lower water-holding capacity of sandy and rocky soils results in a lack of thermal buffering. Soils that hold more water change temperature gradually, since water absorbs and holds large amounts of energy. Dense vegetative cover on mesic and moist soils is greater, further reducing soil temperatures compared to dry soils. Surface temperatures on sandy soils in midsummer can exceed 130 degrees Fahrenheit, so plants endemic to dry prairies have adapted to survive these extremes.

Dry prairies have a higher percentage of early-spring wildflowers and more cool-season grasses than do mesic or moist prairies. Many spring-blooming flowers will produce seed by June or July and essentially go dormant by midsummer. New leaves often appear in late summer or early fall to take advantage of cooler growing conditions to supplement their re-

Wilted Ironweed under midday heat

Wilted Ironweed after rain later that day

Ironweed blooming later in the season

serves for the following year. Some plants completely disappear for the summer, such as Prairie Larkspur (*Delphinium virescens*), only to reappear in fall.

Spiderworts (*Tradescantia* spp.) bloom only in the morning on sunny days in order to conserve precious moisture but may reopen late in the day when temperatures are more moderate. Yellow Coneflower (*Ratibida pinnata*) and Ironweed (*Vernonia* spp.) often wilt in the heat of a hot summer afternoon, causing concern to gardeners. Their leaves will revive after rain or in the cool of night and regain their vitality by morning.

Carbon Dioxide Capture

Warm-season prairie grasses have a significant adaptive advantage for handling hot, dry weather. Plants must open stomata on the undersides of their leaves in order to obtain carbon dioxide from the atmosphere for photosynthesis. Stomata are small apertures that control air and moisture exchange between the interior of the leaf and outside air. Every time stomata open to expel oxygen (a waste product) and take in carbon dioxide, water is lost, especially under conditions of high temperatures and low relative humidity.

Plants have specific molecules that bond with carbon dioxide to extract it from the air and move it into the leaf, where it can be converted into plant sugars. Ninety-five percent of plants (including rice, wheat, oats, and legumes) possess a rather inefficient carbon dioxide capture system which is referred to as a C-3 pathway. C-3 plants, as they are called, lose 97% of the water supplied by their roots in the process of carbon dioxide capture. During hot, sunny weather they are limited as to how long they can keep their stomata open without wilting. As a result, their rate of growth is limited.

A carbon capture system called the C-4 pathway appeared in plants around 25 million to 32 million years ago, during the Oligocene era. This was more efficient and resulted in significantly less water loss from the leaf. C-4 plants have specific cell structures in their stomata that concentrate carbon dioxide in the presence of carbon receptor molecules and enzymes that facilitate the process. C-4 plants are able to continue photosynthesis under hot, dry conditions, making them more productive and competitive in sunny grassland ecosystems. In addition to warm-season prairie grasses and many tropical grasses, some well-known C-4 plants include corn, sorghum, and sugarcane.

Carbon Sequestration in Prairies

The extensive root systems of prairie grasses and flowers incorporate organic matter deeply into the subsoil. The average prairie grass abandons approximately one-third of its root mass (mostly feeder roots) at the end of every growing season. This has created soils rich in organic matter over thousands of years. Some prairie soils in Iowa, for example, have rich black topsoil up to 6 feet deep through the accumulation of root matter; whereas Midwestern forest soils generally have an organic-rich topsoil layer of only 6–12 inches.

About half the biomass of trees resides in the roots, compared to 70% for prairie plants. Standing mature trees contain more carbon in their tissues than do prairie plants per square mile. However, when dead trees decompose, most of their carbon is released into the atmosphere, with less retained in the soil, in contrast to deeper-rooted prairie plants. Over the long term, the total amount of sequestered carbon should theoretically be greater in organic-rich prairie soils than in forest soils. Consequently, reestablishing native grasslands could serve as a mitigating factor for rising carbon dioxide levels in the atmosphere.

Modern Prairie Restoration

The first true prairie restoration was undertaken at the University of Wisconsin Arboretum in Madison in the 1930s. The renowned ecologist Aldo Leopold and his associates came up with the novel idea of restoring a native prairie. Leopold recognized that the Midwestern tallgrass prairie was nearing extirpation. Restoring prairie at the arboretum was a truly radical concept, as arboreta of the time typically highlighted plants from other parts of the world.

At first Leopold and his colleagues painstakingly transplanted large pieces of sod taken from prairie remnants that were under development. This was moderately successful on a small scale but not feasible for creating large, multiacre restorations. They began collecting seeds of various prairie species and sowing them in former agricultural fields. These plots tended to yield tremendous competition from cool-season grasses and weeds, especially Bluegrass and Quackgrass.

The ecologist John T. Curtis, of the University of Wisconsin, knew that prairie fires were common in presettlement times and wondered whether perhaps the power of fire could be harnessed to favor prairie plants over nonnative weeds. In the 1940s he experimented with burning prairies at various times of year to determine the effects. His conclusion was that burning in mid- to late spring suppressed nonnative cool-season weeds and grasses by removing new leaf growth, thus weakening them. Soil also warmed up rapidly after a spring burn, greatly favoring growth of predominantly warm-season prairie plants over nonnative cool-season competitors. The effectiveness of mid- to late-spring burning in controlling cool-season perennial weeds was extraordinary. Controlled burning officially entered the tool kit of prairie restorationists following publication of this research in 1948.[1]

Other universities soon followed Wisconsin's lead and began experimenting with restoring prairies. Basic methods of prairie establishment using seeds were developed and refined over the years. Thanks to these early prairie-planting pioneers, reliable procedures now exist for establishing prairies on a large scale.

3

Understanding
Your Soil

A major factor in success with growing native plants and propagating seeds is their compatibility with your soil. Every species grows within a certain range of soil types. Some thrive in a broad range of soil conditions while others grow only in specific soils. It is very important to choose your seeds and plants to match your soil.

Soil Types

Soils can be divided into three basic classifications: sand, loam, and clay. There are gradations within these groups. However, for the purpose of describing where a given plant will grow best, these categories suffice.

Sandy soils, referred to as "light" soils, consist of large-sized particles that are loose and easy to work. They allow water to drain readily and tend to be low in nutrients. Sandy soils are often more acid than more fertile loams and clays, especially when derived from granitic rock. Certain plants grow only in extremely well-drained sandy soil and will not tolerate any excess moisture. Soils with a history of conifer and oak growth tend to have a low pH, due to the acidic nature of needles and oak leaves. If your soil has a pH lower than 5.5, consider adding lime or wood ash to bring it up to a pH between 6.0 and 7.0.

Clay soils, commonly known as "heavy" soils, consist of small, tightly packed soil particles. Clays tend to be dense and hard to work. However, they are generally rich in nutrients, have a high water-holding capacity, and can be very productive.

Loamy soils are intermediate between sand and clay. A mix of sand, silt, and clay, they combine fertility and moisture-holding capacity with good drainage. Easier to work than clays and better consolidated than sands, loamy soils make an excellent medium for growing most plants. Many prairie plants do best in loam soils.

Determining Your Soil Type

The "feel test" can help determine soil types. Take just enough moist soil to rub between thumb and fingers. As the soil dries between your fingers, rub it into dust. Rub back and forth several times and really *feel* it very carefully. Clay soils will be slick and smooth, and may also have a floury feeling, indicating silt content, but they lack the gritty sand component found in loams. Sandy soils will be gritty and not stick together. Loamy soils feel moderately gritty and stick together, but not tenaciously like clay, and will also have a component that feels like flour. This is silt, a soil particle size somewhere between sand and clay.

If you have difficulty determining your soil type using the feel test, dig into your soil when it is dry. Sandy soils seldom exhibit clods. Any that do form will crumble easily. Loamy soils will have clods that can be sliced cleanly with a shovel. Clay soils tend to form hard, persistent clods. Rather than slicing easily, a shovel will get stuck or will shatter them into hard little blocks. If still in doubt, take a soil sample to your local county extension agent or send to a soils lab for analysis.

Organic Soils

Average garden soils typically contain 3%–10% organic matter. "Organic soils" are defined as those that contain more than 20% organic matter — these are commonly found in low-lying swamps and marshes. They behave much differently from upland "mineral soils," which contain less than 20% organic matter.

Organic soils with a high water table will support plants of wet prairies and sedge meadows. Plants that prefer medium to dry soils will often rot when planted in moist soils. Organic soils that have been drained will support plants that do best in medium to moist soils. They are often rich and productive. Weeds grow profusely, so thorough site preparation and post-planting maintenance are critical to success.

Great care must be taken when burning prairies on organic soils. Organic peats and mucks can catch fire if burned when dry. Prairies on these soils should be burned only when saturated to prevent peat fires and loss of plantings. Fire can creep down into the soil until it eventually reaches the water table, consuming plants and soil in the process.

Soil Moisture

Soil moisture is an essential factor in plant selection. Dry soils include sandy, rocky, and gravelly soils that drain readily and never have standing water, even after a heavy rain. Medium (mesic) soils include well-drained loams and clays. These may have standing water for short periods after a hard rain. Moist soils have access to water in the subsoil throughout the growing season. They may have periods of standing water in winter and spring, but the soil surface usually dries out in summer into fall.

Soil pH and Nutrients

Soil pH and nutrients strongly influence plant growth. Soil pH is a measure of the acidity or alkalinity of soil. The pH scale ranges from 1.0 (most acid) to 14.0 (most alkaline). A neutral soil has a pH of 7.0. The ideal pH for most soils is 6.5, slightly acid. All soil nutrients are readily available at this pH, especially phosphorus (P), an important primary nutrient. Most prairie plants will grow well in soils with a pH between 6.0 and 7.5. Many can grow in alkaline soils with a pH of 8.0 and higher, but some can survive in very acid soils with pH levels of 5.0 and lower. Plant growth is usually slower in extremely acid or alkaline soils, as there is reduced access to phosphorus and other nutrients.

Testing your soil before selecting your plants and seeds is a good investment to ensure your success. Simple test kits for soil pH and nutrients are readily available and reasonably priced. Soil samples can be sent to private or state-run soil testing labs. If the pH is less than 5.5, then lime should be added at the recommended rate for your soil type to raise the pH to 6.5. It is rare for a soil to be excessively alkaline (pH greater than 8.5), although this condition can occur in the Great Plains and intermountain western states where high levels of alkali (sodium) are common.

If the pH is 6.5 or higher, liming is not necessary. Excessive addition of lime can elevate the pH to levels at which phosphorus becomes increasingly unavailable and plant growth diminished.

The tables at the end of this book provide the soil texture and moisture conditions in which each species is known to grow, along with pH ranges for some. Data are limited or nonexistent for the pH range of many of these plants because of a dearth of research to date.

Soil Nutrients

Soil nutrients are divided into three groups: primary nutrients, secondary nutrients, and micronutrients.

Primary Nutrients

Primary nutrients are essential to plant growth. Most prairie plants receive sufficient nitrogen from rainwater, especially during thunderstorms. Phosphorus availability is often low, but usually adequate. Potassium is chronically low in most soils, as it is easily leached. The three primary nutrients are the following:

- Nitrogen (N) — Stimulates growth of leaves and stems, and builds proteins, enzymes, and chlorophyll
- Phosphorus (P) — Promotes root growth and seed formation; essential to photosynthesis
- Potassium (K) — Improves fruit and seed quality, and enhances disease resistance

Secondary Nutrients

Secondary nutrients are required in modest amounts and are present in sufficient quantities in most soils. Extremely sandy soils, especially acid sands, benefit greatly from the addition of dolomitic lime to increase calcium and magnesium levels and to correct soil acidity problems. Secondary nutrients are the following:

- Calcium (Ca) — Essential to cell wall structure, and transfer and regulation of other elements
- Magnesium (Mg) — Essential element of the chlorophyll molecule and many plant enzymes
- Sulfur (S) — Essential in protein formation; improves root growth and seed production

Micronutrients

Micronutrients are trace minerals required in small quantities for enzymes, plant metabolism and energy transfer. They generally occur in levels sufficient for growth of prairie plants, with the exception of subsoil sands. A complete micronutrient package is often necessary to promote good plant growth on these problem soils. Excessive levels of individual micronutrients can adversely affect plant growth. This is seldom a problem in the eastern United States, but it can occur in western states with high alkaline soils. Micronutrients include the following:

- Boron (B)
- Manganese (Mn)

- Copper (Cu)
- Zinc (Zn)
- Iron (Fe)
- Molybdenum (Mo)
- Chlorine (Cl)

Fertilizing

Fertilization is rarely required for proper growth of prairie plants, with the exception of two primary nutrients, phosphorus (P) and potassium (K), and the secondary nutrients calcium (Ca) and magnesium (Mg). The soil chemistry of phosphorus is complicated, and it is often rare in plant-available forms. If your soil tests low for phosphorus, consider adding an organic fertilizer such as bonemeal (4-17-0), or inorganic forms such as triple phosphate (0-45-0). Because phosphorus transfer occurs primarily in soil organic matter, adding compost and similar organic materials provides a reservoir of phosphorus for plants to draw on.

Plants consume potassium (K) in large quantities, and it is readily leached from soil. Soils low in potassium can be amended using organic sources such as wood ash (0-0-6) or greensand (0-0-17). Readily available forms of inorganic potassium include muriate of potash (0-0-60) and K-Mag, a naturally occurring mineral (0-0-22, with 10% magnesium and 21% sulfur).

Fertilizing and liming creates proper conditions for germination and growth of prairie plants. For best results, lime and organic fertilizers should be incorporated into the top 6 inches of soil. Inorganic fertilizers can be spread on the soil surface. Once established, further fertilization should not be necessary, provided that appropriate plants were selected to match soil conditions.

Improving Your Soil with Organic Matter

Sand and clay soils can be improved by adding large amounts of organic matter. Composted manure, household or mushroom compost, and dead leaves are all excellent choices. Do not use sawdust, wood chips, or similar materials. These require a long time to break down, and they rob the soil of nitrogen. Avoid uncomposted manure, as it typically contains lots of weed seeds.

Organic matter holds more water and nutrients than any other soil constituent. It helps break up heavy soils, improving water intake and air exchange at root level. Organic matter also firms sandy soils, making them richer and less drought-prone. In each case, adding organic matter modifies a soil so it behaves more like a loam. Benefits of adding organic matter include increased seedling survival, better root development, and faster plant growth.

"Green manure" crops, such as Buckwheat and Winter Wheat, are another effective method for improving poor soils. Their roots help break up clay. The crop is tilled under while actively growing, usually at flowering stage, to incorporate roots and leaves into the soil. This is a cheap, ecologically sound way to build soil organic matter. See the discussion of smother crops on page 347 for more detail on using green manures.

Tips for Working with Clay Soils

Clay soils with low levels of organic matter have poor soil structure and can be difficult to work. Fine soil particles pack together tightly, impeding drainage and transportation of water and air to and from plant roots. Clay soils warm up slowly in spring due to their high moisture content. They should never be worked when wet. They are easily compacted, and long-term damage can result from walking or driving equipment on them. In the heat of summer, clay soils harden and prevent downward root growth, retarding root development and plant growth.

For garden beds, optimal growth conditions can be created by "double digging." This involves removing two spade depths (approximately 1.5 feet) of soil to create a trench, reserving the topsoil separately from the subsoil. Place six inches of organic matter in the bottom of the trench, and then backfill with subsoil, which should be forked into the organic matter in the bottom of the trench. Finally, replace stockpiled topsoil on top. When all the soil has been replaced, you will be left with a small mound. This process of double digging creates a loose, rich soil that promotes excellent root and plant growth. Fall is the best time to double dig, because soils are drier and easier to work, and the mound settles over the winter, helping prepare soil for spring planting.

Many prairie plants thrive in heavy clay soils (see the tables at the end of the book). With good initial care, they will flourish even on difficult sites. Their roots will gradually work their way down into the clay, opening and improving it, just as they have for thousands of years.

To help maintain soil moisture and encourage seed germination when seeding on clay soils, mulch with 1–2 inches of weed-free straw, such as Winter Wheat (see chapter 6 for details).

Regular light watering for the first 4–8 weeks after seeding will greatly increase germination and seedling survival. Water when the soil surface begins to dry out, and only in early morning to prevent fungal disease problems and seedling mortality. Mulched plantings require less frequent watering.

Inorganic Soil Amendments

An inorganic method of opening up a tight clay soil is to add gypsum (calcium sulfate). This naturally occurring compound has the ability to flocculate (or make loose and disaggregated) clay minerals and break them apart so that air and water can pass through the soil. However, studies have indicated that gypsum is effective only on soils with a high sodium content (sodic soils typical of the western United States) and is of limited utility in improving clay soils in the Midwest and eastern states.

Research has indicated that practically all plants have symbiotic relationships with soil fungi called endomycorrhizae. These fungi develop networks of hyphae — threads that make up mycelium — that extend far into the soil to obtain nutrients. Plants serve as hosts for the fungus while benefiting from nutrients supplied to it by mycorrhizae. Products are available for inoculating seed with endomycorrhizae fungi to enhance plant growth and seedling survival on problem soils such as sterile subsoil clays and sands.

Other amendments that can be applied during the seeding process include formulations containing plant rooting hormones, humic acids, seaweed extracts, *Rhizobium* bacteria inocula, and water-retaining polymer gels. These products foster availability of plant nutrients, maintain soil moisture, and stimulate plant growth by improving soil microflora. On normal soils these specialized soil amendments are not a requirement for success, but they have been shown to improve seed germination and growth on difficult sites with low organic matter and poor soil structure.

4

Designing, Planting, and Maintaining Prairie Gardens

Many of the showiest prairie flowers and grasses of the tallgrass prairie make excellent garden plants. They require little to no fertilizer; few, if any, pesticides; and no irrigation once established. They are adapted to growing in states in the Midwest and Northeast that receive 20–40 inches of annual precipitation. They are an economical and ecologically sound alternative to high-maintenance traditional landscapes.

If the prairie is like a tapestry, then grasses form the fabric, their dense root systems holding everything together, and flowers provide color, textural variation, and embellishment. Together they create a unique plant community that provides habitat for a variety of birds, mammals, reptiles, and other life forms.

Prairie plants can be as "at home" in a formal garden setting as their nonnative counterparts most often used in American gardens. With attention to plant selection and knowledgeable maintenance techniques, you can design a complete formal garden using these wonderful plants.

Prairie plants are a great choice for smaller, more formal areas, as they provide immediate results and are tidier than seeded plantings. Areas of

Front-yard prairie meadow replaces a high-maintenance lawn

A densely planted prairie garden deters weeds in both the rooting and foliar zones

1,000 square feet or less can be established using transplants rather than seed. Plants offer a number of advantages over seeds:

- Most native perennials will bloom the year they are transplanted, whereas seeds typically require three years or longer to bloom.
- Transplants can be planted according to a design to create a desired effect, such as borders and masses.
- Weeds can be readily distinguished from desirable plants in the garden, but it can be difficult to tell the good guys from the bad guys in a seeded planting.

Low-profile prairie garden with Purple Coneflower and Prairie Blazing Star

Tall prairie flowers planted in masses for effect: (*left to right*) Tall Joe Pye Weed, Culver's Root, Prairie Blazing Star, Sweet Blackeyed Susan, and Queen of the Prairie

Naturalistic prairie gardens are typically planted more densely, so plants overlap one another as they would in a natural setting. In more formal flower beds, plants are placed farther apart to prevent crowding and highlight each individual plant.

Plant-spacing recommendations for both prairie gardens and flower beds are listed separately in the plant tables found at the end of this book.

Site Selection and Preparation

Prairie gardens do best when sited in full sun to part shade (at least half a day of direct light). Many prairie plants can also grow in semishaded areas that receive less than a half day of sun. It is essential to select plants matched to your sun, soil, and moisture conditions. Failure to observe these basic guidelines can reduce plant vigor, increase the chance of plant mortality, and necessitate higher maintenance. Sun and soil requirements are listed for each species in the plant descriptions in chapter 5, as well as in the plant tables at the end of this book.

As with any garden, the area to be planted must be completely free of weeds. Areas with a history of weedy growth will require extensive weed control prior to planting, as detailed in chapter 6.

A Quick Guide to Converting Your Lawn to a Prairie Garden

Lawns are easy to convert into prairie gardens. A beautiful, fully developed garden can be established in two years by following these steps.

First, smother the lawn using cardboard, black plastic, a tarp, an old carpet, or something similar. Secure the material with rocks, logs, fenceposts, or similar items to hold it in place. Cover for two months or longer, until the grass is completely dead. Lawn can also be killed by spraying herbicide in midspring or early fall when lawn grasses are actively growing. Wait one week after spraying before installing your plants. It is important to note that recent research has implicated glyphosate herbicides in the precipitous decline of some insect and amphibian populations.[1] Although glyphosate herbicides offer an efficient and cost-effective method for eliminating weeds, other options are available, as described in chapter 6.

If perennial weeds are present in the lawn, they will require smothering for a full growing season, from April through October, or repeated applications of herbicide three to four times every eight weeks from late spring to early fall. If pernicious rhizomatous perennial weeds such as Canada Thistle (*Cirsium arvense*), Carolina Horsenettle (*Solanum carolinense*), Crownvetch (*Securigera varia*), or Field Bindweed (*Convolvulus arvensis*) are present, the area will require smothering for a full two growing seasons, or the use of specific broadleaf herbicides to kill these plants completely. These four species are resistant to glyphosate, and an appropriate broadleaf herbicide must be added to the glyphosate solution to kill them, along with other vegetation.

Second, if the soil is poor heavy clay or dry sand, and requires amendment with organic matter or lime, dig up or till the dead sod and add soil amendments. Sod tends to clump together and does not mix into soil well.

If a spring planting is desired, kill the lawn in fall and allow it to rot over winter before digging or tilling in spring. The sod will break up easily.

Third, if no soil amendment is required, transplants can be installed directly into the dead sod using a small shovel or hand trowel. Weed seeds from below are not exposed, thus reducing weed germination in the new garden. If cardboard is used for smothering, cut a small hole in it and install transplants through it. Leave the cardboard in place and cover with the desired mulch. Weed-free Winter Wheat straw and shredded hardwood are common choices.

Winter Wheat straw mulch in a spring-planted prairie garden

Some people use cocoa bean hulls as they are easy to work with, but they can be toxic to pets, especially dogs. Bark chips are never recommended; they can reduce nutrient levels in the soil and some are toxic to plants. If desired, a preemergent herbicide can be applied after plants have been installed and prior to mulch application. This will further reduce weed germination and garden maintenance.

Finally, water spring- or early-summer-planted gardens in the first two months, when soil under the mulch begins to dry out. Expect to water once a week in the absence of rainfall. A single deep soaking is better than numerous light sprinklings. Once plants are well established, watering should not be necessary except during periods of extended drought. Clay soils hold large amounts of moisture and are susceptible to overwatering by well-meaning gardeners. Gardens planted in clay seldom require watering more than once a week and benefit from periods of no watering to facilitate air exchange to the roots. (For more detailed information on site preparation on existing and former agricultural fields and other situations, see chapter 6.)

Twelve Tips for Designing Prairie Gardens

The following tips on designing your prairie garden combine principles of plant ecology with garden design. You can select the ideas you wish to apply in your garden and express your own style using prairie plants.

1   Plant flowers and grasses together to create a naturalistic meadow effect. The dense root systems of grasses dominate the upper rooting zone and help squeeze out weeds. Grasses will do much of the "weeding" for you by eliminating open soil in which weed seeds germinate. Grasses help support wildflowers, reducing the need for staking. They also take up soil nitrogen, reducing plants' height and their tendency to flop.

2   Select plants to match the scale of your landscape. Use short flowers and grasses (1–5 feet tall) in small prairie gardens. Short grasses, such as Little Bluestem, Sideoats Grama, and Prairie Dropseed, are clump formers that leave room between them for flowers. Use tall flowers and grasses in back borders and areas where bold plants are desired. Tall prairie grasses and flowers can also be used for screening undesirable views in late summer and fall when these plants are at their peak.

Clump-forming Prairie Dropseed with lower-growing Foxglove Beardtongue, Butterflyweed, and Golden Alexanders

3 Avoid aggressive plants that creep by rhizomes or self-sow prolifically. These squeeze out less assertive plants and can take over the garden. These may include Cupplant, Rosinweed, Bergamot, Sunflowers, Oxeye Sunflower, Canada Anemone, Big Bluestem, Indiangrass, Switchgrass, and Prairie Cordgrass.

Rhizome: Downy Sunflower (*Helianthus mollis*)

4 Plant flowers in masses and drifts of color to create drama and impact. Mass plantings of only one or two species of flowers will likely experience weed problems due to lack of grasses to squeeze out weeds. In a closely tended or heavily mulched garden, this may not be a concern.

5 Interplant flowers with taproots, bulbs, and corms with prairie grasses and fibrous-rooted flowers. Taproots, bulbs, and corms do not provide sufficient soil cover to prevent weed growth around them, while grasses and fibrous-rooted flowers do.

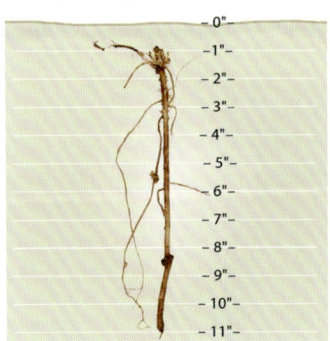

Taproot: Lupine (*Lupinus perennis*)

Bulb: Nodding Pink Onion (*Allium cernuum*)

Corm: Blazing Star (*Liatris* spp.)

Fibrous: Shooting Star (*Dodecatheon meadia*)

6   Arrange plants to complement one another, with texture and color. For instance, plant flowering spikes of Blazing Star (*Liatris*) in front of the bold foliage of Prairie Dock. Most flowers look good mixed with grasses of similar stature. Their green foliage helps highlight flowers planted in the foreground.

7   Select plants for succession of bloom throughout the growing season. This ensures something interesting is always going on in your garden. Prairie grasses provide a great show in fall and winter after the flowers are gone.

8   Create cohesion and interest in the garden by echoing color patterns. Place groups of individual species in various locations to lead the eye across the garden.

9   Include spring-blooming flowers. Many shorter spring bloomers are among the most attractive and delicate of prairie flowers. Most spring prairie flowers go dormant by midsummer, making them good companions for a variety of other flowers and grasses, both tall and short.

10   Plant taller plants toward the back, shorter in front when planting against a background such as a wall, hedge, or wooden fence. This prevents larger plants from obscuring shorter ones. Early-blooming short flowers can be planted throughout, as they will bloom before others overtop them. As the season progresses and taller plants increase in size, they will cover the foliage of shorter plants as they die back, reducing the need for deadheading. In island beds, tall plants should generally be located centrally, with occasional specimens in other strategic locations.

Prairie island beds planted into an existing lawn create interest and reduce mowing

11 Use large "specimen" plants as architectural focal points. Surround individual specimens with lower-growing flowers and grasses to enhance them and control weeds by covering the soil, especially around taprooted specimens.

12 Use groundcover plants for interplanting among taller flowers and grasses and in areas where low-growing cover is desired. Creeping plants will fill in gradually by runners sent out from the main plant. Once established, occasional weeding is all that should be necessary to remove the few interlopers which may find their way into the garden.

By integrating the principles of ecology with those of garden design, you can create attractive prairie gardens. They require few fertilizers, pesticides, and little, if any, irrigation to keep them healthy and vibrant. Even during severe heat and drought, prairie gardens continue to perform while other plants often succumb to the elements.

Planting the Prairie Garden

Planting can commence once all weeds have been eliminated and any necessary soil amendments added. Plantings in mid- to late spring and early to midautumn have higher survival rates and require less watering than summer plantings. Summer plantings are riskier because of high temperatures and increased water requirements. In northern climates, late-fall plantings are not recommended, as they allow limited time for plant establishment prior to the onset of cold weather. In southern latitudes, the planting season can be extended into late fall.

To plant plugs or plants from containers, dig holes twice as wide as the container and at least as deep as the plant root mass. If soil is dry, fill with water and allow to drain. Place plants in the holes, backfill with soil, firm into place, and water in thoroughly.

If container plants are pot-bound, with their roots circling inside the pot, pull the roots apart by hand and spread them outward and downward. Build a cone of soil in the center of the planting hole and spread roots over the cone to encourage proper growth. Backfill with soil, and water in thoroughly. If roots are tightly bound together and cannot be separated by hand, use a knife to cut the root mass into three or four sections, from top to bottom. Spread the roots out over the soil cone, backfill, and water.

After planting plugs or plants from small containers, add a half inch of garden soil over the original soil around the plant to prevent desiccation. For the first month, water the newly planted garden deeply once a week in the absence of rain, until plants become well established.

Prairie Garden Maintenance

Prairie gardens can be maintained similarly to traditional perennial gardens. This includes weed control, deadheading, cutting back, mulching, and watering. Fertilizing is seldom necessary, as it promotes excessive plant growth which encourages leggy, floppy plants. An established prairie garden almost never requires watering, as the plants are well adapted to surviving extended heat and drought. A properly designed prairie garden containing plants in proper densities and proportions will completely occupy the rooting zone and discourage weeds. See tables at the end of the book for guidelines on planting distance.

Foundation garden planted to attract birds, butterflies, and pollinators

Prairie flowers and grasses fill most of this small urban lot with prime habitat

A school prairie garden celebrating the life of an early Illinois pioneer girl

Big bold prairie garden: (*left to right*) Prairie Blazing Star, Pale Purple Coneflower, Big Bluestem, Yellow Coneflower, and Tall Joe Pye Weed

Wetland prairie flowers without grasses in a rain garden include (*front to back*) Swamp Milkweed, Culver's Root, and Yellow Coneflower

As with any garden, some plants may be more successful than others. If a certain species begins to infringe on its neighbors, it should be controlled to ensure proper balance. Rhizomes of spreading species can be cut back to the main plant every few years, as needed. Seedlings of self-sowing species should be removed in their first year when small before they develop deep root systems, which are harder to extract.

Year-End Maintenance

Despite temptation to tidy up the garden in fall, you will do yourself and wildlife a favor by leaving plants standing over winter. Birds will eat seeds and insects in the prairie garden all winter long. Many butterflies, moths, and other beneficial insects overwinter as adults, pupae, and eggs in plant stems and leaves. Cutting the plants down in fall or early winter harms these important invertebrates. Cutting and removing debris results in fewer butterflies and other pollinators in future years. Cut plants down in midspring just as they resume new growth. Cut material should be left on

Remnant vegetation is left standing over winter to preserve invertebrates and feed the birds in this rain garden designed and planted in 1986 by Neil Diboll

The rain garden blooming in its second year, 1987

the ground so that beneficial insects remaining on the plants can emerge and flourish.

An excellent reference for more detailed information on perennial garden maintenance is *The Well-Tended Garden*, by Tracy DiSabato-Aust, published by Timber Press.

Prairie Garden Designs

The variety of prairie garden styles is virtually unlimited. Combining a multitude of flowers and grasses, one can create unique native-plant gardens. Refer to the plant tables at the end of this book for information on which plants to use to create some of the following design ideas. Some garden options include the following:

1 Mixed prairie flowers and grasses
2 Prairie flowers only
3 Prairie grasses with no flowers
4 Rain gardens
5 Formal gardens using prairie plants
6 Butterfly gardens
7 Pollinator gardens
8 Songbird gardens
9 Hummingbird gardens
10 Medicinal plant gardens
11 Deer-resistant gardens
12 Low-growing gardens
13 Big, bold plant gardens
14 White gardens (or moon gardens)

A native prairie pollinator garden installed at a McDonald's restaurant in Westfield, WI, sports (*left to right*) Hoary Vervain, Purple Coneflower, Prairie Blazing Star, Blackeyed Susan (*center rear*), Purple Coneflower again, and Purple Poppymallow growing low along the sidewalk in late July

5

Prairie Species
Field Guide

Monocots

Tradescantia bracteata
Prairie Spiderwort

FAMILY: SPIDERWORT (COMMELINACEAE)

Best behaved of spiderworts, this forms a low-growing carpet of color in spring. Thrives in good soil as well as hot dry spots where most other plants struggle. Rhizomes creep slowly to form a nice patch. Tricky from seed but easily divided.

Habitat: Dry to moist prairies
Garden Uses: Informal gardens, bee gardens, prairie meadows

USDA Hardiness Zones: 3–7
Light Requirements: Full sun to partial shade
Soil Types: Sand, loam, clay
Soil Moisture: Dry to moist
Bloom Time: May–June
Flower Color: Blue to pink
Height: 1–2 feet
Life Expectancy: 20+ years
Root Type: Rhizomatous, fibrous
Propagation: Moist stratify seed 90 days, or plant fresh for fall germination; root division in fall or spring

Aggressiveness: Medium by rhizome
Attracts: Bees
Deer Palatability: Medium
Distinguishing Characteristics:
- Forms low-growing clumps and does not spread rapidly like Western Spiderwort (p. 46)
- Sepals and pedicels have glandular hairs (under the flower petals and on the stems)
- Like most *Tradescantia*, flowers bloom in the morning and close by early afternoon, except on overcast days

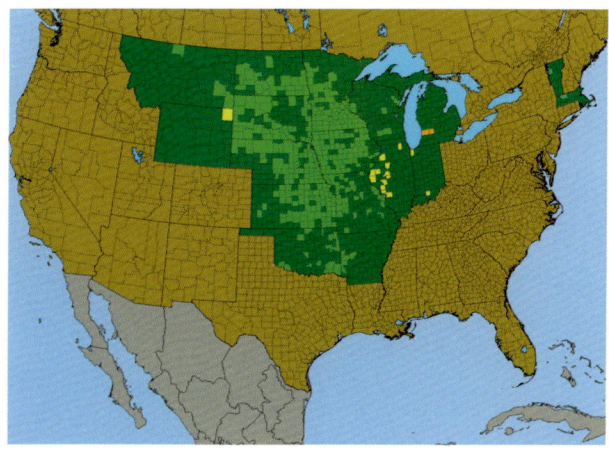

Seedling

Emerging mature plant in spring

Entire plant. Plant is low growing (1–1.5 ft.), forming a clump that does not spread by rhizomes.

Leaf. Leaves tend to stay green later into the season than those of *T. occidentalis* and *T. ohiensis*.

Flower

Early seed

Mature seed. Seedheads covered with dense hairs (similar to *T. occidentalis*).

LOOK-ALIKE PLANTS:

Natives: Similar to other *Tradescantia*, but lower growing than Ohio Spiderwort (p. 48) and usually has blue rather than pink flowers as in Western Spiderwort (p. 46). *T. virginiana* is a similar but taller species at 1.5–3 feet and has hairy but nonglandular sepals.

Tradescantia occidentalis
Western Spiderwort

FAMILY: SPIDERWORT (COMMELINACEAE)

Hot pink to purple flowers explode in a riot of can't-miss-it color in mid- to late spring. Spreads rapidly by rhizomes to form a colony that squeezes out most weeds. Grows in superdry sandy and rocky soil, and other well-drained sites. Somewhat tricky from seed: divide rhizomes for rapid results.

Habitat: Dry to medium prairies, sand prairies
Garden Uses: Groundcovers, informal gardens, bee gardens, prairie meadows

USDA Hardiness Zones: 3–9
Light Requirements: Full sun to partial shade
Soil Types: Sand, loam
Soil Moisture: Dry to medium
Bloom Time: April–August
Flower Color: Pink, purple
Height: 1–2 feet
Life Expectancy: 20+ years
Root Type: Rhizome

Propagation: Moist stratify seed 90 days, or plant fresh for fall germination; root division in fall or spring
Aggressiveness: High by rhizome
Attracts: Bees
Deer Palatability: Medium
Distinguishing Characteristics:
• Spreads rapidly by rhizomes

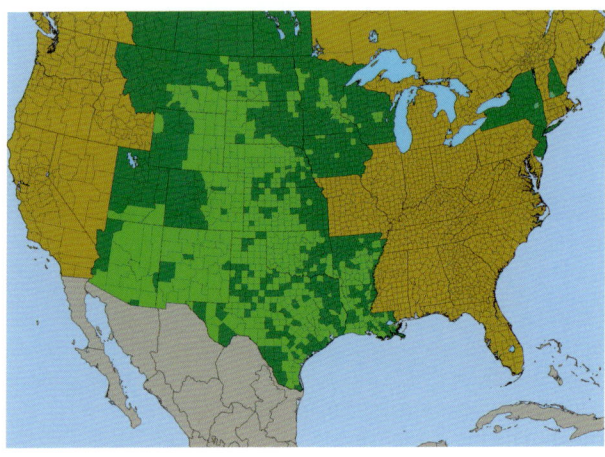

Seedling

Emerging mature plant in spring

Entire plant. Similar in height to *T. occidentalis* but spreads rapidly by rhizomes. Best used as a groundcover, not a garden companion.

Leaf

Flower

Seed. Seedheads covered with dense hairs (similar to *T. bracteata*).

LOOK-ALIKE PLANTS:

Natives: Lower growing than Ohio Spiderwort (p. 48). Prairie Spiderwort (p. 44) is of similar stature but does not spread rapidly by rhizomes. *T. virginiana* is an eastern species that also spreads quickly by rhizomes but is taller, at 1.5–3 feet.

Tradescantia ohiensis
Ohio Spiderwort, Bluejacket

FAMILY: SPIDERWORT (COMMELINACEAE)

Deep-blue flowers light up prairie meadows in late spring. White and pink forms also occur. Best for informal gardens, as it readily self-sows. Will grow in incredibly dry sand to moist clay. Seed requires extended cold treatment to germinate; sow in fall for best results. Easy to divide, one mature plant yields dozens of new ones.

Habitat: Dry to moist prairies, savannas, barrens, glades, dry open woods, bluffs, sand dunes, old fields, roadsides — especially on sandy soils
Garden Uses: Informal gardens, bee gardens, prairie meadows

USDA Hardiness Zones: 3–9
Light Requirements: Full sun to partial shade
Soil Types: Sand, loam, clay
Soil Moisture: Dry to moist
Bloom Time: April–July
Flower Color: Blue, purple
Height: 2–4 feet
Life Expectancy: 20+ years
Root Type: Fibrous
Propagation: Moist stratify seed 90 days, or plant fresh for fall germination; root division in fall or spring

Aggressiveness: High by seed
Attracts: Bees
Deer Palatability: High
Distinguishing Characteristics:
- Taller than most other *Tradescantia*
- Leaves have a bluish tinge
- Like most *Tradescantia*, flowers bloom in the morning and close by early afternoon, except on overcast days

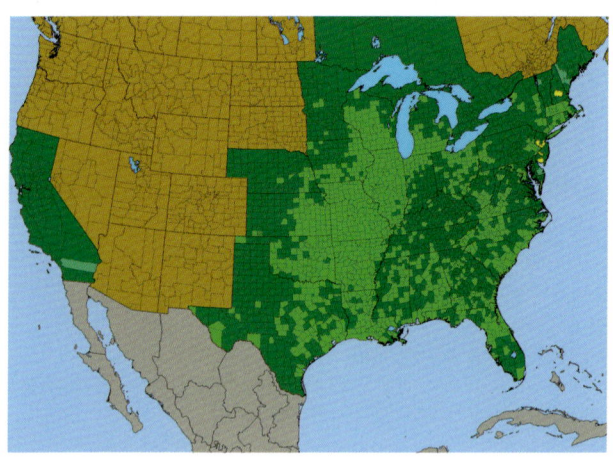

Seedling

Emerging mature plant in spring

Entire plant. Taller than *T. bracteata* and *T. occidentalis* at 2–3 ft. Spreads easily by seed but not by rhizomes.

Leaf. Leaves turn brown after flowering and seed formation by July, resprouting in late summer and early fall.

Flower

Mature seed. Seedheads hairless or nearly so, unlike *T. bracteata* and *T. occidentalis*.

Early seed

LOOK-ALIKE PLANTS:

Natives: Similar to other *Tradescantia* but taller than Prairie Spiderwort and Western Spiderwort (pp. 44–46). *T. virginiana* is similar but does not have blue-tinged foliage, flower pedicels are hairy, and bracts subtending the flowers are larger compared to *T. ohiensis* (1–1.5 inches versus 0.5–1 inches).

Iris virginica var. *shrevei* (*Iris shrevei*)
Wild Iris, Shreve's Iris

FAMILY: IRIS (IRIDACEAE)

This beautiful native iris is at its best in moist soils and pond edges. Even thrives in shallow water but tolerates dry periods in summer. Tricky from seed: sow fresh or in fall for spring germination. Corms are easily divided. It is almost impossible to kill this plant!

Habitat: Wet prairies, marshes, sedge meadows, fens, swamps, lakeshores, streamsides, riverbanks, ponds, wet thickets, wet woods
Garden Uses: Rain gardens, water gardens, pond edges, moist stream banks, wet meadows; flowers are fleeting but daggerlike leaves provide season-long interest

USDA Hardiness Zones: 3–6
Light Requirements: Full sun to partial shade
Soil Types: Sand, loam, clay
Soil Moisture: Wet
Bloom Time: June–July
Flower Color: Blue
Height: 2–3 feet
Life Expectancy: 10–20 years
Root Type: Rhizome, corm
Propagation: Moist stratify seed 90 days, or plant fresh for germination the following spring; root division in fall or spring

Aggressiveness: Low by rhizome
Attracts: Bees, butterflies, game birds, hummingbirds, songbirds
Deer Palatability: Low
Distinguishing Characteristics:
- Emerging leaves are often purple tinged
- Taller than *Iris vernalis* at 1 foot and *Iris cristata* at 10 inches

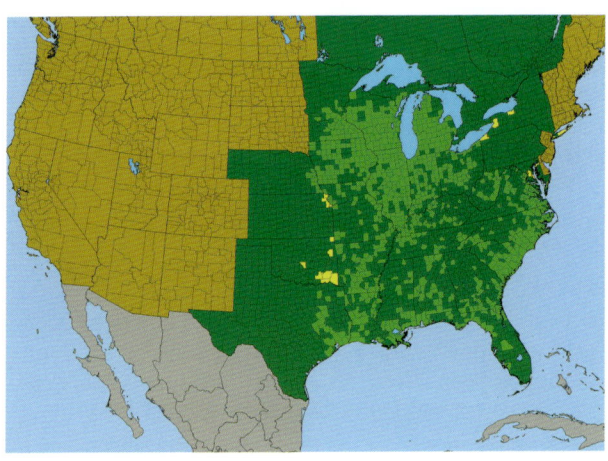

Seedling

Emerging mature plant in spring. Previous year's stringy, dried leaves are often retained on the ground at plant base over winter.

Entire plant

Leaf

Flower

Early seed

Mature seed. Seed is ripe when pods turn brown and feel papery.

LOOK-ALIKE PLANTS:

Natives: Virtually identical to Blue Flag (*Iris versicolor*), which has a zig-zag flowering stem. Leaves resemble Sweet Flag (*Acorus calamus*), a wet-land plant with yellow flower spikes (spathes) whose leaves are fragrant when crushed.

Weeds: Nonnative Yellow Flag Iris (*Iris pseudacorus*) has similar leaves, with yellow flowers, and is commonly planted in water gardens.

Sisyrinchium albidum
White Blue-Eyed Grass

FAMILY: IRIS (IRIDACEAE)

The blue-eyed grasses appear like miniature irises, with starlike flowers at the tips of leaflike stalks. White Blue-Eyed Grass is the only species with pure white flowers instead of blue (rarely occurring in a blue form). A truly charming plant. Easily divided in early spring before new growth appears. Plant seeds immediately after harvesting in mid to late summer. Seedlings typically emerge early fall or the following spring.

Habitat: Medium to moist prairies and meadows, savannas, open woods, limestone glades, railroad edges
Garden Uses: Rock gardens, front of borders, rain gardens

USDA Hardiness Zones: 3–10
Light Requirements: Full sun
Soil Types: Sand, loam
Soil Moisture: Dry, medium
Bloom Time: March–June
Flower Color: White
Height: 0.5–1 feet
Life Expectancy: 3–5 years
Root Type: Fibrous
Propagation: Moist stratify seed 90 days, or plant fresh; root division in early spring or midsummer (cut back 50% of top growth if dividing in summer)

Aggressiveness: Low by seed
Attracts: Bees
Deer Palatability: Low
Distinguishing Characteristics:
- Flowers are almost always white
- Flower stalks have narrow ridges
- Flowers are borne on thin stalks (pedicels) originating from pairs of sessile spathes (flattened bracts from which the flower stalks originate)

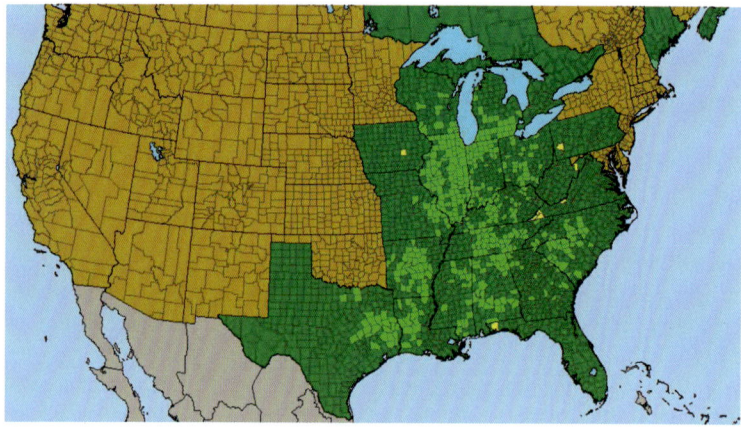

Seedling

Emerging mature plant in spring Entire plant

Flower

Early and mature seed. *S. albidum* Mature seed
has two sessile spathes at the
apex of each flowering stalk, one
of which becomes visible only
after flowering.

LOOK-ALIKE PLANTS:

Natives: Prairie Blue-Eyed Grass (p. 56) and Narrowleaf Blue-Eyed
Grass (p. 54) are very similar but typically have blue to lavender flowers.
Prairie Blue-Eyed Grass flowers and seedpods are borne from single sessile spathes. The spathes of Narrowleaf Blue-Eyed Grass that contain the
flowers and seedpods are produced on peduncled stems.

Sisyrinchium angustifolium
Narrowleaf Blue-Eyed Grass

FAMILY: IRIS (IRIDACEAE)

Showy and robust, this delicate beauty produces deep-blue flowers from large grasslike clumps. Widely adaptable, it is one of the best *Sisyrinchium* for gardens. Easily divided in early spring before new growth appears. Plant seeds immediately while fresh when harvested in midsummer. Seedlings typically emerge in early fall or the following spring.

Habitat: Medium to moist prairies and meadows, open woods, thickets
Garden Uses: Rock gardens, front of borders, rain gardens

USDA Hardiness Zones: 3–10
Light Requirements: Full sun to partial shade
Soil Types: Sand, loam, clay
Soil Moisture: Medium, moist
Bloom Time: April–June
Flower Color: Blue
Height: 0.5–2 feet
Life Expectancy: 3–5 years
Root Type: Fibrous
Propagation: Moist stratify seed 90 days, or plant fresh; root division in early spring or midsummer (cut back 50% of top growth if dividing in summer)
Aggressiveness: Low by seed

Attracts: Bees, butterflies, game birds
Deer Palatability: Low
Distinguishing Characteristics:
- Despite its name, leaves are wider at 0.1875 inch than most other species of blue-eyed grass
- Leaves have "wings" on either side of the midrib, up to 0.25 inches across
- Flowers are borne from flattened bracts called spathes at the ends of secondary stalks branching from the main stem

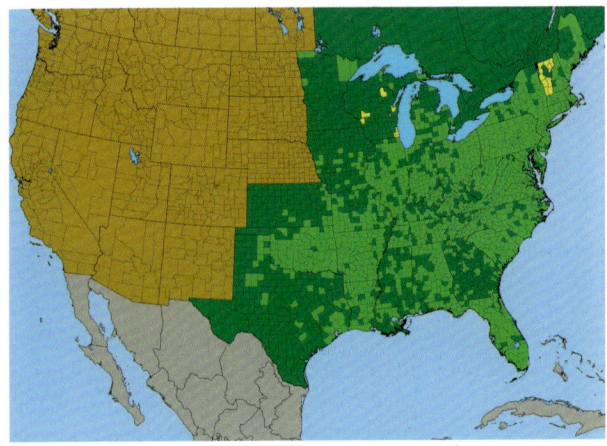

Seedling

Emerging mature plant in spring. Emerging clump is much larger than other *Sisyrinchium*.

Entire plant. Plants grow from large clumps. Leaf blade is wider than most other *Sisyrinchium* at 0.25–0.5 in.

Flower

Early seed

Mature seed

LOOK-ALIKE PLANTS:

Natives: Prairie Blue-Eyed Grass (p. 56) and White Blue-Eyed Grass (p. 52) have narrower leaves, and their flowers are borne from spathes located directly on the main stem, not from secondary stalks.

Sisyrinchium campestre
Prairie Blue-Eyed Grass

FAMILY: IRIS (IRIDACEAE)

A bit daintier than its cousin Narrowleaf Blue-Eyed Grass, with flowers that range from robin's-egg blue to nearly white. Like all blue-eyed grasses, it can be readily divided in early spring before emergence of new leaves. Plant seeds fresh when harvested in midsummer for fall or next spring germination.

Habitat: Dry to medium prairies, sandy fields, savannas, barrens, glades, bluffs, dry open woods
Garden Uses: Fronts of borders, rock gardens, dry prairie gardens

USDA Hardiness Zones: 3–8
Light Requirements: Full sun to partial shade
Soil Types: Sand, loam
Soil Moisture: Dry to medium
Bloom Time: April–June
Flower Color: Blue to white
Height: 0.5–1 feet
Life Expectancy: 3–5 years
Root Type: Fibrous
Propagation: Moist stratify seed 90 days, or plant fresh; root division in early spring or midsummer (cut back 50% of top growth if dividing in summer)

Aggressiveness: Low by seed
Attracts: Bees, butterflies, game birds
Deer Palatability: Low
Distinguishing Characteristics:
- Has narrowest leaves of the prairie *Sisyrinchium*, at 0.0625–0.125 inches wide
- Flowers originate in a single bunch from a flat, widened bract (spathe) located on the stem near the top of each flower stalk

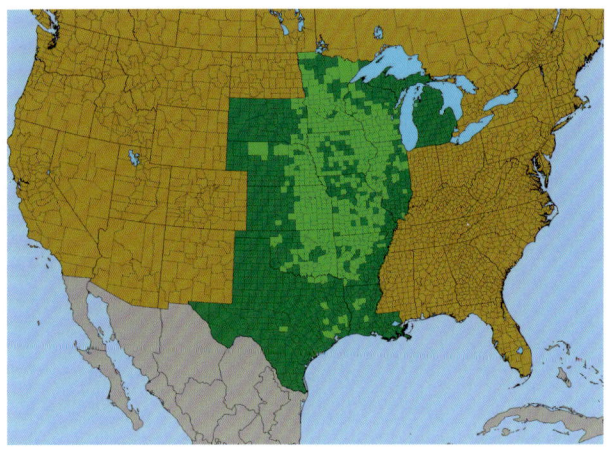

Seedling

Emerging mature plant in spring

Entire plant

Flower

Early seed

Mature seed

LOOK-ALIKE PLANTS:

Natives: White Blue-Eyed Grass (p. 52) is nearly identical, except its flowers are usually white and borne in two individual bunches on the stem. Narrowleaf Blue-Eyed Grass (p. 54) has wider leaves (0.1875 inch), with flowers borne from spathes at the ends of thin stalks that branch off the main stem.

Allium cernuum
Nodding Pink Onion

FAMILY: LILY (LILIACEAE)

This adaptable plant is widely distributed across much of North America. Grows readily from seed, and bulbs are easily divided. Very long-lived, it is excellent in prairie gardens and meadows. Virtually deer-proof due to its onion flavor.

Habitat: Medium to moist prairies, dry to medium open woods
Garden Uses: Large or small groupings, alone or mixed with short grasses; border edges, rain gardens, prairie meadows

USDA Hardiness Zones: 3–8
Light Requirements: Full sun to partial shade
Soil Types: Sand, loam, clay
Soil Moisture: Medium to slightly moist
Bloom Time: June–August
Flower Color: White, pink
Height: 1–2 feet
Life Expectancy: 10–20 years
Root Type: Bulb
Propagation: Moist stratify seed 30 days; root division in fall or spring

Aggressiveness: Medium by seed
Attracts: Bees, hummingbirds
Deer Palatability: Low
Distinguishing Characteristics:
- White to pink flower globes nod downward at maturity; Prairie Onion (p. 60) flowers are up-right and seldom nod
- Leaves smell mildly of onion when bruised
- Globelike seedheads contain angular black seeds

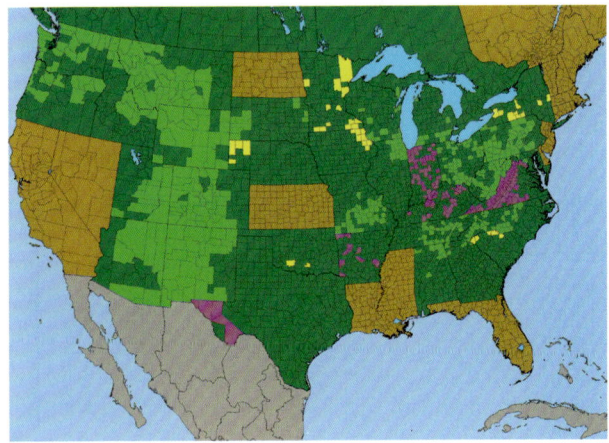

Seedling. Note rounded leaves of seedling and red color at plant base.

Entire plant

Leaf. Leaves are fleshier than grass leaves, with onion odor when bruised.

Emerging mature plant in spring. Note flat leaves and distinct onion odor when bruised.

Flower. Mature flowers nod downward.

Early seed (*left*) and mature seed (*right*)

LOOK-ALIKE PLANTS:

Natives: Prairie Onion; Wild Garlic (*Allium canadense*) has narrow, chive-like leaves, with pinkish-white bulblets and white flowers at the top of the stems. Some nonnative cultivated species of *Allium* are similar in appearance.

Weeds: Wild Garlic is considered by many to be a weed.

Allium stellatum
Prairie Onion

FAMILY: LILY (LILIACEAE)

This western relative of Nodding Pink Onion is similar in appearance, occurring primarily on dry sandy and rocky prairies west of the Mississippi River.

Habitat: Dry to medium prairies, sand prairies, barrens, rocky hillsides, bluffs

Garden Uses: Large or small groupings mixed with short grasses. Looks great planted with Sideoats Grama and Prairie Dropseed. Also mixes well with other low-growing flowers, especially blues, pinks, and purples. Excellent for rock gardens.

USDA Hardiness Zones: 3–7
Light Requirements: Full sun
Soil Types: Sand, loam
Soil Moisture: Dry to medium
Bloom Time: July–October
Flower Color: White to lavender
Height: 1–2 feet
Life Expectancy: 10–20 years
Root Type: Bulb
Propagation: Moist stratify seed 30 days; root division in fall or spring

Aggressiveness: Low by seed
Attracts: Bees, game birds, hummingbirds, songbirds
Deer Palatability: Low
Distinguishing Characteristics:
- Flowers are upright at maturity
- Leaves smell mildly of onion when bruised
- Globelike seedheads contain angular black seeds

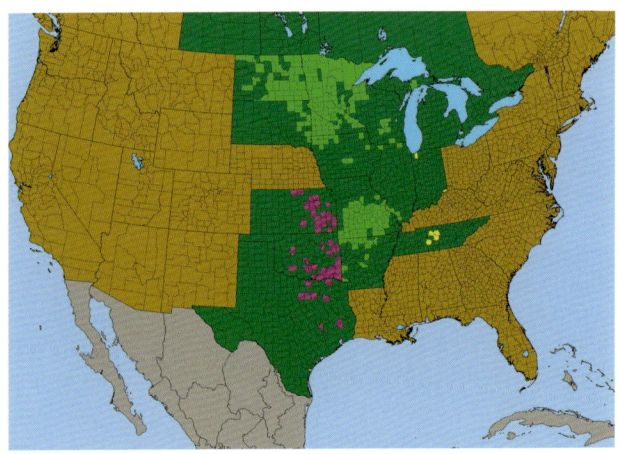

Seedling. Note rounded leaves of seedling; usually not red at plant base.

Emerging mature plant in spring. Note flattened leaves, virtually indistinguishable from *Allium cernuum*.

Entire plant

Flower. Flowers are upright, not nodding, as in *Allium cernuum*, and tend to be darker purple.

Mature seed. Seed is ripe when pods open to reveal black seeds.

LOOK-ALIKE PLANTS:

Natives: Very similar to Nodding Pink Onion (p. 58) except flowers are erect rather than nodding at maturity. Grows primarily in dry soils; Nodding Pink Onion prefers rich, medium to moist soil.

Lilium michiganense
Michigan Lily

FAMILY: LILY (LILIACEAE)

Large and showy, this is a gem for moist gardens. Requires moist soil with lots of organic matter. Tricky from seed and very slow growing. Needs up to five years to reach blooming stage. Bulbs can be easily divided into individual scales when dormant to produce new plants. Deer are fond of the flower buds.

Habitat: Moist to wet prairies and meadows, fens, sedge meadows, seeps, streamsides, wet thickets, open wet woods
Garden Uses: Hummingbird gardens, borders, formal and informal gardens. Lilies like their "heads in the sun and feet in the shade," so surround with low-growing flowers and grasses.

USDA Hardiness Zones: 3–7
Light Requirements: Full sun to partial shade
Soil Types: Sand, loam, clay
Soil Moisture: Moist to wet
Bloom Time: June–August
Flower Color: Orange
Height: 4–6 feet
Life Expectancy: 20+ years
Root Type: Bulb
Propagation: Moist stratify seed 60 days; root division (divide bulb scales and plant 2 inches deep in fall)

Aggressiveness: Low by seed
Attracts: Bees, butterflies, hummingbirds
Deer Palatability: High
Distinguishing Characteristics:
- Large, showy, pendulant orange flowers
- Smooth glossy leaves are borne in whorls around the stem
- Upright, paper-thin seedpods have three chambers in cross section

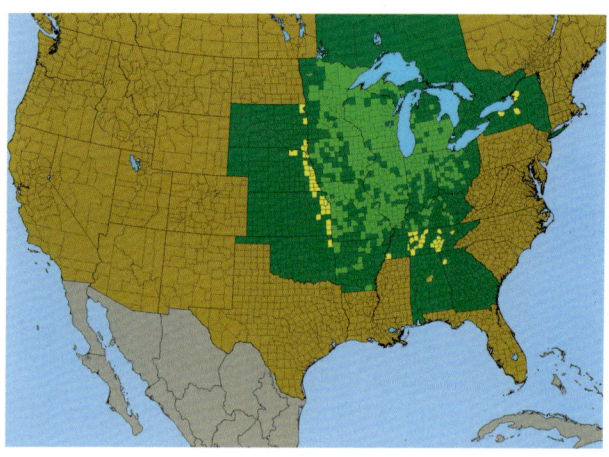

Seedling

Emerging mature plant in spring

Entire plant

Leaf. Leaves are borne in whorls of four or five.

Flower

Early seed. Seed capsule is three-parted and often tinged with red when immature.

Mature seed

LOOK-ALIKE PLANTS:

Natives: *Lilium superbum* is a southern variant with similar flowers and growth form. Wood Lily (*Lilium philadelphicum*) has upright flowers and occurs in dry to mesic prairies and savannas.

Weeds: Certain ornamental Asian Lilies have similar flowers.

Polygonatum biflorum var. *commutatum* (*Polygonatum canaliculatum*)

Great Solomon's Seal, Smooth Solomon's Seal

FAMILY: LILY (LILIACEAE)

Largest of Solomon's Seals. Creeps gradually by rhizomes to form a patch. Creamy-white flowers become bright-blue berries suspended under gracefully arching stems. Does well in any decent well-drained soil. Foliage turns a brilliant yellow in fall. Seeds are double-dormant and require two years in the ground to germinate. Easy to propagate by root division.

Habitat: Savannas, oak barrens, open woods, woodland edges, roadsides and railroads

Garden Uses: Specimens, informal gardens, prairie meadows. Plant with grasses to accentuate its form and in contained spaces. Combine with other aggressive plants in the garden or consign it to the meadow.

USDA Hardiness Zones: 3–9
Light Requirements: Full sun to partial shade
Soil Types: Sand, loam, clay
Soil Moisture: Dry to medium
Bloom Time: May–June
Flower Color: Greenish white
Height: 2–4 feet
Life Expectancy: 10–20 years
Root Type: Rhizome
Propagation: Seed is double-dormant (see chapter 9); root division in fall

Aggressiveness: Low by seed and rhizome
Attracts: Bees, game birds, songbirds
Deer Palatability: Medium
Distinguishing Characteristics:
- Tallest of all Solomon's Seals
- Leaves wider than other Solomon's Seals
- Forms colonies by creeping rhizomes

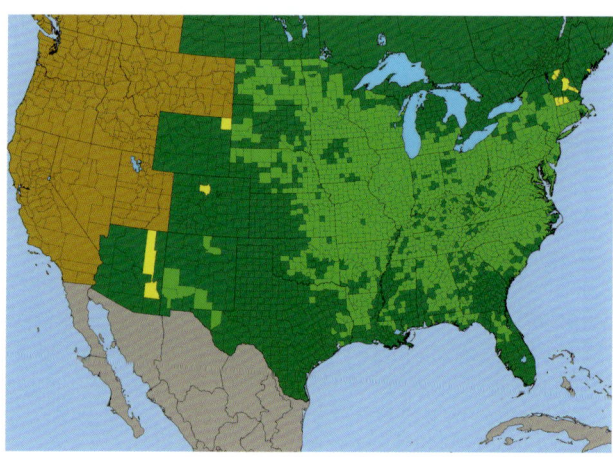

Seedling

Emerging mature plant in spring. Emerging stems are thick and bluish in color, resembling *Baptisia alba*, but they have individual leaves rather than compound leaflets.

Entire plant

Leaf. Leaves are thick and smooth, with numerous prominent parallel veins.

Flower. Flowers are greenish yellow to white, resembling an upside-down urn.

Early seed

Mature seed. Seeds are ripe when they turn deep purplish black; they strip off the stem more easily as they age and soften.

LOOK-ALIKE PLANTS:

Natives: The woodland species Solomon's Seal (*P. biflorum*) is almost identical but grows half as tall. Emerging leaves are similar to Bellwort (*Uvularia* spp.) and Solomon's Plume (*Smilacina racemosa*). Yellow Fairy-bells (*Prosartes lanuginosa*) is an eastern woodland species with similar leaves, half the size of Great Solomon's Seal.

Dicots

Ruellia humilis
Wild Petunia, Fringeleaf Wild Petunia

FAMILY: ACANTHUS (ACANTHACEAE)

Don't let the lovely blue flowers fool you: this is the toughest petunia you'll ever encounter. The delicate trumpetlike flowers resemble cultivated petunias. Grows in dry sandy and rocky soils, as well as loamy clay. Relatively easy from seed; not readily divided.

Habitat: Dry to medium prairies, open dry areas, bluffs, dry open woods
Garden Uses: Hummingbird gardens, informal gardens, prairie meadows. Best planted in foreground areas with other low-growing grasses and flowers, such as Prairie Smoke and Shooting Star.

USDA Hardiness Zones: 4–9
Light Requirements: Full sun to partial shade
Soil Types: Sand, loam
Soil Moisture: Dry to medium
Bloom Time: June–September
Flower Color: Lavender
Height: 1–2 feet
Life Expectancy: 5–10 years
Root Type: Fibrous
Propagation: Moist stratify seed 30 days; easy from stem cuttings early to midseason
Aggressiveness: Low by seed

Attracts: Bees, butterflies, hummingbirds
Deer Palatability: Low
Distinguishing Characteristics:
- Leaves, stems, and flower bases are typically densely hairy, although some ecotypes are nearly hairless
- Seedheads are surrounded by long, narrow upright bract leaves
- Lavender flowers usually have fine purple lines which act as nectar guides to insects

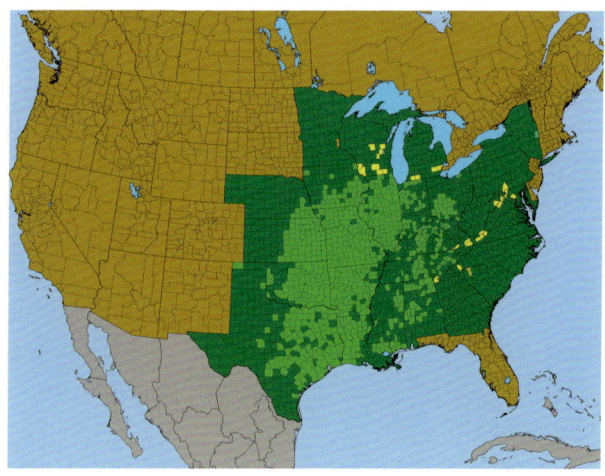

Seedling

Emerging mature plant in spring. Uppermost pair of emerging leaves are upright and directly opposite one another, like praying hands.

Entire plant

Northern strain leaf. Leaves of northern strains have densely hairy leaves and stems.

Southern strain leaf. Leaves of southern strains typically are hairless and smooth.

Flower. Lavender- to violet-colored flowers are similar to garden petunias.

Early seed

Mature seed. Seed is ripe when pods turn from green to tan.

LOOK-ALIKE PLANTS:

Natives: Other *Ruellia* spp. that occur in woodlands are usually taller.

Eryngium yuccifolium
Rattlesnake Master

FAMILY: CARROT (APIACEAE, UMBELLIFERAE)

A truly fantastical plant with Yucca-like leaves and flower clusters of pure white pompoms. Prefers limy soil with adequate nutrition. Pollinated by an amazing diversity of bees and wasps.

Habitat: Dry to moist prairies, barrens, often in calcareous soils
Garden Uses: Specimens and small groups in just about any garden situation. Plant with Prairie Blazing Star for a great midsummer color combo.

USDA Hardiness Zones: 4–9
Light Requirements: Full sun
Soil Types: Sand, loam, clay
Soil Moisture: Dry to moist
Bloom Time: June–August
Flower Color: White
Height: 3–5 feet
Life Expectancy: 5–10 years
Root Type: Fibrous
Propagation: Moist stratify seed 30 days; root division in spring or fall

Aggressiveness: Medium by seed
Attracts: Bees, butterflies, predaceous wasps
Deer Palatability: Low
Distinguishing Characteristics:
- Clusters of white, golf ball–like flower heads
- Long, blue-green leaves resemble Yucca (*Yucca filamentosa*) but do not have sharp points

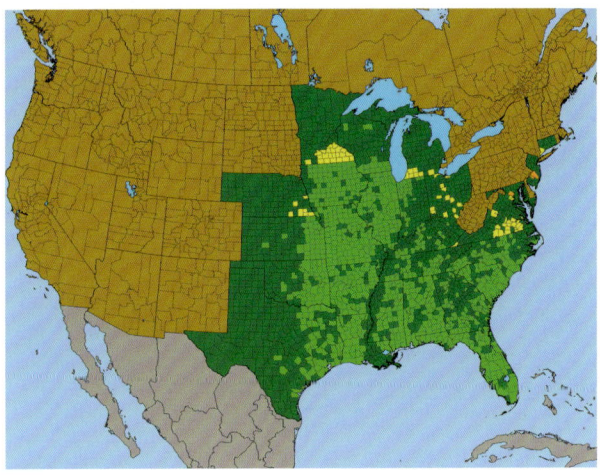

Seedling. Seedling leaves have distinct hairs on margins.

Emerging mature plant in spring

Entire plant

Leaf. Leaves are thick, smooth, and Yucca-like, with scattered long, soft hairs.

Flower

Early and mature seed

LOOK-ALIKE PLANTS:

Natives: Yucca (*Yucca filamentosa* and *Y. glauca*), a western species that seldom occurs in the same habitat as Rattlesnake Master, with large upright spikes of creamy-white flowers. Leaves of Virginia Agave (*Manfreda virginica*) are similar when emerging in spring but are soft and succulent.

Zizia aptera
Heartleaf Golden Alexanders, Meadow Zizia

FAMILY: CARROT (APIACEAE, UMBELLIFERAE)

Among the earliest-blooming flowers of medium and moist soil prairies. Smooth, glossy, heart-shaped leaves possess small silvery-red teeth. Plant seed in fall for spring germination. Division is not recommended.

Habitat: Medium to moist prairies, savannas, barrens
Garden Uses: Foreground areas, informal gardens, prairie meadows

USDA Hardiness Zones: 3–8
Light Requirements: Full sun to partial shade
Soil Types: Sand, loam, clay
Soil Moisture: Medium to moist
Bloom Time: April–June
Flower Color: Yellow
Height: 1–2 feet
Life Expectancy: 5–10 years
Root Type: Taproot
Propagation: Moist stratify seed 30 days

Aggressiveness: Low to medium by seed
Attracts: Bees, butterflies
Deer Palatability: Low
Distinguishing Characteristics:
- Heart-shaped leaves have small smooth teeth
- Stem leaves have no stalks, while stem leaves of Golden Alexanders (p. 74) are stalked (have pedicels)

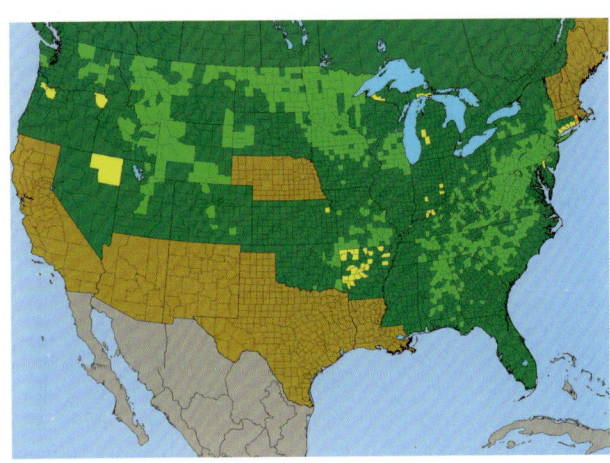

Seedling. First true leaves only minimally notched on edges, compared to *Z. aurea*.

Emerging mature plant in spring

Entire plant. Plant is shorter than *Z. aurea*.

Leaf. Leaves occur as rounded individuals, with no divisions into leaflets, and leaf edges have distinctive teeth, often with a silvery hue.

Flower. Flowers visually indistinguishable from *Z. aurea*.

Early seed. Seedheads visually indistinguishable from *Z. aurea*.

Mature seed

LOOK-ALIKE PLANTS:

Natives: Golden Alexanders (p. 74) have essentially identical flowers, but with three- to five-parted stem leaves, instead of one-or three-parted stem leaves on *Z. aptera*. Yellow Pimpernel (*Taenidia integerrima*) has very similar flowers and leaves but occurs in dry to mesic woods. Its leaves are usually three times divided and smooth-edged with no teeth. Meadow Parsnips (*Thaspium barbinode* and *T. trifoliatum*) have similar flowers. *T. barbinode* leaflets have large blunt teeth and occur in pairs or triplets, the terminal leaflet being largest. Grows up to 3.5 feet tall, twice the average height of *Z. aptera*. *T. trifoliatum* has three-lobed stem leaves on pedicels, while *Z. aptera* stem leaves are without pedicels.

Zizia aurea
Golden Alexanders, Golden Zizia

FAMILY: CARROT (APIACEAE, UMBELLIFERAE)

Taller than Heartleaf Golden Alexanders, has three-lobed leaves borne up stems rather than at the base. Often self-sows. Grows well in clay loam and damp sand, in full to moderate shade. Like Heartleaf Golden Alexanders, seed requires cold treatment for good germination.

Habitat: Medium to wet prairies, wet meadows, fens, open rich woods
Garden Uses: Informal gardens, prairie meadows, woodlands

USDA Hardiness Zones: 3–8
Light Requirements: Full sun to medium shade
Soil Types: Sand, loam, clay
Soil Moisture: Medium to wet
Bloom Time: April–July
Flower Color: Yellow
Height: 1–2 feet
Life Expectancy: 5–10 years
Root Type: Fibrous

Propagation: Moist stratify seed 30 days
Aggressiveness: Low by seed
Attracts: Bees, butterflies
Deer Palatability: Low
Distinguishing Characteristics:
- Basal leaves are three- to five-parted
- Stem leaves are stalked, with pedicels

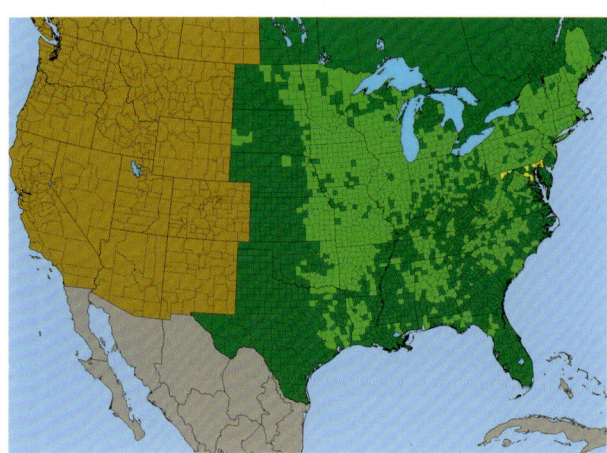

Seedling. First true leaves are more deeply notched on edges than *Z. aptera*.

Emerging mature plant in spring

Entire plant. Plant grows 2–4 ft. tall; *Z. aptera* grows 1–2 ft. high.

Leaf. Leaves are divided into five leaflets, sometimes into three leaflets at the base of the plant.

Flower. Flowers visually indistinguishable from *Z. aptera*.

Mature seed. Seedheads visually indistinguishable from *Z. aptera*.

LOOK-ALIKE PLANTS:

Natives: Heartleaf Golden Alexanders (p. 72) have nearly identical flowers but with single- to three-parted stem leaves, as opposed to three- to five-parted stem leaves on *Z. aurea*. Yellow Pimpernel (*Taenidia integerrima*) has very similar flowers and leaves, and occurs in similar woodland habitats. The leaves are usually three times divided and smooth-edged with no teeth. Meadow Parsnips (*Thaspium barbinode* and *T. trifoliatum*) have similar flowers. *T. barbinode* leaflets have large blunt teeth and occur in pairs or triplets, the terminal leaflet being the largest. Grows up to 3.5 feet tall, twice the average height of *Z. aurea*. *T. trifoliatum* is nearly identical but can be distinguished by close examination of the flower. The central flower of each umbellet is borne on a short stalk (pedicel), and seeds have wings when ripe.

Amsonia tabernaemontana
Eastern Bluestar

FAMILY: DOGBANE (APOCYNACEAE)

An eastern species best known for its pastel blue flowers and compact growth form. Common in savannas and open woodlands, it grows well in full sun. An extraordinary variety of butterflies utilize the plant as a nectar source.

Habitat: Medium to moist meadows, damp woodland edges, moist woods, seeps
Garden Uses: Borders, rain gardens, specimens, and swaths in just about any garden situation

USDA Hardiness Zones: 4–8
Light Requirements: Full sun to partial shade
Soil Types: Sand, loam, clay
Soil Moisture: Medium to moist
Bloom Time: May–June
Flower Color: Blue, white
Height: 2–3 feet
Life Expectancy: 10–20 years
Root Type: Fibrous
Propagation: Moist stratify seed 30 days; tricky from cuttings: use first early growth
Aggressiveness: Medium by seed
Attracts: Bees, butterflies, hummingbirds

Deer Palatability: Low
Distinguishing Characteristics:
- Flowers are present within the leaves as plants emerge from ground in spring
- Five-pointed petals form star-like flowers that are robin's-egg blue
- Plants have white, milky sap
- Leaves are wider than other similar plants: 6 inches long and 2.5 inches wide
- Foliage turns dramatic bright gold in fall

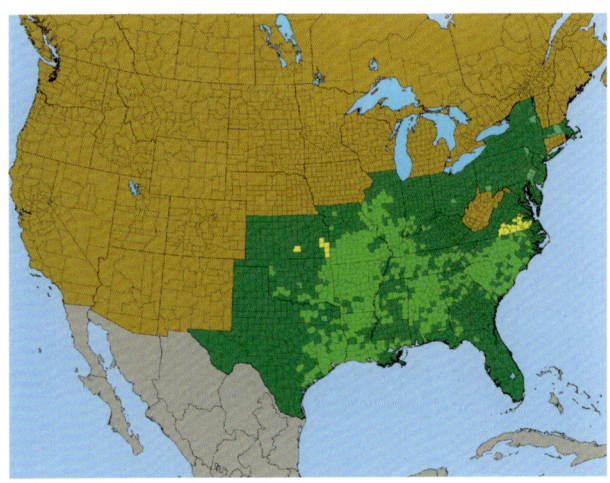

Seedling

Emerging mature plant in spring

Entire plant

Leaf. Leaf is smooth and glossy, with distinct midrib.

Flower. Flower resembles a five-pointed star.

Early seed

Mature seed. Mature seeds are chocolate brown inside seedpod.

LOOK-ALIKE PLANTS:

Natives: Arkansas Bluestar (*Amsonia hubrichtii*) is similar in form and flower but has very narrow leaves, only 0.0625 inches wide. Dogbanes (*Apocynum cannabinum* and *A. androsaemifolium*) have similar leaves with white, milky sap that are opposite on the stem, whereas Bluestars' are alternate.

Asclepias incarnata

Swamp Milkweed, Red Milkweed, Marsh Milkweed

FAMILY: MILKWEED (ASCLEPIADACEAE)

Flowers range from light pink to red, rare hues among prairie flowers. A favorite larval food source of monarch butterflies, *A. incarnata* grows on any damp soil and is impervious to most diseases. Excellent for rain gardens because of its ability to tolerate both moist and dry conditions. Easy to grow from seed, and the roots are easily divided.

Habitat: Wet prairies, marshes, fens, lakeshores, wet swales
Garden Uses: Specimens and small groups in rain gardens, wet meadow seedings; along creek banks, lakeshores, and pond edges

USDA Hardiness Zones: 3–9
Light Requirements: Full sun to partial shade
Soil Types: Sand, loam, clay
Soil Moisture: Moist to wet
Bloom Time: June–August
Flower Color: Pink, red
Height: 3–5 feet
Life Expectancy: 5–10 years
Root Type: Fibrous, sometimes a taproot in its southern range
Propagation: Moist stratify seed 30 days; root division in fall or spring
Aggressiveness: Low by seed
Attracts: Bees, butterflies, hummingbirds
Deer Palatability: Low
Distinguishing Characteristics:
  • One of only two deep-pink- to red-flowered milkweed in United States and Canada (*A. amplexicaulis* is also red; *A. purpurascens* is purple)
  • Long, narrow seedpods stand upright (though not unique among milkweeds)
  • Only true wetland milkweed — *Asclepias hirtella* will sometimes grow in wet-mesic prairies but has white flowers and long, narrow leaves (0.125–0.25 inches wide versus 1–1.5 inches wide for *A. incarnata*)
  • Leaf thickness is less than other sun-loving species of *Asclepias*; *A. exaltata* also has thin leaves but grows only in woodlands
  • One of few fibrous-rooted milkweeds; most other species are taprooted or rhizomatous

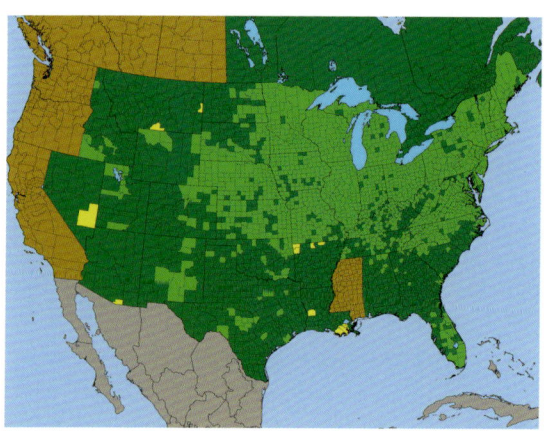

Seedling

Emerging mature plant in spring

Entire plant

Leaf

Flower

Mature seed

LOOK-ALIKE PLANTS:

Natives: Other milkweed species appear similar when emerging.

Weeds: Dogbane (*Apocynum androsaemifolium*) and Indianhemp (*Apocynum cannabinum*) are native, weedy species that spread aggressively by rhizomes. They have short, narrow leaves and smaller flowers.

Asclepias sullivantii
Sullivant's Milkweed, Prairie Milkweed

FAMILY: MILKWEED (ASCLEPIADACEAE)

Very similar to Common Milkweed but does not spread as rapidly by rhizomes. This better-behaved milkweed is an excellent plant for hosting monarch butterfly larvae; its flowers attract many species of butterflies.

Habitat: Medium to moist prairies, fens
Garden Uses: Butterfly gardens, bee and predaceous-wasp gardens, prairie meadows

USDA Hardiness Zones: 4–7
Light Requirements: Full sun
Soil Types: Loam, clay
Soil Moisture: Medium to moist
Bloom Time: June–August
Flower Color: Lavender
Height: 3–5 feet
Life Expectancy: 10–20 years
Root Type: Rhizome
Propagation: Moist stratify seed 30 days
Aggressiveness: Low by rhizomes
Attracts: Bees, butterflies, hummingbirds
Deer Palatability: Low

Distinguishing Characteristics:
- Central vein of each leaf is pink to red
- Seedpods and undersides of leaves are hairless (*A. syriaca* has hairs)
- Flower buds are generally twice the size of Common Milkweed buds
- Seedpods are usually smooth or have only a few bumps compared to Common Milkweed, which have many bumps
- Seedlings grow tall and elongated; the first true leaves are thin and narrow

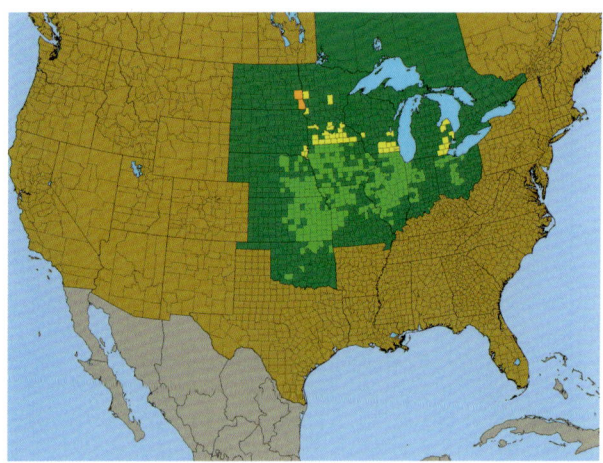

Seedling. Note long, narrow seedling leaves.

Emerging mature plant in spring. Emerging leaves are narrower than *Asclepias syriaca*.

Entire plant

Leaf

Flower

Early seed. Note lack of bumps on pods, compared to *Asclepias syriaca*.

Mature seed. Seeds can be harvested when pods are still green and closed, but seeds within are brown.

LOOK-ALIKE PLANTS:

Natives: Common Milkweed (p. 82) are similar in form and flower, with differences noted above.

Weeds: Dogbane (*Apocynum androsaemifolium*) and Indianhemp (*Apocynum cannabinum*) are native, weedy species that spread aggressively by rhizomes. They have short, narrow leaves and smaller flowers.

Asclepias syriaca
Common Milkweed

FAMILY: MILKWEED (ASCLEPIADACEAE)

Leaves are a favorite food of monarch butterfly larvae. Pollinated by a tremendous variety of insects attracted to the sweet-scented flowers. Creeps rapidly by rhizomes, and is best used in meadows with robust prairie flowers and grasses to keep it in check.

Habitat: Dry to medium prairies, old fields, pastures, waste areas, roadsides, open woodlands, dunes along the Great Lakes
Garden Uses: Use sparingly in meadow seedings. Not recommended for gardens.

USDA Hardiness Zones: 3–8
Light Requirements: Full sun to partial shade
Soil Types: Sand, loam, clay
Soil Moisture: Dry to medium
Bloom Time: June–August
Flower Color: Lavender
Height: 2–4 feet
Life Expectancy: 10–20 years
Root Type: Rhizome
Propagation: Moist stratify seed 30 days; root division using 8-inch rhizomes in spring or fall
Aggressiveness: High by seed and rhizome

Attracts: Bees, butterflies, hummingbirds
Deer Palatability: Low
Distinguishing Characteristics:
- Often forms large patches via rhizomes
- Flowers are aromatic and sweet scented
- Leaf undersides are covered with fine hairs, while most other *Asclepias* spp. are not
- Seedpods are typically covered with bumps

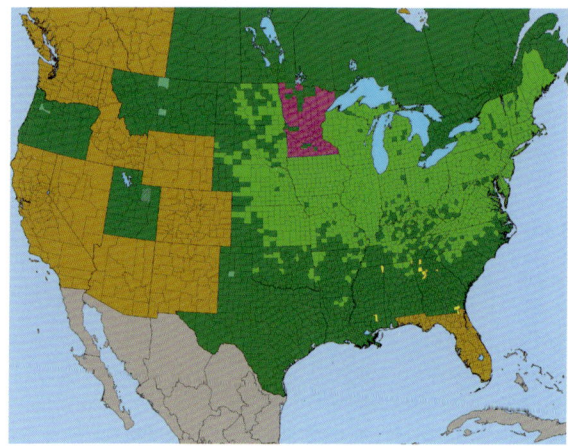

Seedling

Emerging mature plant in spring. Emerging leaves are wider than *Asclepias sullivantii*.

Entire plant

Leaf

Flower

Early seed. Note pronounced bumps on seedpods.

Mature seed. Seeds can be harvested when pods are still green and closed but seeds within are brown.

LOOK-ALIKE PLANTS:

Natives: Sullivant's Milkweed (p. 80) leaves have pink midribs and essentially no hairs on the underside, and seedpods have fewer bumps. Emerging plants are similar to Flowering Spurge (p. 192) and members of the genus *Apocynum*, all of which have white, milky sap.

Weeds: Dogbane (*Apocynum androsaemifolium*) and Indianhemp (*Apocynum cannabinum*) are native, weedy species that spread aggressively by rhizomes. They have short, narrow leaves and smaller flowers.

Asclepias tuberosa
Butterflyweed, Butterfly Milkweed

FAMILY: MILKWEED (ASCLEPIADACEAE)

Its orange flowers are unique among milkweeds. Pollinated by a diversity of butterflies and insects. Drought resistant, tough, and noninvasive. An outstanding plant for gardens and meadows on well-drained soils. Mix with short prairie grasses to set off and keep weeds at bay. Easy from seed. Taproot cannot easily be divided.

Habitat: Dry to medium prairies, glades, barrens, especially in sandy soils
Garden Uses: Specimens and small groups, dry prairie meadows

USDA Hardiness Zones: 3–10
Light Requirements: Full sun
Soil Types: Sand, loam
Soil Moisture: Dry to medium
Bloom Time: June–August
Flower Color: Orange
Height: 2–3 feet
Life Expectancy: 10–20 years
Root Type: Taproot
Propagation: Moist stratify seed 30 days
Aggressiveness: Low by seed
Attracts: Bees, butterflies, hummingbirds

Deer Palatability: Low
Distinguishing Characteristics:
- Only orange-flowered milkweed in United States and Canada
- Sap is clear rather than the milky, white sap typical of most other milkweeds
- Seedpods stand upright (although this is not unique among milkweeds)
- Roots have a distinctive fragrance

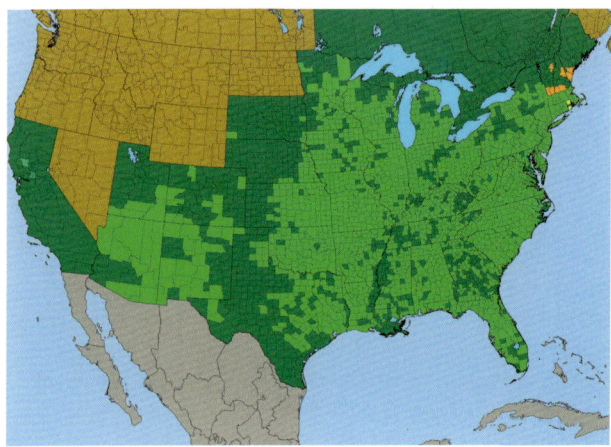

Seedling. Seedling stems are hairy; most other *Asclepias* are hairless.

Emerging mature plant in spring. Emerging stems are hairy; most other *Asclepias* are hairless.

Entire plant

Leaf. Sap is clear, not white, as with other milkweeds.

Flower close-up. One of only two orange flowers of the prairie, along with Hoary Puccoon.

Flower

Mature seed. Seeds can be harvested when pods are still green and closed but seeds within are brown.

LOOK-ALIKE PLANTS:

Natives: Emerging plant is similar to Flowering Spurge (p. 192), which has milky, white sap rather than clear sap.

ASTERACEAE

Antennaria neglecta
Pussytoes, Cat's Foot

FAMILY: ASTER (ASTERACEAE, COMPOSITAE)

One of the lowest-growing prairie plants, seldom taller than 1 foot. Light-green leaves are tinged with silvery-white fuzz on their undersides. White to pinkish seedheads give the appearance of a cat's toe, hence the name.

Habitat: Dry prairies, savannas, barrens, glades, dry open woods
Garden Uses: Groundcover on dry sandy or rocky soils, rock gardens, interplanting between taprooted and corm-producing flowers to provide soil cover

USDA Hardiness Zones: 3–7
Light Requirements: Full sun to light shade
Soil Types: Sand, gravel
Soil Moisture: Dry
Bloom Time: April–May
Flower Color: White
Height: 0.5–1 feet
Life Expectancy: 5–10 years
Root Type: Rhizome
Propagation: Moist stratify seed 30 days; root division in early spring
Aggressiveness: Low by seed
Attracts: Bees, butterflies

Deer Palatability: Low
Distinguishing Characteristics:
- Leaf upper side is silvery light green and covered with hairs; underside is covered with dense white hairs
- Leaves grow close to the ground, with a few small leaves on the stem
- Flower buds often have a light rosy hue prior to the flowers opening
- Flower stalks seldom exceed 1 foot

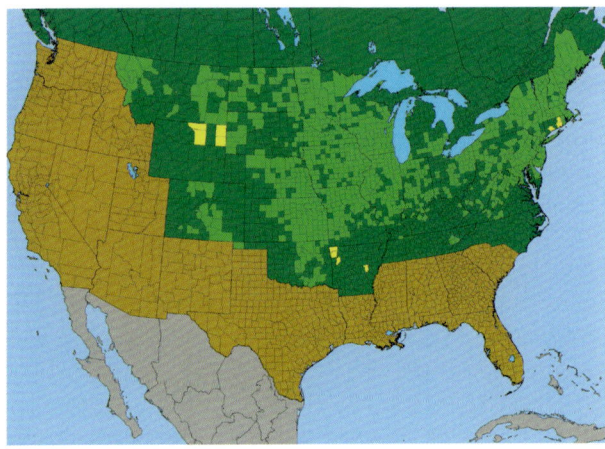

Seedling

Emerging mature plant in spring

Entire plant

Leaf

Male flower. Reddish-pink male stamens with white tips are thicker than the female pistils.

Female flower. Translucent white female pistils are thinner than the whitish male stamens.

Mature seed. Seed is ripe when it begins to fly!

LOOK-ALIKE PLANTS:

Natives: Numerous other species of *Antennaria* grow in dry habitats in full sun to medium shade. *A. parlinii* has bright green leaves on the upper side with distinct white margins, and more rosy-pink flower buds. *A. plantaginifolia* grows in open woods and has rounded leaves. *A. howellii* has green leaves that lack hairs on the upper side.

Arnoglossum atriplicifolium
(Cacalia atriplicifolia)
Pale Indian Plantain

FAMILY: ASTER (ASTERACEAE, COMPOSITAE)

Deep-green leathery leaves adorn this statuesque specimen. More notable for their structure and foliage than for their small white flowers. Grows in almost any soil. Appears to behave as a biennial in northern climates, while more long-lived in southern zones.

Habitat: Dry to wet prairies, savannas, open woods, swamps, thickets
Garden Uses: Specimen, small groups, combines well with shrublike species such as Blue False Indigo and White False Indigo; rain gardens

USDA Hardiness Zones: 4–8
Light Requirements: Full sun to partial shade
Soil Types: Sand, loam, clay
Soil Moisture: Dry to wet
Bloom Time: July–September
Flower Color: White
Height: 5–15 feet
Life Expectancy: 2–3 years (often biennial)
Root Type: Fibrous
Propagation: Moist stratify seed 30 days
Aggressiveness: High in northern latitudes, lower in southern latitudes
Attracts: Bees, predaceous wasps
Deer Palatability: Low
Distinguishing Characteristics:
- Thick, triangular leathery leaves are green on the upper side, white on the underside
- First-year plants form large basal clumps
- Plants typically grow 5–12 feet tall, much taller than other species
- Distinctly purple stems

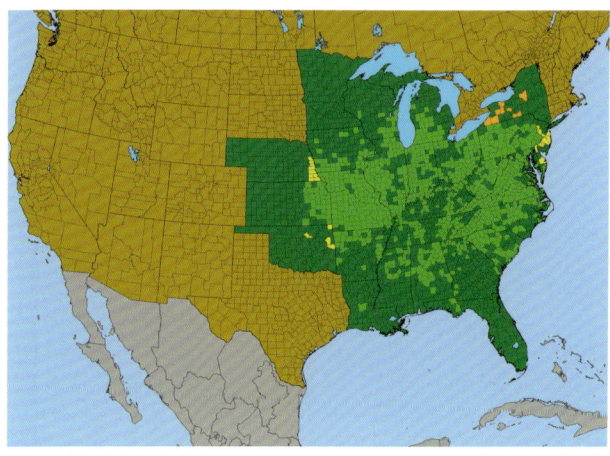

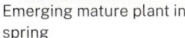
Seedling

Emerging mature plant in spring

Entire plant. Stems are often purple.

Leaf. Leaves are thick, smooth, and rubbery to the touch.

Flower close-up

Flower

Early and mature seed. Mature seeds are sticky; they actually move by themselves in the hand!

LOOK-ALIKE PLANTS:

Natives: Other Indian Plantains (*Arnoglossum reniforme, A. suaveolens*) have similar form and flowers but grow only 3–5 feet and have smaller, narrower leaves and green stems. *A. plantagineum* also has purple stems but grows only 3–7 feet, with kidney-shaped basal leaves, green on the underside, not white.

Weeds: Burdock (*Arctium minus*) is similar as it emerges in spring, but has densely hairy leaves. Elecampane (*Inula helenium*) has large, glossy leaves as it emerges from the ground.

Conoclinium coelestinum
(Eupatorium coelestinum)
Blue Mistflower

FAMILY: ASTER (ASTERACEAE, COMPOSITAE)

It's a groundcover. It's a butterfly magnet. And it's downright gorgeous! This is a great plant for troublesome damp spots that won't grow lawn or traditional garden plants. A few plants go a long way and will rapidly form an attractive, low-maintenance groundcover. One cultivar, "Cory," is more compact and blooms later than the species.

Habitat: Wet meadows, stream banks, woodland edges, wet woods, ditches

Garden Uses: Groundcover in moist soils in sun or part shade, companion to taprooted or corm-forming plants to occupy surrounding soil and reduce weeds

USDA Hardiness Zones: 4–10
Light Requirements: Full sun to partial shade
Soil Types: Sand, loam, clay
Soil Moisture: Medium to moist
Bloom Time: July–October
Flower Color: Blue, purple
Height: 1–3 feet
Life Expectancy: 3–5 years
Root Type: Rhizome
Propagation: Moist stratify seed 30 days; root division in spring; easy from stem cuttings early to midseason

Aggressiveness: Medium by rhizome
Attracts: Bees, butterflies, hummingbirds, songbirds
Deer Palatability: Low
Distinguishing Characteristics:
- Creeps rapidly by rhizomes to form a patch
- Light-green leaves are heavily textured
- Plants are covered with filamentous purplish-pink flowers in late summer, often creating the appearance of a ground mist

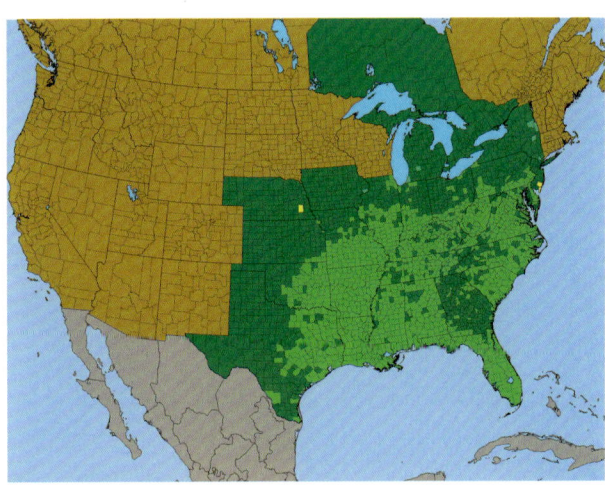

Seedling

Emerging mature plant in spring

Entire plant

Leaf

Flower

Early and mature seed. Mature seeds are white.

LOOK-ALIKE PLANTS:

Natives: Emerging leaves are similar to *Eupatorium* spp. but do not develop individual tall stems.

Weeds: Leaves resemble Clearweed (*Pilea pumila*), a native plant of moist, shady habitats.

Coreopsis lanceolata
Lanceleaf Coreopsis

FAMILY: ASTER (ASTERACEAE, COMPOSITAE)

Showy and easy to grow, thrives in dry, sandy soil but does well in any well-drained soil. Somewhat short-lived, self-sows readily onto open soil. Blooms for an extended period in late spring into summer. Utilized by a wide variety of butterflies and other pollinators.

Habitat: Dry sandy and rocky areas, sand barrens, sand dunes
Garden Uses: Butterfly gardens, bird gardens, borders, dry gardens, prairie meadows. Can be deadheaded to extend bloom well into autumn.

USDA Hardiness Zones: 3–9
Light Requirements: Full sun to light shade
Soil Types: Sand, loam
Soil Moisture: Dry to medium
Bloom Time: May–August
Flower Color: Yellow
Height: 1–2 feet
Life Expectancy: 3–5 years
Root Type: Fibrous
Propagation: Moist stratify seed 30 days; root division spring or fall; tricky from cuttings: use first early growth

Aggressiveness: Medium by seed
Attracts: Bees, butterflies
Deer Palatability: Low
Distinguishing Characteristics:
- Ends of petals have a prominent center tooth
- Leaves have a deep-green glossy cast
- Rounded seeds are black in the center, surrounded by curled brown "wings"

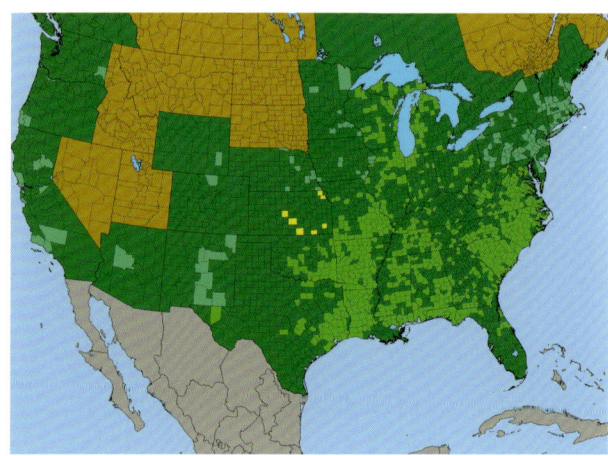

Seedling

Emerging mature plant in spring

Entire plant

Leaf. Some leaves may be three lobed, especially at base of plant.

Flower. Petals overlap, tips visibly indented.

Early seed

Mature seed. Seeds are ripe when pods turn black.

LOOK-ALIKE PLANTS:

Natives: *Coreopsis tinctoria* (Plains Coreopsis) has narrow, bipinnately divided leaves; inner petals and sepals are usually bright red. Leaves of Ohio Goldenrod (p. 134) are similar in spring but generally more elongated.

Coreopsis palmata
Stiff Coreopsis

FAMILY: ASTER (ASTERACEAE, COMPOSITAE)

Taller and less showy than Lanceleaf Coreopsis, Stiff Coreopsis creeps by rhizomes to form a patch. Highly drought tolerant. Somewhat difficult to grow from seed.

Habitat: Dry to medium prairies, sand prairies, sand barrens, savannas, dry open woods
Garden Uses: Prairie meadows. Not recommended for gardens because of its rhizomatous nature.

USDA Hardiness Zones: 3–8
Light Requirements: Full sun
Soil Types: Sand, loam
Soil Moisture: Dry to medium
Bloom Time: June–August
Flower Color: Yellow
Height: 2–3 feet
Life Expectancy: 20+ years
Root Type: Rhizome
Propagation: Moist stratify seed 30 days; root division using 6-inch rhizomes in spring or fall; tricky from cuttings: use first early growth
Aggressiveness: Medium by rhizome

Attracts: Bees, butterflies
Deer Palatability: Low
Distinguishing Characteristics:
- Leaves are narrow and distinctly three lobed
- Tips of petals have less pronounced tooth compared to Lanceleaf Coreopsis (p. 92)
- Only a few flowers are borne atop thin stems, compared to many flowers closer to the ground on Lanceleaf Coreopsis
- Seedpods are smaller in diameter, at 0.5–0.75 inches, than Lanceleaf Coreopsis, at 1–1.5 inches

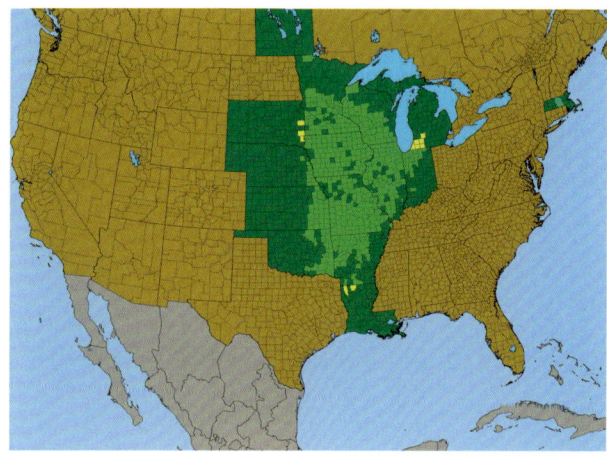

Seedling

Emerging mature plant in spring

Entire plant

Leaf. All leaves are strongly three lobed, with distinct midribs.

Flower. Petals distinct and not overlapping; tips only slightly indented.

Early seed

Mature seed. Seeds are ripe when pods turn black.

Coreopsis tripteris
Tall Coreopsis

FAMILY: ASTER (ASTERACEAE, COMPOSITAE)

Bright lemon-yellow flowers borne in late summer and early fall on smooth, greenish-blue stems. Grows in almost any good garden soil and makes a strong late-season statement.

Habitat: Medium to wet prairies, damp open woods, wet thickets
Garden Uses: As a specimen, in small groups, back of borders

USDA Hardiness Zones: 3–8
Light Requirements: Full sun to partial shade
Soil Types: Sand, loam, clay
Soil Moisture: Medium to wet
Bloom Time: July–September
Flower Color: Yellow
Height: 3–8 feet
Life Expectancy: 10–20 years
Root Type: Fibrous
Propagation: Moist stratify seed 30 days; tricky from cuttings: use first early growth

Aggressiveness: Low by seed
Attracts: Bees, butterflies, songbirds
Deer Palatability: Low
Distinguishing Characteristics:
- Tallest and latest-blooming of North American *Coreopsis* spp.
- Narrow, three-parted leaves are smooth and hairless, often with a bluish cast
- Lightly winged striations occur along the waxy stems

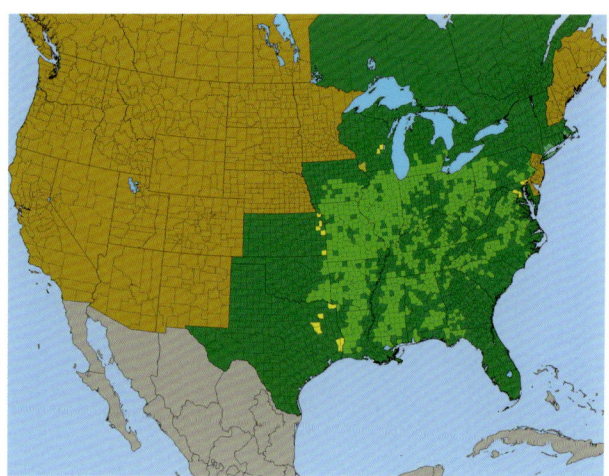

Seedling

Emerging mature plant in spring

Entire plant. Plant much taller than other native *Coreopsis*, up to 8 ft.

Leaf

Flower

Early seed

Mature seed. Seeds are ripe when pods turn black.

LOOK-ALIKE PLANTS:

Natives: Leaves are similar in appearance to Cowbane (*Oxypolis rigidior*), a toxic native plant that also grows in wet habitats. This Parsley family member has white flowers in midsummer rather than yellow flowers in late summer.

Echinacea pallida
Pale Purple Coneflower

FAMILY: ASTER (ASTERACEAE, COMPOSITAE)

Taller, longer-lived, and more drought tolerant than Purple Coneflower, flowers appear a full month earlier. Very popular with butterflies! Birds use seedstalks for perches and food. Seed germinates the following spring when planted in fall.

Habitat: Dry to medium prairies
Garden Uses: Butterfly gardens, bird gardens, informal borders and prairie meadows. Pastel purple flowers mix well with white flowers such as Wild Quinine and prairie grasses.

USDA Hardiness Zones: 4–8
Light Requirements: Full sun
Soil Types: Sand, loam, clay
Soil Moisture: Dry to medium
Bloom Time: May–July
Flower Color: Purple
Height: 3–5 feet
Life Expectancy: 10–20 years
Root Type: Taproot
Propagation: Moist stratify seed 30 days
Aggressiveness: Low by seed
Attracts: Bees, butterflies, game birds, hummingbirds, songbirds

Deer Palatability: Medium
Distinguishing Characteristics:

- Petals droop, in contrast to more rigid petals of Purple Coneflower (p. 102)
- Most leaves occur at base of plant, with only a few on the flower stalks
- Leaves are covered with small hairs; Ozark Coneflower leaves are hairless and shiny
- Emerging leaf buds pointed and strongly tinged with red

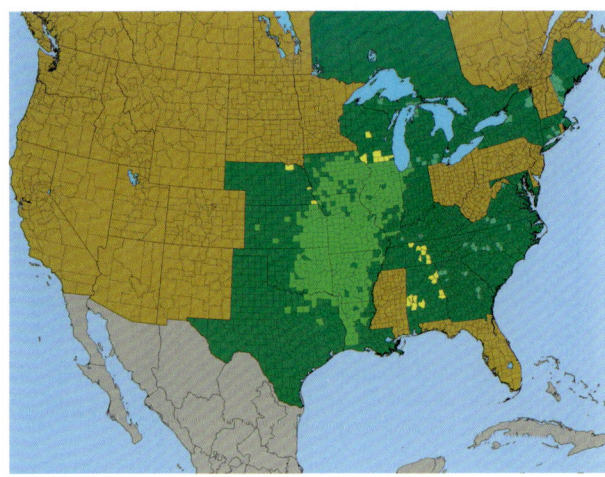

Seedling. Seedling leaves have distinct hairs on margins.

Emerging mature plant in spring. Emerging shoots have reddish bases.

Entire plant

Leaf. Leaves have pronounced midrib and veins.

Flower. Drooping petals distinguish *Echinacea pallida* from *Echinacea purpurea*.

Mature seed. Seed is ripe when heads turn black.

LOOK-ALIKE PLANTS:

Natives: Ozark Coneflower (p. 100) has glossy leaves and yellow flowers; Purple Coneflower (p. 102) has wider, arrowlike leaves and flower petals do not droop. *E. angustifolia*, a Great Plains species, is similar but shorter. *E. simulata* is identical but has yellow pollen instead of white, occurring primarily in limestone glades of southeastern Missouri, and scattered in Arkansas, Illinois, Kentucky, Tennessee, southern Indiana, northern Georgia, and Alabama.

Echinacea paradoxa
Ozark Coneflower, Bush's Coneflower

FAMILY: ASTER (ASTERACEAE, COMPOSITAE)

Nearly identical to Pale Purple Coneflower except for its yellow petals and glossy green leaves. Plant seed in fall for spring germination. Does not self-sow as readily as Purple Coneflower.

Habitat: Dry to medium prairies, barrens, glades
Garden Uses: Butterfly gardens, bird gardens, informal borders, prairie meadows. Mixes well with prairie grasses.

USDA Hardiness Zones: 4–7
Light Requirements: Full sun to partial shade
Soil Types: Sand, loam, clay
Soil Moisture: Dry to medium
Bloom Time: June–August
Flower Color: Yellow
Height: 3–5 feet
Life Expectancy: 10–20 years
Root Type: Taproot
Propagation: Moist stratify seed 30 days

Aggressiveness: Low by seed
Attracts: Bees, butterflies, game birds, hummingbirds, songbirds
Deer Palatability: Medium
Distinguishing Characteristics:
- Yellow drooping petals on a large cone head
- Deep-green, glossy, hairless leaves

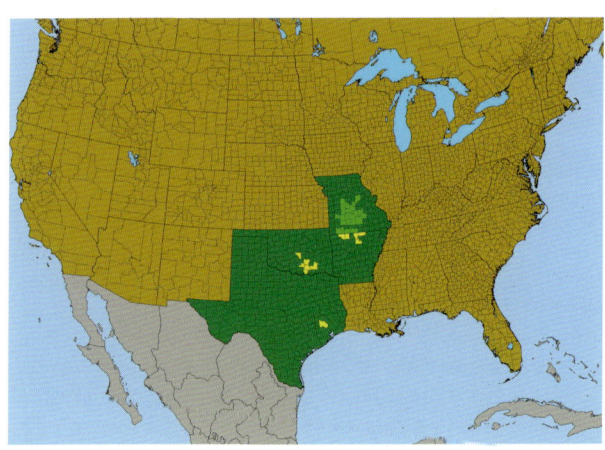

Seedling. Seedlings are virtually indistinguishable from *Echinacea pallida*.

Emerging mature plant in spring

Entire plant

Leaf. Leaves are similar to *Echinacea pallida*, except smoother and deeper green.

Flower

Mature seed. Seed is ripe when heads turn black.

LOOK-ALIKE PLANTS:

Natives: Pale Purple Coneflower (p. 98) is similar in form but has purple flowers and hairy, nonglossy leaves.

Echinacea purpurea
Purple Coneflower

FAMILY: ASTER (ASTERACEAE, COMPOSITAE)

Not a true prairie plant, this showy denizen of woodland edges blooms from midsummer into autumn. Highly adaptable, grows in well-drained and moist soils. Very easy from seed. Attracts many species of butterflies. Songbirds relish its seeds.

Habitat: Medium to moist prairies, open woods
Garden Uses: Excellent for almost any garden and meadow application, especially butterfly and bird gardens. Although not as long-lived as other Echinacea, it self-sows readily and typically increases in the garden. In prairie meadows does best with morning or afternoon shade.

USDA Hardiness Zones: 4–8
Light Requirements: Full sun to partial shade
Soil Types: Sand, loam, clay
Soil Moisture: Medium to moist
Bloom Time: June–September
Flower Color: Purple
Height: 3–4 feet
Life Expectancy: 5–10 years
Root Type: Fibrous
Propagation: Moist stratify seed 30 days
Aggressiveness: Medium by seed
Attracts: Bees, butterflies, game birds, hummingbirds, songbirds

Deer Palatability: Medium
Distinguishing Characteristics:
- Large purple to pink flowers have orange centers (some other *Echinacea* also have orange centers)
- Petals extend out horizontally from center; *E. pallida* and *E. paradoxa* droop (some other less common Coneflower species have horizontal petals)
- Leaves are wide and arrow-shaped (some less common *Echinacea* have similar leaves)

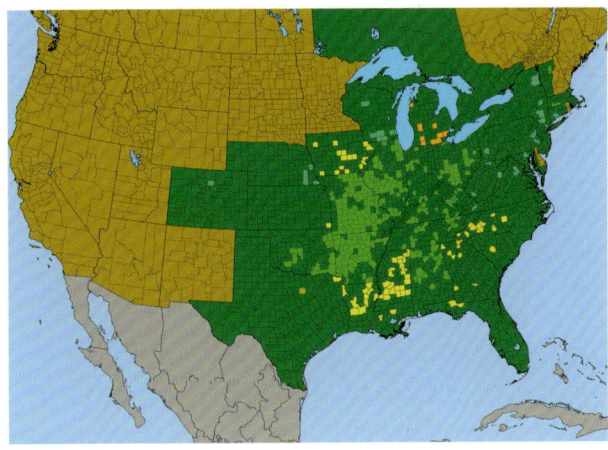

Seedling. Seedling leaves are wider than other *Echinacea* species, with short hairs on margins.

Emerging mature plant in spring

Entire plant

Basal leaf. Leaf surface is essentially hairless; *Rudbeckia triloba* leaf surface is hairy.

Stem leaf

Flower. Flower petals are horizontal to slightly drooping.

Early and mature seed

LOOK-ALIKE PLANTS:

Natives: Pale Purple Coneflower (p. 98) and Ozark Coneflower (p. 100) are similar but have thicker, narrower leaves. Two rare species, *Echinacea laevigata* and *E. tennesseensis*, which occur naturally in the southeastern coastal states and Tennessee respectively, have similar flowers but are shorter in stature. Leaves are similar to Orange Coneflower (p. 142) and Sweet Blackeyed Susan (p. 148).

Eupatorium perfoliatum
Boneset, Common Boneset

FAMILY: ASTER (ASTERACEAE, COMPOSITAE)

Crinkled lime-green leaves surround the stem. Pure white flowers distinguish this from other Joe Pyes. Self-sows prolifically. Long stems with leaves look great in cut flower bouquets. It was once believed this plant helped set broken bones.

Habitat: Wet prairies, marshes, sedge meadows, fens, lakeshores, streamsides, wet woods and swamps
Garden Uses: Plant with Dense Blazing Star for a lavender and white combo. Remove flowers prior to seed set to prevent proliferation. Reliable from seed in wet meadow restorations.

USDA Hardiness Zones: 3–8
Light Requirements: Full sun to part shade
Soil Types: Sand, loam, clay
Soil Moisture: Medium, moist, wet
Bloom Time: July–October
Flower Color: White
Height: 3–4 feet
Life Expectancy: 5–10 years
Root Type: Fibrous
Propagation: Moist stratify seed 30 days; root division in spring or fall; tricky from cuttings: use first early growth

Aggressiveness: Medium by seed
Attracts: Bees, birds, butterflies
Deer Palatability: Low
Distinguishing Characteristics:

- Only white *Eupatorium* that grows in full sun (*E. rugosum* has white flowers but grows in dry to medium woodlands)
- Leaves are heavily textured and pierced completely by the stem (perfoliate)
- Shorter than other *Eupatorium* that grow in open wetlands, seldom exceeding 4 feet in height

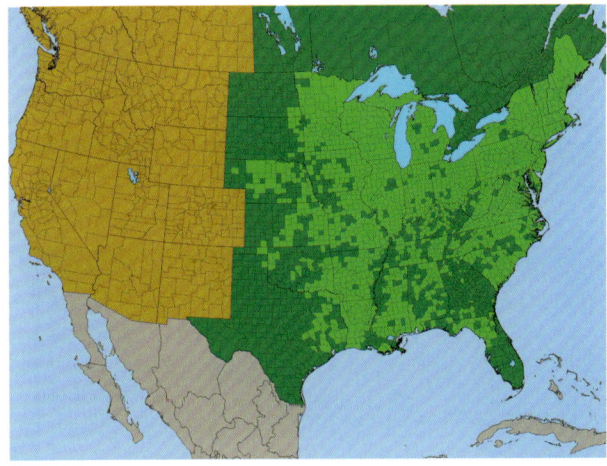

Seedling

Emerging mature plant in spring. Emerging leaves are furry, with numerous dense hairs.

Entire plant. Plants are usually shorter at 3–4 ft. compared with *Eutrochium maculatum* at 5–7 ft.

Leaf. Leaves are joined, with the stem penetrating through them.

Flower

Mature seed. Seed is ripe when it begins to fluff out.

LOOK-ALIKE PLANTS:

Natives: Leaves are similar to *Eupatorium purpureum*, a purple-flowered woodland species. *E. rugosum* is similar but grows in shaded uplands and does not have perfoliate leaves.

Eutrochium fistulosum (*Eupatorium fistulosum, Eupatoriadelphus fistulosus*)

Tall Joe Pye Weed, Trumpetweed

FAMILY: ASTER (ASTERACEAE, COMPOSITAE)

King of the Joe Pyes, tallest of stature and largest of flowers, a sentinel of the garden, great for back of the border or centerpiece for island beds. Attracts a plethora of butterflies. Requires rich soil with reasonable organic matter. A truly stunning plant!

Habitat: Wet prairies and meadows, streamsides, wet woods and woodland edges

Garden Uses: Rain gardens, butterfly gardens, bird gardens, wet meadow seedings

USDA Hardiness Zones: 4–9
Light Requirements: Full sun to partial shade
Soil Types: Sand, loam, clay
Soil Moisture: Medium to moist
Bloom Time: July–September
Flower Color: Purple, pink
Height: 5–8 feet
Life Expectancy: 5–10 years
Root Type: Fibrous
Propagation: Moist stratify seed 30 days; root division in spring or fall; tricky from cuttings: use first early growth

Aggressiveness: Low by seed
Attracts: Bees, butterflies, songbirds
Deer Palatability: Low
Distinguishing Characteristics:
- Largest and tallest of North American *Eutrochium*
- Very large domed flower clusters are up to 12 inches across
- Stems are streaked with lines of purple

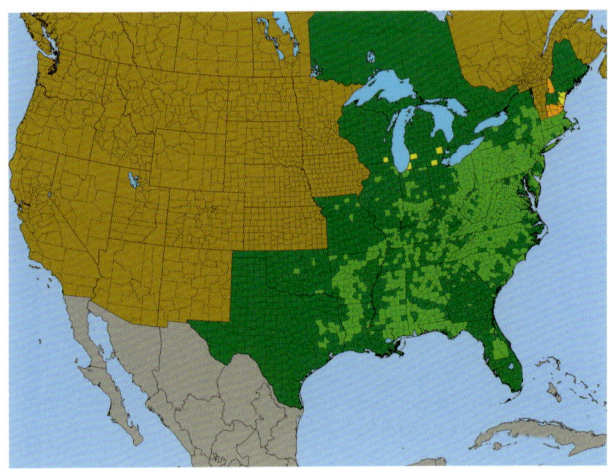

Seedling

Emerging mature plant in spring.
Emerging stems have numerous
purple spots.

Entire plant

Leaf. Leaves are produced in whorls on
thick, hollow purple stems.

Flower. Flowers are about twice the size of
Eupatorium maculatum at 6–12 in. wide.

Mature seed. Seed is ripe when it begins
to fluff out.

LOOK-ALIKE PLANTS:

Natives: *E. maculatum* (p. 108) and *E. purpureum* (a woodland plant) have
similar leaves and flowers but are shorter with smaller flower heads.

Eutrochium maculatum (*Eupatorium maculatum, Eupatoriadelphus maculatus*)
Joe Pye Weed, Spotted Joe Pye Weed

FAMILY: ASTER (ASTERACEAE, COMPOSITAE)

About half the size of Tall Joe Pye, possesses a similar magnetism for butterflies. Bumblebees are fond of roosting on the flowers on cool autumn nights.

Habitat: Wet prairies, sedge meadows, marshes, fens, streamsides, lakeshores, wet thickets, wet woods and woodland edges

Garden Uses: Specimens, massing, rain gardens, butterfly gardens, bird gardens, and wet meadows. Mix with Blue Mistflower for a late-season show.

USDA Hardiness Zones: 3–6
Light Requirements: Full sun to partial shade
Soil Types: Sand, loam, clay
Soil Moisture: Moist to wet
Bloom Time: July–September
Flower Color: Pink
Height: 4–6 feet
Life Expectancy: 5–10 years
Root Type: Fibrous
Propagation: Moist stratify seed 30 days; root division in spring or fall; tricky from cuttings: use first early growth

Aggressiveness: Low by seed
Attracts: Bees, butterflies, songbirds
Deer Palatability: Low
Distinguishing Characteristics:
- Stems are reddish purple, and half the diameter of *E. fistulosum* (p. 106), which has hollow, greenish stems streaked with purple
- Stems are filled with pith, never completely hollow

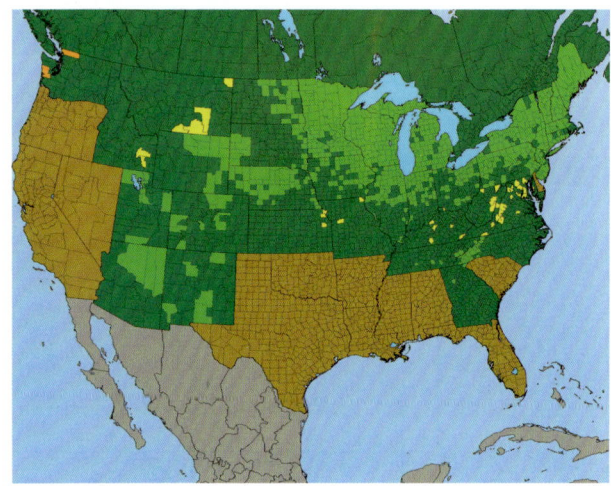

Seedling

Emerging mature plant in spring

Entire plant. Stems are often reddish tinged.

Leaf

Flower. Flowers are about half the size of *Eutrochium fistulosum* at 3–6 in. wide.

Mature seed. Seed is ripe when it begins to fluff out.

LOOK-ALIKE PLANTS:

Natives: *E. fistulosum* (p. 106) is taller with larger, domed flower heads. *E. purpureum* is similar but has green stems and occurs primarily in shady, mesic woodlands rather than open wetlands.

Helenium autumnale
Sneezeweed, Dogtooth Daisy

FAMILY: ASTER (ASTERACEAE, COMPOSITAE)

This floriferous daisy relative has petals with distinctively indented tips. Flowers have been crushed and sniffed to induce sneezing for relief from colds and headaches, hence the name. It does not cause sneezing in the garden!

Habitat: Wet prairies and meadows, marshes, sedge meadows, lakeshores, open wet woods, thickets

Garden Uses: Bird gardens, rain gardens, pond edges and stream banks, wet meadows, informal gardens, borders (deadheading helps reduce self-sowing)

USDA Hardiness Zones: 3–9
Light Requirements: Full sun
Soil Types: Sand, loam, clay
Soil Moisture: Moist to wet
Bloom Time: August–October
Flower Color: Yellow
Height: 3–6 feet
Life Expectancy: 5–10 years
Root Type: Fibrous
Propagation: Moist stratify seed 30 days; root division in spring or fall; tricky from cuttings: use first early growth

Aggressiveness: Moderate by seed
Attracts: Bees, butterflies
Deer Palatability: Low
Distinguishing Characteristics:
- Petals have two pronounced indentations on their tips
- Stems have ridges that protrude from the main stem

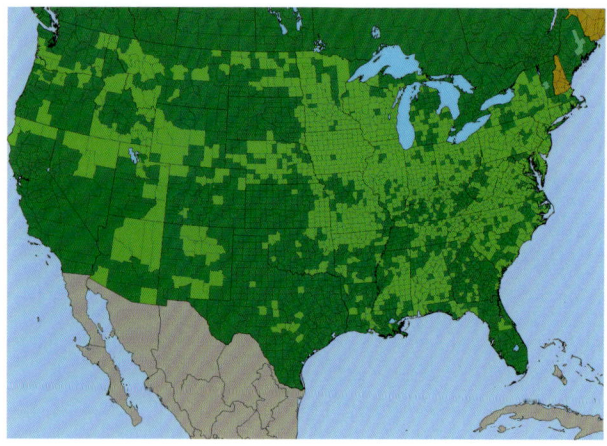

Seedling. Seedling leaves have small teeth, with scattered hairs on leaf surfaces.

Emerging mature plant in spring

Entire plant

Leaf

Flower. Tips of flower petals are indented.

Early and mature seed. Seeds are ripe when heads turn brown.

LOOK-ALIKE PLANTS:

Natives: White Turtlehead (p. 278) and Obedient Plant (p. 234) have similar leaves, but plants are shorter and do not have ridged stems. Obedient Plant also looks similar when emerging in spring. Wingstem (*Verbesina alternifolia*) grows in wet habitats too, but has more pronounced stem wings and only one indentation, not two, on the petal tip.

Helianthus ×laetiflorus (Helianthus pauciflorus, Helianthus rigidus)

Showy Sunflower, Prairie Sunflower, Cheerful Sunflower

FAMILY: ASTER (ASTERACEAE, COMPOSITAE)

Sunny and cheerful it may be, but this is not a friendly garden plant. Birds love the seeds, which makes the plant excellent for wildlife plantings. Eastern pocket gophers eat the rhizomes. Easy to propagate from seed and rhizomes.

Habitat: Dry to medium prairies, sand prairies, barrens, roadsides
Garden Uses: Not recommended due to aggressive spread by rhizomes; seed at low rates in prairie meadows and wildlife plantings

USDA Hardiness Zones: 3–8
Light Requirements: Full sun
Soil Types: Sand, loam, clay, gravel
Soil Moisture: Dry to medium
Bloom Time: August–September
Flower Color: Yellow
Height: 3–6 feet
Life Expectancy: 20+ years
Root Type: Rhizome
Propagation: Moist stratify seed 30 days; root division using 6-inch rhizomes in spring or fall; tricky from cuttings: use first early growth
Aggressiveness: High by rhizome

Attracts: Bees, butterflies, game birds, songbirds
Deer Palatability: Medium
Distinguishing Characteristics:
- Leaves are thick, rough, and "raspy"
- Stems are rough textured
- Large flowers have darker centers than most other native sunflowers
- Creeps aggressively by rhizomes
- "Fairy rings" can result as the inner colony dies out due to plant toxins produced by the allelopathic roots

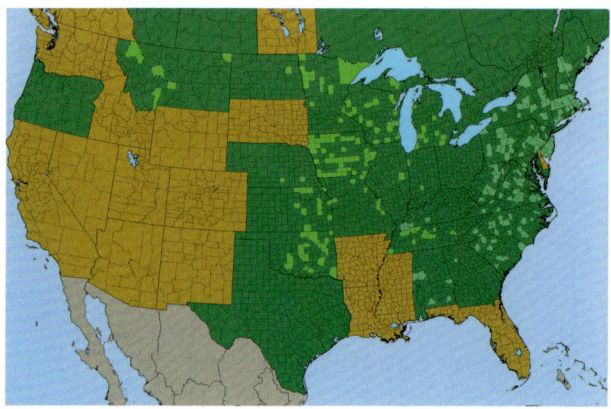

Seedling. Seedling stems extend above the soil soon after germinating.

Emerging mature plant in spring

Entire plant

Leaf. Leaves are rough and feel like sandpaper.

Flower

Mature seed. Seeds are ripe when the interior portion of the head turns brown.

LOOK-ALIKE PLANTS:

Natives: Similar to Downy Sunflower (p. 114) whose leaves are lighter, soft, and more closely spaced along the stem. Leaves are larger and plants taller than Western Sunflower (p. 116). *H. decapetalus*, *H. divaricatus*, and *H. strumosus* are woodland species that may mingle with Showy Sunflower but have smoother leaves and stems, and their flowers have yellow centers.

Helianthus mollis
Downy Sunflower, Ashy Sunflower

FAMILY: ASTER (ASTERACEAE, COMPOSITAE)

Less aggressive than other native sunflowers, creeping to form a patch. Birds savor the seeds, making it a good choice for wildlife gardens and meadows. Dig up leading-edge rhizomes occasionally to prevent excessive spread in the garden. Grows on any well-drained soil.

Habitat: Dry to medium prairies, sand prairies
Garden Uses: Bird gardens, informal gardens, prairie meadows

USDA Hardiness Zones: 4–9
Light Requirements: Full sun
Soil Types: Sand, loam, clay
Soil Moisture: Dry to medium
Bloom Time: July–October
Flower Color: Yellow
Height: 4–6 feet
Life Expectancy: 20+ years
Root Type: Rhizome
Propagation: Moist stratify seed 30 days; root division using 6-inch rhizomes in spring or fall; tricky from cuttings: use first early growth

Aggressiveness: Medium by rhizome
Attracts: Bees, butterflies, game birds, songbirds
Deer Palatability: Low
Distinguishing Characteristics:
- Leaves are thick, downy and soft, neither smooth nor raspy
- Leaves occur close together on hairy stems
- Leaves clasp the stem
- Creeps slowly by rhizomes
- Flowers sometimes face away from sun

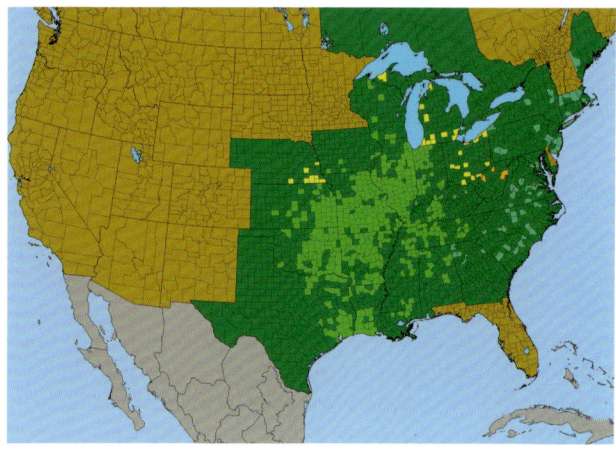

Seedling. Seedlings hug the ground after germinating.

Emerging mature plant in spring

Entire plant

Leaf. Leaves clasp the stems and are covered with fine, fuzzy hairs.

Flower

Mature seed. Seeds are ripe when heads turn brown.

LOOK-ALIKE PLANTS:

Natives: Showy Sunflower (p. 112) has similar stature and flowers, but its leaves and stems are raspy rather than soft, with darker flower centers.
Weeds: Rose Campion (*Lychnis coronaria*), a garden plant, has similarly hairy, grayish silver leaves in spring.

Helianthus occidentalis

Western Sunflower, Naked Sunflower, Fewleaf Sunflower

FAMILY: ASTER (ASTERACEAE, COMPOSITAE)

Shortest of our native sunflowers, with smaller flowers. Creeps by rhizomes to form a patch. Grows in extremely dry, sandy soils. Birds devour the seeds, making it a valuable component of wildlife plantings on dry sites.

Habitat: Dry to medium prairies, sand prairies, barrens, dry and sandy open woods
Garden Uses: Bird gardens, dry gardens, prairie meadows

USDA Hardiness Zones: 3–9
Light Requirements: Full sun
Soil Types: Sand, loam
Soil Moisture: Dry to medium
Bloom Time: July–September
Flower Color: Yellow
Height: 2–5 feet
Life Expectancy: 20+ years
Root Type: Rhizome
Propagation: Moist stratify seed 30 days; root division using 6-inch rhizomes in spring or fall; tricky from cuttings: use first early growth

Aggressiveness: Medium by rhizome
Attracts: Bees, butterflies, game birds, songbirds
Deer Palatability: Low
Distinguishing Characteristics:
- Most leaves are basal; stems nearly leafless
- Flowers are smaller than other native sunflowers at 2 inches across

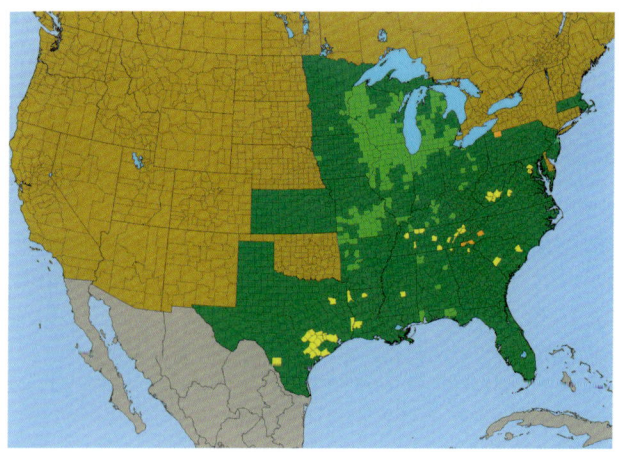

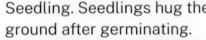

Seedling. Seedlings hug the ground after germinating.

Emerging mature plant in spring. Emerging leaves have pronounced white veins.

Entire plant. Stems have a few small leaves, hence the common name Naked Sunflower.

Leaf

Flower

Early seed. Early seedheads are bright yellow.

Mature seed. Seeds are ripe when the interior portion of the head turns brown.

LOOK-ALIKE PLANTS:

Natives: Showy Sunflower (p. 112) is taller; it has larger flowers, and raspy leaves and stems.

Weeds: Basal leaves resemble those of many introduced Hawkweeds (*Hieracium* spp.), which often colonize dry, sandy soils.

Heliopsis helianthoides
Oxeye Sunflower, Smooth Oxeye, Early Sunflower

FAMILY: ASTER (ASTERACEAE, COMPOSITAE)

One of summer's earliest yellow composites. Grows on almost any soil, often aggressive by seed. An early successional perennial, it dominates seeded prairies in years 2–5. Recedes as longer-lived species mature. Use sparingly in seed mixes!

Habitat: Dry to moist prairies, savannas, open woods, woodland edges, streamsides, lakeshores, roadsides; can be a bit weedy
Garden Uses: Not recommended for most gardens due to invasiveness by seed. Seed at moderate rates in prairie meadows and wildlife plantings.

USDA Hardiness Zones: 3–8
Light Requirements: Full sun to part shade
Soil Types: Sand, loam, clay
Soil Moisture: Dry, medium, moist
Bloom Time: June–October
Flower Color: Yellow
Height: 3–6 feet
Life Expectancy: 5–10 years
Root Type: Fibrous
Propagation: Moist stratify seed 30 days; root division in spring; tricky from cuttings: use first early growth
Aggressiveness: High by seed

Attracts: Bees, butterflies, game birds, hummingbirds, songbirds
Deer Palatability: Low
Distinguishing Characteristics:
- Blooms earlier than most true sunflowers
- Flowers are mostly held aloft, whereas true sunflowers are often nodding
- Does not creep by rhizomes as do true sunflowers
- Emerging leaves have smooth, glaucous upper surfaces; *Helianthus* are hairy or raspy
- Seeds have square tops; true sunflower seeds are rounded

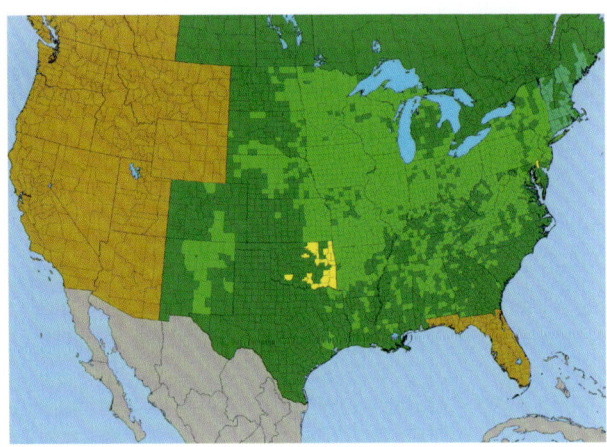

Seedling. Seedling stems and leaves are not densely hairy as with true sunflowers (*Helianthus*).

Emerging mature plant in spring. Emerging leaves often strongly red or purple tinged.

Entire plant

Leaf. Leaves are mildly raspy and not as hairy as true sunflowers (*Helianthus*).

Flower

Mature seed. Seeds are ripe when seedheads turn completely black.

LOOK-ALIKE PLANTS:

Natives: Leaves somewhat resemble those of Purple Coneflower (p. 102).

Liatris aspera
Rough Blazing Star, Button Blazing Star, Tall Blazing Star

FAMILY: ASTER (ASTERACEAE, COMPOSITAE)

One of the most drought tolerant of all blazing stars, with buttonlike pink flowers spaced along tall stalks. Even the buds are attractive! Grows in very dry sand and well-drained loam. Birds prefer its seeds over all other Liatris except Meadow Blazing Star. Easy from seed. Corms cannot be divided.

Habitat: Dry to medium prairies, sand prairies, sand barrens, dry open woods, bluffs
Garden Uses: Butterfly gardens, bird gardens, formal and informal gardens, prairie meadows. Plant with grasses or fibrous-rooted forbs to reduce weed growth.

USDA Hardiness Zones: 3–8
Light Requirements: Full sun
Soil Types: Sand, loam
Soil Moisture: Dry to medium
Bloom Time: August–September
Flower Color: Purple, pink
Height: 2–5 feet
Life Expectancy: 10–20 years
Root Type: Corm
Propagation: Moist stratify seed 30 days; hard from cuttings: use early growth in spring
Aggressiveness: Low by seed

Attracts: Bees, butterflies, hummingbirds, songbirds
Deer Palatability: Low
Distinguishing Characteristics:
- Flowers borne individually on stems, rather than in spikes or clusters, occasionally branching out on short side stems
- Individual flowers are among the largest in the genus
- Flower buds are often greenish white to whitish pink
- Grows in extremely dry sand and inland sand dunes

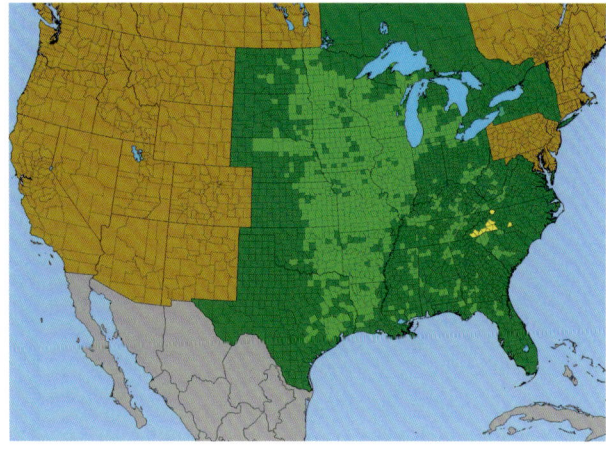

Seedling

Emerging mature plant in spring. Emerging leaves are hairless; leaves of other *Liatris* spp. are densely hairy or have hairs on the margins.

Entire plant

Leaf

Flower close-up. Individual flowers are larger than most other *Liatris* at 1–2 in. across, with exception of *L. ligulistylis.*

Flower. Flowers are borne as distinct individuals along a long stalk, 1–4 ft. long.

Early seed

Mature seed

LOOK-ALIKE PLANTS:

Natives: Meadow Blazing Star (p. 122) is similar in height and flower size but does not grow in dry sand. Its flowers are crimson when in bud and extend out from the stalk on long pedicels.

Liatris ligulistylis
Meadow Blazing Star, Rocky Mountain Blazing Star

FAMILY: ASTER (ASTERACEAE, COMPOSITAE)

Crimson flower buds open into deep purple-pink flowers borne individually on branching stems. Monarch butterflies flock to the blooms, and songbirds devour the seeds. Requires rich loamy soil. Unusually short-lived for a *Liatris*, only three to five years. Reliable from seed; corms cannot be readily divided.

Habitat: Medium to moist prairies
Garden Uses: Butterfly gardens, bird gardens, formal and informal gardens, prairie meadows. Mix with short grasses and flowers to help minimize weed growth.

USDA Hardiness Zones: 3–6
Light Requirements: Full sun
Soil Types: Loam
Soil Moisture: Medium to moist
Bloom Time: August–September
Flower Color: Purple, pink
Height: 3–5 feet
Life Expectancy: 3–5 years
Root Type: Corm
Propagation: Moist stratify seed 30 days; hard from cuttings: use first flush growth
Aggressiveness: Low by seed

Attracts: Bees, butterflies, hummingbirds, songbirds
Deer Palatability: Low
Distinguishing Characteristics:
- Flowers are borne on long pedicels
- Flowers crimson red to deep pink, darker than other *Liatris* species
- Flowers emit a pheromone that specifically attracts monarch butterflies

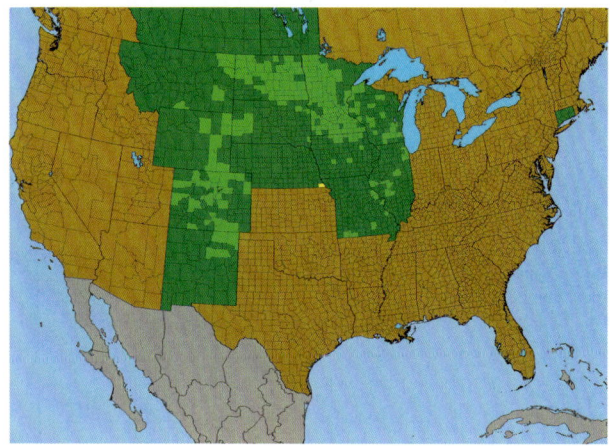

Seedling

Entire plant

Emerging mature plant in spring.
Emerging leaves are noticeably hairy.

Leaf

Flower. Flower stems are widely
branching, typically with pedicels of
1 in. or longer. Flowers are spaced
farther apart along stem than similar
species *L. aspera*.

Early seed. Bracts surrounding base of   Mature seed
flower are often reddish; flower buds
are often deep crimson red.

LOOK-ALIKE PLANTS:

Natives: Rough Blazing Star (p. 120) has a similar growth form and lighter
pink flowers that seldom extend out from the stem on pedicels.

Liatris punctata
Dotted Blazing Star

FAMILY: ASTER (ASTERACEAE, COMPOSITAE)

Native to the western tallgrass prairie and Great Plains, this low-growing *Liatris* tolerates drought and high alkalinity. The root is a unique combination of corm and taproot, ideally adapted to dry conditions. Blooms the latest of all *Liatris* spp.

Habitat: Dry to medium prairies, sand prairies, bluffs, more common in the Great Plains
Garden Uses: Butterfly gardens, bird gardens, rock gardens, dry gardens, and prairie meadows. Plant in the foreground as this diminutive *Liatris* blooms late in fall.

USDA Hardiness Zones: 3–9
Light Requirements: Full sun
Soil Types: Sand, loam
Soil Moisture: Dry to medium
Bloom Time: July–October
Flower Color: Purple, pink
Height: 1–2 feet
Life Expectancy: 10–20 years
Root Type: Taprooted corm
Propagation: Moist stratify seed 30 days; hard from cuttings: use first flush growth
Aggressiveness: Low by seed

Attracts: Bees, butterflies, hummingbirds, songbirds
Deer Palatability: Medium
Distinguishing Characteristics:
- Flower spikes are tightly bunched and bloom starts very close to the ground
- Flower spikes are narrower than most *Liatris* spp.
- Grows only in well-drained sandy soils and calcareous clay in the dry climates of the Great Plains

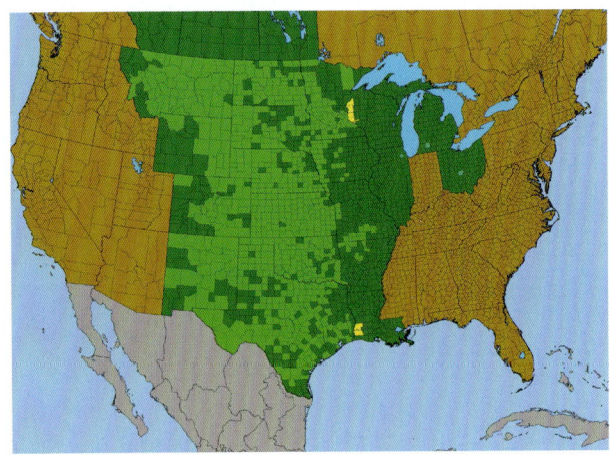

Seedling

Emerging mature plant
in spring

Entire plant. Plants are shorter than most other *Liatris*
spp. of the prairie at 12–24 in., except for *L. squarrosa*,
L. cylindracea, and *L. hirsuta*. Can live up to 35 years,
much longer than most species of *Liatris*.

Leaf. Leaves very narrow,
0.125–0.25 in. wide, very
rigid, with hairs scattered
along the margins.

Flower close-up

Flower. Flower stalks resemble a
miniature *Liatris pycnostachya*, 6–
12 in. long.

Mature seed. Mature
seeds have a whitish-
silvery glint.

LOOK-ALIKE PLANTS:

Natives: *Liatris cylindracea* and *L. scariosa* have similar stature and form
but with looser, more open inflorescences. The former is a Midwestern
species and the latter an eastern US native. *L. punctata* is most common
on the Great Plains.

Liatris pycnostachya
Prairie Blazing Star, Kansas Gayfeather

FAMILY: ASTER (ASTERACEAE, COMPOSITAE)

A most striking blazing star, with stunning wands of purple-pink flowers. Thrives in clay as well as rich, medium to moist soil. Do not fertilize with nitrogen, as taller *Liatris* species tend to flop. Reliable from seed and often self-sows on open soil. Corms are difficult to divide.

Habitat: Medium to moist prairies, fens, edges of sunny open wetlands, open wet woods
Garden Uses: Butterfly gardens, bird gardens, rain gardens, informal gardens, prairie meadows. Effective in masses. Plant with grasses and sturdy flowers to provide support and prevent flopping.

USDA Hardiness Zones: 3–9
Light Requirements: Full sun
Soil Types: Sand, loam, clay
Soil Moisture: Medium to moist
Bloom Time: July–September
Flower Color: Purple, pink
Height: 3–5 feet
Life Expectancy: 10–20 years
Root Type: Corm
Propagation: Moist stratify seed 30 days; hard from cuttings: use first flush growth

Aggressiveness: Low by seed
Attracts: Bees, butterflies, hummingbirds, songbirds
Deer Palatability: Medium
Distinguishing Characteristics:
- Flowers are borne on long spikes, similar to *L. spicata*, which is often slightly taller
- Among the earliest of *Liatris* to begin blooming, often starting in early July

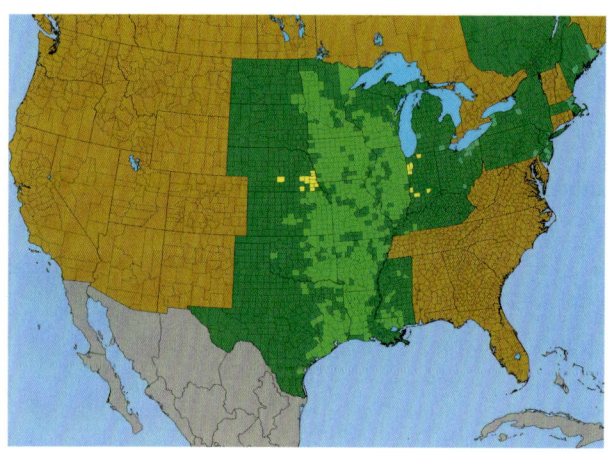

Seedling

Emerging mature plant in spring

Entire plant

Leaf

Flower

Early seed

Mature seed

LOOK-ALIKE PLANTS:

Natives: Dense Blazing Star (p. 128) is similar in form and appearance. The only way to tell the two species apart is that *L. pycnostachya* has re-curved bracts at the base of each individual flower, whereas the bracts of *L. spicata* are straight.

Liatris spicata
Dense Blazing Star, Marsh Blazing Star

FAMILY: ASTER (ASTERACEAE, COMPOSITAE)

Similar to Prairie Blazing Star but will grow in even damper soils. A number of cultivars are available. Like Prairie Blazing Star it tends to flop, so fertilizing is not recommended. Reliable from seed, self-sows on open soil. Unlike most other species of *Liatris*, corms can be divided.

Habitat: Medium to moist prairies, occasional on dry limestone barrens, bluffs, and outcrops

Garden Uses: Butterfly gardens, bird gardens, rain gardens, informal gardens, prairie meadows. Effective in masses. Mix with grasses and robust flowers to lend support.

USDA Hardiness Zones: 4–10
Light Requirements: Full sun
Soil Types: Sand, loam, clay
Soil Moisture: Medium to moist
Bloom Time: June–October
Flower Color: Purple, pink
Height: 3–6 feet
Life Expectancy: 10–20 years
Root Type: Corm
Propagation: Moist stratify seed 30 days; root division (divide corms in early spring); hard from cuttings: use first flush growth

Aggressiveness: Low by seed
Attracts: Bees, birds, butterflies, hummingbirds, sphinx moths
Deer Palatability: Medium
Distinguishing Characteristics:
- Flower spikes are long and narrow, usually taller than Prairie Blazing Star (p. 126)
- Certain ecotypes bloom as late as October, while others may begin blooming as early as late June

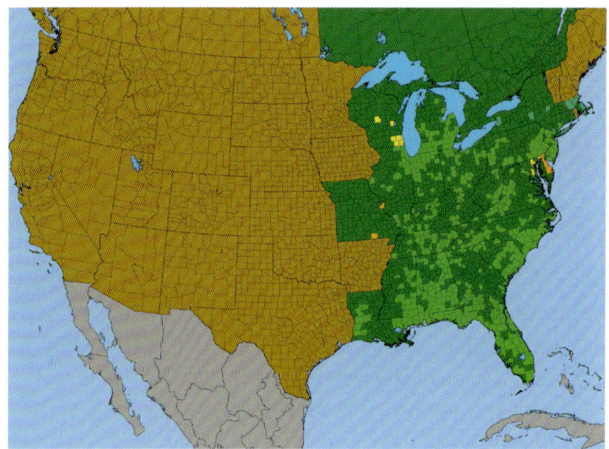

Seedling

Emerging mature plant in spring

Entire plant

Leaf

Flower close-up

Flower

Mature seed

LOOK-ALIKE PLANTS:

Natives: Prairie Blazing Star (p. 126) is similar in form and appearance. The two species can be distinguished in that *L. pycnostachya* has recurved bracts at the base of each individual flower, whereas the bracts of *L. spicata* are straight.

Liatris squarrosa
Scaly Blazing Star

FAMILY: ASTER (ASTERACEAE, COMPOSITAE)

Low-growing mounds of purple flowers make this *Liatris* a perfect foil for lance leaves and low-growing grasses. Tolerates extremely dry, sandy and rocky soils but thrives in any well-drained garden soil. Reliable from seed and the corms can be divided, unlike most other Liatris.

Habitat: Dry to medium prairies, especially dry, sandy prairies, barrens
Garden Uses: Dry gardens, informal gardens, butterfly gardens, bird gardens, prairie meadows

USDA Hardiness Zones: 4–8
Light Requirements: Full sun
Soil Types: Sand, loam
Soil Moisture: Dry to medium
Bloom Time: June–September
Flower Color: Purple, pink
Height: 1–2 feet
Life Expectancy: 10–20 years
Root Type: Corm
Propagation: Moist stratify seed 30 days; hard from cuttings: use first flush growth
Aggressiveness: Low by seed
Attracts: Bees, butterflies, hummingbirds, songbirds

Deer Palatability: Medium
Distinguishing Characteristics:
- Flowers borne on highly branched stalks, often creating a mounded appearance
- Individual flowers resemble starbursts, unlike most other more compact *Liatris* spp. flowers
- Scaly bracts at the base of the flowers are more prominent than any other *Liatris* spp.
- One of the shortest Liatris, no taller than 18 inches

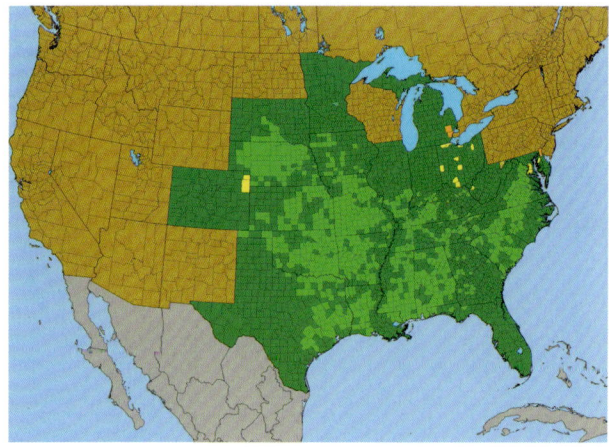

Seedling

Emerging mature plant
in spring

Entire plant. Plants are shorter than other *Liatris* spp. of the prairie at
10–18 in., except for *L. punctata, L. cylindracea,* and *L. hirsuta.*

Leaf

Flower close-up. Individual
florets curve outward and are
starlike; bracts at base are
large, scaly, and recurved.

Flower. Flowers are borne
on short pedicels in small
clusters at the top of plant
stems.

Early seed

Mature seed

LOOK-ALIKE PLANTS:

Natives: *L. cylindracea* flowers are also starlike but borne on narrow, individual, unbranched stalks with long green bracts subtending the flowers.

Solidago ptarmicoides (*Aster ptarmicoides, Oligoneuron album*)

Prairie Goldenrod (White Aster)

FAMILY: ASTER (ASTERACEAE, COMPOSITAE)

Neat, sweet, and petite, a great plant for dry gardens. White flowers cover the plant in early autumn. Seedheads are an attractive silvery gray. Easy from seed, grows only in dry, sandy, and rocky soils.

Habitat: Dry prairies, sand barrens, bluffs, particularly in dry, calcareous soils
Garden Uses: Rock gardens, hot, dry, sandy locations, pockets in rock walls, dry prairie meadows

USDA Hardiness Zones: 3–7
Light Requirements: Full sun
Soil Types: Sand, gravel
Soil Moisture: Dry
Bloom Time: August–September
Flower Color: White
Height: 1–2 feet
Life Expectancy: 5–10 years
Root Type: Fibrous
Propagation: Moist stratify seed 30 days; root division in spring; easy from stem cuttings early to midseason
Aggressiveness: Low by seed

Attracts: Bees, butterflies, songbirds
Deer Palatability: Low
Distinguishing Characteristics:
- Flowers are borne in a single terminal head
- Large plants may appear as small "bushes" often entirely covered in flowers
- Seed receptacles are shiny silver after seeds are gone
- Hairless leaves are narrow and "straplike"

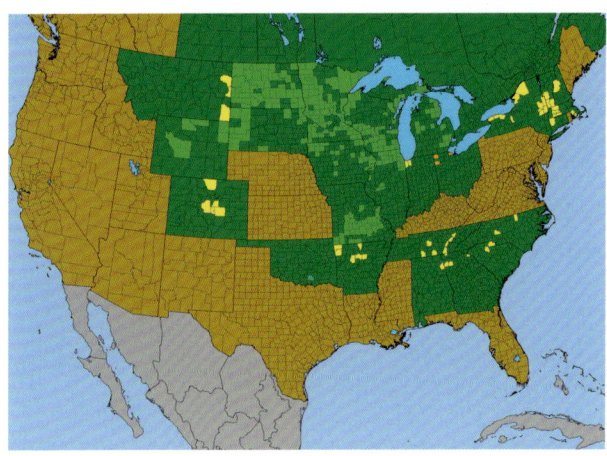

Seedling. Seedling leaves long, narrow, and smooth, with only a few hairs on margins.

Emerging mature plant in spring

Entire plant. Plant short and compact, 0.50–1.50 ft.

Leaf. Leaves long, narrow, and smooth, nearly hairless except on margins, up to ten times longer than wide.

Flower

Mature seed

LOOK-ALIKE PLANTS:

Natives: White Panicle Aster (*Symphyotrichum lanceolatum* or *Aster lanceolatus*) has narrow hairless leaves and white flowers that are borne on multiple stems along the stalk. Grows in moist areas and creeps by rhizomes to form large patches.

Oligoneuron ohioense (Solidago ohioensis)
Ohio Goldenrod

FAMILY: ASTER (ASTERACEAE, COMPOSITAE)

One of the showiest yet underutilized native plants. Has large, compact golden flower heads that hum with butterflies, bees, and a multitude of other pollinators. Performs beautifully in any good soil, especially clay. Easy from seed and can be readily divided.

Habitat: Moist to wet prairies, fens, moist beaches, especially in calcareous soils
Garden Uses: Specimens, small groups, butterfly gardens, bird gardens, bee gardens, rain gardens, informal gardens, prairie meadows

USDA Hardiness Zones: 4–6
Light Requirements: Full sun to partial shade
Soil Types: Sand, loam, clay
Soil Moisture: Medium to wet
Bloom Time: August–September
Flower Color: Yellow
Height: 3–4 feet
Life Expectancy: 5–10 years
Root Type: Fibrous
Propagation: Moist stratify seed 30 days; root division in fall or spring; easy from stem cuttings early to midseason

Aggressiveness: Low by seed
Attracts: Bees, butterflies, songbirds
Deer Palatability: Low
Distinguishing Characteristics:

- Large, flat-topped, domelike flower heads are among the largest of all goldenrods, covering most of the top of the plant
- Compact growth form with sturdy stems
- Nearly hairless, straplike leaves are smooth to the touch

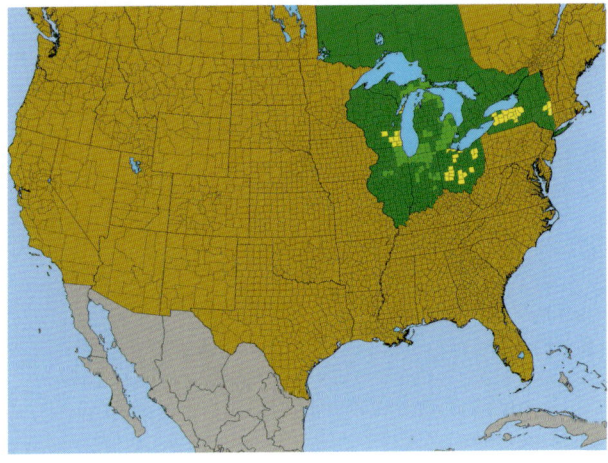

Seedling

Emerging mature plant in spring

Entire plant. Plant is lower growing than most other goldenrods.

Leaf. Leaves are smooth, narrow, and straplike, 0.75–2 in. wide.

Flower. Flowers in flat-topped bunches, usually broader (4–8 in.) than *Oligoneuron rigidum* (2–5 in.).

Early seed

Mature seed

LOOK-ALIKE PLANTS:

Natives: Stiff Goldenrod (p. 136) has smaller flat-topped flowers and wider, raspy leaves. *Euthamia graminifolia* and *Oligoneuron riddellii* are usually taller at 4–6 feet, as opposed to 3–4 feet for *O. ohioense*. *E. graminifolia* is clonal and forms patches. *O. riddellii* has smaller, more open flower clusters. Emerging leaves of Showy Goldenrod (p. 160) are similar, as are those of Ozark Coneflower (p. 100), which has three distinctive parallel veins the length of the leaf.

Oligoneuron rigidum (*Solidago rigida*)
Stiff Goldenrod

FAMILY: ASTER (ASTERACEAE, COMPOSITAE)

Stateliest of goldenrods, long-blooming and sturdy. Adapted to a wide variety of soil conditions, from dry sand to heavy clay. Spreads readily by seed if birds don't get them first. Seed germinates readily and plants are easily divided. Monarch butterflies love the flowers.

Habitat: Dry to moist prairies, dry open places, pastures, old fields
Garden Uses: Butterfly gardens, bird gardens, bee gardens, rain gardens, informal gardens, prairie meadows

USDA Hardiness Zones: 3–9
Light Requirements: Full sun
Soil Types: Sand, loam, clay
Soil Moisture: Dry to moist
Bloom Time: August–September
Flower Color: Yellow
Height: 3–5 feet
Life Expectancy: 5–10 years
Root Type: Fibrous
Propagation: Moist stratify seed 30 days; root division in fall or spring; easy from stem cuttings early to midseason
Aggressiveness: Medium by seed
Attracts: Bees, butterflies, game birds, songbirds

Deer Palatability: Medium
Distinguishing Characteristics:
- Flat-topped flower heads are closely bunched or slightly branching
- Individual flowers are the largest of our native goldenrods
- Leaves have a yellowish-green cast and are rough to the touch
- Silvery, starlike seed receptacles that appear after seed dispersal are larger than most other goldenrods; good for dried arrangements

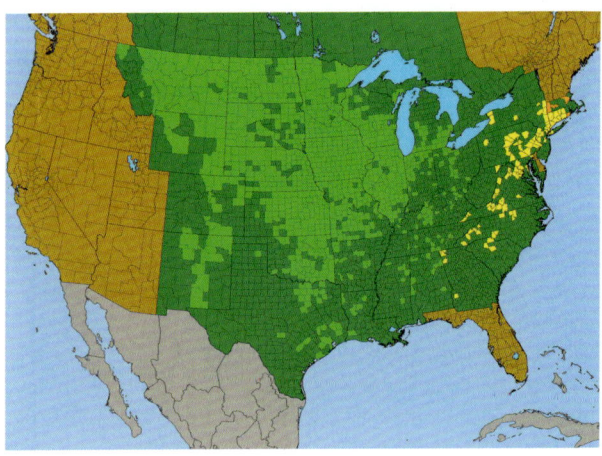

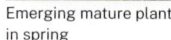
Seedling

Emerging mature plant
in spring

Entire plant

Leaf. Leaves are raspy
to the touch, thicker and
wider (1–3 in.) than most
other goldenrods.

Flower. Flower heads are nearly flat
topped.

Early seed

Mature seed

LOOK-ALIKE PLANTS:

Natives: Ohio Goldenrod (p. 134) has larger flattened flower heads.
O. riddellii and *Euthamia graminifolia* have similar-sized and flattened
flower heads and occur in wet to wet-mesic prairies. *E. graminifolia* is
strongly rhizomatous and forms large clones. *O. riddellii* has widely
branching, flat-topped flower clusters.

Parthenium integrifolium
Wild Quinine

FAMILY: ASTER (ASTERACEAE, COMPOSITAE)

One of the best white-flowered plants for prairie gardens, blooming for up to three months. Arrowlike leaves hold up well all season long. Requires rich soil and tolerates moisture. Somewhat tricky from seed, so plant in fall for spring germination. Its fibrous taproot cannot be easily divided. Looks great with Prairie Blazing Star (p. 126).

Habitat: Medium to moist prairies, open dry woods in rich soil
Garden Uses: Rain gardens, informal gardens, prairie meadows

USDA Hardiness Zones: 4–8
Light Requirements: Full sun to partial shade
Soil Types: Sand, loam, clay
Soil Moisture: Medium to moist
Bloom Time: May–October
Flower Color: White
Height: 3–5 feet
Life Expectancy: 10–20 years
Root Type: Taproot; may have rhizomes
Propagation: Moist stratify seed 30 days

Aggressiveness: Low by seed
Attracts: Songbirds, bees
Deer Palatability: Low
Distinguishing Characteristics:

- Five-pointed white flowers bloom for two to three months
- Thick, deep-green "crinkly" leaves are strongly veined
- Seeds emit an astringent but not unpleasant odor when crushed

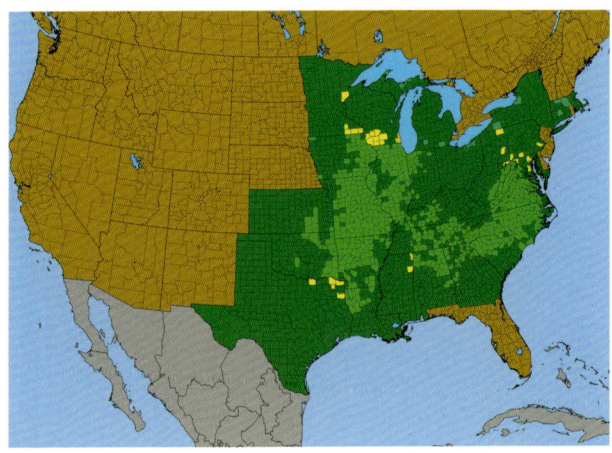

Seedling

Emerging mature plant in spring

Entire plant

Leaf. Leaves are thick and crinkly, often with a strong white midrib.

Flower close-up. Flowers have five mouse-ear-like petals that protrude from the center.

Flower. Flowers often retain their white color for three to four months from late June to early October

Early and mature seed

LOOK-ALIKE PLANTS:

Natives: Early leaves of Rosinweed (p. 152) and Whorled Rosinweed (*Silphium trifoliatum*) are similar when emerging in spring but have fewer and less pronounced serrations.

Weeds: Curly Dock (*Rumex crispus*) is a taprooted perennial with similar leaves in spring but soon attains a reddish-brown color and has green flower spikes that produce brown membranous seeds.

Ratibida pinnata
Yellow Coneflower, Grayhead Coneflower

FAMILY: ASTER (ASTERACEAE, COMPOSITAE)

A widely distributed and highly ornamental plant, looks good all year. Blooms for an extended period in summer and is a butterfly favorite. Plants often wilt on hot afternoons to conserve water but will perk up the following morning. Seedheads turn grayish black and make great dried arrangements. Self-sows with abandon.

Habitat: Dry to wet prairies
Garden Uses: Butterfly gardens, informal gardens, prairie meadows

USDA Hardiness Zones: 3–9
Light Requirements: Full sun to partial shade
Soil Types: Sand, loam, clay
Soil Moisture: Dry to wet
Bloom Time: July–September
Flower Color: Yellow
Height: 3–6 feet
Life Expectancy: 10–20 years
Root Type: Fibrous
Propagation: Moist stratify seed 30 days
Aggressiveness: Medium by seed

Attracts: Bees, butterflies, game birds, songbirds
Deer Palatability: Low
Distinguishing Characteristics:
- Drooping yellow flower petals are wider and bloom later than Ozark Coneflower (p. 100)
- Seedheads are smaller and more conical than *Echinacea* spp.
- Seedheads turn gray as they mature and have a spicy, minty fragrance when crushed

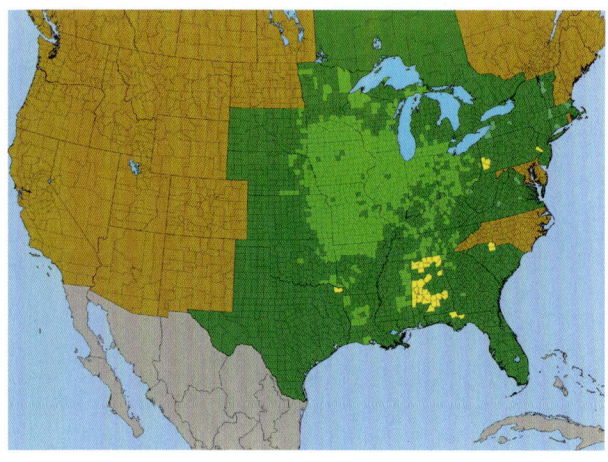

Seedling. Spadelike true leaves are borne on long, upright stems. *Rudbeckia* seedling leaves are usually on short stalks and not toothed (*Rudbeckia laciniata* seedlings may be slightly toothed).

Emerging mature plant in spring. Similar to *Rudbeckia laciniata*, except emerging leaves are narrower and not as intricately divided or fernlike.

Entire plant

Leaf. Leaves are deeply divided, usually with five to seven lobes.

Flower. Long, narrow, drooping yellow petals and black flower centers are distinctive.

Mature seed. Seed is ripe when seedheads break apart when pulled by hand.

LOOK-ALIKE PLANTS:

Natives: Leaves are similar to, but smaller than, Green-Headed Coneflower (p. 146) and Sweet Blackeyed Susan (p. 148). Seedlings are similar to Purple Coneflower (p. 102) and Sweet Blackeyed Susan.

Rudbeckia fulgida
Orange Coneflower

FAMILY: ASTER (ASTERACEAE, COMPOSITAE)

Prized for fine foliage, compact form, and long bloom period. Native to woodland edges, it is intolerant of hot sun and drought. Does best in heavy soils that retain moisture. *R. fulgida* var. *sullivantii* "Goldsturm" is a well-known selection. Easy from seed, it can also be divided. Can be deadheaded to extend bloom period.

Habitat: Moist meadows, woodland edges, open woods
Garden Uses: Specimens, or mass for effect; butterfly gardens, bird gardens, informal gardens, woodland edges

USDA Hardiness Zones: 4–8
Light Requirements: Full sun to partial shade
Soil Types: Loam, clay
Soil Moisture: Medium to moist
Bloom Time: June–September
Flower Color: Yellow
Height: 2–4 feet
Life Expectancy: 5–10 years
Root Type: Fibrous, rhizomatous

Propagation: Moist stratify seed 30 days; easy from root divisions in spring or fall
Aggressiveness: Low by seed
Attracts: Bees, birds, butterflies
Deer Palatability: Low
Distinguishing Characteristics:
- Leaves are broader than other *Rudbeckia* spp., and more corrugated and textured
- Distinctly winged leaf petioles

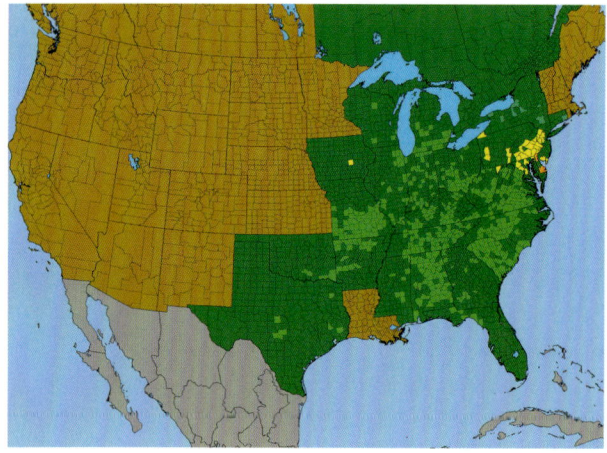

Seedling

Emerging mature plant in spring

Entire plant. Plants are usually fairly short and compact, growing 2–3 ft. tall.

Leaf. Large, undivided dark-green glossy leaves differentiate this plant from other *Rudbeckia* species.

Flower. Flowers are produced primarily at the top of the plant.

Mature seed. Seed is ripe when heads turn black.

LOOK-ALIKE PLANTS:

Natives: Leaves are similar to Purple Coneflower (p. 102) and Sweet Blackeyed Susan (p. 148). Blackeyed Susan (p. 144) leaves are conspicuously hairy. Missouri Coneflower (*R. missouriensis*) has similar flowers but differs by having narrow, hairy leaves.

Rudbeckia hirta
Blackeyed Susan

FAMILY: ASTER (ASTERACEAE, COMPOSITAE)

This irrepressible biennial grows almost anywhere with little assistance. Provides early color in prairie meadow seedings. Predominant in years 2 and 3, it fades as longer-lived perennials take over. Not a great garden plant; it dies after blooming and foliage turns prematurely gray. Propagation is by seed only. Self-sows readily, and can be allowed to self-sow in gardens.

Habitat: Dry to moist prairies, savannas, barrens, glades, roadsides; a variety of open, sometimes disturbed areas
Garden Uses: Butterfly gardens, bird gardens, informal gardens, prairie meadows

USDA Hardiness Zones: 3–10
Light Requirements: Full sun to partial shade
Soil Types: Sand, loam, clay
Soil Moisture: Dry to moist
Bloom Time: June–September
Flower Color: Yellow
Height: 1–3 feet
Life Expectancy: 2 years (biennial)
Root Type: Fibrous
Propagation: Dry or moist stratify seed 30 days
Aggressiveness: Medium by seed
Attracts: Bees, birds, butterflies

Deer Palatability: Low
Distinguishing Characteristics:
- Leaves are covered with fine, fuzzy hairs, and never have lobes
- Mid to upper leaves clasp the stem
- Shallow-rooted biennial, usually blooms in year 2 and dies, but can bloom in year 1 from seed
- Foliage is subject to fungus mid- to late summer, often turning gray and dying back

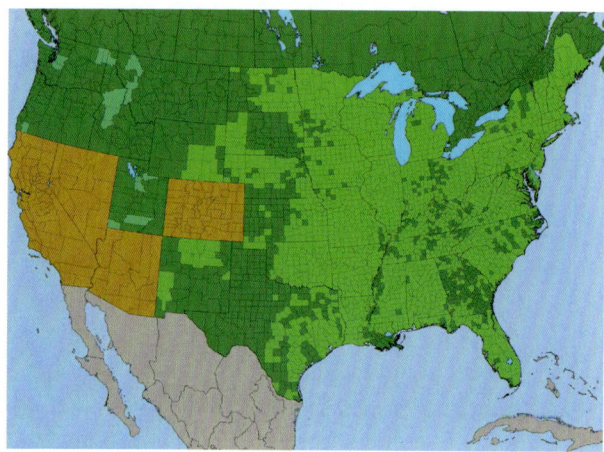

Seedling

Emerging mature plant in spring

Entire plant. Being biennial, plants turn gray after flowering and die.

Leaf. Leaves are soft, covered with hairs, and never divided.

Flower

Mature seed. Seed is ripe when heads turn black.

LOOK-ALIKE PLANTS:

Natives: Browneyed Susan (p. 150) has hairy leaves that are often deeply divided and smaller flowers with jet-black centers. Missouri Coneflower (*R. missouriensis*) is similar in appearance but has narrower leaves.

Weeds: The fuzzy leaves of young plants may resemble seedlings of Common Mullein (*Verbascum thapsus*).

Rudbeckia laciniata
Green-Headed Coneflower, Cutleaf Coneflower

FAMILY: ASTER (ASTERACEAE, COMPOSITAE)

Among the tallest *Rudbeckia*, commonly occurring in wetlands subject to periodic disturbance due to a relatively short life cycle. Requires constant soil moisture for good growth. Due to its fast growth habit, provides rapid soil stabilization when used in wetland seed mixes. Cultivar "Golden Glow" has double flowers.

Habitat: Wet prairies and meadows, stream banks, swales, ditches, wet woods, thickets
Garden Uses: Specimens, small groups, rain gardens

USDA Hardiness Zones: 3–8
Light Requirements: Full sun to partial shade
Soil Types: Sand, loam, clay
Soil Moisture: Moist to wet
Bloom Time: August–October
Flower Color: Yellow
Height: 5–8 feet
Life Expectancy: 3–5 years
Root Type: Fibrous, rhizomatous
Propagation: Moist stratify seed 30 days
Aggressiveness: Low by seed

Attracts: Bees, butterflies, songbirds
Deer Palatability: Low
Distinguishing Characteristics:
- Large, deeply cut laciniate leaves
- Flowers have light-green to yellow centers instead of black or brown, as with other *Rudbeckia*, *Ratibida*, and related genera
- Grows in wet soils, often in shade, where other *Rudbeckia* rarely venture

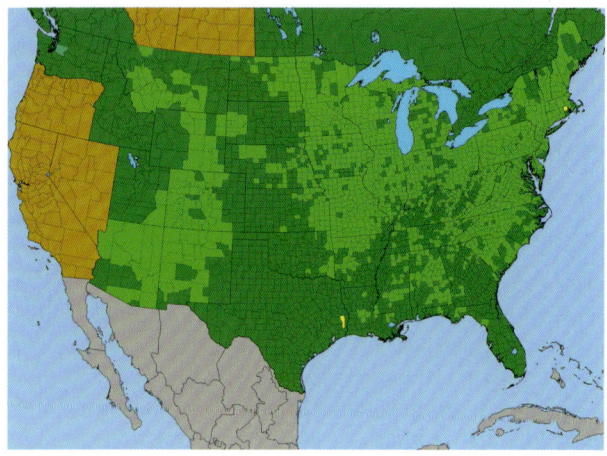

Seedling

Emerging mature plant in spring. Similar to *Ratibida pinnata*, except emerging leaves are typically wider and have a fernlike appearance.

Entire plant. Plant is far taller than other *Rudbeckia* and *Ratibida*, often up to 8–10 ft.

Leaf. Leaves are deeply divided, usually with three lobes.

Flower. Green flower centers separate this species from *Ratibida* and other *Rudbeckia*.

Early seed

Mature seed. Seed is ripe when heads turn black.

LOOK-ALIKE PLANTS:

Natives: Leaves are approximately 50% larger than Yellow Coneflower (p. 140) and Sweet Blackeyed Susan (p. 148). The lower pair of leaflets are cut all the way down to the midrib.

Weeds: Leaves and plant may resemble Giant Ragweed (*Ambrosia trifida*), which grows in similar moist habitats.

Rudbeckia subtomentosa
Sweet Blackeyed Susan, Sweet Coneflower

FAMILY: ASTER (ASTERACEAE, COMPOSITAE)

A most garden-worthy *Rudbeckia*. Foliage holds up well all season long, rarely afflicted by leaf spot. Flowers bloom for an extended period and attract a specialist (oligolectic) bee. Sweet-scented foliage and flowers add a sensory element to the garden. Longest-lived of the *Rudbeckia*, it thrives in heavy soils and damp areas. Easy from seed, it can also be divided.

Habitat: Medium to moist prairies, river bottoms, open wet woods
Garden Uses: Butterfly gardens, bird gardens, rain gardens, informal gardens, prairie meadows

USDA Hardiness Zones: 4–8
Light Requirements: Full sun to partial shade
Soil Types: Sand, loam, clay
Soil Moisture: Medium to moist
Bloom Time: August–October
Flower Color: Yellow
Height: 4–6 feet
Life Expectancy: 10–20 years
Root Type: Fibrous
Propagation: Moist stratify seed 30 days; root division in spring
Aggressiveness: Low by seed
Attracts: Bees, butterflies, songbirds

Deer Palatability: Low
Distinguishing Characteristics:
- Deep-green leaves are smooth and covered with very fine white hairs
- Leaves are often three lobed, with the two lower lobes appearing like wings
- Flower centers are iridescent reddish black
- The plant exudes a pleasant light fragrance
- Grows up to 6 feet, taller than most other *Rudbeckia*, except *R. laciniata* and *R. maxima*

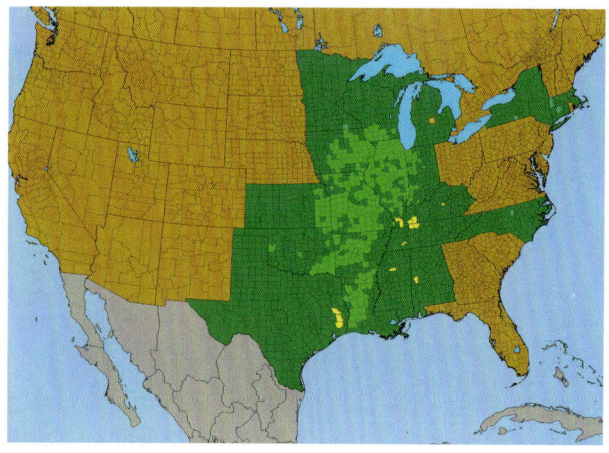

Seedling

Emerging mature plant in spring

Entire plant

Leaf. Leaves commonly exhibit rust spots in summer and fall; upper leaves typically not divided, though some lower leaves are often three lobed.

Flower

Mature seed. Seed is ripe when heads turn black; seedheads are aromatic, hence the name Sweet Blackeyed Susan.

LOOK-ALIKE PLANTS:

Natives: Leaves are similar to Yellow Coneflower (p. 140) and Purple Coneflower (p. 102).

Rudbeckia triloba
Browneyed Susan

FAMILY: ASTER (ASTERACEAE, COMPOSITAE)

Plants that are literally buried under myriad deep-yellow and jet-black blooms make an incredible late-summer show. Although short-lived, it reseeds readily on open soil. Widely adaptable, grows in a variety of soil and sun conditions.

Habitat: Medium to moist prairies (often disturbed), edges of sunny wetlands, stream banks, riverbanks, thickets, open woods
Garden Uses: Specimens, short-lived hedges, butterfly gardens, bird gardens, informal gardens, prairie meadows

USDA Hardiness Zones: 4–9
Light Requirements: Full sun to partial shade
Soil Types: Sand, loam
Soil Moisture: Medium to moist
Bloom Time: July–October
Flower Color: Yellow
Height: 2–5 feet
Life Expectancy: 2–3 years (usually biennial)
Root Type: Fibrous
Propagation: Moist stratify seed 30 days
Aggressiveness: Medium by seed
Attracts: Bees, butterflies, songbirds

Deer Palatability: Low
Distinguishing Characteristics:
- Smaller flowered than other *Rudbeckia*
- Flowers have coal-black centers with shiny, almost waxy, butter-yellow petals
- Emerging leaves have slightly indented margins and are less fuzzy than Blackeyed Susan (p. 144)
- Blooms later than most other *Rudbeckia*, except for Sweet Blackeyed Susan (p. 148)

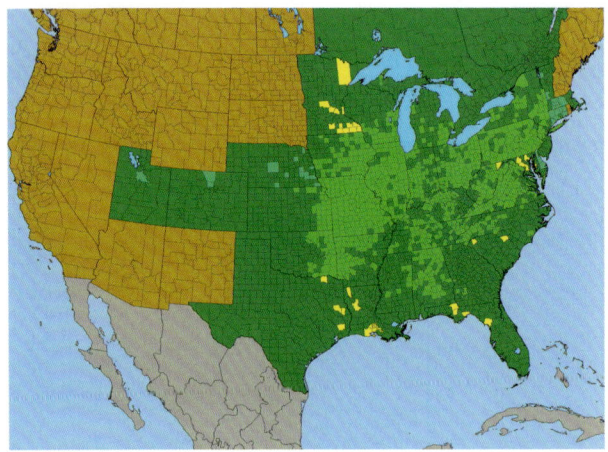

Seedling

Emerging mature plant in spring

Entire plant. Flowers are often produced in profusion on all stems, from top to bottom. Plant height varies depending on growing conditions and light intensity; *R. triloba* blooms from late summer into fall, well after *R. hirta* has finished flowering.

Leaf. Leaves are highly variable, with numerous sizes and forms.

Single leaf

Tri-lobed leaf

Flower. Flower diameter is smaller than other *Rudbeckia* at 1.5–2 in.; flower centers are usually pure black.

Mature seed. Seed is ripe when heads turn black.

LOOK-ALIKE PLANTS:

Natives: Distinguishable from other *Rudbeckia* by its smaller flowers, deep-black centers, shiny butter-yellow petals, and late bloom time.
Weeds: Leaves are similar to early-season leaves of Giant Ragweed (*Ambrosia trifida*).

Silphium integrifolium
Rosinweed, Wholeleaf Rosinweed

FAMILY: ASTER (ASTERACEAE, COMPOSITAE)

Although smaller than other *Silphium*, it still forms large clumps and can grow to 6 feet tall. Many birds savor the seeds. Best for prairie meadows and wildlife plantings, as it tends to self-sow aggressively. Easy from seed; clumps can be divided with some effort.

Habitat: Dry to wet prairies
Garden Uses: Not recommended for gardens. Seed sparingly in prairie meadows and wildlife plantings.

USDA Hardiness Zones: 4–8
Light Requirements: Full sun
Soil Types: Sand, loam, clay
Soil Moisture: Dry to wet
Bloom Time: July–September
Flower Color: Yellow
Height: 2–6 feet
Life Expectancy: 20+ years
Root Type: Taproot
Propagation: Moist stratify seed 30 days; root division in spring or fall (labor intensive)
Aggressiveness: High by seed
Attracts: Bees, butterflies, game birds, songbirds
Deer Palatability: Low

Distinguishing Characteristics:
- Smaller, narrower leaves (1.5–3 inches) than most other *Silphium*, except *S. trifoliatum* (0.75–1.5 inches)
- Lower growing (6 feet) than most other *Silphium*
- Flowers and seedheads are smaller (1.5 inches) than other *Silphium*, except for *S. trifoliatum*
- Exudes a resin when flower stems are plucked, which can be chewed like gum, hence the name Rosinweed (same as Compassplant, p. 154.)

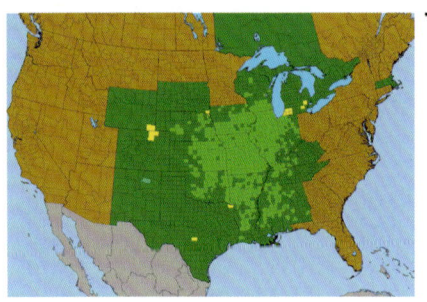

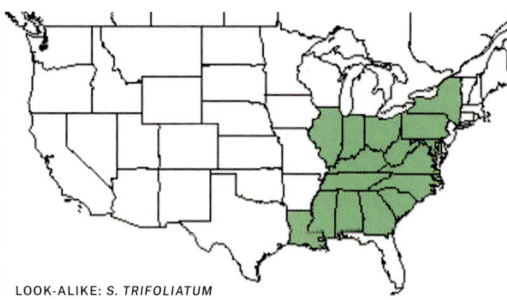

LOOK-ALIKE: *S. TRIFOLIATUM*

Seedling. Seedlings are very similar to *Silphium laciniatum*; typically half as wide as *Silphium terebinthinaceum*.

Emerging mature plant in spring. Emerging leaves are not lobed and originate at ground level without stems.

Entire plant

Leaf. Leaves occur along the length of the stalk and are not toothed as in *S. perfoliatum*.

Leaf

Flower

Early seed. Seedheads resemble those of *Silphium perfoliatum*, but bracts around the seeds are more open.

Mature seed. Seeds are ripe when they turn brown and begin spreading across the receptacle.

S. trifoliatum, entire plant. Plants have whorls of four narrow leaves all along the stem.

S. trifoliatum, leaf. Leaves in whorls along length of stems are much narrower than other *Silphium* (1–2 in.).

S. trifoliatum, flower. Flower petals narrower than other *Silphium* at 0.5–1 in. at widest point.

LOOK-ALIKE PLANTS:

Natives: *S. trifoliatum* of the eastern United States has similar stature and flowers, but its leaves are whorled along the stem rather than in pairs. Emerging leaves are almost identical to those of Rosinweed.

Weeds: Emerging leaves may resemble Curly Dock (*Rumex crispus*).

Silphium laciniatum
Compassplant

FAMILY: ASTER (ASTERACEAE, COMPOSITAE)

Large, deeply divided leaves and towering flower stalks place this plant in a class of its own. Not prone to roaming, it is the best behaved of *Silphium*. Extremely long-lived, its life span is measured in decades, and it can live for up to 100 years. Slow from seed, sometimes requiring up to ten years to bloom. Easily established using young transplants, it cannot be moved or divided at maturity because of its deep taproot.

Habitat: Dry to medium prairies
Garden Uses: Specimens, butterfly gardens, bird gardens, informal gardens, prairie meadows. Interplant with tall prairie grasses for late-season effect.

USDA Hardiness Zones: 4–9
Light Requirements: Full sun
Soil Types: Sand, loam, clay
Soil Moisture: Dry to medium
Bloom Time: June–September
Flower Color: Yellow
Height: 3–10 feet
Life Expectancy: 20+ years
Root Type: Taproot
Propagation: Moist stratify seed 30 days
Aggressiveness: Low by seed
Attracts: Bees, butterflies, game birds, songbirds

Deer Palatability: Low
Distinguishing Characteristics:
- Unique, deeply indented leaves evident from early leaf emergence in spring
- Red emerging bud sheaths are typical of most *Silphium*
- Leaves often orient in a north-south direction, hence the plant's name
- Seedheads are larger than other *Silphium* at 2–3 inches wide, and bracts are covered with bristly hairs

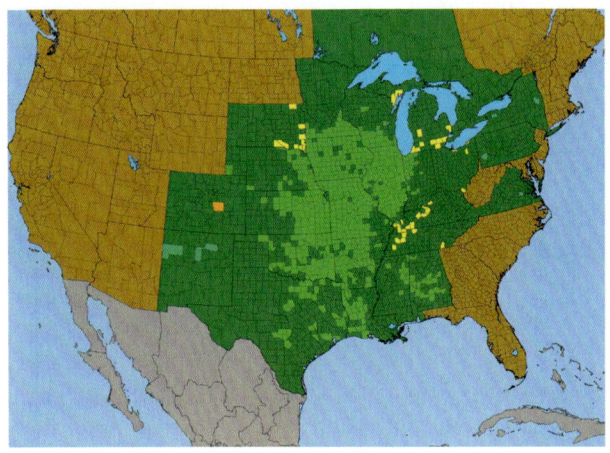

Seedling. Seedlings' first leaves are not indented; later leaves exhibit characteristic indentations.

Emerging mature plant in spring. Emerging leaves are deeply indented.

Entire plant

Leaf. Leaves originate from the ground and are deeply and distinctively indented.

Flower

Early seed. Seedpods are surrounded by thick, bristly bracts.

Mature seed. Seed is ripe when tips of individual seeds display a silver-colored point in their centers.

LOOK-ALIKE PLANTS:

Natives: Flowers are similar to other *Silphium*, but the plant is easily distinguished by its deeply cleft leaves and large, hairy seedheads.

Silphium perfoliatum
Cup Plant

FAMILY: ASTER (ASTERACEAE, COMPOSITAE)

Largest of the *Silphium*, attaining heights of 12 feet or more. Leaves cup the stems, holding water after rain or heavy dew. Birds come to eat seeds and have a drink. Loves clay soil and tolerates high moisture. Spreads wildly by seed on favorable sites. Plant with other big sturdy plants, as smaller plants will be engulfed. Seed sparingly in wet prairie meadows.

Habitat: Moist to wet prairies and meadows, moist woodland edges, wet woods, bottomlands

Garden Uses: "Big and tall gardens" (with other ruffians!). Great for bird gardens but be prepared to remove numerous seedling volunteers.

USDA Hardiness Zones: 3–8
Light Requirements: Full sun to partial shade
Soil Types: Sand, loam, clay
Soil Moisture: Medium to wet
Bloom Time: July–September
Flower Color: Yellow
Height: 3–10 feet
Life Expectancy: 20+ years
Root Type: Fibrous taproot; may have short rhizomes
Propagation: Moist stratify seed 30 days; root division in spring or fall (labor intensive)

Aggressiveness: High by seed
Attracts: Bees, butterflies, game birds, hummingbirds, songbirds
Deer Palatability: Low
Distinguishing Characteristics:
- Biggest and tallest of all *Silphium*
- Large square stems are 0.75–2 inches thick
- Leaves are pierced by the stems, creating "cups"
- Leaves have creases and are not as thick as other *Silphium*

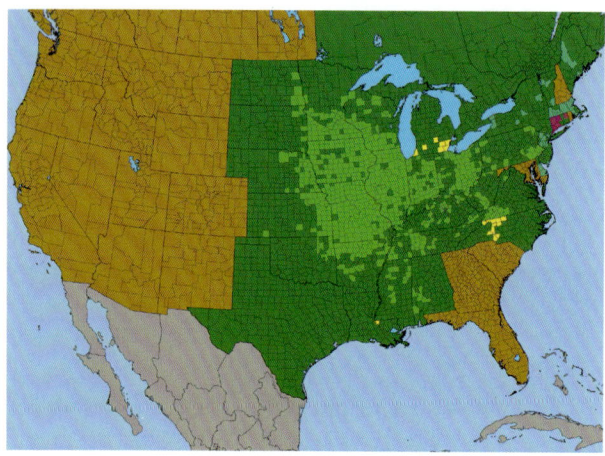

Seedling. Lower half of the seedling's first true leaves are slightly toothed.

Emerging mature plant in spring. Emerging leaves are toothed but not deeply indented as with *S. laciniatum*.

Entire plant

Leaf. Large, toothed leaves occur along the length of the square stems.

Flower

Early seed. Seedheads resemble those of *Silphium terebinthinaceum*, but bracts of *S. terebinthinaceum* curl upward and over the seeds when immature.

Mature seed. Seed is ripe when it turns brown and is wide spreading across the receptacle.

LOOK-ALIKE PLANTS:

Natives: Flowers are similar to other *Silphium*, but leaves are creased and not as thick and raspy as other species. Emerging leaves may resemble Wild Quinine (p. 138).

Silphium terebinthinaceum
Prairie Dock

FAMILY: ASTER (ASTERACEAE, COMPOSITAE)

Gigantic elephant-ear leaves produce a tropical effect in the garden. Ten-foot-tall flower stalks attract birds, butterflies, and other pollinators. Seed germinates readily but grows slowly, requiring five to ten years to reach maturity. Thrives in heavy clay and moist soils. Large taproot cannot be divided or easily transplanted once established.

Habitat: Medium to wet prairies, roadside ditches, railroads
Garden Uses: Specimens, small groups, informal gardens, butterfly gardens, bird gardens, prairie meadows

USDA Hardiness Zones: 4–7
Light Requirements: Full sun
Soil Types: Sand, loam, clay
Soil Moisture: Medium to wet
Bloom Time: July–September
Flower Color: Yellow
Height: 3–10 feet
Life Expectancy: 20+ years
Root Type: Taproot
Propagation: Moist stratify seed 30 days
Aggressiveness: Low by seed

Attracts: Bees, butterflies, game birds, hummingbirds, songbirds
Deer Palatability: Low
Distinguishing Characteristics:
- Large, raspy leaves resemble elephant ears
- Seedheads are smooth and hairless (as opposed to Compassplant)
- Seedstalks are completely leafless; other *Silphium* have at least some small leaves on their stalks

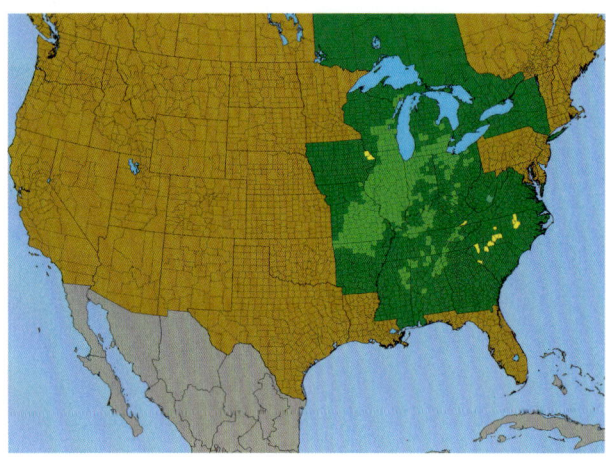

Seedling. Seedling leaves are not lobed and about twice as wide as *S. integrifolium*.

Emerging mature plant in spring. Emerging arrowhead-like leaves are not lobed and are borne on stems above the ground, compared to *S. integrifolium*.

Entire plant

Leaf. Leaves originate from the ground and are very broad, unlobed, and resemble elephant ears.

Flower

Early seed. Early seeds are nearly encompassed by green bracts.

Mature seed. Seed is ripe when tips of individual seeds display a silver-colored point in their centers and heads just begin to spread open.

LOOK-ALIKE PLANTS:

Natives: Flowers are similar to other *Silphium*, but the plant is easily distinguished by its large, wide leaves.

Weeds: Emerging leaves are similar to Common Burdock (*Arctium minus*); Common Burdock's basal leaves are hairy and lie flat, Prairie Dock's are raspy and upright. Early leaves are also similar to Elecampane (*Inula helenium*), whose leaves are not as thick and raspy as those of Prairie Dock.

Solidago speciosa
Showy Goldenrod

FAMILY: ASTER (ASTERACEAE, COMPOSITAE)

Aptly named, the elegantly sculpted, conical flower heads stand out in the garden. Red stems add to the effect. Attracts an astounding diversity of pollinators. Prefers a loose, well-drained soil. Reliable from seed. Can be divided.

Habitat: Dry to medium prairies, sand barrens, old fields, roadsides
Garden Uses: Specimens, small groups, informal herbaceous hedges, butterfly gardens, bird gardens, bee gardens, informal gardens, prairie meadows

USDA Hardiness Zones: 3–8
Light Requirements: Full sun to partial shade
Soil Types: Sand, loam
Soil Moisture: Dry to medium
Bloom Time: August–September
Flower Color: Yellow
Height: 1–4 feet
Life Expectancy: 5–10 years
Root Type: Fibrous, rhizomatous
Propagation: Moist stratify seed 30 days; root division in fall or spring; easy from stem cuttings early to midseason

Aggressiveness: Low by seed
Attracts: Bees, butterflies, songbirds
Deer Palatability: Medium
Distinguishing Characteristics:
- Conical, nonarching flower heads
- Stems are often reddish in color
- Flowering stalks typically "spray" outward evenly from base of plant

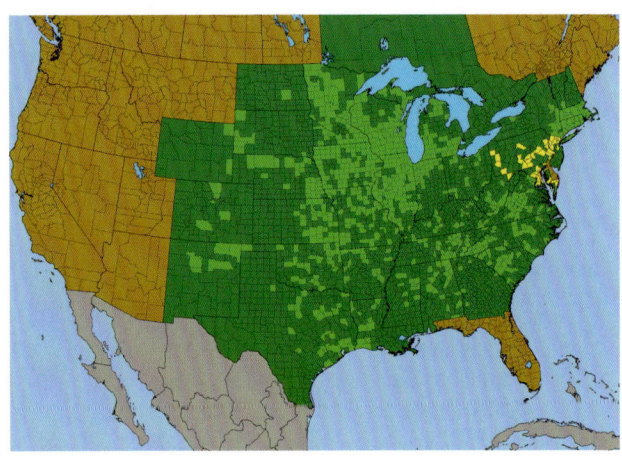

Seedling

Emerging mature plant in spring

Entire plant

Leaf. Leaves are smooth to the touch and narrower than *Oligoneuron ohioense*, whose leaves are also smooth.

Flower. Flower head is conical, without downward arching tips typical of *Solidago nemoralis*, *S. canadensis*, and others.

Mature seed

LOOK-ALIKE PLANTS:

Natives: Sweet-Scented Goldenrod (*S. speciosa* var. *jejunifolia*) is almost identical, but blooms a month earlier and has a semisweet fragrance. Emerging leaves are similar to Ohio Goldenrod (p. 134), as well as Ozark Coneflower, which has three distinct parallel leaf veins.

Symphyotrichum ericoides (*Aster ericoides*)
Heath Aster, White Heath Aster

FAMILY: ASTER (ASTERACEAE, COMPOSITAE)

This low-growing, mounding aster creeps gradually by rhizomes to form an attractive "bush." Masses of tiny white flowers appear in mid- to late fall, pollinated by myriad insects.

Habitat: Dry prairies, barrens, dry open woods, old fields, disturbed areas, especially dry sandy or gravelly soil, former quarries
Garden Uses: Groundcover, border foregrounds and edges, small groups; with grasses and taller flowers as background, esp. blue asters

USDA Hardiness Zones: 3–9
Light Requirements: Full sun
Soil Types: Sand, loam, gravel
Soil Moisture: Dry to medium
Bloom Time: August–October
Flower Color: White
Height: 1–3 feet
Life Expectancy: 10–20 years
Root Type: Rhizome
Propagation: Moist stratify seed 30 days; root division using 4–6 inches rhizomes in spring; easy from stem cuttings early to mid-season

Aggressiveness: Medium by rhizome
Attracts: Bees, butterflies, game birds
Deer Palatability: Low; medium on new growth
Distinguishing Characteristics:
- Creeps gradually by rhizomes to form compact "bushes"
- Leaves are very short and narrow
- Individual flowers are smaller than most other asters: 0.5 inch across
- Stems almost woody

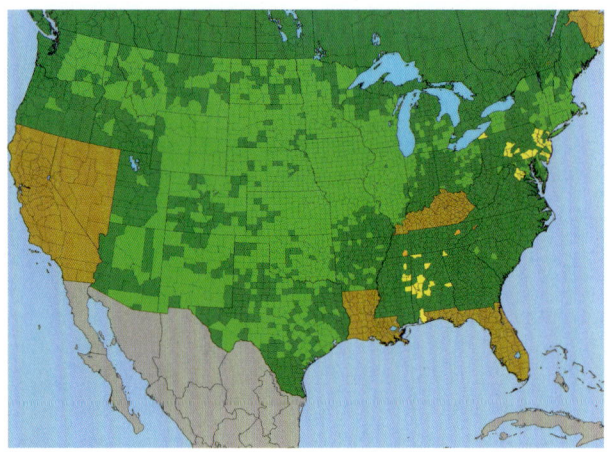

Seedling. Seedlings grow upright and tall; most other aster seedlings stay lower to the ground initially.

Emerging mature plant in spring

Entire plant. Plant spreads by rhizomes to form patches.

Leaf. Plant stems and leaf margins hairy; small, stiff leaves are four to five times longer than wide, and seldom longer than 1 in., with distinctive hairs on margins.

Flower close-up

Flower

Early seed

Mature seed

LOOK-ALIKE PLANTS:

Natives: Frost Aster (*Symphyotrichum pilosum*) is very similar but usually taller (2–3 feet), and it spreads more rapidly by rhizomes. Emerging leaves are similar to those of New England Aster (p. 166).

Symphyotrichum laeve (Aster laevis)
Smooth Aster

FAMILY: ASTER (ASTERACEAE, COMPOSITAE)

Very adaptable, this long-lived aster provides late-season color and nectar for pollinators. Reliable from seed; easy to transplant and divide. Flowers for an extended period and foliage holds up well all season.

Habitat: Dry to moist prairies, barrens, glades, open woods, and along roadsides and railroads
Garden Uses: Butterfly gardens, bird gardens, prairie meadows

USDA Hardiness Zones: 4–8
Light Requirements: Full sun
Soil Types: Sand, loam, clay
Soil Moisture: Dry to moist
Bloom Time: August–October
Flower Color: Blue
Height: 2–4 feet
Life Expectancy: 5–10 years
Root Type: Fibrous
Propagation: Moist stratify seed 30 days; root division in spring; easy from stem cuttings early to midseason
Aggressiveness: Medium by seed

Attracts: Bees, butterflies, game birds, songbirds
Deer Palatability: High
Distinguishing Characteristics:
- Emerging leaves are edged with red
- Mature leaves are bluish and smooth to the touch
- Pronounced leaf veins are distinctive
- One of the latest-blooming asters (New England Aster and Aromatic Aster are also late)

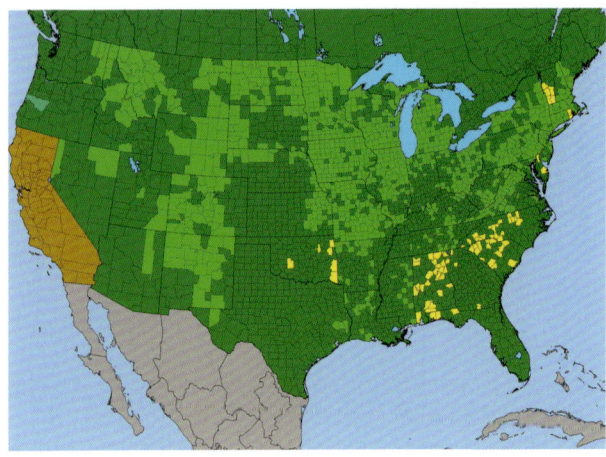

Seedling

Emerging mature plant in spring

Entire plant

Leaf. Leaves and stems smooth and hairless.

Flower close-up

Flower

Early seed

Mature seed close-up

Mature seed

LOOK-ALIKE PLANTS:

Natives: Skyblue Aster (p. 170) has similar flowers but blooms a few weeks earlier. Aromatic Aster (p. 168) blooms at the same time but has narrow, hairy leaves and stems and a spreading growth habit.

Symphyotrichum novae-angliae (Aster novae-angliae)

New England Aster

FAMILY: ASTER (ASTERACEAE, COMPOSITAE)

One of the best and most adaptable asters, its blue, lavender, and pink flowers bloom well into late fall. Mix with earlier-blooming flowers to serve as a foundation for any garden. Butterflies depend on it for a late-season nectar source, and it is a staple resource for migrating monarchs. Loves clay. Reliable from seed, often self-sows. Easy to divide. Many cultivars available.

Habitat: Medium to moist prairies, wetland edges, open woods, roadsides
Garden Uses: Perennial borders, butterfly gardens, bird gardens, rain gardens, prairie meadows

USDA Hardiness Zones: 3–7
Light Requirements: Full sun to partial shade
Soil Types: Sand, loam, clay
Soil Moisture: Medium to moist
Bloom Time: August–October
Flower Color: Blue, purple, pink
Height: 3–6 feet
Life Expectancy: 5–10 years
Root Type: Fibrous

Propagation: Moist stratify seed 30 days; root division in spring; easy from stem cuttings early to midseason
Aggressiveness: Low by seed
Attracts: Bees, butterflies
Deer Palatability: Medium
Distinguishing Characteristics:
- Leaf bases clasp the stem
- Large flowers are borne on tall plants (up to 6 feet)
- One of the latest-blooming asters

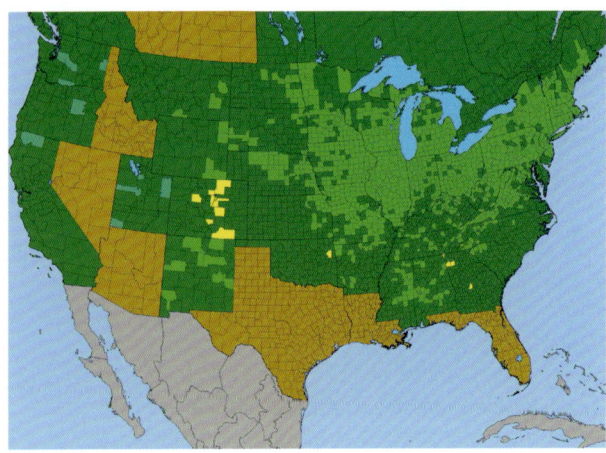

Seedling

Emerging mature plant in spring

Entire plant. One of the tallest asters at 3–6 ft.

Leaf. Leaves four to five times longer than wide; 2–3 in. long.

Flower close-up

Flower

Early seed (*top*) and mature seed (*bottom*)

Mature seed

LOOK-ALIKE PLANTS:

Natives: Swamp Aster (*Symphyotrichum puniceum*, also called Purple Stem Aster) has purple stems with flower heads clustered more toward the top of the plant. Canada Goldenrod (*Solidago canadensis*) has similar leaves that do not clasp the stem.

Symphyotrichum oblongifolium
(Aster oblongifolius)
Aromatic Aster

FAMILY: ASTER (ASTERACEAE, COMPOSITAE)

One of the last asters to bloom, it is covered with stunning sky-blue flowers from September into November. Resistant to aster yellows disease, the foliage and flowers often extend to the base. Foliage is pleasantly aromatic on warm summer days. Similar to New England Aster, this is a garden stalwart. Mixes well with most other flowers and grasses, makes an outstanding seasonal hedge or focal point.

Habitat: Dry rocky and sandy prairies, sand barrens, bluffs
Garden Uses: Perennial borders, bee gardens, butterfly and moth gardens

USDA Hardiness Zones: 3–8
Light Requirements: Full sun to partial shade
Soil Types: Sand, loam, clay
Soil Moisture: Dry to medium
Bloom Time: September–November
Flower Color: Blue, purple
Height: 2–4 feet
Life Expectancy: 5–10 years
Root Type: Rhizomatous
Propagation: Moist stratify seed 30 days; root division in spring; easy from stem cuttings early to midseason

Aggressiveness: Medium by rhizome
Attracts: Bees, butterflies
Deer Palatability: Low
Distinguishing Characteristics:
- Stems become woody as season advances
- Leaves are coarsely hairy
- Plant creeps rapidly by rhizomes
- Plants are weakly erect and spreading

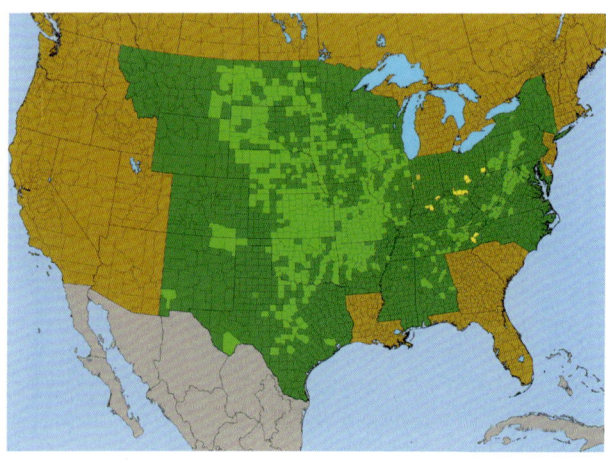

Seedling

Emerging mature plant in spring

Entire plant. Plants are low growing and compact at 1–2 ft. tall.

Leaf

Flower close-up. Flowers are deep blue, covering the plant in fall.

Flower

Early seed

Mature seed close-up

Mature seed

LOOK-ALIKE PLANTS:

Natives: Spreading Aster (*Symphyotrichum patens*) and Willowleaf Aster (*Symphyotrichum praealtum*) have blue flowers, are of similar height, and creep by rhizomes. *S. praealtum* has an upright growth form with tall flower spikes. *S. patens* is delicate, with oval, untoothed leaves that clasp the stem.

Symphyotrichum oolentangiense
(Aster azureus)
Skyblue Aster

FAMILY: ASTER (ASTERACEAE, COMPOSITAE)

Bright-blue flowers cover this plant in autumn. Highly adaptable, grows on any loose, well-drained soil. Tolerates light shade. Plant with grasses, White Asters, goldenrods, and earlier-blooming flowers. Short-lived, it requires periodic replanting. For a longer-lived blue aster, consider using Smooth Aster (p. 164).

Habitat: Dry to moist prairies, savannas, barrens, bluffs, dry open woods, Lake Michigan dunes, especially in sandy and rocky soils
Garden Uses: Butterfly gardens, bird gardens, prairie meadows

USDA Hardiness Zones: 3–8
Light Requirements: Full sun to partial shade
Soil Types: Sand, loam
Soil Moisture: Dry to medium
Bloom Time: August–October
Flower Color: Blue
Height: 2–3 feet
Life Expectancy: 5–10 years
Root Type: Fibrous
Propagation: Moist stratify seed 30 days; easy from stem cuttings early to midseason

Aggressiveness: Low by seed
Attracts: Bees, butterflies, songbirds
Deer Palatability: Medium
Distinguishing Characteristics:
- Arrowlike blue-green leaves
- Most leaves occur at base of plant
- Leaf surface is rough to the touch, whereas that of Smooth Aster is not

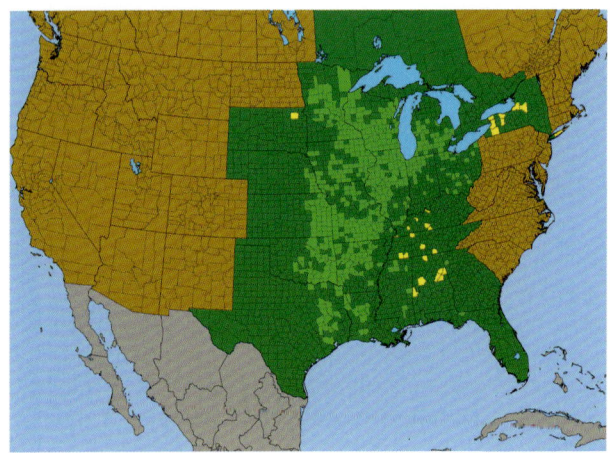

Seedling. Seedling stems are usually reddish; leaf margins slightly indented.

Emerging mature plant in spring. Emerging leaves are covered with fine hairs, and stems are typically reddish.

Entire plant. Plant growth form is compact, seldom exceeding 2 ft. tall.

Leaf

Flower close-up

Flower

Early seed

Mature seed close-up

Mature seed

LOOK-ALIKE PLANTS:

Natives: Smooth Aster (p. 164) has similar flowers but blooms later and has smooth, rubbery leaves. Blue Wood Aster (*Symphyotrichum cordifolium*) and White Arrowleaf Aster (*Symphyotrichum urophyllum*) are true woodland asters with similar flowers.

Vernonia fasciculata
Prairie Ironweed, Smooth Ironweed, Common Ironweed

FAMILY: ASTER (ASTERACEAE, COMPOSITAE)

Gorgeous, deep-violet-pink flowers attract flocks of butterflies. Named for its tough roots. Birds feast on its seeds. Prefers damp soil and shines when given a home in clay. Easy from seed. Can be divided using a sharp spade, machete, or chain saw!

Habitat: Medium to moist prairies, streamsides, riverbanks, lakeshores, moist pastures

Garden Uses: Butterfly gardens, bird gardens, rain gardens, informal gardens, prairie meadows

USDA Hardiness Zones: 3–7
Light Requirements: Full sun
Soil Types: Sand, loam, clay
Soil Moisture: Medium to moist
Bloom Time: July–September
Flower Color: Violet to deep pink
Height: 4–6 feet
Life Expectancy: 5–10 years
Root Type: Fibrous
Propagation: Moist stratify seed 30 days; root division in fall or spring; tricky from cuttings: use first early growth
Aggressiveness: Low by seed

Attracts: Bees, butterflies, songbirds
Deer Palatability: Low
Distinguishing Characteristics:
- One of few prairie plants with reddish-pink flowers with a flat-topped flower head
- Underside of leaves are hairless and have a prominent midvein
- *Vernonia* have alternate leaves, whereas the often similar *Eutrochium* have opposite leaves
- Leaves are 2–5 inches long and 0.25–0.75 inches wide

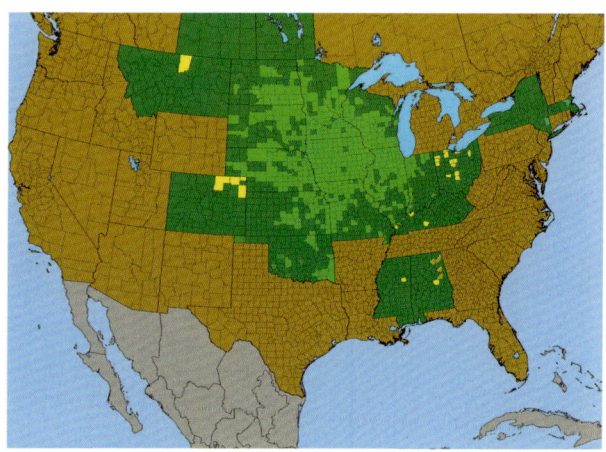

Seedling. First true leaves about half as wide as those of *V. gigantea* (see *V. gigantea* captions for details).

Emerging mature plant in spring

Entire plant

Leaf. Leaves generally possess minimal texturing and fissuring.

Flower

Early seed

Mature seed

LOOK-ALIKE PLANTS:

Natives: Similar to Tall Ironweed (p. 174) but shorter, with smaller flower heads (6 inches versus 12 inches across). Leaves seldom exceed 5 inches long and 0.5 inches wide, while Tall Ironweed leaves are up to 10 inches long and usually wider. *V. noveboracensis* is an eastern native with similar stature and flower heads. Its leaves are wider at 0.5–2 inches wide. *V. missurica* is a lower-growing native to the central Midwest and US South, grows 2–4 feet with stems and leaves covered in fine white hairs and leaves wider at 0.75–2 inches. *V. baldwinii* is a Midwestern and Great Plains native that grows 2.5–5 feet tall with broader leaves that are 0.75–2 inches wide; and can be considered weedy.

Vernonia gigantea (Vernonia altissima)
Tall Ironweed, Giant Ironweed

FAMILY: ASTER (ASTERACEAE, COMPOSITAE)

Huge purple flowers are an absolute butterfly magnet. Birds love the seeds. Grows naturally in moist areas but thrives in any good garden soil. Easy from seed; roots can be divided with a sharp machete or pickax!

Habitat: Wet meadows, moist to wet woodlands, woodland edges
Garden Uses: Specimens, butterfly gardens, bird gardens, rain gardens, informal gardens, wet prairie meadows

USDA Hardiness Zones: 4–9
Light Requirements: Full sun to partial shade
Soil Types: Sand, loam, clay
Soil Moisture: Moist to wet
Bloom Time: July–September
Flower Color: Purple, pink
Height: 5–8 feet
Life Expectancy: 10–20 years
Root Type: Fibrous
Propagation: Moist stratify seed 30 days; root division in fall or spring; tricky from cuttings: use first early growth

Aggressiveness: Medium by seed
Attracts: Bees, butterflies, songbirds
Deer Palatability: Low
Distinguishing Characteristics:
- Very large purplish-pink flower heads up to 1 foot across
- Leaves are up to 10 inches long and 2–2.5 inches wide, largest of all the ironweeds
- *Vernonia* have alternate leaves, whereas the often similar *Eutrochium* have opposite leaves

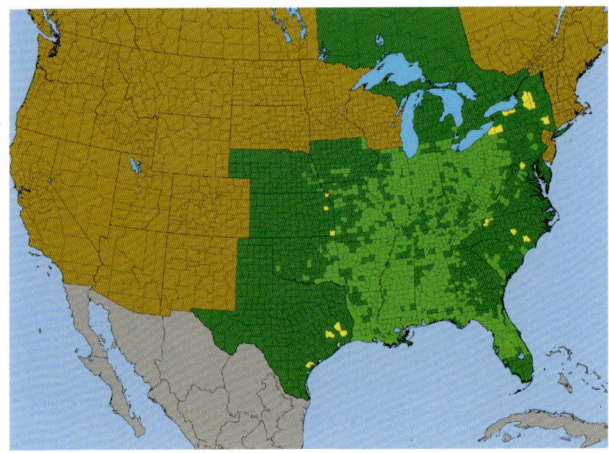

Seedling. First true leaves about twice as wide as *V. fasciculata* seedling leaves, at 0.50–0.75 in. versus 0.2–0.4 in. — leaf width varies as plants develop.

Emerging mature plant in spring

Entire plant

Leaf. Leaves generally more textured than *V. fasciculata*.

Flower. Flower heads about twice as wide as *V. fasciculata*, at 0.5–1 ft. across versus 0.25–0.5 ft.

Early seed

Mature seed

LOOK-ALIKE PLANTS:

Natives: Prairie Ironweed (p. 172) is usually shorter at 4–6 feet, with smaller flower heads 6 inches across. Its leaves seldom exceed 5 inches long and 1 inch wide. *V. noveboracensis* is an eastern native, usually shorter at 3–6 feet with typically smaller flower heads 6 inches across, and leaves 0.5–2 inches wide. *V. missurica* is a lower-growing native to the central Midwest and US South, 2–4 feet tall, with stems and leaves covered in fine white hairs; leaves similar in width at 0.75–2 inches wide but only 2–5 inches long. *V. baldwinii* is a Midwestern and Great Plains native that grows 2.5–5 feet tall with broader leaves that are 0.75–2 inches wide; and can be considered weedy.

Lithospermum canescens
Hoary Puccoon

FAMILY: BORAGE (BORAGINACEAE)

Puccoons are much loved but seldom grown due to a lack of seed availability and difficulty in transplanting. Harvest your own seed to start these plants in your garden. Seeds should be sown immediately in the desired location. They will germinate the following spring. Collect Hoary Puccoon seeds when they turn from green to chocolate brown and are easily removed from the stem. Sow in a semishaded location. Does best with a half day of direct light.

Habitat: Dry to medium prairies, savannas, barrens, open woods, bluffs; prefers calcareous soils, often in partial sun on the north sides of oak-hickory woods
Garden Uses: Shaded rock gardens and forest edges

USDA Hardiness Zones: 3–7
Light Requirements: Full sun to partial shade
Soil Types: Sand, loam
Soil Moisture: Dry to medium
Bloom Time: April–June
Flower Color: Orange to yellow
Height: 1 foot
Life Expectancy: 5–10 years
Root Type: Taproot
Propagation: Scarify seed and moist stratify 90 days, or plant fresh for spring germination
Aggressiveness: Low by seed
Attracts: Bees, butterflies
Deer Palatability: Low
Distinguishing Characteristics:

- Flowers are orange to yellow, while those of Hairy Puccoon (p. 178) and Fringed Puccoon are bright yellow
- Leaf hairs are soft; Hairy Puccoon has spiny, prickly leaf hairs
- Leaf tips are blunt; Hoary Puccoon and Fringed Puccoon leaves are pointed
- Seeds are smaller than those of Hairy Puccoon, turning chocolate brown when ripe; Hairy Puccoon seeds turn grayish white when ripe, usually a month later; Fringed Puccoon seeds turn brownish white when ripe in late summer

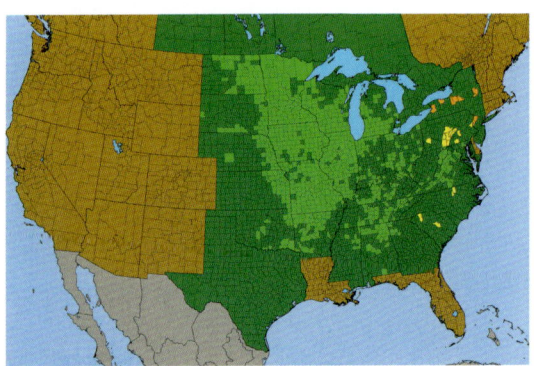

Seedling

Emerging mature plant in spring. Plants emerge as single stems with flower buds formed, which bloom almost immediately.

Entire plant. Plants are small in stature at 6–12 in. with upright stems.

Leaf. Leaves have soft hairs, never rough or raspy to the touch.

Flower. Flowers are commonly light orange, sometimes yellow.

Early seed

Mature seed. Seeds turn white when fully ripe.

LOOK-ALIKE PLANTS:

Natives: Hairy Puccoon (p. 178) and Fringed Puccoon (p. 180)

Lithospermum caroliniense
Hairy Puccoon, Carolina Puccoon

FAMILY: BORAGE (BORAGINACEAE)

Brightest and largest-flowered of puccoons, Hairy Puccoon grows only in dry, sandy soil. Puccoon seeds ripen indeterminately at different times, so collect only the ripe seeds. Hairy Puccoon seeds turn pure white when ripe. Sow seeds in desired location. Plants develop taproots 3 feet or longer in their first growing season, making transplant impossible.

Habitat: Dry sand prairies, sand barrens, oak barrens, sandy bluffs
Garden Uses: Rock gardens, dry gardens, and dry prairie meadows

USDA Hardiness Zones: 3–8
Light Requirements: Full sun
Soil Types: Sand
Soil Moisture: Dry
Bloom Time: May–July
Flower Color: Yellow-orange
Height: 1–2 feet
Life Expectancy: 10–20 years
Root Type: Taproot
Propagation: Scarify seed and moist stratify 90 days, or plant fresh for spring germination
Aggressiveness: Low by seed
Attracts: Butterflies
Deer Palatability: Medium
Distinguishing Characteristics:
- Bright-yellow to orange flowers bloom after Hoary Puccoon (p. 176), and before Fringed Puccoon (p. 180)
- Leaf hairs are sharp and bristly; Hoary Puccoon has soft leaf hairs on blunt-tipped leaves; Fringed Puccoon has short leaf hairs on long, narrow leaves
- Hairy Puccoon grows in bush-like form; Hoary Puccoon produces a few upright short stems; Fringed Puccoon has narrow leaves scattered on trailing, often prostrate, stems
- Hairy Puccoon seeds are larger than both Hoary Puccoon and Fringed Puccoon and turn grayish white when ripe; Hoary Puccoon seeds turn chocolate brown when ripe; Fringed Puccoon seeds are similar in size to Hoary Puccoon but turn light gray rather than brown when ripe

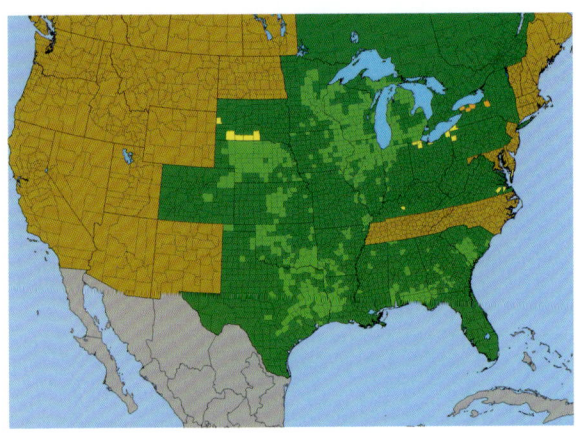

Seedling

Emerging mature plant in spring. Mature plants emerge as large clumps with numerous individual stems.

Entire plant

Leaf. Leaves covered in long, bristly hairs and are rough and raspy to the touch.

Flower. Flowers are bright yellow.

Early seed

Mature seed. Seeds turn gray-brown when fully ripe.

LOOK-ALIKE PLANTS:

Natives: Hoary Puccoon and Fringed Puccoon

Lithospermum incisum
Fringed Puccoon

FAMILY: BORAGE (BORAGINACEAE)

Fringed lemon-yellow flowers distinguish this from other puccoons. Growth form is more relaxed and often nearly prostrate. Grows in dry sandy and rocky soils. As with other puccoons, individual seeds should be collected as they ripen and planted immediately in the desired location.

Habitat: Dry limestone prairies and open hillsides, limestone bluffs
Garden Uses: Rock gardens, dry gardens, rock walls, dry prairie meadows

USDA Hardiness Zones: 3–9
Light Requirements: Full sun
Soil Types: Sand, gravel
Soil Moisture: Dry
Bloom Time: April–June
Flower Color: Pale yellow
Height: 1–2 feet
Life Expectancy: 5–10 years
Root Type: Taproot
Propagation: Scarify seed and moist stratify 90 days, or plant fresh for spring germination
Aggressiveness: Low by seed
Attracts: Butterflies, skippers, maybe ruby-throated hummingbird (cross-pollination); otherwise is self-pollinating
Deer Palatability: Low

Distinguishing Characteristics:
- Flowers are pale yellow and distinctly fringed; Hoary Puccoon (p. 176) has orange-yellow flowers, and Hairy Puccoon (p. 178) has large, yellow-orange flowers
- Narrow grayish-green leaves (0.4 inches) are scattered along stems, often nearly prostrate on the ground; Hairy Puccoon grows in an upright, compact bushlike form and produces a few short, upright stems with wider, deep-green leaves
- Leaf hairs are much less pronounced than in Hairy Puccoon; Hoary Puccoon has soft, hairy leaves

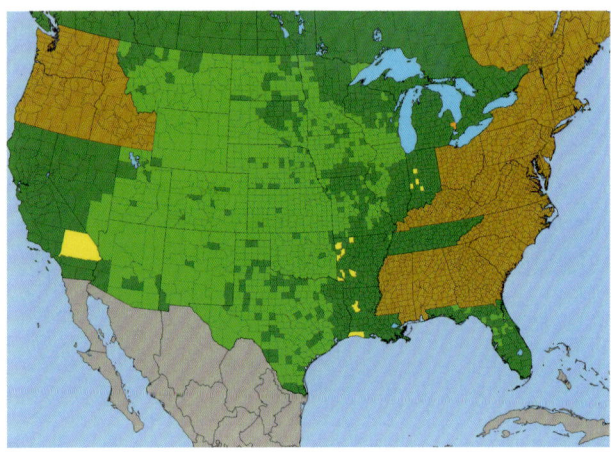

Seedling

Emerging mature plant
in spring

Entire plant. Plants are small in stature at 9–16 in., with
numerous arching stems.

Leaf. Numerous
hairs on upper leaf
surface often lend
a silvery sheen to
the narrow leaves
(0.125–0.25 in.).

Flower. Long tubular flowers are light
yellow with distinctly fringed petals.

Early seed

Mature seed. Ripe seeds
are easily dislodged by the
slightest movement!

LOOK-ALIKE PLANTS:
Natives: Hoary Puccoon and Hairy Puccoon

Onosmodium bejariense
(Onosmodium molle)
Marbleseed

FAMILY: BORAGE (BORAGINACEAE)

Marbleseed has fabulous foliage and adds a nice architectural element to gardens and prairies. Its subtle white flowers peek out from under deeply furrowed, dark-green leaves. Collect seeds when they turn from green to white and plant directly in the ground in fall. Cuttings can also be taken in late summer and rooted out for fall transplanting.

Habitat: Dry limestone prairies, dry calcareous hillsides, bluffs, dry open woods
Garden Uses: Specimens, informal gardens, prairie meadows

USDA Hardiness Zones: 3–8
Light Requirements: Full sun
Soil Types: Sand, loam
Soil Moisture: Dry to medium
Bloom Time: May–July
Flower Color: White
Height: 1–3 feet
Life Expectancy: 10–20 years
Root Type: Taproot
Propagation: Moist stratify seed 90 days, or plant fresh for germination in following spring
Aggressiveness: Low by seed
Attracts: Butterflies, game birds, songbirds
Deer Palatability: Low

Distinguishing Characteristics:
- Thick, deep-green leaves are heavily veined and larger than those of puccoons
- Upper stems and leaves typically droop downward
- Drooping white flowers occur on terminal stems at the top of the plant
- Large brown seeds turn white at maturity, borne in long rows along the upper stems; seeds are similar in shape to puccoons but larger, darker, and do not drop immediately upon ripening

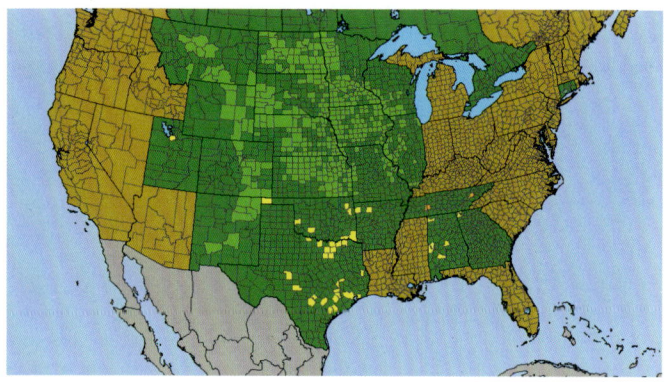

Seedling

Emerging mature plant in spring

Entire plant

Leaf. Leaves are deeply veined and densely hairy.

Flower. Distinctive drooping white flowers are shrouded within a green calyx.

Early seed. Early seeds are chocolate brown.

Mature seed. Seeds turn cream to light brown when ripe; unlike puccoons, seeds on each stem ripen simultaneously.

LOOK-ALIKE PLANTS:

Natives: Leaves are similar to American Gromwell (*Lithospermum latifolium*), native to undisturbed rich woods, with small yellow flowers borne on terminal stems.

Lobelia cardinalis
Cardinal Flower

FAMILY: BELLFLOWER (CAMPANULACEAE)

Brilliant red blooms are a magnet for hummingbirds. Requires rich moist soil with plenty of humus. Short-lived (three to five years), it rarely self-sows in a garden situation. Prefers light shade but thrives in full sun with adequate moisture. Plants can be divided every year to maintain viability. Mulch in fall with clean straw or leaves to prevent winter damage.

Habitat: Wet meadows, marshes, seeps, shaded riverbanks and stream-sides, wet woods and thickets
Garden Uses: Hummingbird gardens, borders, formal and informal gardens, annual beds

USDA Hardiness Zones: 3–9
Light Requirements: Full sun to partial shade
Soil Types: Sand, loam, clay
Soil Moisture: Moist to wet
Bloom Time: July–September
Flower Color: Red
Height: 2–5 feet
Life Expectancy: 3–5 years
Root Type: Fibrous
Propagation: Moist stratify seed 30 days; root division of two-year-old plants in early spring; easy from stem cuttings early to mid-season

Aggressiveness: Low by seed
Attracts: Bees, butterflies, hummingbirds
Deer Palatability: Medium
Distinguishing Characteristics:
- Bright red flowers borne in long wands
- Basal leaves often evergreen over winter
- Stems exude white milky sap when cut
- Small, fine seeds are orange

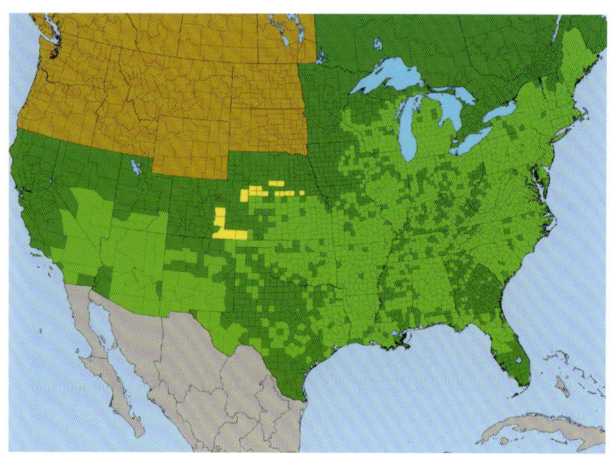

Seedling

Emerging mature plant in spring

Entire plant

Leaf

Flower

Early seed. Immature seedpods are tinged with red.

Mature seed. Seed is ripe when heads turn from green to brown; individual seeds are tiny and orange.

LOOK-ALIKE PLANTS:

Natives: Leaves are essentially indistinguishable from those of Great Blue Lobelia (p. 186).

Lobelia siphilitica
Great Blue Lobelia

FAMILY: BELLFLOWER (CAMPANULACEAE)

Longer-lived than Cardinal Flower, grows in similar moist habitats. Deep-blue flowers attract hummingbirds, bees, and other pollinators. White-flowered forms may appear when grown from seed. Self-sows readily and can be safely transplanted at any time. Mature plants easily divided.

Habitat: Wet prairies and meadows, fens, streamsides, wet swales, wet open woods
Garden Uses: Hummingbird gardens, bee gardens, rain gardens, informal gardens, moist meadows

USDA Hardiness Zones: 3–9
Light Requirements: Full sun to medium shade
Soil Types: Sand, loam, clay
Soil Moisture: Moist to wet
Bloom Time: July–September
Flower Color: Blue
Height: 1–4 feet
Life Expectancy: 5–10 years
Root Type: Fibrous
Propagation: Moist stratify seed 30 days; root division in early spring; easy from stem cuttings early to midseason

Aggressiveness: Medium by seed
Attracts: Bees, butterflies, hummingbirds
Deer Palatability: Low
Distinguishing Characteristics:
- Flowers are blue (Cardinal Flower is red)
- Basal leaves often evergreen over winter
- Seeds are grayish white; Cardinal Flower seeds are orange

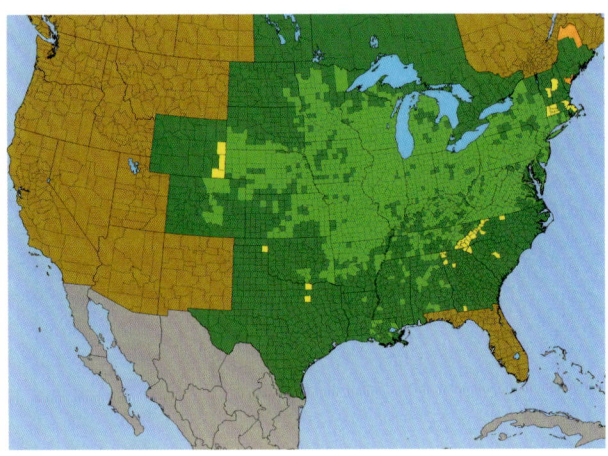

Seedling

Emerging mature plant in spring

Entire plant

Leaf

White flower. Albinism is not uncommon in the genera *Lobelia*, *Liatris*, *Echinacea*, and *Physostegia*. *Lobelia siphilitica* and *Lobelia cardinalis* can hybridize and produce a purple-flowering form.

Flower

Early and mature seed. Immature seedpods have no red coloration, as is typical in *Lobelia cardinalis*.

LOOK-ALIKE PLANTS:

Natives: Leaves are essentially indistinguishable from those of Cardinal Flower (p. 184). In flower, it resembles Tall Bellflower (*Campanulastrum americanum*), a biennial woodland-edge plant that grows in medium to moist habitats.

Silene regia
Royal Catchfly

FAMILY: PINK (CARYOPHYLLACEAE)

One of few red flowers of the North American prairie. Plant with short grasses and flowers to help control weeds. Its nectar is a hummingbird favorite; sticky flower petals actually catch flies that hummingbirds eat! Requires rich, well-drained soil. The taproot cannot be divided; relatively easy to grow from seed. Although native to the southern Midwest, a strain that is reliably hardy to USDA Zone 4 has been bred by Prairie Nursery of Westfield, Wisconsin.

Habitat: Medium prairies, sunny open woods
Garden Uses: Specimens, small groups, hummingbird gardens, informal gardens, prairie meadows

USDA Hardiness Zones: 5–9
Light Requirements: Full sun to partial shade
Soil Types: Loam
Soil Moisture: Medium
Bloom Time: July–October
Flower Color: Red
Height: 2–4 feet
Life Expectancy: 5–10 years
Root Type: Taproot
Propagation: Moist stratify seed 30 days

Aggressiveness: Low by seed
Attracts: Butterflies, hummingbirds
Deer Palatability: Low
Distinguishing Characteristics:
- Bright red, five-petaled starlike flowers
- Flower bracts and sepals are sticky
- Seeds borne in brown "urns" with six little points on top

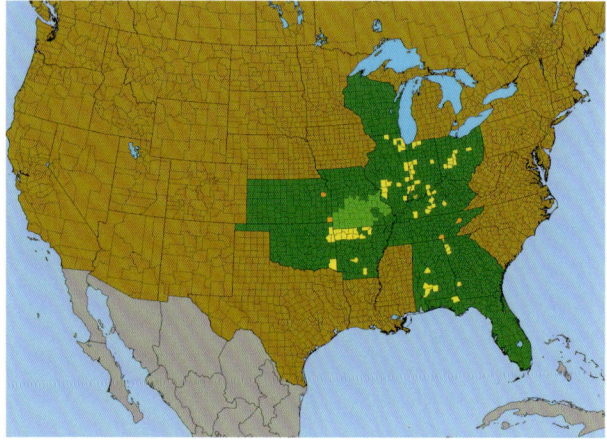

Seedling

Emerging mature plant in spring

Entire plant

Leaf

Flower. Five-petaled red flowers are unique among prairie plants; similar to *Silene virginica*, a plant of savannas and open woodlands.

Early seed

Mature seed. Pods resemble tiny tan urns when mature; seed is ripe when pods turn tan color.

LOOK-ALIKE PLANTS:

Natives: *S. virginica* is a woodland species with similar flowers whose petals are notched at the tips and has glossy, oblong leaves. Seedpods hang down on *S. virginica* rather than upright.

Weeds: Basal rosette and stem leaves of the white-flowered Bladder Campion (*Lychnis alba*) are similar, as are the seed capsules.

Hypericum ascyron
(Hypericum pyramidatum)
Great St. Johnswort

FAMILY: ST. JOHNSWORT (CLUSIACEAE)

Showy yellow flowers turn into chocolate-brown seed candelabras. A good companion to Culver's Root (*Veronicastrum virginicum*) and Michigan Lily (*Lilium michiganense*). Seedheads make excellent dried arrangements; be sure to empty the seeds before bringing them indoors!

Habitat: Wet prairies and meadows, sedge meadows, riverbanks, wet thickets, moist woodland edges
Garden Uses: Specimens, rain gardens, summer screen plantings, pond and stream edges, and wet meadows

USDA Hardiness Zones: 3–5
Light Requirements: Full sun to part shade
Soil Types: Sand, loam, clay
Soil Moisture: Moist to wet
Bloom Time: June–August
Flower Color: Yellow
Height: 4–6 feet
Life Expectancy: 5–10 years
Root Type: Fibrous
Propagation: Moist stratify seed 30 days; easy from stem cuttings early to midseason
Aggressiveness: Low by seed
Attracts: Bees, butterflies

Deer Palatability: Low
Distinguishing Characteristics:

- Seedheads resemble brown candelabras
- Large seedpods have distinctive points at the end of each capsule lobe
- Small brown seeds look like cupcake sprinkles
- Flowers have five styles (female pollen receptors), whereas all other native *Hypericum* have between one and three styles, except *H. kalmianum*, a shrubby species that grows 2–4 feet tall

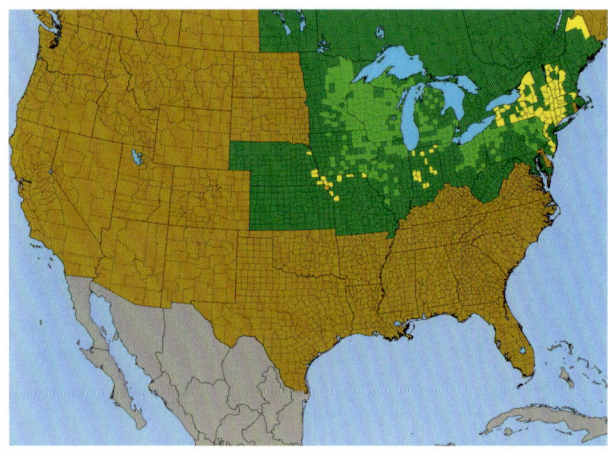

Seedling. Seedling leaves are smooth and hairless.

Emerging mature plant in spring. Emerging stems are reddish brown.

Entire plant

Leaf. Leaves are smooth and opposite on the stem.

Flower. Flower has a large ovary above with a prominent, five-parted stigma.

Mature seed. Seedpods are large and distinctive; ripe seeds are chocolate brown and resemble tiny cupcake sprinkles.

LOOK-ALIKE PLANTS:

Natives: Flowers are similar to many other *Hypericum* species, none of which reaches the same stature, with the exception of Shrubby St. Johnswort (*H. prolificum*), which has three styles rather than five styles on the ovary.

Euphorbia corollata
Flowering Spurge, Prairie Baby's Breath

FAMILY: EUPHORBIACEAE (SPURGE)

Thrives in extremely dry sandy soil, as well as medium to moist sand and well-drained loam. Often creeps by rhizomes to form a patch. Small groupings create an airy effect. Mix with low to medium-height flowers and grasses. Seed is best planted in fall for spring germination. Slow from seed but long-lived.

Habitat: Dry to moist prairies, barrens, glades, lakeshores, bluffs, old fields, roadsides, especially in sandy soils
Garden Uses: Prairie meadows, informal gardens

USDA Hardiness Zones: 3–9
Light Requirements: Full sun
Soil Types: Sand, loam
Soil Moisture: Dry to moist
Bloom Time: May–October
Flower Color: White
Height: 2–4 feet
Life Expectancy: 10–20 years
Root Type: Rhizome
Propagation: Moist stratify seed 60 days; root division using 4-inch rhizomes in fall or spring; easy from stem cuttings; use early growth
Aggressiveness: Medium by rhizome

Attracts: Bees, game birds, songbirds
Deer Palatability: Low
Distinguishing Characteristics:
- Plant produces a white, milky sap when cut that can cause a rash in some people
- Flowers are produced in open whorled panicles on long stems, somewhat resembling Baby's Breath (*Gypsophila paniculata*)
- Stems often tinged with red, especially late in the season
- Grayish-white seeds borne in groups of three, inside "exploding" capsules

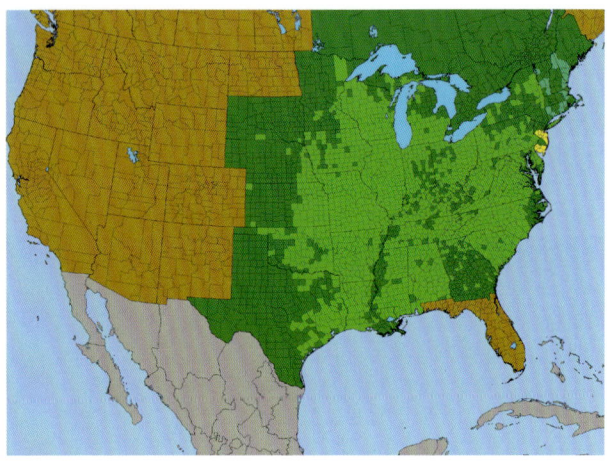

Seedling. Seedlings usually have red stems.

Emerging mature plant in spring

Entire plant. Plants often (not always) have red stems, especially in late summer.

Leaf. Leaves are very smooth and exude white sap similar to milkweed when cut.

Flower. Flowers appear similar to Baby's Breath (*Gypsophila* spp.).

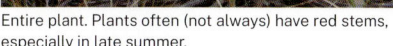
Early seed

Mature seed close-up. Seeds explode when ripe. Harvest seeds shortly after plants have finished flowering and pods are fully filled out but still green.

Mature seed

LOOK-ALIKE PLANTS:

Natives: Emerging plants are similar to Butterflyweed (p. 84), which does not have white milky sap. Also similar to Common Milkweed (p. 82) and members of the genus *Apocynum*, all of which have white, milky sap. Seedlings resemble Roundhead Bush Clover (p. 212), but young leaves are blunt-tipped rather than pointed.

Weeds: Leaves and white sap similar to nonnative spurges, such as *E. esula* and *E. cyparissias*.

Amorpha canescens
Leadplant

FAMILY: PEA (FABACEAE, LEGUMINOSAE)

This lovely small shrub thrives in poor, sandy soil. Although slow growing, it is extremely long-lived. Excellent for dry prairie gardens and meadows. Its deep taproot cannot be divided, mature plants are not easily moved.

Habitat: Dry to medium prairies, sand prairies, savannas, barrens, dry open woods
Garden Uses: Specimens and small groups, alone or mixed with short- to medium-height grasses. Butterfly gardens, prairie meadows.

USDA Hardiness Zones: 3–8
Light Requirements: Full sun to light shade
Soil Types: Sand, loam
Soil Moisture: Dry to medium
Bloom Time: June–July
Flower Color: Purple to violet
Height: 2–3 feet
Life Expectancy: 20+ years
Root Type: Taproot
Propagation: Scarify seed and moist stratify 30 days
Aggressiveness: Low by seed

Attracts: Bees, butterflies
Deer Palatability: Medium
Distinguishing Characteristics:
- Finely divided leaves are grayish green
- Purple flower spires have prominent yellow stamens
- Ripe seedheads are gray, soft to the touch, and aromatic when crushed
- Plants will flower on old stems, as well as on new growth following burning or mowing

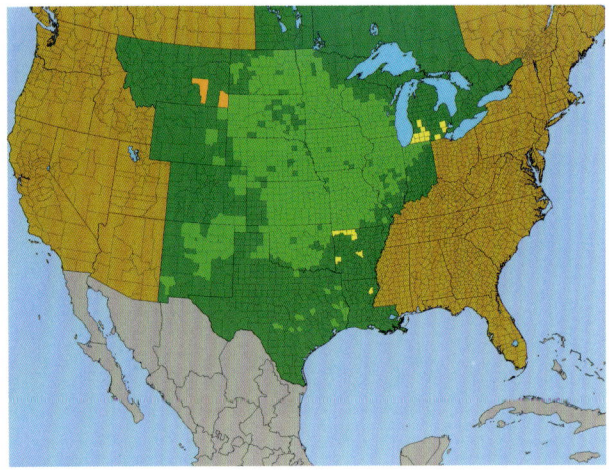

Seedling. Seedling stems usually reddish; leaves broadly rounded.

Emerging mature plant in spring. Whitish-green leaves emerge from woody stalks.

Entire plant. Wide-spreading, shrubby growth form.

Leaf. Leaves are typically silvery green, especially the undersides.

Flower. Orange stamens are prominent against blue petals.

Mature seed. Seeds are ripe when pods turn greenish brown and strip off stem easily — wear gloves!

LOOK-ALIKE PLANTS:

Natives: False Indigo Bush (*Amorpha fruticosa*) has similar leaves and stems but grows up to 10 feet in moist soils along streams. Leaves of Illinois Bundleflower (p. 208) are similar but wider and longer. Leaves of American Wisteria (*Wisteria frutescens*) are similar as they emerge on woody stems.

Astragalus canadensis
Canada Milkvetch

FAMILY: PEA (FABACEAE, LEGUMINOSAE)

Creamy-yellow flowers borne on upright spikes produce seeds relished by many birds. A bit sprawling for most garden applications and works best in a meadow situation. Readily grown from seed; mature plants are taprooted and cannot be easily divided or transplanted.

Habitat: Dry to medium prairies, savannas, glades, open woods, woodland edges, lakeshores, streamsides
Garden Uses: Informal gardens, prairie meadows

USDA Hardiness Zones: 3–8
Light Requirements: Full sun
Soil Types: Sand, loam
Soil Moisture: Dry to medium
Bloom Time: July–October
Flower Color: Yellow
Height: 2–3 feet
Life Expectancy: 10–20 years
Root Type: Taproot
Propagation: Scarify seed and moist stratify 30 days

Aggressiveness: Medium by seed
Attracts: Bees, butterflies, game birds, hummingbirds, songbirds
Deer Palatability: Medium
Distinguishing Characteristics:
- Stems often reddish colored
- Small, creamy-yellow flowers borne in tight clusters on stems above the plant
- Seedpods are blackish brown and pointed on top

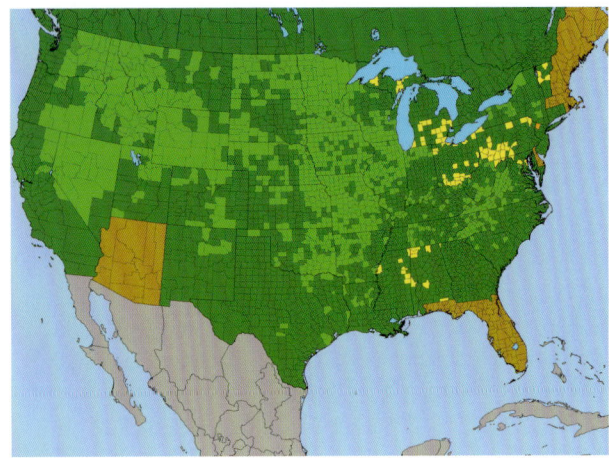

Seedling

Emerging mature plant in spring. Emerging leaves similar to *Tephrosia virginiana* but wider and blunt tips.

Entire plant. Central plant stem is usually red.

Leaf

Flower

Early and mature seed. Seeds are ripe when pods turn black.

LOOK-ALIKE PLANTS:

Natives: Other species of *Astragalus* have similar leaves and flowers. Most are native to the western United States. *A. crassicarpus* has similar leaves and occurs on the tallgrass prairie and Great Plains but has flowers that range from white to purple in color.

Weeds: Leaves are similar to Crownvetch (*Coronilla varia*), which has blue flowers and creeps aggressively by rhizomes.

Baptisia alba *(Baptisia lactea, B. leucantha)*
White False Indigo

FAMILY: PEA (FABACEAE, LEGUMINOSAE)

Prized for its towering spires of pure white flowers and distinctive blue-green foliage. One of the finest specimen plants of the prairie. Unlike Blue False Indigo, there are no lower leaves to shade out weeds; plant short prairie grasses and flowers around base to minimize weed growth. Very tolerant of moisture. Slow from seed, but very long-lived. Transplant one-year-old seedlings before taproot develops. Mature plants cannot be divided or transplanted. Seedstalks with mature pods make a stunning centerpiece in dried arrangements.

Habitat: Medium to moist prairies, savannas, dry open woods
Garden Uses: Specimens and groups of three to five plants, rain gardens

USDA Hardiness Zones: 4–9
Light Requirements: Full sun to partial shade
Soil Types: Sand, loam, clay
Soil Moisture: Medium to moist
Bloom Time: June–August
Flower Color: White
Height: 3–5 feet
Life Expectancy: 20+ years
Root Type: Taproot
Propagation: Scarify seed and moist stratify 90 days; tricky from cuttings: use first early growth
Aggressiveness: Low by seed

Attracts: Bees, butterflies, game birds, songbirds
Deer Palatability: Low
Distinguishing Characteristics:
- Emerging plant resembles asparagus (similar to Blue False Indigo, p. 200)
- Flowers and seedpods borne on long stalks held above the leaves
- Leaves are sparse along the stems, whereas *B. australis* leaves are abundant and surround the stems from the ground up

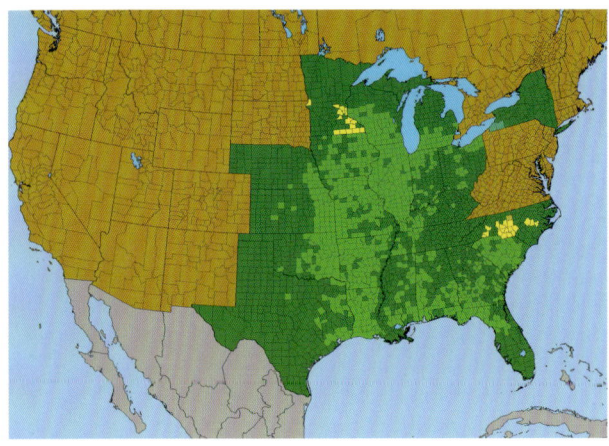

Seedling

Emerging mature plant in spring. Bases of emerging stems tinged with purple.

Entire plant

Leaf

Flower

Mature seed. Tall seedstalks are usually elevated above leaves; "tails" on seedpods are shorter than on *Baptisia australis*.

LOOK-ALIKE PLANTS:

Natives: Blue False Indigo (p. 200) has blue flowers rather than white.

Baptisia australis
Blue False Indigo

FAMILY: PEA (FABACEAE, LEGUMINOSAE)

Reaching up to 5 feet in diameter, makes an excellent specimen. Deep-blue flowers are borne among the leaves: other *Baptisia* species produce flowers outside the leaves. Mixes effectively with other flowers, grasses, and shrubs. Slow from seed. Unlike other taprooted *Baptisia*, mature plants can be divided and transplanted.

Habitat: Medium prairies and open woods
Garden Uses: Individual specimens and small groves of 3–5 plants, perennial gardens and borders as a year-round focal point

USDA Hardiness Zones: 3–10
Light Requirements: Full sun to partial shade
Soil Types: Sand, loam, clay
Soil Moisture: Medium
Bloom Time: May–July
Flower Color: Blue
Height: 3–5 feet
Life Expectancy: 20+ years
Root Type: Fibrous taproot
Propagation: Scarify seed and moist stratify 90 days; root division in spring or fall; tricky from cuttings: use first early growth
Aggressiveness: Low by seed
Attracts: Bees, butterflies, game birds, songbirds

Deer Palatability: Low
Distinguishing Characteristics:
- Emerging plant resembles asparagus (also similar to White False Indigo, p. 198)
- Flowers and seedpods are borne on both upper and lower stems; White False Indigo has flowers only on upper stems
- Seedpod clusters are shorter (8–15 inches) than those of White False Indigo (12–36 inches) and nestled among the leaves; White False Indigo has very long, terminal seedstalks

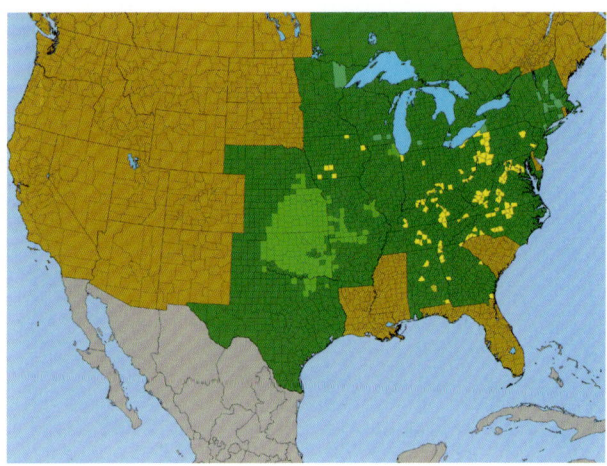

Seedling

Emerging mature plant in spring. Base of emerging stems are usually green, not purple, as in *Baptisia alba*.

Entire plant

Leaf

Flower

Mature seed. Seedstalks are usually produced within leafy stems, not above them; "tails" on seedpods are longer than on *Baptisia alba*.

LOOK-ALIKE PLANTS:

Natives: White False Indigo (p. 198) has white flowers rather than blue.

Baptisia bracteata (*B. leucophaea*)
Cream False Indigo

FAMILY: PEA (FABACEAE, LEGUMINOSAE)

Long racemes of cream-yellow flowers make this one of the showiest prairie flowers. Pollinated primarily by bumblebees. Mix with short prairie grasses to control weeds around its base. Slow from seed but extremely long-lived. Transplant one-year-old seedlings before taproot develops. Mature plants cannot be easily divided or moved.

Habitat: Dry to medium prairies, savannas, dry open woods
Garden Uses: As a low-growing, early-season specimen plant, perennial borders

USDA Hardiness Zones: 4–9
Light Requirements: Full sun to partial shade
Soil Types: Sand, loam
Soil Moisture: Dry to medium
Bloom Time: May–June
Flower Color: Cream
Height: 1–2 feet
Life Expectancy: 20+ years
Root Type: Taproot
Propagation: Scarify Seed and moist stratify 90 days; tricky from cuttings: use first early growth
Aggressiveness: Low by seed

Attracts: Bees, butterflies, game birds, songbirds
Deer Palatability: Low
Distinguishing Characteristics:
- Emerging plant resembles asparagus (similar to other False Indigo)
- One of the lowest growing of *Baptisia*
- Flower stalks and seedpods are 12–18 inches long, borne on arching stems radiating out from the plant, often hidden by the leaves

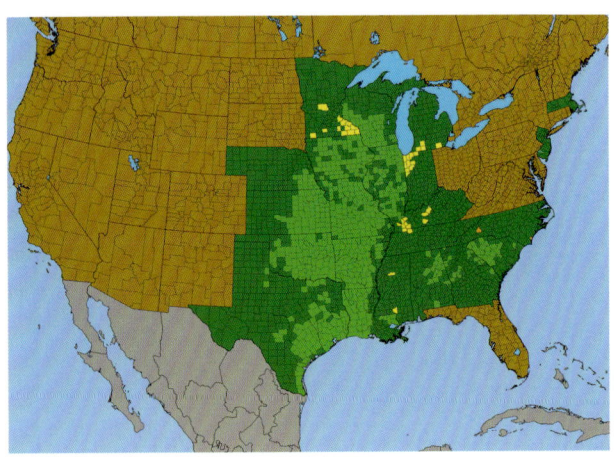

Seedling. Seedling stems minutely hairy, not smooth, as in *B. alba* and *B. australis*.

Emerging mature plant in spring. Emerging shoots are often chocolate brown.

Entire plant

Leaf. Leaves hairy, not smooth, as with *B. alba* and *B. australis*.

Flower

Early and mature seed. Seedstalks pendulant, not borne upright; seeds are ripe when pods turn black.

LOOK-ALIKE PLANTS:

Natives: *B. tinctoria* is an eastern native with short flower stalks (4–5 inches) and bright-yellow to creamy flowers. *B. sphaerocarpa*, native to south-central US, has bright-yellow flowers on upright stalks.

Dalea candida (Petalostemum candidum, Petalostemon candidus)

White Prairie Clover

FAMILY: PEA (FABACEAE, LEGUMINOSAE)

Similar to Purple Prairie Clover, with white flowers and slightly wider leaves. Plant with Purple Prairie Clover for effect, and with short grasses to help control weeds. Easy to grow from seed. Its deep taproot cannot be divided, and mature plants are not easily moved.

Habitat: Dry to medium prairies, glades, dry open woods
Garden Uses: Bird gardens, informal gardens, prairie meadows

USDA Hardiness Zones: 3–9
Light Requirements: Full sun
Soil Types: Sand, loam
Soil Moisture: Dry to medium
Bloom Time: July–October
Flower Color: White
Height: 1–2 feet
Life Expectancy: 10–20 years
Root Type: Taproot
Propagation: Moist stratify seed 30 days
Aggressiveness: Low by seed
Attracts: Bees, butterflies, game birds, songbirds

Deer Palatability: High
Distinguishing Characteristics:
- Individual leaves (leaflets) are broader (0.2 inches) than Purple Prairie Clover leaves (0.08 inches)
- Seedheads are coarse and raspy; Purple Prairie Clover seeds are soft to the touch
- Roots are yellow in color when young, turning browner after 3 years

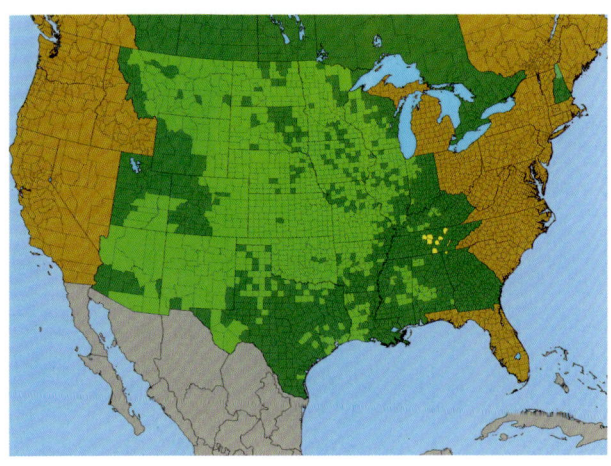

Seedling

Emerging mature plant
in spring

Entire plant

Leaf. Leaf is two to three times
as wide as *Dalea purpurea*.

Flower

Early and mature
seed

LOOK-ALIKE PLANTS:

Natives: Purple Prairie Clover (p. 206) has narrower leaflets. Leafy Prairie Clover (*Dalea foliosa*), a rare species found in a few locations in Illinois, Tennessee, and Alabama, has purple flowers and many more leaflets per leaf. Leaves of *Dalea aurea* are similar, but this western species has yellow flowers.

Dalea purpurea (Petalostemum purpureum, Petalostemon purpureus)

Purple Prairie Clover

FAMILY: PEA (FABACEAE, LEGUMINOSAE)

Highly adaptable, long-lived, and extremely drought tolerant. Fine, lacy foliage is topped with bright purple and yellow flowers. Associate with fibrous-rooted plants to help control weeds around its taproot. Mixes nicely with White Prairie Clover, Butterflyweed, Rattlesnake Master, and low-growing grasses. Easy to grow from seed. Deep taproots cannot be divided, and mature plants cannot be easily moved.

Habitat: Dry to medium prairies, barrens, savannas, dry open woods
Garden Uses: Butterfly gardens, bird gardens, informal gardens, and prairie meadows

USDA Hardiness Zones: 3–8
Light Requirements: Full sun
Soil Types: Sand, loam, clay
Soil Moisture: Dry to medium
Bloom Time: July–October
Flower Color: Purple-yellow
Height: 1–2 feet
Life Expectancy: 10–20 years
Root Type: Taproot
Propagation: Moist stratify seed 30 days
Aggressiveness: Low by seed
Attracts: Bees, butterflies, game birds, songbirds

Deer Palatability: High
Distinguishing Characteristics:

- Individual leaves (leaflets) are narrower (0.08 inches) than White Prairie Clover leaves (0.2 inches)
- Leaves are aromatic when crushed
- Seedheads are soft to the touch, whereas White Prairie Clover seeds are coarse and raspy
- Roots are bright yellow when young, turning browner later

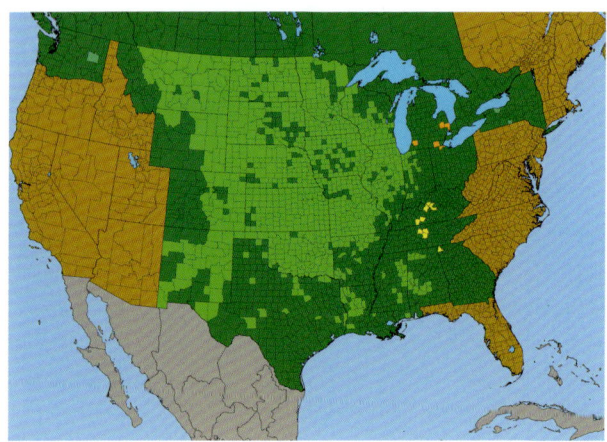

C

A
Seedling

B
Emerging mature plant
in spring

Entire plant

D

Leaf. Leaf one-half to one-third as
wide as *Dalea candida*.

E

Flower

F

Mature seed. Mature seed
is soft to the touch (*Dalea
candida* is rough); ripe seed
strips off without resistance.

LOOK-ALIKE PLANTS:

Natives: White Prairie Clover (p. 204) has wider leaflets; Leafy Prairie Clover (*Dalea foliosa*) has similar flowers but many more leaflets per leaf.

Desmanthus illinoensis

Illinois Bundleflower, Prairie Bundleflower, Prairie Mimosa

FAMILY: PEA (FABACEAE, LEGUMINOSAE)

Notable for its fine, lacy foliage and delicate white flowers, a great garden plant. Curly, light-brown seedpods provide added fall and winter interest and make great dried arrangements.

Habitat: Dry to moist prairies, savannas, dry open woods
Garden Uses: Specimens and small groups, bird gardens, cutting gardens (dried seedheads)

USDA Hardiness Zones: 5–9
Light Requirements: Full sun
Soil Types: Sand, loam, clay
Soil Moisture: Dry to moist
Bloom Time: June–August
Flower Color: White
Height: 2–5 feet
Life Expectancy: 5–10 years
Root Type: Taproot
Propagation: Scarify seed and moist stratify 30 days
Aggressiveness: Low by seed

Attracts: Bees, game birds, songbirds
Deer Palatability: Medium
Distinguishing Characteristics:
- Compound leaves are finely divided, lacy
- Delicate flowers resemble Mimosa Tree (*Albizia julibrissin*) blossoms but white
- Curly, chocolate-brown seedpods are unique among prairie legumes

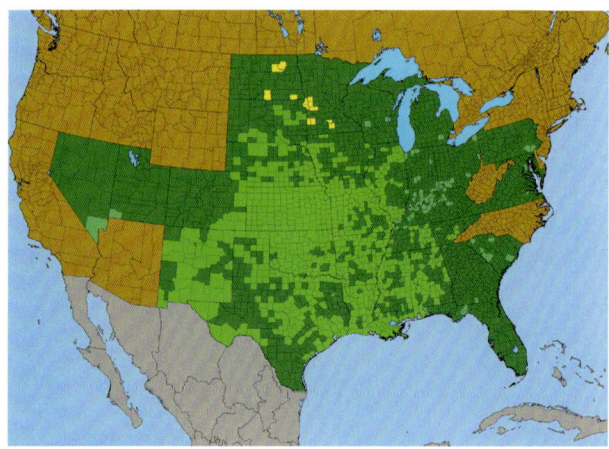

Seedling

Emerging mature plant in spring

Entire plant

Leaf

Flower

Mature seed. Seeds are ripe when they turn brown.

LOOK-ALIKE PLANTS:

Natives: Leaves resemble those of Leadplant (p. 194) but are wider and longer.

Desmodium canadense
Canada Ticktrefoil, Showy Ticktrefoil

FAMILY: PEA (FABACEAE, LEGUMINOSAE)

Showiest of our native Ticktrefoils. Utilized by a wide variety of wildlife, including upland game birds. Excellent for creating habitat when planted with tall prairie grasses and forbs. Easy to grow from seed, the stick-tight seedpods spread far and wide.

Habitat: Dry to wet prairies, savannas, open woods, woodland edges, roadsides
Garden Uses: Prairie meadows and wildlife plantings. Not recommended as a garden plant because of its sticky seeds and proclivity to self-sow.

USDA Hardiness Zones: 3–7
Light Requirements: Full sun to part shade
Soil Types: Sand, loam, clay
Soil Moisture: Dry to moist
Bloom Time: July–October
Flower Color: Pale purple
Height: 3–5 feet
Life Expectancy: 10–20 years
Root Type: Taproot
Propagation: Dry stratify seed 60 days
Aggressiveness: Medium by seed

Attracts: Bees, butterflies, game birds, hummingbirds, songbirds
Deer Palatability: High
Distinguishing Characteristics:
- Leaves are smooth above, hairy below; Illinois Ticktrefoil (*D. illinoense*) has lighter-colored leaves that are hairy on both sides
- Stick-tight seedpods are level on top and indented on the bottom; Illinois Ticktrefoil seeds are indented on both top and bottom

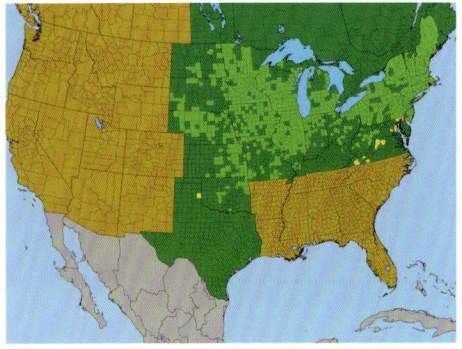

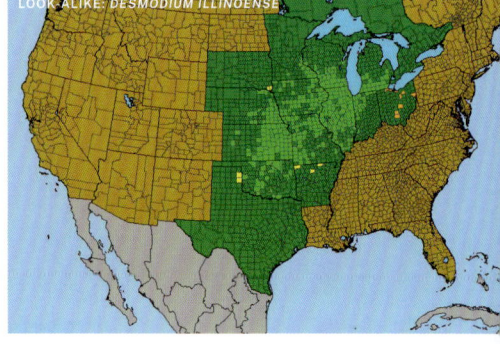

LOOK-ALIKE: *DESMODIUM ILLINOENSE*

Seedling

Emerging mature plant in spring

Entire plant

Leaf. Leaves are smooth with few hairs; leaves of *Desmodium illinoense* are hairy and rough.

Flower. Flowers are showy; much larger than *Desmodium illinoense*.

Early seed. Top of seedpod is almost flat; *Desmodium illinoense* is indented.

Mature seed. Seeds are ripe when they turn brown.

Desmodium illinoense, leaf. Leaves are rough and hairy; leaves of *Desmodium canadense* are nearly smooth.

Desmodium illinoense, flower. Flowers are small and hardly noticeable.

Desmodium illinoense, early seed. Top of seedpod is indented; *Desmodium canadense* is nearly flat.

LOOK-ALIKE PLANTS:

Natives: Illinois Ticktrefoil (*Desmodium illinoense*) occurs in mesic to dry mesic prairies, differing as noted above.

Lespedeza capitata
Roundhead Bush Clover

FAMILY: PEA (FABACEAE, LEGUMINOSAE)

At its best in fall and winter, after insignificant white flowers mature into attractive brown seedheads. Seeds are retained late into winter, making this an important food source for resident songbirds. The stiff stems make excellent perches. Reliable from seed, seldom installed as transplants.

Habitat: Dry to medium prairies, sand prairies, barrens, savannas, glades, open woods, bluffs, sandy old fields
Garden Uses: Prairie meadows, wildlife plantings

USDA Hardiness Zones: 3–8
Light Requirements: Full sun to partial shade
Soil Types: Sand, loam, gravel
Soil Moisture: Dry to medium
Bloom Time: August–September
Flower Color: White
Height: 3–5 feet
Life Expectancy: 10–20 years
Root Type: Taproot
Propagation: Scarify seed and moist stratify 30 days
Aggressiveness: Low by seed

Attracts: Bees, butterflies, game birds, songbirds
Deer Palatability: Medium
Distinguishing Characteristics:
- Small white flowers are inconspicuous and bloom for a short period of time
- Faded flowers appear bright orange as plant goes into seed
- Brown seedheads stand late into winter, retaining seeds as a winter bird food
- Hulled seeds are greenish red in color

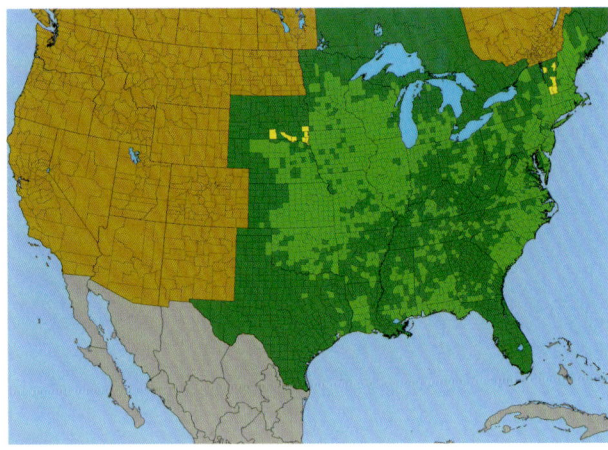

Seedling. Seedling leaves are pointed; hairs are readily visible on stems and leaf margins.

Emerging mature plant in spring. Emerging stems are often leggy and reddish.

Entire plant

Leaf. Leaves are often tinged with silver highlights.

Flower

Early seed

Mature seed

LOOK-ALIKE PLANTS:

Natives: Seedlings resemble Flowering Spurge (p. 192) but leaves are pointed rather than blunt, and stems do not have white, milky sap.

Weeds: Numerous species of *Lespedeza* have been introduced for agricultural and wildlife plantings, but few reach the height of *L. capitata*.

Lupinus perennis
Wild Lupine, Sundial Lupine

FAMILY: PEA (FABACEAE, LEGUMINOSAE)

Blue spires of Lupine are a garden favorite. This native grows only on dry, deep sandy soils. Deeply taprooted, seedlings have roots 12 inches long when new sprouts are only 2–3 inches tall, making it hard to transplant. Will not grow on loam or clay. Plant with short grasses and flowers to minimize weed growth. Leaves are the primary food source for federally endangered Karner blue butterflies. Seedpods twist open and explode, propelling seeds up to 33 feet. Harvest seeds when pods are yellow and rattle when tapped. Dry seedpods loosely in a closed paper bag or in boxes covered with window screen. Seeds germinate in late summer when planted fresh. Plants cannot be divided.

Habitat: Dry sand prairies, sand barrens, oak barrens, dry open sandy woods
Garden Uses: Butterfly gardens, small groupings in informal dry sand gardens, dry prairie meadows

USDA Hardiness Zones: 3–8
Light Requirements: Full sun to partial shade
Soil Types: Sand
Soil Moisture: Dry
Bloom Time: May–June
Flower Color: Blue
Height: 1–2 feet
Life Expectancy: 10–20 years
Root Type: Taproot
Propagation: Moist stratify seed 7 days, or plant fresh for fall germination
Aggressiveness: Medium by seed

Attracts: Bees, butterflies, game birds, hummingbirds, songbirds
Deer Palatability: Low
Distinguishing Characteristics:
- Leaves radiate out from center like a sundial
- Plants grow only 2 feet tall, much shorter than other lupines, such as garden ornamentals like *L. polyphyllus* "Russell Hybrids," which reach 4 feet tall
- Seedpods turn from green to yellow to black and explode when ripe

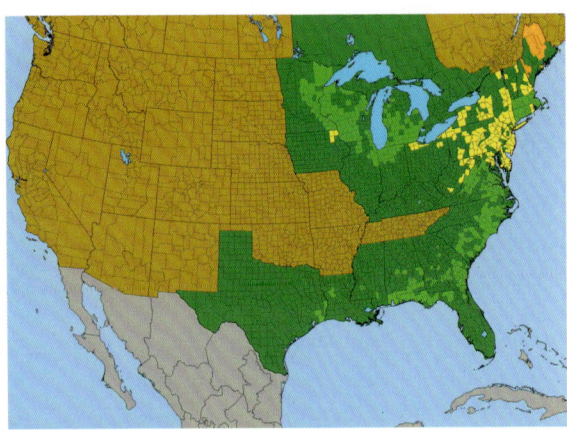

Seedling. Seedlings have large cotyledons and five true leaves with long hairs on margins and stem.

Emerging mature plant in spring

Entire plant

Leaf

Flower

Early and mature seed

LOOK-ALIKE PLANTS:

Natives: Prairie Turnip (*Pediomelum esculentum*) is a western prairie species with similar palmate leaves that are deep green and thick textured. Small, light-blue flowers are borne in clusters among the leaves rather than above them. Seedpods do not explode.

Weeds: "Russell Hybrids" Lupine is a garden selection made from *Lupinus polyphyllus*, native to the West Coast of North America.

Senna hebecarpa (*Cassia hebecarpa*)
Wild Senna

FAMILY: PEA (FABACEAE, LEGUMINOSAE)

Deep-green foliage and butter-yellow flowers on 6 foot stems make this an outstanding specimen plant. Attracts a wide variety of pollinators, especially bees. Easy from seed, grows rapidly and self-sows readily. Associate with other vigorous plants to help control its spread. Great for alkaline clay soils.

Habitat: Wet prairies, moist open woods, woodland edges, in alluvial soils along streamsides
Garden Uses: Specimens, bee gardens, wildlife plantings, prairie meadows

USDA Hardiness Zones: 4–7
Light Requirements: Full sun
Soil Types: Sand, loam, clay
Soil Moisture: Medium to wet
Bloom Time: July–October
Flower Color: Yellow
Height: 4–6 feet
Life Expectancy: 10–20 years
Root Type: Fibrous taproot
Propagation: Scarify seed and moist stratify 30 days; tricky from cuttings: use first early growth

Aggressiveness: Medium by seed
Attracts: Bees, butterflies, game birds, hummingbirds, songbirds
Deer Palatability: Low
Distinguishing Characteristics:
- Yellow, pea-like flowers are borne in large clusters at the ends of stems
- Dark-brown seedpods droop downward from the stems, with large, squarish chocolate-brown seeds inside

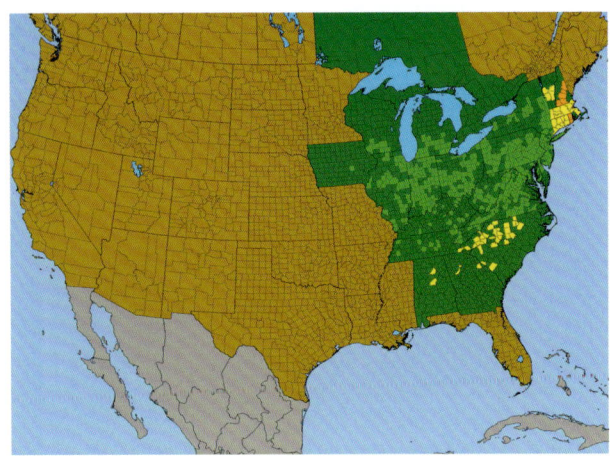

Seedling

Emerging mature plant in spring. Emerging leaves similar to *Astragalus canadensis*, except on longer stems and wider leaves.

Entire plant

Leaf

Flower. Pistils have long white hairs.

Mature seed. Seeds are ripe when pods turn brown.

LOOK-ALIKE PLANTS:

Natives: Maryland Senna (*S. marilandica*) has flowers borne in the axils of the leaves along the stem, with four to eight pairs of leaflets per leaf. *S. hebecarpa* has flowers borne in large terminal clusters, with six to ten pairs of leaflets per leaf. *Chamaechrista fasciculata* (Partridge Pea) is a lower-growing annual with similar leaves and flowers.

Weeds: Crownvetch (*Coronilla varia*) has similar leaves, with a creeping habit rather than growing upright.

Senna marilandica (Cassia marilandica)
Senna

FAMILY: PEA (FABACEAE, LEGUMINOSAE)

Essentially the same in appearance as *Senna hebecarpa*. Attracts a wide variety of pollinators, especially bees.

Habitat: Moist prairies, woodland openings, savannas, riverbanks, limestone glades
Garden Uses: Informal borders, rain gardens, bee gardens, prairie meadows

USDA Hardiness Zones: 4–9
Light Requirements: Full sun
Soil Types: Sand, loam, clay
Soil Moisture: Medium to wet
Bloom Time: July–October
Flower Color: Yellow
Height: 3–6 feet
Life Expectancy: 10–20 years
Root Type: Central taproot, rhizomes
Propagation: Easy from seed
Aggressiveness: Low by seed

Attracts: Birds, bumblebees, butterflies
Deer Palatability: Low
Distinguishing Characteristics:
- Narrow seeds twice as long as broad — Wild Senna (p. 216) has seeds as broad as they are long
- Flowers borne in the axils of the leaves along the stem
- Has four to eight pairs of leaflets per leaf

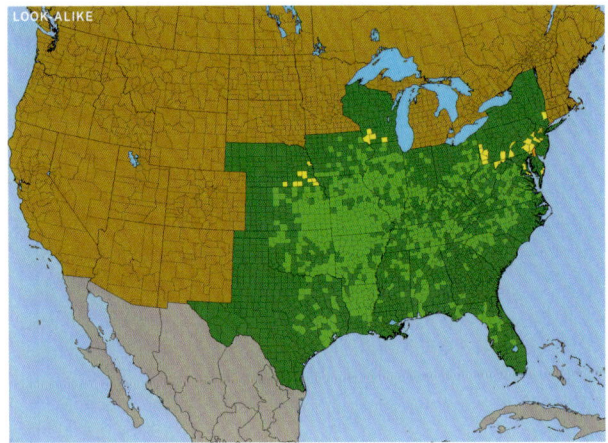

LOOK-ALIKE

Flower. Pistils are shorter and closer together (appressed) than in *S. hebecarpa* flowers.

Mature seed. Seeds are ripe when pods turn brown.

LOOK-ALIKE PLANTS:

Natives: *Strophostyles* spp. (Fuzzybean) as it emerges through the ground, *Senna hebecarpa*

Tephrosia virginiana
Goatsrue, Virginia Tephrosia

FAMILY: PEA (FABACEAE, LEGUMINOSAE)

Unique pink and yellow flowers rise above gray-green, ferny foliage. Grows only on dry sandy soil. Will not grow on loam or clay. Very long-lived. Its deep taproot cannot be divided or moved. The root contains rotenone, a natural insecticide. Seeds should be sown in desired location in fall for spring germination.

Habitat: Sand prairies, sandy barrens, oak barrens, dry open woods on sandy soils
Garden Uses: Butterfly gardens, bee gardens, dry sand gardens, dry prairie meadows and savannas

USDA Hardiness Zones: 4–9
Light Requirements: Full sun to partial shade
Soil Types: Sand, gravel
Soil Moisture: Dry
Bloom Time: June–August
Flower Color: Pink, yellow
Height: 1–2 feet
Life Expectancy: 10–20 years
Root Type: Taproot
Propagation: Scarify seed and moist stratify 30 days
Aggressiveness: Low by seed
Attracts: Bees, butterflies, game birds, songbirds

Deer Palatability: Low
Distinguishing Characteristics:
- Bicolor pink and yellow flowers are unique among native legumes
- Thin, lacy, pea-like leaves often have a grayish cast
- Seedpods are covered with silver hairs and are produced in pendulant bunches close to the ground
- Plant has a low, bushy habit, never creeping or climbing

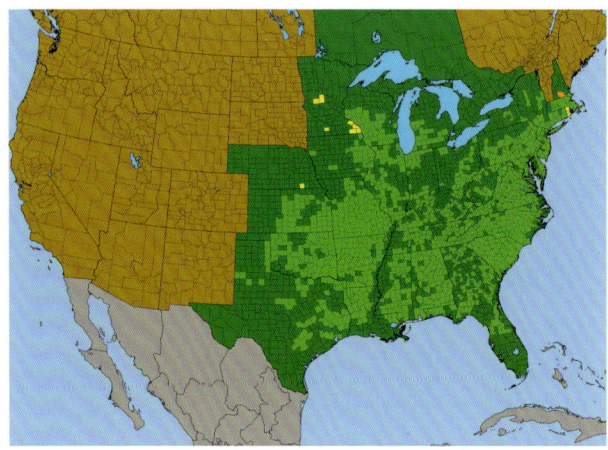

Seedling

Emerging mature plant in spring. Emerging leaves similar to *Astragalus canadensis*, except narrower and sharp tipped.

Entire plant

Leaf. Individual leaves on each leaflet are 0.25–0.375 in. wide, with up to twelve or more leaves on each side of the stem.

Flower. Flowers are a unique pink and yellow.

Early seed

Mature seed with seedpods intact

After natural dispersal. Exploding pods are half the width of Lupine pods, at 0.375 in. versus 0.75 in., and appear two months after Lupine.

LOOK-ALIKE PLANTS:

Natives: Lupine (p. 214) has similar growth form and habitats but larger palmate leaves, blue flowers, and wider upright seedpods. Various species of Vetch (*Vicia* spp.) have similar leaves, but most exhibit a trailing habit with blue to purple flowers. *Strophostyles umbellata* is a southern savanna plant whose flowers may appear similar upon aging but has wider leaves with only three leaflets.

Gentiana alba (Gentiana flavida)

Cream Gentian, Plain Gentian

FAMILY: GENTIAN (GENTIANACEAE)

Similar to Bottle Gentian, with creamy-white instead of blue flowers. Grows in medium rather than moist soil. Foliage is attractive all season. White flowers are produced in profusion in late summer. Growing much larger than Bottle Gentian, clumps can reach 2 feet in diameter.

Habitat: Dry to moist prairies, savannas, open woods, woodland edges, particularly on calcareous soils

Garden Uses: Specimens, borders, informal gardens, prairie meadows

USDA Hardiness Zones: 3–7
Light Requirements: Full sun to partial shade
Soil Types: Sand, loam, clay
Soil Moisture: Dry to moist
Bloom Time: July–October
Flower Color: White, cream
Height: 1–2 feet
Life Expectancy: 10–20 years
Root Type: Fibrous
Propagation: Moist stratify seed 30 days; root division in spring or fall

Aggressiveness: Low by seed
Attracts: Bumblebees
Deer Palatability: Low
Distinguishing Characteristics:
- Creamy-white, vase-shaped flowers do not actually open
- Leaves are thick, smooth, and glossy yellow-green
- Very similar to Bottle Gentian (p. 224), except has white flowers instead of blue

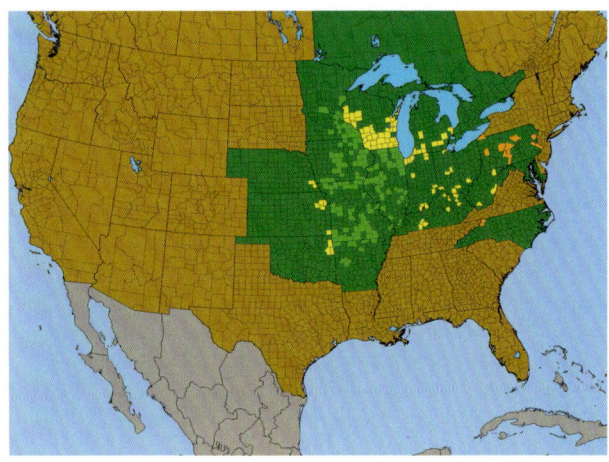

Seedling. Seedlings are indistinguishable from Bottle Gentian (*Gentiana andrewsii*).

Emerging mature plant in spring. Mature plants have numerous stems; *Gentiana andrewsii* typically has a single or only a few stems.

Entire plant

Leaf. Leaves are smooth and fleshy.

Flower. Flowers open only partially.

Early and mature seed. Seed is ripe when seedpods begin to crack open.

LOOK-ALIKE PLANTS:

Natives: Leaves and seedheads are virtually indistinguishable from Bottle Gentian (p. 224).

Gentiana andrewsii
Bottle Gentian, Closed Gentian

FAMILY: GENTIAN (GENTIANACEAE)

One of the last plants to bloom in fall, its indigo-blue flowers are virtually frost-proof. Flowers do not open and are pollinated exclusively by bumblebees, the only insect strong enough to gain entry to the inner sanctum. Plant in foreground, due to its short stature and late blooming time. Never invasive, mature plants are easily divided. Slow but fairly reliable from seed.

Habitat: Moist to wet prairies, sedge meadows, fens, streamsides, lakeshores, wet thickets
Garden Uses: Rain gardens, borders, informal gardens

USDA Hardiness Zones: 3–6
Light Requirements: Full sun to partial shade
Soil Types: Sand, loam, clay
Soil Moisture: Moist to wet
Bloom Time: August–October
Flower Color: Blue
Height: 1–2 feet
Life Expectancy: 5–10 years
Root Type: Fibrous
Propagation: Moist stratify seed 30 days; root division in spring or fall
Aggressiveness: Low by seed
Attracts: Bumblebees
Deer Palatability: Low
Distinguishing Characteristics:
- Deep-blue, vase-shaped flowers never open
- Leaves are thick, smooth, and glossy green

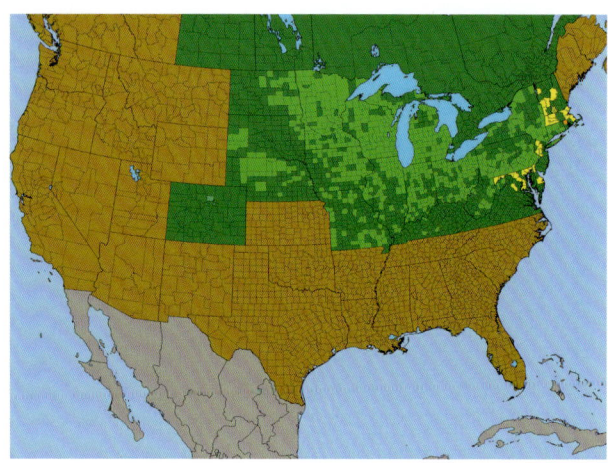

Seedling. Seedlings are indistinguishable from Cream Gentian (*Gentiana alba*).

Emerging mature plant in spring. Margins of emerging leaves are often tinged with red.

Entire plant

Leaf. Leaves are smooth and fleshy.

Flower

Mature seed. Seed is ripe when seedpods begin to crack open.

LOOK-ALIKE PLANTS:

Natives: *Gentiana clausa* is a similar eastern native that does not occur in the tallgrass prairie. Leaves and seedheads are almost identical to Cream Gentian (*Gentiana alba*), which has creamy-white flowers.

Agastache foeniculum
Lavender Hyssop, Anise Hyssop

FAMILY: MINT (LAMIACEAE, MENTHACEAE)

Essentially deer-proof thanks to the anise-flavored oils in its beautifully textured leaves. Its deep purple to lavender flowers are a favorite of honeybees. An excellent choice for savannas and semishaded areas, as well as full sun. Thrives in heavy soils, unlike most other members of this genus.

Habitat: Dry prairies, savannas, barrens
Garden Uses: Bee gardens, prairie meadows; containers in Zones 5–10. Not generally recommended for formal areas as it self-sows prolifically in sun or shade.

USDA Hardiness Zone: 2–6
Light Requirements: Full sun to partial shade
Soil Types: Sand, loam, amended clay
Soil Moisture: Dry to medium
Bloom Time: June–October
Flower Color: Purple
Height: 1–3 feet
Life Expectancy: 3–5 years
Root Type: Fibrous
Propagation: Moist stratify seed 30 days; root division in early spring; easy from stem cuttings early to midseason
Aggressiveness: Medium by seed
Attracts: Bees, butterflies, hummingbirds
Deer Palatability: Low
Distinguishing Characteristics:
- Leaves have a distinctive licorice fragrance when crushed
- Square stems (like most Mint family plants)

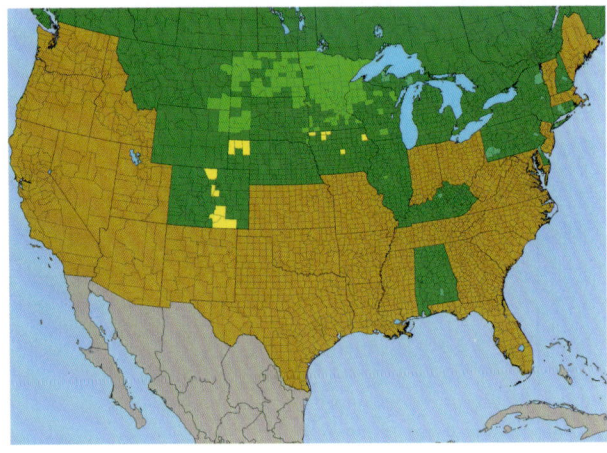

Seedling

Emerging mature plant in spring.
Note square stem and red-purple
undersides of leaves.

Entire plant

Leaf

Flower

Early and mature
seed

LOOK-ALIKE PLANTS:

Natives: Mint family members, including other Agastache species and
Monardas. Leaves also resemble many *Eutrochium*, which do not have
square stems.

Weeds: Mint family members, such as Wild Mint (*Mentha* spp.) and Catnip
(*Nepeta cataria*)

Blephilia ciliata

Downy Wood Mint, Downy Pagoda Plant, Ohio Horsemint

FAMILY: MINT (LAMIACEAE, MENTHACEAE)

Downy Wood Mint is prized for its pagodalike layers of circular flowers that are prominently clustered at the top of its stems. Adaptable to most well-drained soils, in full sun to moderate shade, it mixes nicely with lower-growing white, purple, blue and yellow flowers. Short-lived, it requires open, disturbed soil to regenerate by seeds. Easy from seed. Does not divide well.

Habitat: Dry to medium rich prairies, limestone glades and outcrops, thickets
Garden Uses: Butterfly gardens, borders, prairie gardens, masses, prairie meadows

USDA Hardiness Zones: 5–8
Light Requirements: Full sun to partial shade
Soil Types: Gravel, loam, clay, alkaline soils
Soil Moisture: Dry to moist
Bloom Time: May–September
Flower Color: Lavender to white
Height: 1–4 feet
Life Expectancy: 3–5 years
Root Type: Taproot
Propagation: Moist stratify seed 30 days; easy from stem cuttings early to midseason

Aggressiveness: Low by seed
Attracts: Bees, butterflies, flies
Deer Palatability: Low
Distinguishing Characteristics:
- Flower has pronounced purple spots on the lower lip
- Green to pink flower bracts curve upward
- Flowers arranged atop one another on stem
- Central stem is ridged with white hairs
- Has square stems (like most mints)

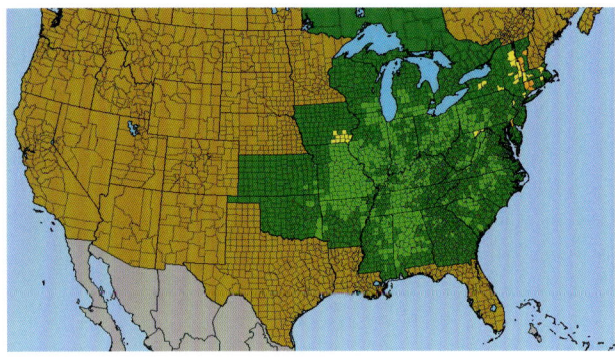

Seedling

Emerging mature plant in spring.
Note dense hairs on emerging leaves.

Entire plant

Leaf

Flower. Flowers stacked above
one another on the stem.

Early seed

Mature seed

LOOK-ALIKE PLANTS:

Natives: Dotted Mint (p. 232) has nearly identical leaves and stature but has showy pink bracts below the true flowers. Dotted Mint grows on sand barrens, whereas Downy Wood Mint grows on rich black soil prairies, often in calcareous soils. The seedheads are easily confused when ripe. Bergamot (p. 230) is two to three times as tall and has much longer flowers, with no spots on the lower lip.

Weeds: Many introduced members of the Mint family have similar flowers, including Spearmint (*Mentha spicata*).

Monarda fistulosa
Bergamot

FAMILY: MINT (LAMIACEAE, MENTHACEAE)

Incredibly adaptable, grows in almost any soil. Hummingbirds frequent the flowers. Spreads by rhizomes, best planted with other vigorous flowers and grasses to prevent it from spreading. Easy from seed and rhizome divisions. Dried seedheads have their own architectural beauty.

Habitat: Dry to moist prairies, savannas, barrens, glades, open woods, pastures, roadsides
Garden Uses: Butterfly gardens, hummingbird gardens, informal borders and gardens, prairie meadows

USDA Hardiness Zones: 3–9
Light Requirements: Full sun to partial shade
Soil Types: Sand, loam, clay
Soil Moisture: Dry to moist
Bloom Time: June–September
Flower Color: Lavender
Height: 2–5 feet
Life Expectancy: 10–20 years
Root Type: Rhizome
Propagation: Moist stratify seed 30 days; root division in spring or fall; easy from stem cuttings early to midseason

Aggressiveness: Medium by seed and rhizome
Attracts: Bees, butterflies, hummingbirds, hummingbird moths, songbirds
Deer Palatability: Low
Distinguishing Characteristics:
- Leaves emit a strong minty aroma when crushed, similar to Dotted Mint (p. 232) and Beebalm (*Monarda didyma*)
- Has square stems (like most mints)
- Emerging leaves are strongly purpled tinted

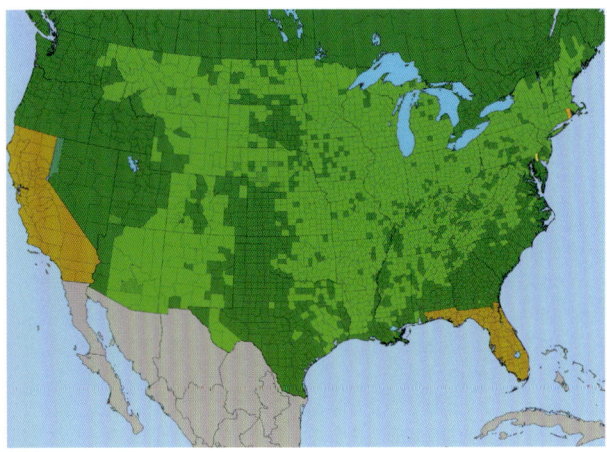

Seedling

Emerging mature plant in spring. Mature plants emerge as clumps, with numerous individual stems; back of leaves are often purple.

Entire plant. Plants are clonal and spread by rhizomes.

Leaf. Leaves are distinctly serrated on margins, exuding a strong, minty fragrance when crushed.

Flower

Early seed

Mature seed. Seeds are ripe when heads turn brown and individual seeds can be easily shaken out.

LOOK-ALIKE PLANTS:

Natives: Leaves and flower structure are similar to Beebalm (*Monarda didyma*), which has bright red rather than lavender flowers. Flowers do not have pink bracts below them, as does Dotted Mint (p. 232). Emerging plants appear similar to Lavender Hyssop (p. 226), which also has purple-tinted but rounded leaves.

Monarda punctata
Dotted Mint, Horsemint, Spotted Beebalm

FAMILY: MINT (LAMIACEAE, MENTHACEAE)

Bright pink and lavender bracts appear like flowers, creating an impressive display. Requires replanting if a consistent presence in the garden is desired. Great bee plant, it attracts a wide variety of other pollinators. A favorite nectar plant of the federally endangered Karner blue butterfly. Leaves are extremely aromatic, one of the strongest of the mints. This self-sowing biennial naturally appears in irregular years in dry prairies. Highly cyclical, it may be abundant one year then not reappear for many years. Easy from seed. Does not divide well.

Habitat: Dry sand prairies, sand barrens, open dry woods, sandy outcrops, sandy old fields
Garden Uses: Butterfly gardens, hummingbird gardens, dry sandy prairie meadows

USDA Hardiness Zones: 3–9
Light Requirements: Full sun
Soil Types: Sand, gravel
Soil Moisture: Dry
Bloom Time: June–September
Flower Color: Lavender
Height: 1–2 feet
Life Expectancy: 2 years (biennial)
Root Type: Fibrous
Propagation: Moist stratify seed 30 days; easy from stem cuttings early to midseason
Aggressiveness: Medium by seed

Attracts: Bees, butterflies, hummingbirds, songbirds
Deer Palatability: Low
Distinguishing Characteristics:
- Lower growing than other Monardas
- Flowers arranged atop one another on stem
- Flowers have pink and white bract "flowers" — colorful modified leaves at the base of the actual less showy flowers
- Has square stems (like most mints)

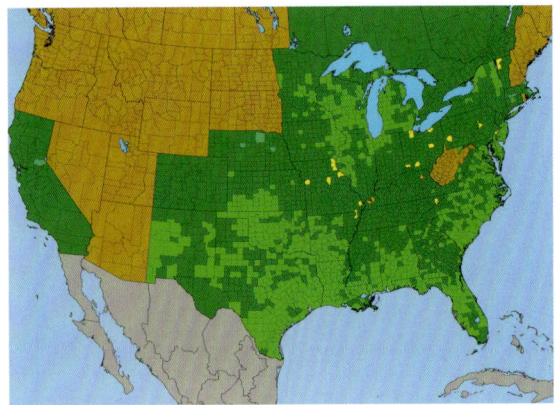

Seedling

Emerging mature plant in spring

Entire plant

Leaf. Leaves are strongly aromatic; leaf margins are not serrated as with Bergamot.

Flower. Colorful pink and white "flowers" are actually leaf bracts.

Early seed

Mature seed. Seeds are ripe when heads turn brown and individual seeds easily shaken out.

LOOK-ALIKE PLANTS:

Natives: Bergamot (p. 230) is two to three times as tall, with only lavender flowers, no pink "bract flowers." Downy Wood Mint (*Blephilia ciliata*) (p. 228) has nearly identical leaves and stature but no showy pink bracts below the true flowers. Dotted Mint grows on sand barrens, whereas Downy Wood Mint grows on clay, loam, or gravel soil, often in calcareous soils. It is easy to confuse the seedheads when ripe.

Physostegia virginiana
Obedient Plant, False Dragonhead

FAMILY: MINT (LAMIACEAE, MENTHACEAE)

Obedient? Not really. Its flowers stay in place on the stem when positioned by hand, hence the common name. Creeping by rhizomes, it makes a nearly impervious groundcover in damp soils. Gorgeous when planted with Great Blue Lobelia. Dark-green toothed foliage looks great all year. Tricky from seed, but unstoppable when installed as plants or root divisions.

Habitat: Moist prairies and meadows, marshes, sedge meadows, moist swales, riverbanks, streamsides, moist thickets
Garden Uses: Groundcover, hummingbird gardens, bee gardens, moist meadows

USDA Hardiness Zones: 3–9
Light Requirements: Full sun to partial shade
Soil Types: Sand, loam, clay
Soil Moisture: Medium to moist
Bloom Time: July–September
Flower Color: Pink
Height: 1–2 feet
Life Expectancy: 10–20 years
Root Type: Rhizome
Propagation: Moist stratify seed 30 days; root division in spring or fall; easy from stem cuttings early to midseason
Aggressiveness: High by rhizome

Attracts: Bees, butterflies, hummingbirds
Deer Palatability: Low
Distinguishing Characteristics:
- Flowers remain in position when moved by hand, hence the name Obedient Plant
- Square stems (like most Mint family)
- Leaves are not aromatic, unlike other Mint family members
- Most genotypes creep aggressively by rhizomes to form patches, although some do not

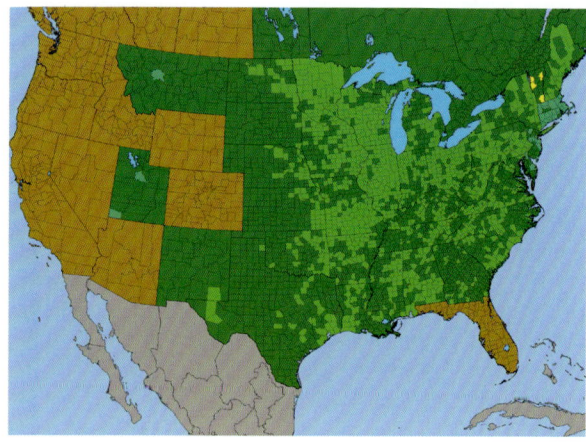

Seedling. Seedling leaves are distinctly toothed.

Emerging mature plant in spring

Entire plant. A member of the Mint family, the stems are square, but the leaves have no fragrance when crushed.

Leaf. Leaves have sawblade-like teeth.

Flower. Individual flowers are among the largest of the native mints.

Early and mature seed. Seed is ripe when they turn dark brown inside the seedpods.

LOOK-ALIKE PLANTS:

Natives: Leaves of White Turtlehead (p. 278) and Sneezeweed (p. 110) are similar. Flowers resemble those of some penstemons but bloom much later.

Pycnanthemum tenuifolium
Narrowleaf Mountainmint

FAMILY: MINT (LAMIACEAE, MENTHACEAE)

White flowers slightly larger but very similar to Virginia Mountainmint. Blooms a little later. Bushy form. Less aromatic foliage. Plant around Rattlesnake Master for a complementary display; and with other flowers and grasses to keep it in check. Easy from seed, grows in any good garden soil.

Habitat: Medium to dry prairies, rich loamy soils, gravelly areas, openings in woodlands
Garden Uses: Bee gardens, informal gardens, medium to dry prairie meadows

USDA Hardiness Zones: 3–7
Light Requirements: Full sun
Soil Types: Sand, gravel, loam, clay
Soil Moisture: Medium to dry
Bloom Time: July–September
Flower Color: White
Height: 1–4 feet
Life Expectancy: 5–10 years
Root Type: Taproot, rhizomes
Propagation: Moist stratify seed 30 days; root division in spring or fall

Aggressiveness: Medium by rhizome
Attracts: Bees, butterflies
Deer Palatability: Low
Distinguishing Characteristics:
- Leaves much narrower than Common Mountainmint
- Leaf aroma not so intense as Common Mountainmint
- White flowers attract many pollinators
- Square stems have no hairs

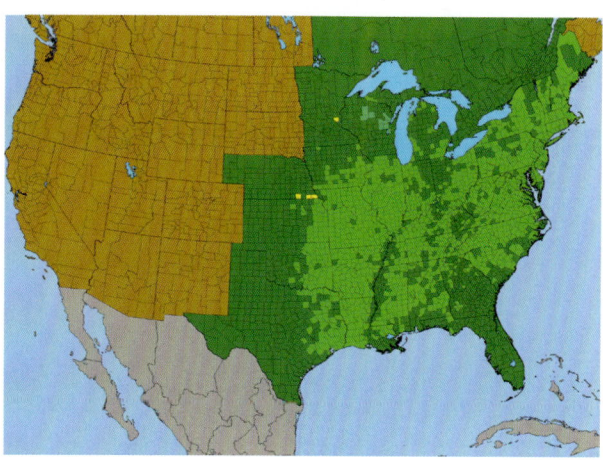

Seedling

Emerging mature plant in spring. Emerging stems can be red or purple.

Entire plant. Stems are hairless, unlike Virginia Mountainmint.

Leaf. Leaves have strong, minty fragrance when crushed.

Leaf close-up

Flower. Flowers usually white, without purple spots, as in *Pycnanthemum virginianum*.

Early and mature seed

Mature seed. Seed is ripe when seedheads turn brown to black.

LOOK-ALIKE PLANTS:

Natives: Common Mountainmint (*Pycnanthemum virginianum*) has less conspicuous white flowers and broader leaves. *P. verticillatum* has stems covered in white hairs, and wider, hairy leaves that are 0.75 inches across.

Pycnanthemum virginianum
Virginia Mountainmint, Common Mountainmint

FAMILY: MINT (LAMIACEAE, MENTHACEAE)

Tiny white flowers appear like little smiling faces on a midsummer day. Attractive form. Extremely aromatic foliage is greenish gray. Plant around Rattlesnake Master for a complementary display, and with other flowers and grasses to keep it in check. Easy from seed, and grows in any good garden soil.

Habitat: Moist to wet prairies, wet meadows, lowland pastures
Garden Uses: Bee gardens, informal gardens, moist prairie meadows

USDA Hardiness Zones: 3–7
Light Requirements: Full sun to partial shade
Soil Types: Sand, loam, clay
Soil Moisture: Moist to wet
Bloom Time: July–September
Flower Color: White
Height: 1–3 feet
Life Expectancy: 5–10 years
Root Type: Rhizome
Propagation: Moist stratify seed 30 days; root division in spring or fall
Aggressiveness: Medium by rhizome

Attracts: Bees, butterflies
Deer Palatability: Low
Distinguishing Characteristics:
- Leaves emit an intensely pungent, spicy mint fragrance when crushed
- White flowers are smaller than most other native mints, and occur in dense clusters at the top of the plant
- Square stems have lines of white hairs
- Leaves are 0.25–0.5 inches wide (see below)

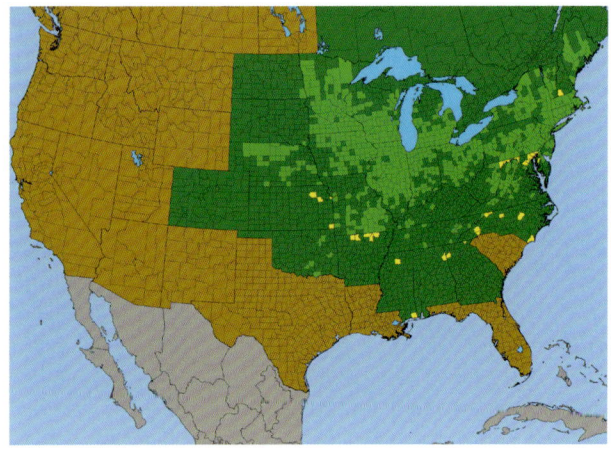

Seedling

Emerging mature plant in spring. Emerging stems can be red or purple.

Entire plant. Plants usually taller than *Pycnanthemum tenuifolium*, at 2–3 ft.

Stem. Stems have white hairs along the ridges.

Leaf. Leaves have strong, minty fragrance when crushed.

Flower. White flowers are usually covered with small purple spots.

Early seed

Mature seed. Seed is ripe when seedheads turn brown.

LOOK-ALIKE PLANTS:

Natives: Slender Mountainmint (*Pycnanthemum tenuifolium*) has more conspicuous white flowers and narrower leaves seldom more than 0.25 inches wide. Whorled Mountainmint (*P. verticillatum*) has stems covered in white hairs, and wider, hairy leaves 0.75 inches across.

Salvia azurea
Blue Sage

FAMILY: MINT (LAMIACEAE, MENTHACEAE)

Notable for its attractive arrowlike foliage and robin's-egg-blue flowers. Provides a strong vertical element to the garden. Does well in any well-drained soil, even dry sand and gravel. Not easily divided.

Habitat: Dry to medium prairies and meadows, savannas, woodland edges, open woods, pastures
Garden Uses: Butterfly gardens, hummingbird gardens, informal gardens, prairie meadows

USDA Hardiness Zones: 4–9
Light Requirements: Full sun to partial shade
Soil Types: Sand, gravel, loam
Soil Moisture: Dry to medium
Bloom Time: June–August
Flower Color: Blue
Height: 3–5 feet
Life Expectancy: 5–10 years
Root Type: Fibrous

Propagation: Moist stratify seed 30 days
Aggressiveness: Low by seed
Attracts: Bees, butterflies, hummingbirds
Deer Palatability: Low
Distinguishing Characteristics:
- Leaves have the aroma and taste of sage
- Sky-blue, snapdragon-like flowers

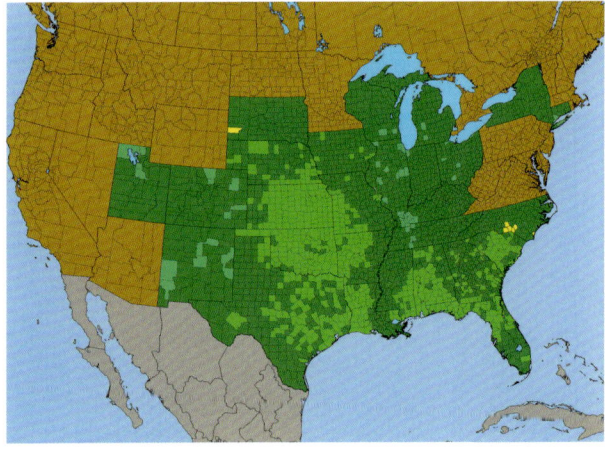

Seedling. Seedlings have square stems; very similar to *Verbena hastata* and *V. stricta* seedlings.

Emerging mature plant in spring

Entire plant. Plants are tall (up to 6 ft.) and leggy.

Leaf. Leaves are narrow, and serrated tips are not pointed on ends.

Flower. Robin's-egg blue, snapdragon-like flowers are distinctive.

Mature seed. Seed is ripe when pods turn silvery white and begin to open.

LOOK-ALIKE PLANTS:

Natives: Leaves resemble Obedient Plant (*Physostegia virginiana*, p. 234) and White Turtlehead (*Chelone glabra*, p. 278), but Blue Sage has small stipules, or axil leaves, where leaf meets stem.

Callirhoe involucrata
Purple Poppymallow

FAMILY: MALLOW (MALVACEAE)

Forming a 4–5 foot circle, this low-growing groundcover sports brilliant magenta flowers from spring through fall. Attractive geranium-like foliage covers the soil and helps reduce weed growth. Flowering stems twine up taller plants, dying back in fall. When planted in containers, spills attractively over the sides. Once an important food of Native Americans. Susceptible to slug damage.

Habitat: Dry to medium prairies, rocky and sandy hillsides, thickets, open woods
Garden Uses: As groundcover, border edging, in short prairie meadows

USDA Hardiness Zones: 4–9
Light Requirements: Full sun to part shade
Soil Types: Sand, loam, clay
Soil Moisture: Dry to medium
Bloom Time: May–August
Flower Color: Magenta
Height: 1 foot
Life Expectancy: 10–20 years
Root Type: Corm-like taproot
Propagation: Scarify seed and moist stratify 60 days; easy from basal stem cuttings in midspring
Aggressiveness: Medium by seed

Attracts: Bees, butterflies, songbirds
Deer Palatability: Medium
Distinguishing Characteristics:
- Palmate leaves are deeply lobed and cleft, unlike the arrow-shaped, undivided leaves of Poppymallow (p. 244)
- Stems may twine up other plants, whereas Poppymallow lies prostrate on the ground and seldom, if ever, twines
- Root is widened near the soil surface similar to a sweet potato, tapering to a deep taproot

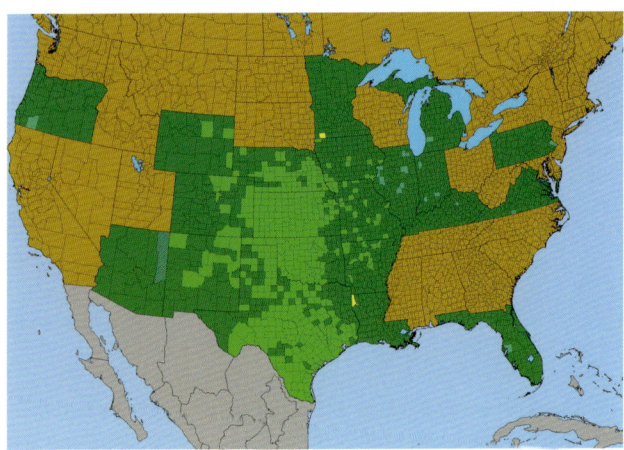

Seedling

Emerging mature plant in spring

Entire plant

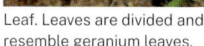

Leaf. Leaves are divided and resemble geranium leaves.

Flower

Early seed

Mature seed

LOOK-ALIKE PLANTS:

Natives: Flowers similar to Poppymallow (p. 244). Leaves resemble Thimbleweed (p. 256), Canada Anemone (p. 254), and Wild Geranium (*Geranium maculatum*). *Callirhoe alcaeoides* (Pale Poppymallow) occurs within the same range as *C. involucrata* but has narrowly divided palmate leaves, and flowers are white, lavender, and pale pink.

Callirhoe triangulata
Poppymallow, Clustered Poppymallow

FAMILY: MALLOW (MALVACEAE)

Spreads prostrate over the ground, seldom exceeding 1 foot in height. Deep-magenta flowers resemble petunias, with distinctive Mallow family "Hollyhock" pistils. Very drought resistant, growing only in extremely well-drained, dry sandy soil. Mix with short prairie grasses and other flowers.

Habitat: Dry sand prairies and sand barrens; rare, occurring in isolated locations
Garden Uses: Groundcover, rock gardens, dry sandy locations

USDA Hardiness Zones: 4–8
Light Requirements: Full sun
Soil Types: Sand
Soil Moisture: Dry
Bloom Time: July–October
Flower Color: Magenta
Height: 1 foot
Life Expectancy: 10–20 years
Root Type: Taproot
Propagation: Scarify seed and moist stratify 60 days
Aggressiveness: Low by seed

Attracts: Bees, butterflies, songbirds
Deer Palatability: Low
Distinguishing Characteristics:
- Plant creeps along the ground, seldom if ever twining up other plants, as does *C. involucrata* (p. 242)
- Greenish-yellow arrowlike leaves are covered with small hairs

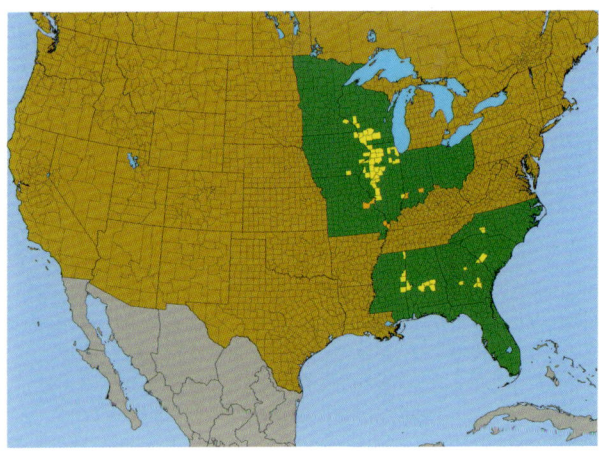

Seedling

Emerging mature plant in spring

Entire plant

Leaf. Leaves are triangular and not divided, as in *Callirhoe involucrata*.

Flower

Early and mature seed. Seeds are ripe when pods turn yellow and seeds inside are brown.

LOOK-ALIKE PLANTS:

Natives: Purple Poppymallow (p. 242) has nearly identical flowers but deeply cleft palmate leaves instead of arrow-shaped leaves.

Weeds: Leaves are similar to the nonnative garden plant *Anemone japonica*.

Hibiscus moscheutos (*Hibiscus palustris*)
Rosemallow, Swamp Rosemallow, Marsh Mallow

FAMILY: MALLOW (MALVACEAE)

Not a true prairie plant, this denizen of moist eastern meadows produces huge red, pink, or white hollyhock-like flowers up to six inches in diameter. Grows naturally in moist soils but does beautifully in rich, well-drained garden soil. Pollinated specifically by the Rosemallow Bee.

Habitat: Wet meadows, marshes, swamps, moist thickets
Garden Uses: Specimen, rain gardens, creeksides, lakesides, and low wet areas

USDA Hardiness Zones: 5–9
Light Requirements: Full sun to partial shade
Soil Types: Sand, loam, clay
Soil Moisture: Medium, moist, wet
Bloom Time: July–September
Flower Color: White, pink, red
Height: 3–7 feet
Life Expectancy: 10–20 years
Root Type: Rhizome
Propagation: Moist stratify seed 30 days; easy from stem cuttings early to midseason

Aggressiveness: Low by seed and rhizome
Attracts: Bees, butterflies, hummingbirds
Deer Palatability: Low
Distinguishing Characteristics:
- Large red, pink, or white flowers each bloom for only one day
- Leaves are heart-shaped without lobes and nearly hairless
- Seedpods are hairless; pods of similar native *Hibiscus* are conspicuously hairy

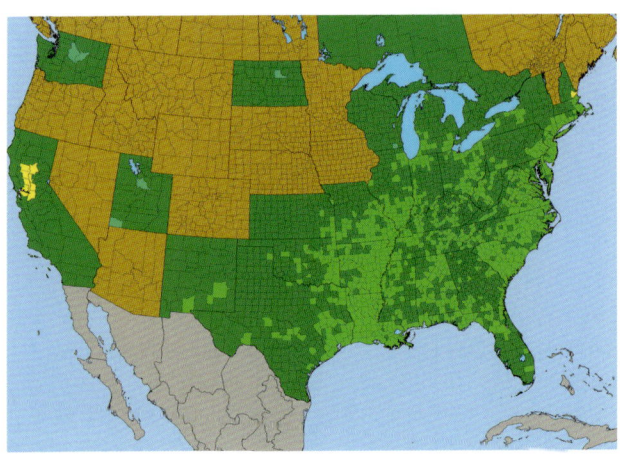

Seedling

Entire plant

Leaf. Leaves are large, 4–6 in. wide and 6–10 in. long.

Emerging mature plant in spring. Leaves emerge atop long stems, 1–2 ft. tall.

Flower. Flowers are large, 4–6 in. across; white, red, and pink are common.

Pink flower

Mature seed. Interior of seedpod is densely hairy; seeds resemble rounded pellets.

LOOK-ALIKE PLANTS:

Natives: Halberdleaf Rosemallow (*Hibiscus laevis*) and Woolly Rose-mallow (*H. lasiocarpa*) are similar, but both have hairy outer seedpods. The leaves of Rosemallow are covered with hairs on both sides. Leaves of Halberdleaf Rosemallow are hairless but have narrow, sharp-pointed lobes.

Oenothera gaura (Gaura biennis)

Biennial Beeblossom, Biennial Gaura

FAMILY: EVENING PRIMROSE (ONAGRACEAE)

Showy white flowers, which resemble butterflies, turn pink as they age. Often blooms for an extended period. This biennial forms a basal rosette in the first year, then blooms and dies in the second. In rich soil it can become large and sprawling. Pruning in midsummer will keep it more compact and further extend bloom period.

Habitat: Dry to moist prairies, open woods, woodland edges
Garden Uses: Butterfly gardens, borders and informal gardens, prairie meadows

USDA Hardiness Zones: 4–8
Light Requirements: Full sun to partial shade
Soil Types: Sand, loam, clay
Soil Moisture: Dry to moist
Bloom Time: June–September
Flower Color: White, turning pink with age
Height: 3–6 feet
Life Expectancy: 2 years (biennial)
Root Type: Fibrous
Propagation: Moist stratify seed 30 days; easy from stem cuttings early to midseason

Aggressiveness: Medium by seed
Attracts: Bees, butterflies
Deer Palatability: Medium
Distinguishing Characteristics:
- Drooping flowers have protruding stamens and stigmas that extend well beyond the petals
- Flowers resemble butterflies in flight
- Flowers occur along entire length of plant
- Forms a large basal rosette in first growing season

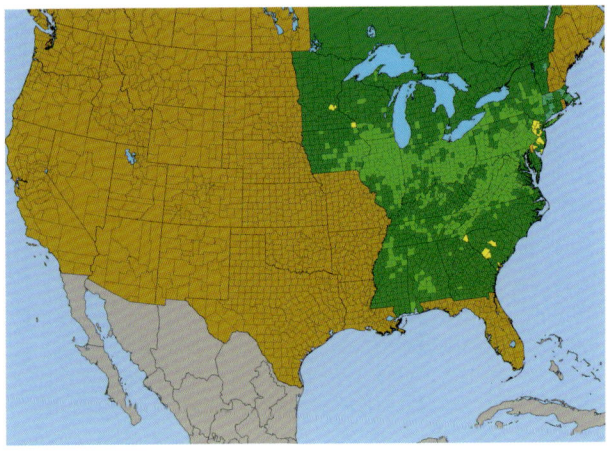

Seedling. Seedling leaves are big and fleshy with distinct midrib; seedlings grow rapidly.

Emerging mature plant in spring. Basal rosettes strongly resemble those of Evening Primrose (*Oenothera biennis*).

Entire plant

Gaura moth (*Schinia gaurae*)

Leaf. Leaves are commonly attacked by leaf-mining larvae of a Momphid moth, maybe resulting in red spots on foliage.

Flower

Early seed

Mature seed close-up

Mature seed. Seeds are ripe when they turn brown.

LOOK-ALIKE PLANTS:

Natives: Evening Primroses (*Oenothera* spp.) form similar basal rosettes with pronounced white midribs in the leaves and reddish-tinged stems, but they have yellow flowers.

Phlox pilosa
Downy Phlox, Prairie Phlox

FAMILY: PHLOX (POLEMONIACEAE)

This delightful plant packs a lot of punch for its small size. Bluish-pink flowers are a real spring showstopper. Very adaptable, grows in dry sand to moist loam. Seeds explode when ripe and are hard to collect. Plant seed fresh for germination the following spring. Most easily propagated by root divisions or stem cuttings.

Habitat: Dry to moist prairies, savannas, barrens, woodland edges
Garden Uses: Massing (mass plantings will require regular weeding), groundcover, border edges, informal gardens, prairie meadows

USDA Hardiness Zones: 3–9
Light Requirements: Full sun to partial shade
Soil Types: Sand, loam
Soil Moisture: Dry to moist
Bloom Time: May–June
Flower Color: Pink
Height: 1–2 feet
Life Expectancy: 3–5 years
Root Type: Taproot
Propagation: Moist stratify seed 60 days; root division in early spring; easy from stem cuttings early to midseason

Aggressiveness: Low by seed
Attracts: Bees, butterflies, hummingbirds
Deer Palatability: High
Distinguishing Characteristics:
- Grows in groups of small individual stems
- Often sprawls on the ground but does not creep by rhizomes
- Leaves are sparse along the stem
- Seedpods explode when ripe

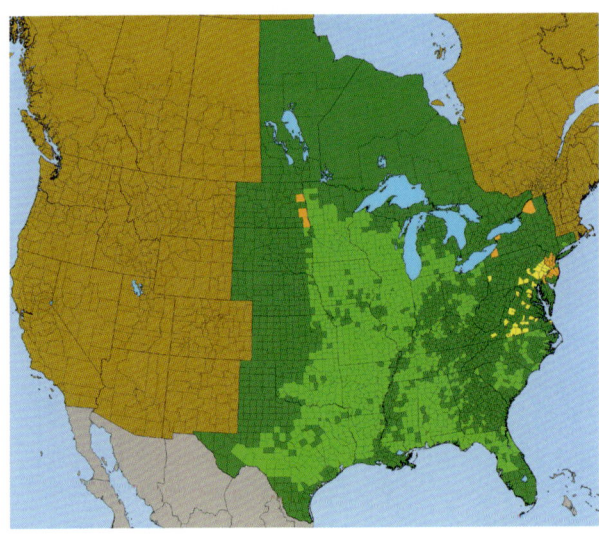

Seedling

Emerging mature plant in spring.
Emerging stems resemble
Pycnanthemum spp. but are
round rather than square.

Entire plant

Leaf. Leaves strongly resemble
those of *Pycnanthemum* spp.
but have no minty fragrance
when crushed.

Flower

Early seed

Mature seed. Seeds explode when
ripe and must be collected when the
spherical pods inside the calyx turn
light brown.

LOOK-ALIKE PLANTS:

Natives: The woodland plant Creeping Phlox (*Phlox stolonifera*) has similar stature and flowers but forms large patches by rhizomes.

Dodecatheon meadia
Shooting Star, Pride of Ohio

FAMILY: PRIMROSE (PRIMULACEAE)

One of the best native plants for gardens. Long-lived, disease-free, and easy to divide when dormant in late summer or fall. Never disturb when actively growing! Thrives in full sun to medium shade in good soil. Sown in late summer or fall, seed germinates readily the following spring. Mix with later-blooming species to cover dormant plants. Extremely slow growing, requires five to ten years to bloom from seed.

Habitat: Dry lime prairies, medium to moist prairies, fens, savannas, open woods, bluffs
Garden Uses: Rock gardens, rain gardens, borders, prairie meadows

USDA Hardiness Zones: 4–8
Light Requirements: Full sun to medium shade
Soil Types: Sand, loam, clay
Soil Moisture: Dry to moist
Bloom Time: May–June
Flower Color: White to pink
Height: 1–2 feet
Life Expectancy: 20+ years
Root Type: Fibrous
Propagation: Moist stratify seed 30 days and plant in early spring; root division in late summer or fall when dormant
Aggressiveness: Low by seed

Attracts: Bees
Deer Palatability: Medium
Distinguishing Characteristics:
- Decumbent flowers resemble a shooting star
- Early growth looks similar to lettuce leaves
- Leaves are soft, smooth, and hairless
- Seedpods appear like small brown urns
- Plants go dormant in mid-summer
- Roots have a distinctive fragrance

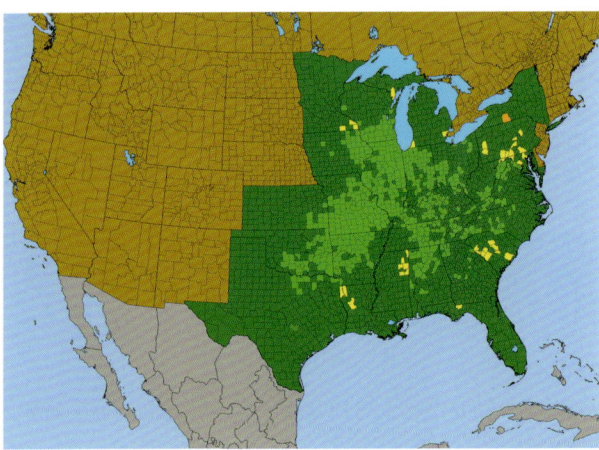

Seedling. Seedlings grow slowly and enter dormancy by midsummer.

Emerging mature plant in spring. Leaves are fleshy and smooth, resembling Bibb lettuce.

Entire plant

Leaf

Flower. Flower color ranges from pure white to deep pink.

Mature seed. Seeds are ripe when pods turn brown and top of pod begins to open.

LOOK-ALIKE PLANTS:

Natives: Amethyst Shooting Star (*Dodecatheon amethystinum*) is similar in form but slightly smaller with bright pink flowers.

Anemone canadensis
Canada Anemone

FAMILY: BUTTERCUP (RANUNCULACEAE)

Pure white flowers cover the lovely deep-green foliage in late spring. Beware, however, as this seemingly demure denizen of wet prairies and lightly shaded woodlands can creep aggressively by rhizomes. Makes an excellent groundcover in damp soil situations.

Habitat: Moist prairies, sedge meadows, wet open woods, wet swales
Garden Uses: Moist meadow seedings. Not recommended for gardens, as it self-seeds aggressively in sun or shade.

USDA Hardiness Zones: 3–6
Light Requirements: Full sun to partial shade
Soil Types: Sand, loam, clay
Soil Moisture: Medium to moist
Bloom Time: April–July
Flower Color: White
Height: 1–2 feet
Life Expectancy: 10–20 years
Root Type: Rhizome
Propagation: Moist stratify seed 90 days and plant in early spring; easy from root divisions in spring or fall

Aggressiveness: High by rhizome
Attracts: Bees
Deer Palatability: Low
Distinguishing Characteristics:
- Forms large patches by rhizomes
- No central stem: leaves arise from rhizomes
- Flat, hairless seeds have pointed ends
- Leaves turn black in fall

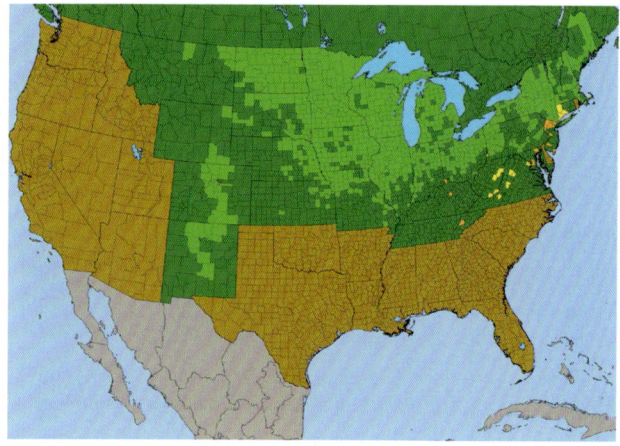

Seedling

Emerging mature plant in spring

Entire plant. Plants typically form large colonies.

Leaf

Flower

Early seed

Mature seed. Seeds are ripe when they turn from green to yellow.

LOOK-ALIKE PLANTS:

Natives: Leaves resemble other *Anemone* spp., *Geranium*, *Delphinium*, and other Buttercup family members. Some of the nonnative cultivated Anemones are similar in appearance.

Weeds: Some gardeners, based on their bitter experience, consider this plant a weed!

Anemone cylindrica
Thimbleweed, Candle Anemone

FAMILY: BUTTERCUP (RANUNCULACEAE)

Demure white flowers with green veins sit atop stiff stems. The seed-heads give the plant its name, appearing like thimbles or small candles. Fluffy white seeds resemble sheep's wool and actually have an oily feel similar to lanolin. Interplant with short to medium-height grasses and low-growing flowers.

Habitat: Dry to medium prairies, sand prairies, barrens, glades
Garden Uses: Small groups, prairie meadows

USDA Hardiness Zones: 3–6
Light Requirements: Full sun
Soil Types: Sand, loam
Soil Moisture: Dry to medium
Bloom Time: May–June
Flower Color: White
Height: 2–5 feet
Life Expectancy: 10–20 years
Root Type: Fibrous
Propagation: Moist stratify seed 60 days
Aggressiveness: Low by seed

Attracts: Bees
Deer Palatability: Low
Distinguishing Characteristics:
- Conical seedheads resemble sheep's wool when ripe
- Small white flowers are borne on long, nearly leafless, stems
- Seedheads are usually one to three times longer than wide; narrower and longer than other native anemones

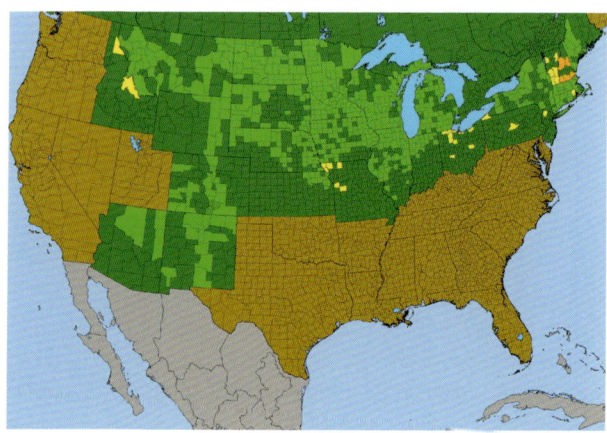

Seedling

Emerging mature plant in spring

Entire plant

Leaf

Flower. Note pink pistils in the center of the flower.

Early seed

Mature seed. Ripe seeds are fluffed out and readily stripped from the seedhead.

LOOK-ALIKE PLANTS:

Natives: Pasque Flower (p. 260) has very similar leaves. Tall Thimbleweed (*Anemone virginiana*) grows in open, dry woods and has similar but shorter seedhead (0.5–1.25 inches).

Delphinium carolinianum subsp. *virescens* (*Delphinium virescens*)

Prairie Larkspur, Carolina Larkspur

FAMILY: BUTTERCUP (RANUNCULACEAE)

Lush, deep-green foliage appears in early spring, topped with pure white flowers in June. Plants go dormant and disappear in midsummer with new foliage reappearing in early autumn.

Habitat: Dry to medium prairies, barrens, glades, bluffs
Garden Uses: Plant with sturdy tall plants to help support the delicate stems. Other plants will cover the soil when larkspur is dormant.

USDA Hardiness Zones: 3–8
Light Requirements: Full sun
Soil Types: Sand, loam
Soil Moisture: Dry to medium
Bloom Time: June–July
Flower Color: White
Height: 1–4 feet
Life Expectancy: 5–10 years
Root Type: Fibrous
Propagation: Moist stratify seed 30 days, or plant fresh for fall germination

Aggressiveness: Low by seed
Attracts: Bees
Deer Palatability: Low
Distinguishing Characteristics:
- Pure white, larkspur-form flowers
- Leaves are light bluish green
- Plant goes dormant in midsummer; new leaves appear in fall

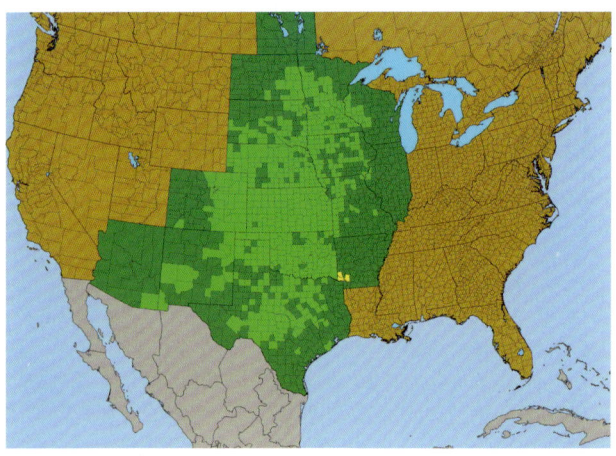

Seedling

Emerging mature plant in spring

Entire plant

Leaf

Flower close-up

Flower

Mature seed. All seeds on a stalk are ripe when upper pods turn brown and begin to open.

LOOK-ALIKE PLANTS:

Natives: Leaves resemble some other members of Buttercup family; and Wild Geranium. Leaves and seedpods are similar to Tall Larkspur (*Delphinium exaltatum*), an eastern woodland plant with deep-blue flowers that does not exhibit summer dormancy.

Pulsatilla patens (*Anemone patens*)
Pasque Flower, Wild Crocus, Windflower

FAMILY: BUTTERCUP (RANUNCULACEAE)

One of the earliest-blooming prairie flowers, often covered in early-spring snow. Delicate blooms herald the advent of a new growing season, and their hoary seedheads are gone before most other species begin to flower. Grows only on extremely well-drained sandy and rocky soils in full sun. Mix with low-growing flowers and grasses, such as Prairie Smoke, Blue-Eyed Grass, and Junegrass.

Habitat: Dry prairies, barrens, glades, rocky open hillsides
Garden Uses: Rock gardens, dry prairie gardens

USDA Hardiness Zones: 3–6
Light Requirements: Full sun
Soil Types: Sand, gravel
Soil Moisture: Dry
Bloom Time: April–May
Flower Color: White
Height: 0.5–1 foot
Life Expectancy: 10–20 years
Root Type: Fibrous
Propagation: Moist stratify seeds 30 days, or plant fresh for fall germination
Aggressiveness: Low by seed

Attracts: Bees
Deer Palatability: High
Distinguishing Characteristics:
- Large, showy flowers appear much earlier than other *Anemone*
- Flowers appear before leaves emerge
- Distinctive seedheads are similar to *Clematis*
- Narrow leaves are seldom more than 0.4 inches wide
- Grows only 1 foot tall

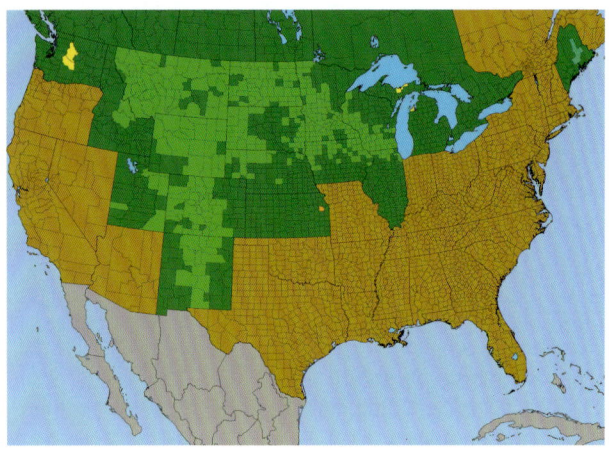

Seedling

Emerging mature plant in spring. Flowers emerge and bloom before leaves appear.

Entire plant

Leaf. Lobes of leaves are thinner than other native anemones.

Flower

Early seed

Mature seed. Seeds are ripe when brown and are easily pulled from the seedhead.

LOOK-ALIKE PLANTS:

Natives: Leaves resemble those of other native *Anemone*, except the lobes of Pasque Flower (*A. patens*) are much narrower.

Thalictrum dasycarpum
Meadow-Rue, Tall Meadow-Rue, Purple Meadow-Rue

FAMILY: BUTTERCUP (RANUNCULACEAE)

Fine foliage and airy white flowers make this wetland plant an excellent backdrop in border gardens. Plants are dioecious, with male and female flowers produced on separate plants. Grows in any damp soil, including heavy clay. Plant seed in fall for spring germination. Can be divided.

Habitat: Moist to wet prairies, sedge meadows, fens, streamsides, lakeshores, swales, ditches, open wet woods, woodland edges, thickets
Garden Uses: Specimens, small groups, butterfly gardens, rain gardens, borders and informal gardens, wet meadows

USDA Hardiness Zones: 3–9
Light Requirements: Full sun to partial shade
Soil Types: Sand, loam, clay
Soil Moisture: Moist to wet
Bloom Time: April–July
Flower Color: White
Height: 3–6 feet
Life Expectancy: 5–10 years
Root Type: Fibrous
Propagation: Moist stratify seed 30 days
Aggressiveness: Low by seed

Attracts: Bees, butterflies
Deer Palatability: Low
Distinguishing Characteristics:

- Male and female flowers are produced on separate plants
- Delicate, divided blue-green foliage very similar to Columbine, and both are in the Buttercup family
- Taller than Early Meadow-Rue, which grows in well-drained soils, usually in deeper shade

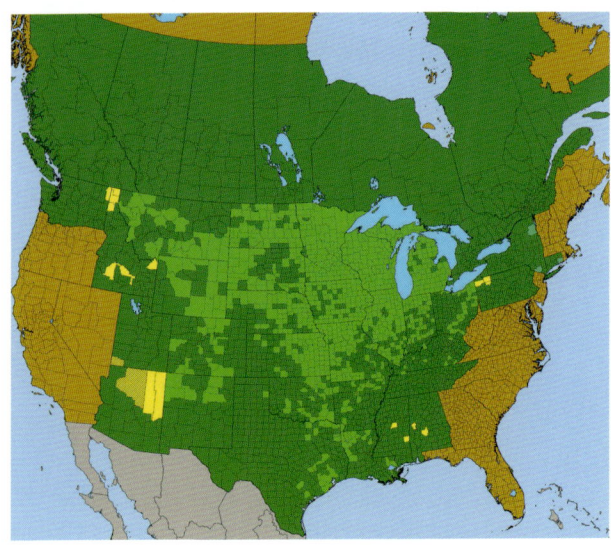

Seedling

Emerging mature plant in spring. Early shoots are often a deep purple.

Entire plant

Leaf

Female flower close-up

Male flower close-up

Male flowers

Early (yellow) and mature (brown to black) seed

LOOK-ALIKE PLANTS:

Natives: Seedlings and early-spring leaves are almost identical to Columbine (*Aquilegia canadensis*), Early Meadow-Rue (*Thalictrum dioicum*), and Rue Anemone (*T. thalictroides*).

Ceanothus americanus
New Jersey Tea

FAMILY: BUCKTHORN (RHAMNACEAE)

This exquisite small shrub sports glossy green leaves and is awash in white flowers in midsummer. Colonists brewed a tea from the leaves during the American Revolution. Hummingbirds eat small insects that come for nectar. Makes a fine low hedge. Difficult to grow from seed; best established using transplants.

Habitat: Dry to medium prairies, savannas, barrens, glades, bluffs, dry open woods, woodland edges
Garden Uses: Specimens, low hedges, hummingbird gardens; plant 2 feet apart for a hedge, or 2 feet apart in staggered rows for groundcover

USDA Hardiness Zones: 3–9
Light Requirements: Full sun to partial shade
Soil Types: Sand, loam
Soil Moisture: Dry to medium
Bloom Time: July–October
Flower Color: White
Height: 2–3 feet
Life Expectancy: 20+ years
Root Type: Taproot
Propagation: Pour boiling water over seeds in a bowl, let cool, moist stratify 30 days; easy from cuttings: use new wood with rooting hormone

Aggressiveness: Low by seed
Attracts: Bees, butterflies, game birds, hummingbirds
Deer Palatability: Medium
Distinguishing Characteristics:
- One of few shrubs on the prairie, with woody, greenish stems
- Smooth, shiny leaves are deeply furrowed
- Rounded seedpods turn black when mature
- Seeds are hard, round, and reddish brown

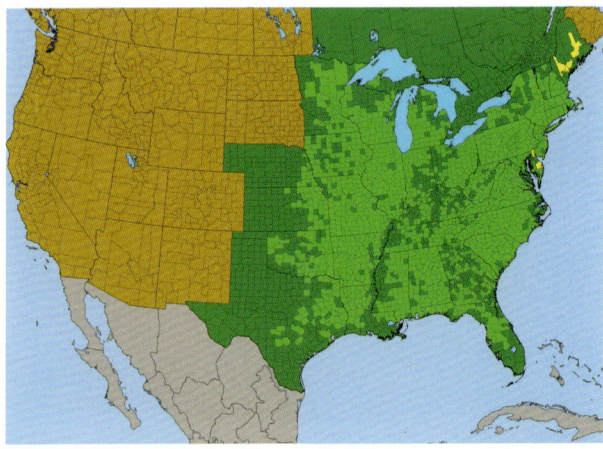

Seedling

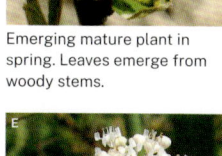

Emerging mature plant in spring. Leaves emerge from woody stems.

Entire plant

Leaf

Flower

Early seed

Mature seed. Seedpods turn black at maturity and then explode to disperse seeds.

LOOK-ALIKE PLANTS:

Natives: Redroot (*Ceanothus ovatus*) grows on very dry, sandy soil in full sun to light shade. It blooms one month earlier and has a more open, looser growth form.

ROSACEAE
Filipendula rubra
Queen of the Prairie

FAMILY: ROSE (ROSACEAE)

Stunning pink plumes are the hallmark of this regal plant. Creeps by rhizomes to form a patch. Give it room to spread, as it will often annex territory on its own. A member of the Rose family, it is subject to Japanese beetles, Rose Chafers, and Powdery Mildew. Somewhat difficult and slow from seed.

Habitat: Wet prairies and meadows, roadside ditches, wet woods and woodland edges: rare
Garden Uses: Specimen clumps, backs of borders, centers of island beds, rain gardens

USDA Hardiness Zones: 3–6
Light Requirements: Full sun to part shade
Soil Types: Rich sand, loam, clay with plenty of organic matter
Soil Moisture: Medium to moist
Bloom Time: June–July
Flower Color: Pink
Height: 4–6 feet
Life Expectancy: 20+ years
Root Type: Rhizome
Propagation: Moist stratify seed 30 days; root division, using 6-inch or longer rhizomes with terminal buds in spring or fall
Aggressiveness: Medium by rhizomes
Attracts: Bees
Deer Palatability: Low
Distinguishing Characteristics:
- Emerging leaves are "crinkly" and tinged with deep red
- Large pink flowers resemble cotton candy
- Roots and rhizomes have a distinctive fragrance, not unlike roses

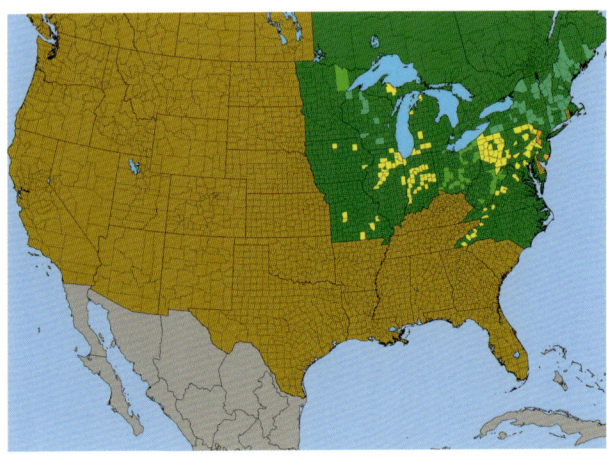

Seedling

Emerging mature plant in spring. Emerging stems are red.

Entire plant

Leaf

Flower. Flowers have a light but distinct rose fragrance.

Mature seed. Seedpods are pink when ripe, turning brown as they age and open up.

LOOK-ALIKE PLANTS:

Weeds: A nonnative Meadowsweet, *Filipendula ulmaria*, has creamy-white fragrant flowers, leaves similar to Queen of the Prairie (*Filipendula rubra*, p. 266) and similar growing requirements; it is considered a "noxious" weed in some Midwestern states.

Geum triflorum
Prairie Smoke, Old Man's Whiskers

FAMILY: ROSE (ROSACEAE)

Noted more for its fluffy pink seedheads than its flowers. Fernlike foliage holds up well all year, turning red in fall. Extremely drought resistant, does best in well-drained soils. Rhizomes establish readily and can easily be divided. Spreads gradually to form a matlike groundcover. Recalcitrant and slow from seed.

Habitat: Dry prairies, sand prairies, barrens, bluffs, dry rocky hillsides, rarely in moist meadows

Garden Uses: Rock gardens, dry gardens, perennial gardens. Plant in foreground between later-blooming flowers and grasses for best effect.

USDA Hardiness Zones: 3–6
Light Requirements: Full sun
Soil Types: Sand, loam
Soil Moisture: Dry to medium
Bloom Time: May–June
Flower Color: Pink
Height: 0.5 feet
Life Expectancy: 20+ years
Root Type: Rhizomes
Propagation: Moist stratify seed 30 days, or plant fresh for fall germination; root division early spring or midsummer

Aggressiveness: Low by seed and rhizome
Attracts: Bees, butterflies
Deer Palatability: Low
Distinguishing Characteristics:
- Finely divided, featherlike leaves
- Distinctive wispy, pink seedheads
- Low growing, never taller than 1 foot
- Rhizomes are black and scaly

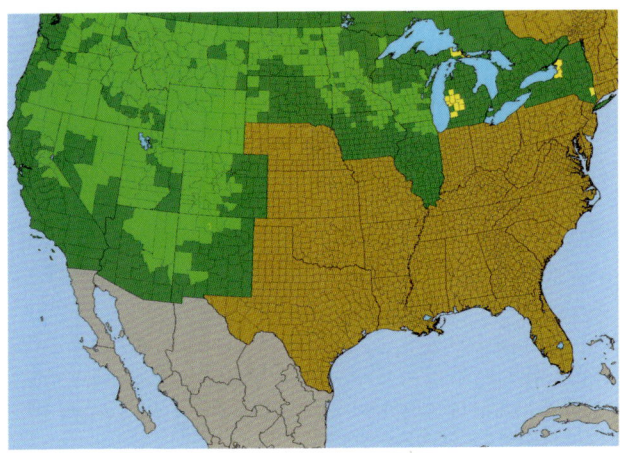

Seedling. Seedlings have distinctive pointed leaves with long hairs on the margins.

Emerging mature plant in spring

Entire plant

Leaf. Leaves are deeply dissected and often have a silvery sheen due to numerous fine hairs.

Flower. Flowers are nondescript; whitish internal flower parts are mostly obscured by the red calyx.

Early seed. Unripe seeds are bright green.

Mature seed. Seed is ripe when it turns from green to brown and can be easily pulled off the seedhead.

LOOK-ALIKE PLANTS:

Natives: Leaves could be mistaken for Silverweed (*Argentina anserina*), which is densely silver-hairy on leaf undersides and occurs near the Great Lakes, in the northern prairies, and in the Rocky Mountains.

Rosa carolina
Pasture Rose, Carolina Rose

FAMILY: ROSE (ROSACEAE)

The lowest growing of North American native roses, often hugging the ground at no more than a foot tall. Large pink flowers are similar to *Rosa arkansana*. Unlike the relatively thornless Meadow Rose, Pasture Rose is armed and dangerous! Its lovely rosehips make a fine tea, loaded with vitamin C. Mix with other low-growing prairie flowers and grasses to attenuate its spread and reduce weed growth. Grows in extremely dry, poor soil but will thrive in any well-drained garden soil.

Habitat: Dry to medium prairies, meadows, barrens, pastures, open hillsides, dry woodlands
Garden Uses: Informal gardens, prairie meadows, groundcover

USDA Hardiness Zones: 3–8
Light Requirements: Full sun to partial shade
Soil Types: Sand, loam
Soil Moisture: Dry to medium
Bloom Time: May–July
Flower Color: Pink
Height: 1–2 feet
Life Expectancy: 10–20 years
Root Type: Rhizome
Propagation: Seed is double-dormant (see chapter 9); root division in fall or spring; easy from cuttings: use new wood with rooting hormone
Aggressiveness: Low by rhizome
Attracts: Bees, butterflies, game birds, songbirds
Deer Palatability: Medium
Distinguishing Characteristics:
- One of the lowest-growing native roses, often only 1 foot tall
- Leaves have three to seven leaflets each
- Stems are well armed with thorns

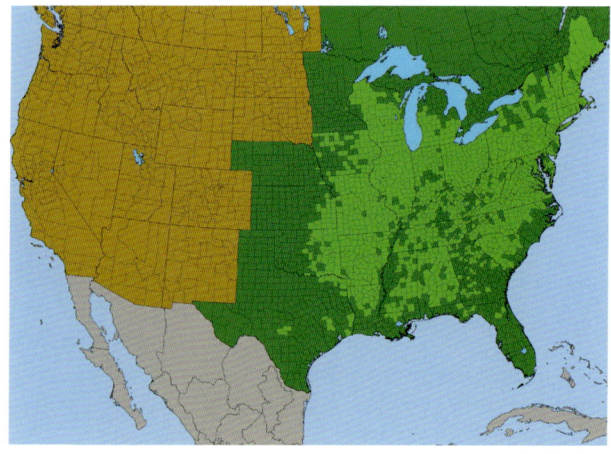

Seedling

Emerging mature plant in spring. Leaves emerge from woody stems.

Thorn. Thorns are long and narrow, scattered along the length of the stem (*Rosa arkansana* has numerous thorns of variable length on the stems, while *Rosa blanda* has no thorns on its new growth).

Entire plant. Plants usually grow only 1–2 ft. tall, shortest of the roses of the North American prairie.

Leaf. Leaves commonly have five to seven leaflets (sometimes three on lower leaves).

Flower

Seed. Seeds are ripe when hips turn red.

LOOK-ALIKE PLANTS:

Natives: Prairie Rose (*Rosa arkansana*) stems are densely covered with thorns and is usually taller; it has five to eleven leaflets per leaf. Climbing Prairie Rose (*Rosa setigera*, p. 272) has fewer thorns, is much taller and has three to five leaflets per leaf. Meadow Rose (*Rosa blanda*) is usually taller, has few thorns, and has five to seven leaflets per leaf.

Rosa setigera
Prairie Rose, Climbing Prairie Rose

FAMILY: ROSE (ROSACEAE)

One of the showiest native roses, with attractive, shiny leaves and large, deep-pink flowers. Climbs on fences, arbors, and adjacent vegetation. Makes a great hedge when planted 3 feet on center. Tea brewed from rosehips is refreshing and full of vitamin C. Grows in good, well-drained soil but also tolerates moisture. Typically resistant to attack by Japanese beetle.

Habitat: Medium to moist prairies, savannas, thickets, woodland edges and clearings, acid gravel seeps, fencerows
Garden Uses: Formal and informal gardens, hedging

USDA Hardiness Zones: 5–9
Light Requirements: Full sun to partial shade
Soil Types: Sand, loam, clay
Soil Moisture: Medium to moist
Bloom Time: May–July
Flower Color: White to pink
Height: 6–15 feet
Life Expectancy: 10–20 years
Root Type: Branching taproot
Propagation: Seed is double-dormant (see chapter 9); root division in fall or spring; easy from cuttings: use new wood with rooting hormone

Aggressiveness: Low by seed
Attracts: Bees, butterflies, game birds, songbirds
Deer Palatability: Medium
Distinguishing Characteristics:
- Taller than other native roses, except *Rosa palustris*, a native rose of wetlands
- Leaves consist of three (sometimes five) sharply pointed leaflets; all other native roses have five or more leaflets per leaf
- Commonly found entwined with small trees and shrubs because of its climbing habit

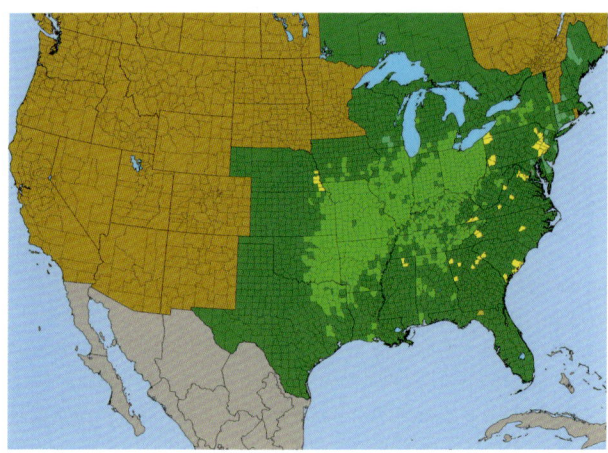

Seedling

Emerging mature plant in spring. Leaves emerge from woody stems.

Thorn. Thorns are robust and triangular in shape, common along the entire length of stem.

Entire plant. Plants climb on adjacent shrubs and fences, reaching up to 10 ft.

Leaf. Leaves commonly have three leaflets (occasionally five).

Flower

Seed. Seeds are smaller than other native roses of the prairie region. They are ripe when the hips turn red.

LOOK-ALIKE PLANTS:

Natives: Carolina Rose (*Rosa carolina*, p. 270) is much shorter and has three to seven leaflets per leaf. Meadow Rose (*Rosa blanda*) has five to seven leaflets per leaf. Prairie Rose (*Rosa arkansana*) is shorter and has five to eleven leaflets per leaf.

Weeds: Nonnative and highly invasive Multiflora Rose (*Rosa multiflora*) is of similar stature but typically grows in large clumps and has seven to nine leaflets per leaf that are blunter and not tapered to a sharp point at the end.

Spiraea alba
Meadowsweet, White Meadowsweet

FAMILY: ROSE (ROSACEAE)

Delicate spires of white to pale pink flowers grace the upper stems of this durable wetland shrub. Their white flowers mix well with blues, pinks, purples, and reds. Deeply veined foliage is attractive all season, turns golden yellow in fall. Its ability to grow in bogs makes this an excellent choice for wet, acid soils.

Habitat: Moist to wet prairies, sedge meadows, bogs, streamsides, swales, swamps, wet thickets
Garden Uses: Specimens, small groups of three to five plants, borders, low hedges, background for shorter flowers

USDA Hardiness Zones: 3–7
Light Requirements: Full sun to partial shade
Soil Types: Sand, loam, clay
Soil Moisture: Moist to wet
Bloom Time: June–September
Flower Color: White to pale pink
Height: 2–4 feet
Life Expectancy: 10–20 years
Root Type: Fibrous
Propagation: Moist stratify seed 30 days; easy from cuttings: use new wood with rooting hormone

Aggressiveness: Low by seed
Attracts: Bees, butterflies, game birds, songbirds
Deer Palatability: Medium
Distinguishing Characteristics:
- Essentially identical to Steeple-bush (p. 276) except flowers are white instead of pink
- Underside of Meadowsweet leaves are smooth and green

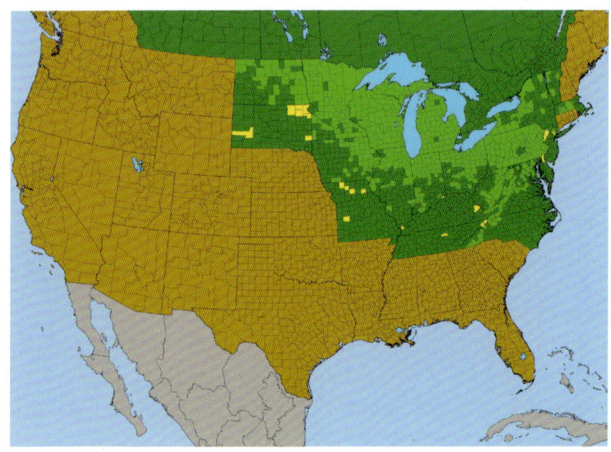

Seedling

Emerging mature plant in spring. Leaves emerge from woody stems.

Entire plant

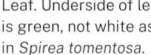

Leaf. Underside of leaf is green, not white as in *Spirea tomentosa*.

Flower

Mature seed

LOOK-ALIKE PLANTS:

Natives: Steeplebush (p. 276) has leaves with minutely pubescent white undersides, whereas Meadowsweet's leaf undersides are green and hairless.

Weeds: Another Meadowsweet, nonnative *Filipendula ulmaria*, has creamy-white fragrant flowers, leaves similar to Queen of the Prairie (*Filipendula rubra*, p. 266) and similar growing requirements; it is considered a "noxious" weed in some Midwestern states.

Spiraea tomentosa
Steeplebush

FAMILY: ROSE (ROSACEAE)

Similar to Meadowsweet, Steeplebush has bright pink flower spires instead of white, which mix well with white, purple, lavender, and other pink flowers. They are otherwise virtually indistinguishable from Meadowsweet, with similar foliage and the same golden fall color. Although native to wetlands, both plants do well in good garden soil with occasional watering during dry periods.

Habitat: Wet prairies, sedge meadows, bogs, lakeshores, seeps, wet woods and thickets, often on sandy soils
Garden Uses: Specimens, small groups of three to five plants, borders, low hedges, background for shorter flowers

USDA Hardiness Zones: 3–8
Light Requirements: Full sun to partial shade
Soil Types: Sand, loam, clay
Soil Moisture: Moist to wet
Bloom Time: August–September
Flower Color: Pink
Height: 2–4 feet
Life Expectancy: 10–20 years
Root Type: Rhizome
Propagation: Moist stratify seed 30 days; easy from cuttings: use new wood with rooting hormone

Aggressiveness: Low by seed and rhizome
Attracts: Bees, butterflies, songbirds
Deer Palatability: Low
Distinguishing Characteristics:
- Undersides of Steeplebush leaves are minutely pubescent and white, whereas the undersides of Meadowsweet leaves are smooth and green
- Leaves have a bitter sap; Meadowsweet leaves do not

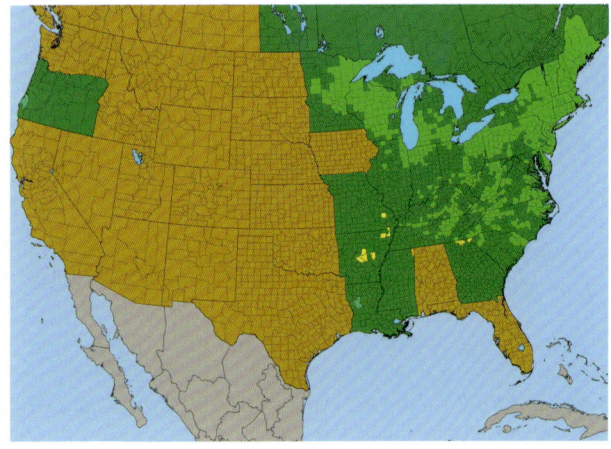

Seedling

Emerging mature plant in spring. Leaves emerge from woody stems.

Entire plant

Leaf

Leaf underside. Underside of leaf is silvery white with a dense cover of small hairs.

Flower

Mature seed

LOOK-ALIKE PLANTS:

Natives: Meadowsweet (*Spiraea alba*, p. 274) has leaves whose undersides are green and hairless.

Weeds: Another Meadowsweet, nonnative *Filipendula ulmaria* has creamy-white fragrant flowers and similar growing requirements; it is considered a "noxious" weed in some Midwestern states.

Chelone glabra
White Turtlehead

FAMILY: FIGWORT (SCROPHULARIACEAE)

A unique plant for moist gardens and wetland plantings. Mass for flowering effect. Baltimore checkerspot caterpillar depends almost exclusively on this plant as a food source. A bit tricky from seed but easily established by transplants.

Habitat: Wet prairies, marshes, sedge meadows, fens, streamsides, wet thickets
Garden Uses: Rain gardens, shaded borders, pond, stream and wetland edges

USDA Hardiness Zones: 3–8
Light Requirements: Full sun to partial shade
Soil Types: Sand, loam, clay, muck
Soil Moisture: Moist to wet
Bloom Time: August–September
Flower Color: White
Height: 2–4 feet
Life Expectancy: 5–10 years
Root Type: Fibrous
Propagation: Moist stratify seed 30 days; easy from stem cuttings early to midseason

Aggressiveness: Low by seed
Attracts: Bees, butterflies, hummingbirds
Deer Palatability: Low
Distinguishing Characteristics:
- Distinctive turtle-head-shaped white flowers
- Narrow, distinctly toothed leaves
- Plant may sprawl along the ground or among adjacent plants

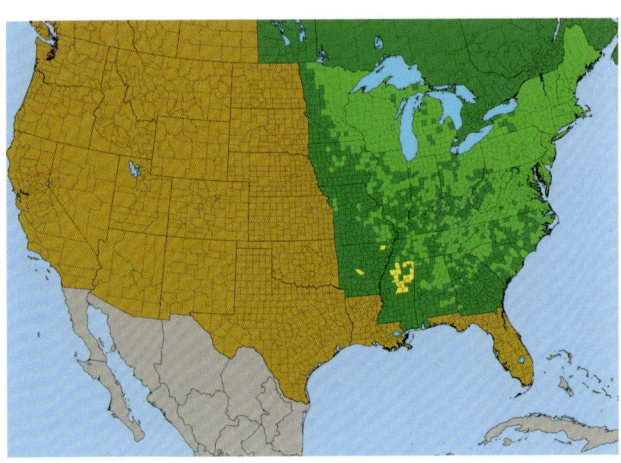

Seedling

Emerging mature plant
in spring

Entire plant

Leaf. Leaves often have a
crinkly, corrugated texture.

Flower. Flower resembles a turtle's head.

Early seed

Mature seed. Seeds are
ripe when pods turn brown
and top pods begin to
crack open.

LOOK-ALIKE PLANTS:

Natives: Pink Turtlehead (*Chelone lyonii*) is native to the southern Appalachian Mountains in seeps and moist, north-facing hillsides. Leaves of *Physostegia virginiana* (Obedient Plant, p. 234) and *Helenium autumnale* (Sneezeweed, p. 110) are similar in appearance.

Penstemon cobaea

Cobaea Beardtongue, Cobaea Penstemon, Wild Foxglove

FAMILY: FIGWORT (SCROPHULARIACEAE)

Large purple flowers and shiny deep-green foliage make this a great garden plant. Prefers well-drained soil but tolerates moisture and clay soil. Easy from seed.

Habitat: Dry to medium prairies, rocky or sandy hillsides, bluffs
Garden Uses: Borders, hummingbird gardens, informal gardens, prairie meadows

USDA Hardiness Zones: 5–8
Light Requirements: Full sun
Soil Types: Sand, gravel
Soil Moisture: Dry to medium
Bloom Time: May–June
Flower Color: White, pink, purple
Height: 1–4 feet
Life Expectancy: 5–10 years
Root Type: Fibrous
Propagation: Moist stratify seed 30 days
Aggressiveness: Low by seed

Attracts: Bees, birds, butterflies, hummingbirds
Deer Palatability: Medium
Distinguishing Characteristics:
- Large red-lavender to purple flowers have distinct white hairs (a sterile stamen) inside the bottom lip
- Flower stigmas often remain attached to ripe seedpods rather than dropping off

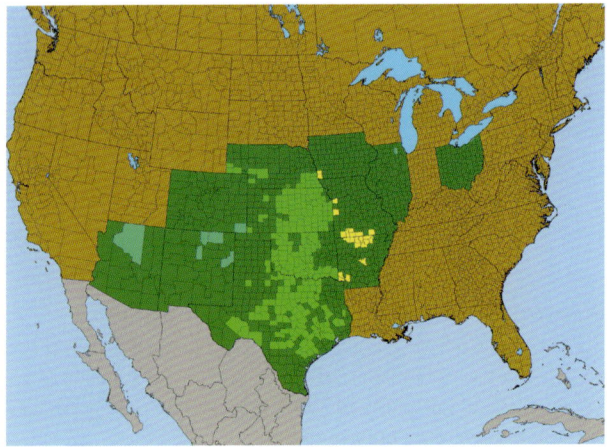

Seedling

Emerging mature plant in spring

Entire plant

Leaf. Smooth, glossy serrated leaves are deep green and clasp the stem.

Flower. Flowers are large and intensely reddish purple.

Flower close-up. Note prominent stigma with long hairs on the floor of the flower tube.

Mature seed. Seeds are ripe when pods turn brown.

LOOK-ALIKE PLANTS:

Natives: Large-Flowered Beardtongue (p. 286) is similar but has thicker leaves, lavender flowers, and no hairs on the bottom lip.

Penstemon digitalis

Smooth Penstemon, Foxglove Beardtongue, Talus Slope Penstemon

FAMILY: FIGWORT (SCROPHULARIACEAE)

One of only few penstemons that grow in clay. Tolerates high moisture conditions and moderate shade. Cheerful white flowers attract humming-birds, moths, bees, and other pollinators. Mixes well with the blues and pinks of spiderworts. Generally well behaved in the garden, it is easy to grow from seed, and plants can be divided. Like most penstemons it is relatively short-lived. A well-known cultivar is "Husker Red."

Habitat: Medium to moist prairies, savannas, open woods, old fields
Garden Uses: Borders, hummingbird gardens, bee gardens, rain gardens, formal and informal gardens, prairie meadows

USDA Hardiness Zones: 3–8
Light Requirements: Full sun to partial shade
Soil Types: Sand, loam, clay
Soil Moisture: Medium to moist
Bloom Time: May–July
Flower Color: White
Height: 2–3 feet
Life Expectancy: 5–10 years
Root Type: Fibrous
Propagation: Moist stratify seed 30 days; root division in early spring

Aggressiveness: Low by seed
Attracts: Bees, butterflies, hummingbirds
Deer Palatability: Low
Distinguishing Characteristics:
- One of few white-flowered penstemons in eastern and central North America
- Grows in medium to moist soils, unlike most penstemons which require well-drained soils
- Ripe seeds smell like musty gym socks

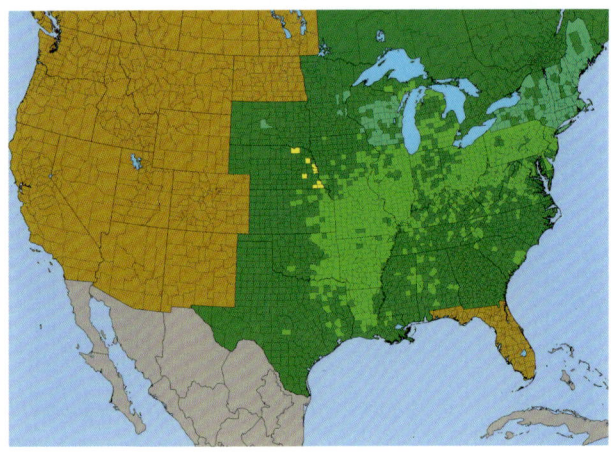

Seedling

Emerging mature plant in spring. Previous year's dark-green to reddish basal leaves are often retained over winter.

Entire plant

Leaf. Basal leaves are often reddish and rounded, with little serration; stem leaves are usually lighter and more serrated.

Flower

Flower close-up. Flowers are pure white with thin purple veins.

Mature seed. Seeds are ripe when pods turn brown.

LOOK-ALIKE PLANTS:

Natives: *Penstemon tubiflorus* has similar white flowers and growth form; more common in Arkansas, Kansas, Missouri, and Oklahoma.

Penstemon gracilis
Slender Beardtongue, Slender Penstemon

FAMILY: FIGWORT (SCROPHULARIACEAE)

This smallest of prairie penstemons, seldom more than 1 foot tall, produces clusters of tiny white-streaked lavender flowers. Foliage is an unusual grayish green. Requires well-drained sandy or rocky soil.

Habitat: Sand prairies, barrens, dry open woods, bluffs, sandy old fields
Garden Uses: Mass in hummingbird gardens, rock gardens, dry gardens, dry prairie meadows

USDA Hardiness Zones: 3–6
Light Requirements: Full sun to partial shade
Soil Types: Sand, gravel
Soil Moisture: Dry
Bloom Time: April–July
Flower Color: Lavender
Height: 1–2 feet
Life Expectancy: 5–10 years
Root Type: Fibrous
Propagation: Moist stratify seed 30 days

Aggressiveness: Low by seed
Attracts: Bees, butterflies, hummingbirds
Deer Palatability: Low
Distinguishing Characteristics:
- Small lavender flowers streaked with white
- Thin stems rise from small basal leaves
- Small seedpods are often reddish green prior to full maturity

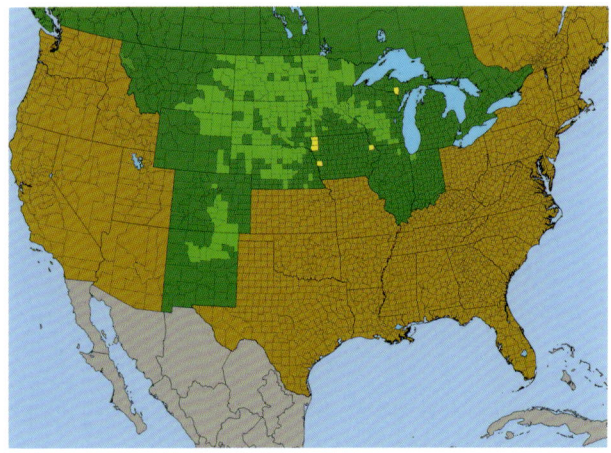

Seedling

Entire plant

Emerging mature plant in spring. Basal leaves are smooth and hairless; *Penstemon hirsutus* basal leaves are covered in fine hairs.

Leaf. Leaves are thin and narrow, without serrations on margins; leaves of *Penstemon hirsutus* are similar in size but distinctly serrated.

Flower

Flower close-up. Flowers are flattened from top to bottom, compared to most other penstemons.

Early and mature seed. Immature seeds often have reddish-green tinge; seeds are ripe when pods turn brown.

LOOK-ALIKE PLANTS:

Natives: *Penstemon hirsutus* (Hairy Penstemon, p. 288) has a shorter flower tube and dense hairs on its stems. *P. pallidus* (Pale Beardtongue) is similar in form but has white flowers with purple lines in the throat and occurs south of the natural range of *P. gracilis*.

Penstemon grandiflorus

Large-Flowered Beardtongue, Beardtongue, Shell-Leaf Penstemon

FAMILY: FIGWORT (SCROPHULARIACEAE)

Largest and showiest of North American prairie penstemons. Waxy, blue-green leaves make their own foliage statement even without flowers. Cut seedstalks when pods are green for dried arrangements. Regular visitors include hummingbirds and bumblebees. Grows only on dry sandy soils. Easy to grow from seed but short-lived. Sow seed in fall for spring germination.

Habitat: Dry sand prairies, sand barrens, oak barrens, open sandy woods, sandy roadsides
Garden Uses: Specimens, small groupings, hummingbird gardens, bee gardens, dry gardens, dry prairie meadows

USDA Hardiness Zones: 3–7
Light Requirements: Full sun
Soil Types: Sand, gravel
Soil Moisture: Dry
Bloom Time: May–June
Flower Color: Lavender
Height: 2–4 feet
Life Expectancy: 3–5 years
Root Type: Fibrous
Propagation: Moist stratify seed 30 days; root division of 2-year-old plants in early spring
Aggressiveness: Low by seed
Attracts: Bees, hummingbirds

Deer Palatability: Low
Distinguishing Characteristics:
- Larger flowers than most other eastern penstemon, except *P. cobaea* and *P. calycosus*
- Thick, sturdy plants are taller than most other eastern penstemon
- Large thick, smooth, blue-green leaves resemble those of a succulent plant
- Big brown seedheads contain large, cubelike seeds

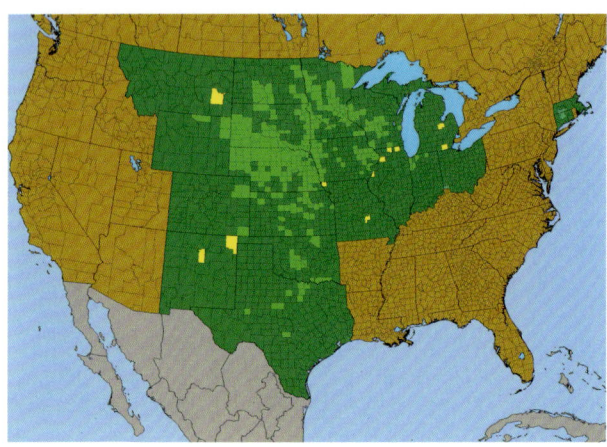

Seedling

Emerging mature plant in spring. Plants emerge as tightly packed whorls of smooth, light-green leaves.

Entire plant

Leaf. Leaves are thick, smooth, and completely hairless, somewhat resembling Eucalyptus leaves.

Flower

Flower close-up. Flowers are similar in size to *Penstemon cobaea* but lavender rather than purplish red and without fine hairs.

Mature seed. Seeds are ripe when pods turn brown.

LOOK-ALIKE PLANTS:

Natives: Cobaea Penstemon (*Penstemon cobaea*, p. 280) is almost as big but has smaller seedpods and thinner leaves. Longsepal Beardtongue (*P. calycosus*) has similar flowers but narrower, pointed serrated leaves.

Penstemon hirsutus
Hairy Penstemon

FAMILY: FIGWORT (SCROPHULARIACEAE)

Multiple spikes of two-toned blue-and-white flowers borne above lemon-green leaves. Seedpods turn attractive purple in late summer, excellent for dried arrangements. Self-sows readily. Native to dry prairies, yet can be successfully grown in well-drained heavier soils in garden situations.

Habitat: Dry prairies, rocky, often calcareous, hillsides, open woods, bluffs
Garden Uses: Specimens, small groups, hummingbird gardens, bee gardens, informal gardens, prairie meadows

USDA Hardiness Zones: 3–6
Light Requirements: Full sun to partial shade
Soil Types: Sand, gravel
Soil Moisture: Dry
Bloom Time: May–July
Flower Color: Lavender to blue
Height: 1–2 feet
Life Expectancy: 5–10 years
Root Type: Fibrous

Propagation: Moist stratify seed 30 days; root division in early spring
Aggressiveness: Medium by seed
Attracts: Bees, butterflies, hummingbirds
Deer Palatability: Low
Distinguishing Characteristics:
- Stems covered in fine hairs
- Delicate lavender flowers, similar to Slender Penstemon (p. 284)

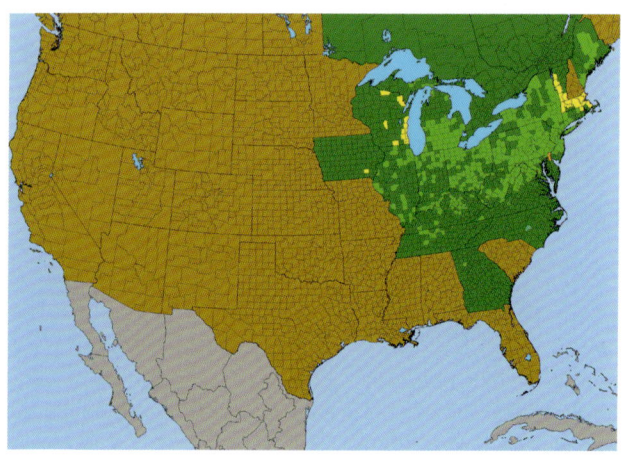

Seedling

Entire plant

Leaf. Stem leaves are serrated on margins, whereas *Penstemon gracilis* leaf margins are not.

Emerging mature plant in spring. Basal leaves are hairy; *Penstemon gracilis* basal leaves are smooth.

Flower

Flower close-up

Early seed. Immature seeds often have reddish-green tinge.

Mature seed. Seeds are ripe when pods turn brown.

LOOK-ALIKE PLANTS:

Natives: Flowers are similar to Slender Penstemon (*Penstemon gracilis*, p. 284). The flower throat of Hairy Penstemon (*P. hirsutus*) is nearly closed by the lower lobe; Slender Penstemon has a more open flat throat. Pale Penstemon (*P. pallidus*) is similar in form but has pure white flowers.

Veronicastrum virginicum (*Veronica virginica*)
Culver's Root

FAMILY: FIGWORT (SCROPHULARIACEAE)

Distinctive white flower spires borne above deep-green, whorled leaves. An excellent architectural plant. If deadheaded will flower late into the season. Prefers a rich moist soil and tolerates moderate shade. Creeps ever so slowly by rootstocks which can be easily divided. Tiny dust-like seeds germinate readily but grow slowly. Seed should be harvested when seedheads are yellow, before they turn brown.

Habitat: Medium to moist prairies, fens, savannas, open woods, woodland edges
Garden Uses: Specimens, masses, formal and informal gardens, rain gardens, bee gardens, prairie meadows

USDA Hardiness Zones: 3–8
Light Requirements: Full sun to medium shade
Soil Types: Sand, loam, clay
Soil Moisture: Medium to moist
Bloom Time: June–August
Flower Color: White
Height: 3–6 feet
Life Expectancy: 10–20 years
Root Type: Fibrous rhizome
Propagation: Moist stratify seeds 30 days; root division in fall or spring; tricky from cuttings: use first early growth

Aggressiveness: Low by seed
Attracts: Bees, butterflies
Deer Palatability: Low
Distinguishing Characteristics:
- Spires of white flowers with red stamens are unique among prairie plants
- Leaves occur in whorls around the stem
- Extremely fine, dark-orange seeds (750,000 per ounce)
- Root is black and gnarly

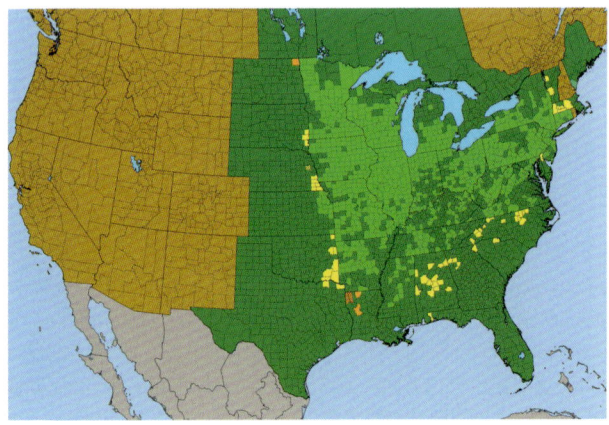

Seedling. Seedling similar to *Verbena hastata* but has fine hairs on stems and leaf margins.

Emerging mature plant in spring has densely hairy stems

Entire plant. Plant is strongly vertical and serves as an excellent structural element in the garden.

Leaf. Leaves occur in whorls of five along the stem.

Flower. Flowers similar in appearance to *Actaea racemosa*, but the flowers of *A. racemosa* extend well above the leaves on a stalk.

Mature seed. Seedheads similar to *Verbena hastata* but occur in whorls along stem; *V. hastata* seedheads emanate from a branched stem.

LOOK-ALIKE PLANTS:

Natives: Some *Veronica* species have similar flower spires, but plants are shorter and most have blue flowers. Ironweed species (*Vernonia* spp.) often have similar leaves but not in whorls around the stem. The seedheads strongly resemble Blue Vervain (*Verbena hastata*, p. 292) when ripe, and both occur in similar moist habitats. Individual seedstalks originate from a whorl on the main stem of Culver's Root, whereas Blue Vervain has multiple branching seedstalks.

Verbena hastata

Blue Vervain, Swamp Verbena

FAMILY: VERVAIN (VERBENACEAE)

Striking blue-violet flowers appear in profusion on conical heads. Arrow-like leaves are "crinkled" with pointed ends. A fast-growing biennial, it usually self-sows to ensure its future presence. An important component of wetland seed mixes, it provides quick vegetation cover and early color. Propagation by seed or cuttings.

Habitat: Moist to wet prairies, sedge meadows, marshes, streamsides, lakeshores, wet thickets, moist pastures
Garden Uses: Butterfly gardens, rain gardens, bee gardens, informal gardens, moist meadows

USDA Hardiness Zones: 3–8
Light Requirements: Full sun to partial shade
Soil Types: Sand, loam, clay
Soil Moisture: Medium to wet
Bloom Time: July–September
Flower Color: Blue
Height: 3–6 feet
Life Expectancy: 2 years (biennial) or annual
Root Type: Fibrous
Propagation: Moist stratify seed 30 days; easy from stem cuttings early to midseason
Aggressiveness: Medium by seed

Attracts: Bees, butterflies, hummingbirds, songbirds
Deer Palatability: Low
Distinguishing Characteristics:
- Small conical flower heads bloom in a ring around the center cone
- Flower heads are shorter and more conical than the longer heads of Hoary Vervain (p. 294)
- "Crinkly" pointy leaves are similar to Hoary Vervain but are usually darker green and less textured

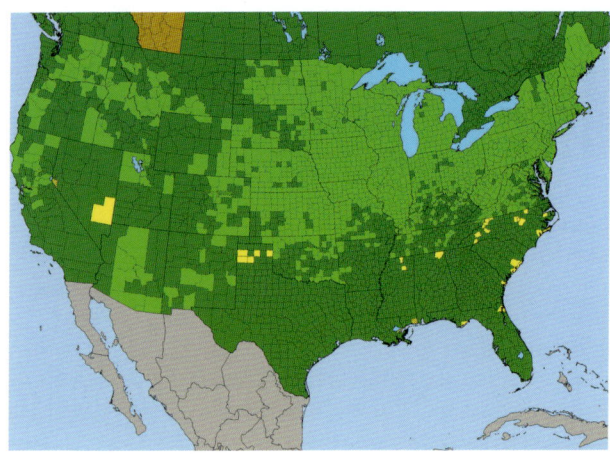

Seedling

Emerging mature plant in spring

Entire plant

Leaf. Leaves are not deeply indented, as with *V. stricta*.

Flower. Individual flower heads and seedstalks about half as long as *V. stricta*, at 0.25–0.5 ft.

Mature seed. Seedheads similar to *Veronicastrum virginicum* but emanate from a branched stem, not whorls.

LOOK-ALIKE PLANTS:

Natives: Taller than Hoary Vervain (*Verbena stricta*, p. 294) and grows in moist, rather than dry sites. Narrowleaf Vervain (*V. simplex*) has showy white to lavender flowers, leaves only 0.5 inch wide, and grows in dry alkaline sites. White Vervain (*V. urticifolia*) has similar but wider leaves (about 2 inches versus 1.5 inches) and tiny white flowers.

Weeds: Leaves resemble the native Stinging Nettle (*Urtica dioica*) but are deeper green and thicker textured. The difference becomes obvious upon contact!

Verbena stricta
Hoary Vervain, Hoary Verbena

FAMILY: VERVAIN (VERBENACEAE)

Robin's-egg-blue flower spikes stand out in the garden. Relatively short-lived, it often self-sows onto loose, open soil. Prefers dry soils and well-drained loams. Easy from seed.

Habitat: Dry to medium prairies, edges of dry woodlands, sandy old fields and pastures
Garden Uses: Butterfly gardens, informal gardens, prairie meadows

USDA Hardiness Zones: 3–8
Light Requirements: Full sun
Soil Types: Sand, gravel, loam
Soil Moisture: Dry to medium
Bloom Time: July–September
Flower Color: Blue
Height: 2–4 feet
Life Expectancy: 3–5 years
Root Type: Central taproot with tillers
Propagation: Moist stratify seed 30 days; easy from stem cuttings early to midseason

Aggressiveness: Medium by seed
Attracts: Bees, butterflies, game birds, hummingbirds, songbirds
Deer Palatability: Low
Distinguishing Characteristics:
- Lavender-blue flowers bloom in a ring around a spike that is longer and not as conical as Blue Vervain (*Verbena hastata*, p. 292)
- Leaves are heavily crenellated and textured with deep fissures

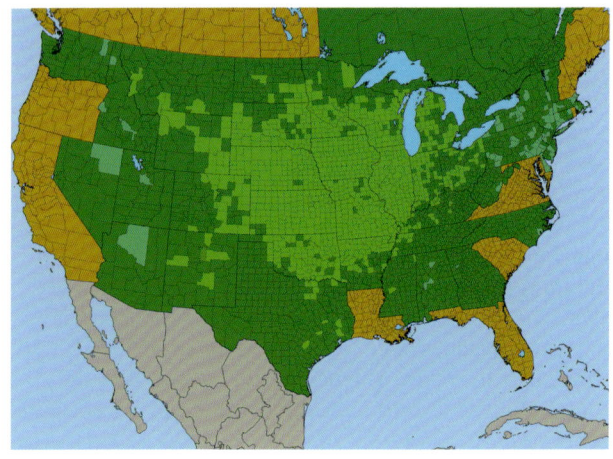

Seedling

Emerging mature plant in spring

Entire plant

Leaf. Deeply fissured leaves have crinkly indentations.

Flower. Individual flower heads and seedstalks generally twice as long as *V. hastata* (0.5–1 ft.).

Mature seed

LOOK-ALIKE PLANTS:

Natives: Shorter than Blue Vervain and grows in dry rather than damp soils. Narrowleaf Vervain (*V. simplex*) also grows in dry sites but has showy white to lavender flowers, and its leaves are only 0.5 inch wide. White Vervain (*V. urticifolia*) has similar wide leaves (2–2.5 inches) but tiny white flowers rather than showy blue flowers.

Viola pedata
Birdfoot Violet

FAMILY: VIOLET (VIOLACEAE)

A gem of dry prairies with the largest flowers of North American native violets. Its leaves resemble a bird's foot, hence the name. Grows in prohibitively dry, rocky, and sandy environments. A recessive color gene produces a bicolor flower with the two upper petals deep velvety purple. Seedpods explode when ripe; harvest when pods are horizontal with the ground or higher to ensure ripeness. Easily divided in early spring and midsummer after seeds have ripened.

Habitat: Dry prairies, barrens, glades, bluffs, dry open woods; restricted to dry, sandy and rocky soils
Garden Uses: Rock gardens, rock walls, dry gardens, dry prairie meadows

USDA Hardiness Zones: 3–9
Light Requirements: Full sun to partial shade
Soil Types: Sand, gravel
Soil Moisture: Dry
Bloom Time: April–June
Flower Color: Blue, purple (bicolor variety)
Height: 3–6 inches
Life Expectancy: 5–10 years
Root Type: Fibrous
Propagation: Moist stratify seeds 30 days, or plant fresh for fall germination; root division early spring or midsummer
Aggressiveness: Low by seed
Attracts: Bees, butterflies, songbirds
Deer Palatability: Low
Distinguishing Characteristics:
- Leaves are deeply divided
- Flowers are larger than other violets
- Flowers lack hairs at the throat
- Seeds have sugary gel that attracts ants, which harvest and distribute them

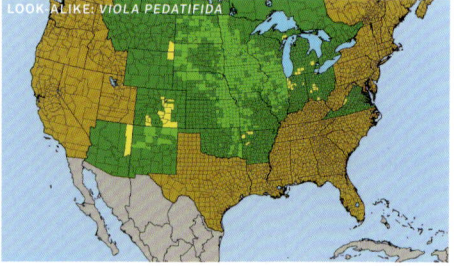

LOOK-ALIKE: *VIOLA PEDATIFIDA*

Seedling

Emerging mature plant in
spring

Entire plant

Leaf. Leaves narrower than
most other violets but almost
indistinguishable from
V. pedatifida.

Flower. Large blue flowers are unique
among native violets and much wider
at 1.0–1.5 in.

Mature seed

Viola pedata var. *bicolor*, entire plant.
Plant is the same species as regular
V. pedata; only the flowers are
different.

Viola pedata var. *bicolor*,
flower. Upper two
flower lobes are deep
purple instead of blue.

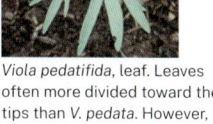
Viola pedatifida, leaf. Leaves
often more divided toward the
tips than *V. pedata*. However,
this is variable and not a
reliable difference.

Viola pedatifida, entire plant

Viola pedatifida,
flower. Flower
only about half as
wide, at 0.5–0.75
in., as *V. pedata*, at
1.0–1.5 in.

LOOK-ALIKE PLANTS:

Natives: Prairie Violet (*V. pedatifida*) grows in similar habitats and often
has deeply divided leaves. It has smaller, deep-blue flowers with noticeable hairs at the throat.

Grasses and Sedges

Carex brevior
Shortbeak Sedge, Plains Oval Sedge

FAMILY: SEDGE (CYPERACEAE)

One of the most widely distributed upland prairie sedges, thriving in dry and moist habitats. Forms tight clumps with strongly upright stems and flower stalks. Like all sedges, flowers are reduced and not prominent. Mix with short prairie flowers in low-growing areas to control weeds without overshadowing flowers. Tightly bunched seedheads are not showy. Shortbeak Sedge is usually seeded as an element of prairie restorations.

Habitat: Dry to moist prairies, sand prairies, hillsides and bluffs
Garden Uses: Excellent in rain gardens in sandy soils that dry out between rains

USDA Hardiness Zones: 3–7
Light Requirements: Full sun to partial shade
Soil Types: Sand, loam
Soil Moisture: Dry to moist
Bloom Time: May–June
Fall Color: Straw
Height: 1–2 feet
Life Expectancy: 10–20 years
Root Type: Fibrous, bunch former
Propagation: Moist stratify seed 30 days; root division in fall or spring
Aggressiveness: Low by seed

Attracts: Butterflies, game birds, songbirds
Deer Palatability: Low
Distinguishing Characteristics:
- Has triangular stems, as with all sedges
- Seeds are flattened, with a central germ surrounded by nearly orbicular, thin, papery "wings"
- A member of the *Ovales* (*Scoparia*) group in the genus *Carex*, its seeds (perigynia) are larger than most other members of this group at 0.16 inches wide and 0.24 inches long

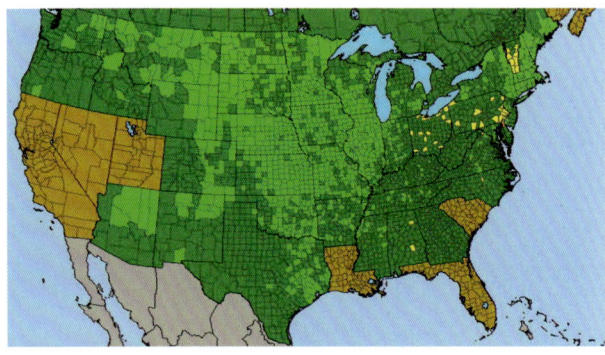

Seedling. Like all sedges, stems are triangular from the time of seedling emergence.

Emerging mature plant in spring

Entire plant

Flower. Larger creamy-yellow male flowers are borne at the base of each spikelet; small white female flowers with two stigmas each occur at tips.

Early seed

Mature seed. Individual spikelets (flower clusters and seedheads) are separated on the stem, unlike many other members of the Scoparia (Ovales) sedge subgroup that are clustered together at the stem tip.

Perigynium. A member of the difficult Scoparia (Ovales) subgroup of sedges, the mature seeds (perigynia) of *C. brevior* and *C. bicknellii* are differentiated from similar members by being nearly circular, not elongated.

LOOK-ALIKE PLANTS:

Natives: Very similar to Bicknell's Sedge (*Carex bicknellii*), which has slightly less-rounded wings on its perigynia. Sedges are best known as plants of wetlands and woodlands, but about a dozen upland species occur in upland prairies and meadows. The taxonomy of sedges is complex and identification is based upon seed structure. Details of sedge identification cannot be adequately addressed here.

Weeds: Yellow Nutsedge (*Cyperus esculentus*), with prominent golden bristly seedheads, has triangular stems and does not form clumps but creeps aggressively by rhizomes to form patches.

Andropogon gerardii
Big Bluestem, Turkeyfoot

FAMILY: GRASS (POACEAE, GRAMINEAE)

King of the prairie grasses: big, beautiful, and adaptable. Grows in clumps but forms a sod. Competes aggressively with shallow-rooted flowers, so plant with taprooted species or spring bloomers that go dormant by late summer, when this grass is most active. Excellent habitat for grassland birds. Seeds readily onto open soil. Seed at low rates of 0.5 pound per acre or less, in prairie meadows; or sow at 2–5 pounds per acre with other prairie grasses to create dense wildlife cover.

Habitat: Dry to wet prairies, savannas, sand barrens, glades, dry open woods, roadsides
Garden Uses: Specimen, otherwise not recommended for gardens due to its tendency to self-sow and dominate

USDA Hardiness Zones: 3–9
Light Requirements: Full sun
Soil Types: Sand, loam, clay
Soil Moisture: Dry to wet
Bloom Time: August–October
Fall Color: Bronze, red
Height: 5–8 feet
Life Expectancy: 20+ years
Root Type: Fibrous, sod former
Propagation: Dry stratify seed 60 days; root division in fall or spring
Aggressiveness: Medium by seed

Attracts: Butterflies, game birds, songbirds
Deer Palatability: Low
Distinguishing Characteristics:
- Stems and foliage turn bronze-red in fall
- Three- to five-parted seedhead resembles a turkey foot
- Emerging stalks are round; those of Little Bluestem (p. 326) are flat
- The clumps form a dense sod but do not creep by rhizomes

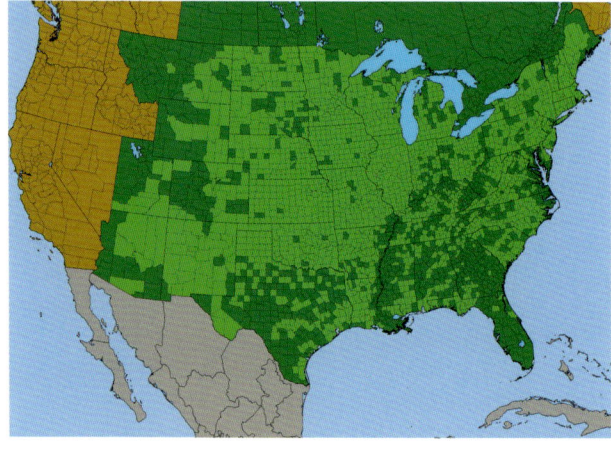

Seedling

Flower close-up

Flower

Emerging mature plant in spring. New shoots are reddish at the base, with hairy lower stems (similar to Indiangrass).

Entire plant. Among the tallest prairie grasses, on average 1 ft. taller than Indiangrass; only Prairie Cordgrass and Eastern Gamagrass are as tall or taller.

Seed. Seeds are borne in groups of three, resembling a turkey's foot, a common name sometimes used for this grass.

Ligule. The papery ligule extends above the sheath, with a somewhat ragged upper edge.

LOOK-ALIKE PLANTS:

Natives: Leaves resemble Switchgrass (*Panicum virgatum*, p. 324) and Indiangrass (*Sorghastrum nutans*, p. 328) but can be distinguished prior to flowering by differences in ligules. Sand Bluestem (*Andropogon hallii*) is a similar species of the Great Plains but usually grows only 2–6 feet tall, as opposed to 5–8 feet for *A. gerardii. A. hallii* is strongly rhizomatous, and the seeds are copiously hairy. Elliott's Bluestem (*A. gyrans*) is native to the southeastern and southern Midwest, and has similar seedheads, but it grows only 1–4 feet tall. Bushy Bluestem (*A. glomeratus*) is a southeastern native that can appear similar prior to seed formation but is easily distinguished by its fluffy, densely agglomerated, seedheads when ripe. *A. glomeratus* grows in low wet areas, ditches, swales, and boggy areas.

Weeds: King Ranch Bluestem (*Bothriochloa ischaemum* var. *songarica*) is an introduction from southern Eurasia planted for forage in the South. The seedheads may resemble Big Bluestem, but the plant only grows 1.5–4 feet tall.

Andropogon virginicus
Broomsedge

FAMILY: GRASS (POACEAE, GRAMINEAE)

Very similar to Little Bluestem, it is widely distributed in eastern and southeastern United States, whereas Little Bluestem extends farther west. Thrives in heavy clay soil, even in damp sites. Beware, as it self-sows readily and is often considered a weed. Use seed to establish dry to moist eastern meadows and for reclamation of disturbed acid clay soils. Seeds aggressively onto open disturbed soil. Dead plant material is reported to be allelopathic, discouraging growth of other species. Provides good cover for songbirds and upland game birds.

Habitat: Dry to wet prairies and meadows, open woods, old fields, especially on eroded acid clay soils
Garden Uses: Meadows, may be used as a replacement for Little Bluestem on clay soils

USDA Hardiness Zones: 5–10
Light Requirements: Full sun
Soil Types: Sand, loam, clay
Soil Moisture: Dry to moist
Bloom Time: August–October
Fall Color: Reddish brown
Height: 2–5 feet
Life Expectancy: 20+ years
Root Type: Fibrous, bunch former
Propagation: Dry stratify seed 60 days; root division in fall or spring
Aggressiveness: High by seed
Attracts: Butterflies, game birds, songbirds
Deer Palatability: Low

Distinguishing Characteristics:
- Flat leaves are similar to Little Bluestem (*Schizachyrium scoparium*, p. 326)
- Fall color is more reddish brown compared to the bright red of Little Bluestem
- Smaller, "fluffier" seeds typically range farther down stem than Little Bluestem, whose seeds are plumper and more clustered toward top of plant
- Tends to grow on acid clay soils, whereas Little Bluestem prefers dry, sandy soils
- Generally tolerates wetter soils than Little Bluestem

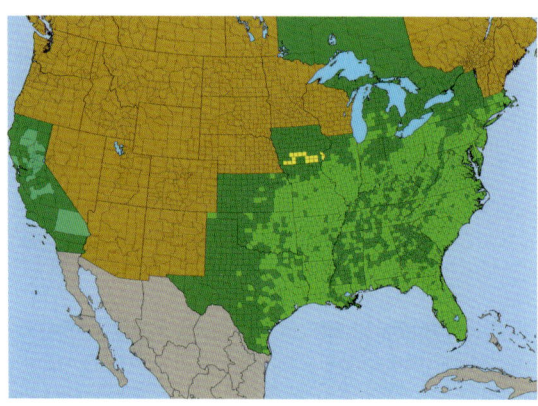

Seedling

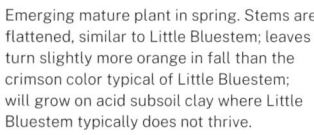

Emerging mature plant in spring. Stems are flattened, similar to Little Bluestem; leaves turn slightly more orange in fall than the crimson color typical of Little Bluestem; will grow on acid subsoil clay where Little Bluestem typically does not thrive.

Flower

Seed. Seed grains are smaller than Little Bluestem; seeds are clustered in bunches along the stem, with many leaflike bracts at base of seedstalks; Little Bluestem seeds are produced on longer seedstalks, with few or no bracts at base.

Ligule

LOOK-ALIKE PLANTS:

Natives: Little Bluestem (p. 326) is virtually identical until seed forms in fall.

Bouteloua curtipendula
Sideoats Grama

FAMILY: GRASS (POACEAE, GRAMINEAE)

Excellent for gardens and meadows, this short bunchgrass does not compete strongly with other flowers and grasses, making it a good neighbor and an excellent companion for low-growing flowers. Requires good drainage and thrives in dry, limy soils. Spreads slowly by stolons to form a sod that helps keep weeds at bay. Not recommended for clay soils in areas with more than 20 inches of annual rainfall. Birds love the seeds. Excellent for bobwhite quail habitat.

Habitat: Dry to medium prairies, savannas, glades, bluffs, especially on alkaline soils
Garden Uses: Border edges, informal gardens

USDA Hardiness Zones: 3–9
Light Requirements: Full sun
Soil Types: Sand, loam
Soil Moisture: Dry to medium
Bloom Time: July–September
Fall Color: Straw
Height: 2–3 feet
Life Expectancy: 10–20 years
Root Type: Fibrous, rhizomatous, bunch former
Propagation: Dry stratify seed 60 days; root division in spring

Aggressiveness: Low by seed
Attracts: Butterflies, game birds, songbirds
Deer Palatability: Low
Distinguishing Characteristics:
• Leaf edges have small hairs on lower half of leaf blades
• Small, oatlike seeds are suspended on one side of the seedstalk

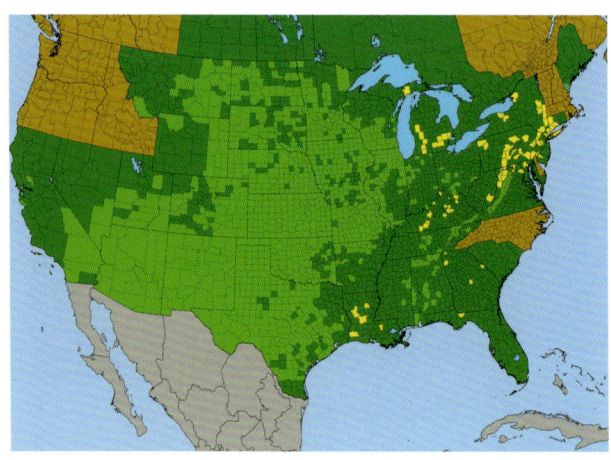

Seedling

Emerging mature plant in spring

Entire plant. Forms bunches, often appearing tuftlike with curly leaves when young; very fine, long hairs extending about halfway up the leaf margins.

Flower close-up

Flower stalk

Seed. Distinctive oatlike seeds are borne on one side of the seedstalk.

Ligule. Thin hairs occur around the ligule and extend about halfway up the leaf along the edges of the leaf blade.

LOOK-ALIKE PLANTS:

Natives: Sideoats Grama is the easternmost species of the Grama grasses (*Bouteloua* spp.), but is most common in the Great Plains and high desert. Other Grama grasses, such as Blue Grama (*B. gracilis*) and Hairy Grama (*B. hirsuta*), have leaf hairs that extend more than halfway up the grass blade. Buffalograss (*B. dactyloides*) is planted in dry climates as a low-maintenance lawn. The seedheads of these three species are longer in length at 0.5–1.5 inches and appear like little mustaches.

Bromus kalmii
Prairie Bromegrass, Arctic Brome

FAMILY: GRASS (POACEAE, GRAMINEAE)

Gracefully arching seedheads are a focal point, as it does not have great fall color. Although short-lived, it often self-sows.

Habitat: Dry to moist prairies and meadows, open woods, thickets
Garden Uses: Dry gardens, rock gardens, grass gardens

USDA Hardiness Zones: 3–6
Light Requirements: Full sun to part shade
Soil Types: Sand, gravel, loam
Soil Moisture: Dry to moist
Bloom Time: June–July
Fall Color: Straw
Height: 2–4 feet
Life Expectancy: 3–5 years
Root Type: Fibrous
Propagation: Dry stratify seed 60 days

Aggressiveness: Low by seed
Attracts: Game birds, songbirds
Deer Palatability: Medium
Distinguishing Characteristics:
- Seeds are covered with fine hairs
- Leaves are covered with fine hairs
- Broad leaves are up to 0.75 inches wide
- Seedheads droop at maturity

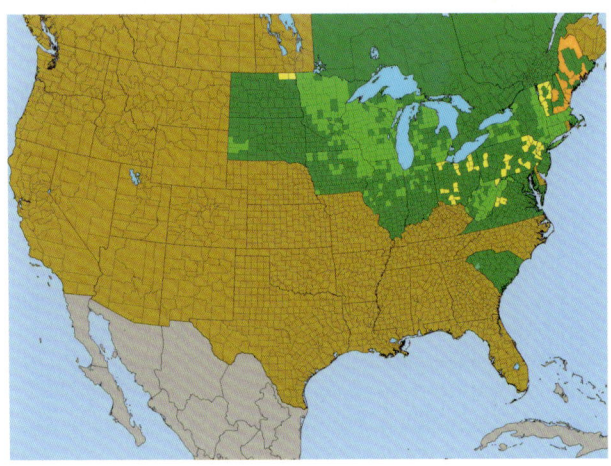

Seedling

Emerging mature plant in spring.
Leaves are covered in dense hairs
from emergence to the end of the
growing season.

Entire plant. Plants grow in small clumps, with
arching seedstalks.

Flower

Mature seed. Seedheads are covered
with dense hairs; *Bromus ciliatus*,
a wetland grass, is similar, but the
hairs are shorter (less than 0.04 in.)
and less dense.

Ligule

LOOK-ALIKE PLANTS:

Natives: Similar to Fringed Brome (*B. ciliatus*) of wet prairies, meadows,
and thickets. Arctic Brome (*B. kalmia*) grows only in well-drained sites.

Calamagrostis canadensis
Canada Bluejoint, Bluejoint Grass

FAMILY: GRASS (POACEAE, GRAMINEAE)

A fine-textured wetland grass for moist locations. Leaves have a slight bluish cast. Very erect in form, it combines well with flowers of various heights and colors. Its small seeds should be planted in fall or early spring.

Habitat: Moist to wet prairies, sedge meadows, marshes, fens, bogs, swales, streamsides, riverbanks, lakeshores, open wet woods
Garden Uses: Rain gardens, pond edges, wet swales, streamsides, wet prairie meadows

USDA Hardiness Zones: 3–6
Light Requirements: Full sun
Soil Types: Sand, loam, clay
Soil Moisture: Moist to wet
Bloom Time: June–July
Fall Color: Straw
Height: 4–6 feet
Life Expectancy: 10–20 years
Root Type: Fibrous, rhizomatous
Propagation: Moist stratify seed 30 days; root division in fall or spring
Aggressiveness: Low by rhizome and seed; can be high in some regions
Attracts: Game birds, songbirds
Deer Palatability: Medium
Distinguishing Characteristics:
  • Long, narrow, paper-thin ligule

• Leaves are relatively narrow compared to other tall prairie grasses, usually 0.25 inches wide
• Flowering stems are 0.0625 inches wide, half the diameter of most other prairie grasses, except Junegrass (*Koeleria macrantha*, p. 322) and Prairie Dropseed (*Sporobolus heterolepis*, p. 332) at 0.04 inch
• Seedheads are fluffy and airy; seeds are hidden inside white "chaff"
• Seeds are much smaller than other prairie grasses (except Junegrass) at 0.03 inch in diameter

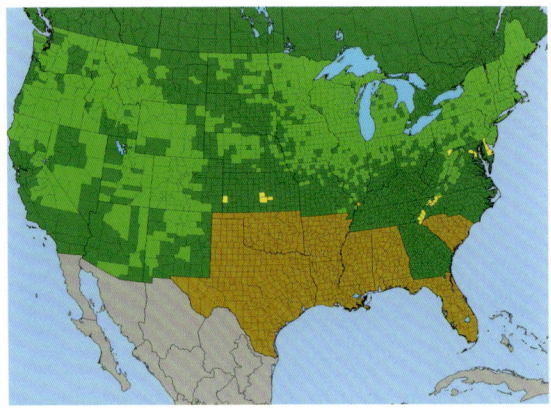

Seedling

Emerging mature plant in spring. Emerging stems narrow (less than 0.125 in. wide) with red bases and short lower leaves.

Entire plant. Stems are narrow (0.125 in.) compared to other taller native grasses.

Flower close-up

Flower. Individual flower stems are borne in whorls on a delicate, multiple-branched flower stalk.

Seed. Small seeds are borne in a narrow, soft-to-the-touch seedhead.

Ligule. Thin, nearly translucent ligule extends up from base approximately 0.125 in.

LOOK-ALIKE PLANTS:

Natives: Leaves and stems resemble Fowl Bluegrass (*Poa palustris*), a wetland grass with a more open inflorescence and leaves with curled "boat tips" on their ends.

Weeds: Reed Canary Grass (*Phalaris arundinacea*) grows in disturbed wetland habitats but has a more compact seedhead, wider leaves (0.5 inch), thicker stems (0.1 inch), and red rhizomes close to the soil surface.

Elymus canadensis
Canada Wildrye, Nodding Wildrye

FAMILY: GRASS (POACEAE, GRAMINEAE)

This short-lived, cool-season grass is commonly used as an early successional nurse crop in prairie meadow seedings. Not a great garden plant due to its short-lived nature. Best used as a fast-growing, early successional "nurse crop" in prairie meadow seedings. Sow at a rate of 3–5 pounds per acre in conjunction with other prairie flowers and grasses to provide quick vegetative cover. Grows in almost any soil except wet. Seedheads look great in dried arrangements. Many birds relish the large seeds.

Habitat: Dry to moist prairies, savannas, sand dunes, lakeshores, riverbanks, dry open woods
Garden Uses: Informal gardens, meadows

USDA Hardiness Zones: 2–9
Light Requirements: Full sun
Soil Types: Sand, loam, clay
Soil Moisture: Dry to moist
Bloom Time: July–August
Fall Color: Straw
Height: 4–5 feet
Life Expectancy: 3–5 years
Root Type: Fibrous, bunch former
Propagation: Dry stratify seed 60 days

Aggressiveness: Medium by seed
Attracts: Butterflies, songbirds
Deer Palatability: Low
Distinguishing Characteristics:
- Awns of mature seeds are curved (awns of other *Elymus* species are straight)
- Leaves often have a bluish cast
- Common on Great Lakes sand dunes

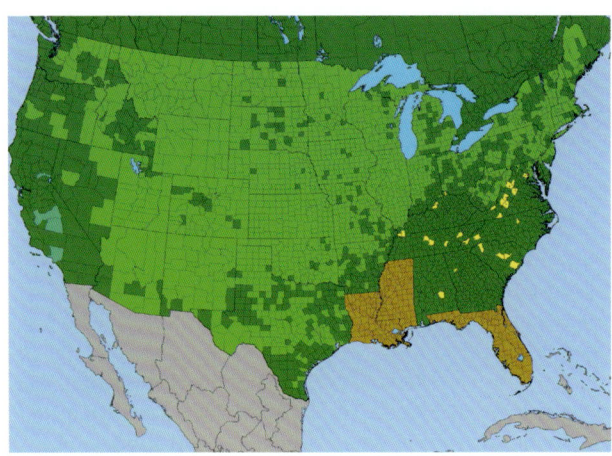

Seedling. Seedling virtually identical to *E. virginicus.*

Emerging mature plant in spring

Entire plant

Flower close-up. Flower not obvious and hidden within flower heads.

Flower

Seed. Awns at tips of seeds are curved backward, unlike the many other species of *Elymus.*

Ligule. Ligule is flat at the leaf, with no upward extension. It is easily confused with the nonnative weedy Quackgrass (*Elymus repens*).

LOOK-ALIKE PLANTS:

Natives: Virtually indistinguishable from Virginia Wildrye (*Elymus virginicus,* p. 314) until seeds mature. Seeds of Canada Wildrye (*Elymus canadensis*) exhibit recurved awns, whereas Virginia Wildrye has straight awns. Hairy Wildrye (*E. villosus*) and Riverbank Wildrye (*E. riparius*) are woodland grasses of similar form and stature. Hairy Wildrye leaves are covered with fine white hairs. The seeds of Riverbank Wildrye have straight awns rather than curved.

Weeds: Leaves and ligules are very similar to Quackgrass (*Elytrigia repens*). Canada Wildrye leaves are usually 50% wider at about 0.5 inch at the widest point; and its ligule will sometimes have a reddish hue. Wildryes are bunch-forming grasses, whereas Quackgrass has distinctive, widely ranging white rhizomes in the upper 4 inches of soil.

Elymus virginicus
Virginia Wildrye

FAMILY: GRASS (POACEAE, GRAMINEAE)

Similar to Canada Wildrye (*Elymus canadensis*) except awns of seeds are straight instead of curved. Short-lived, it is often used as a nurse crop in wet prairie seedings. As with Canada Wildrye, this short-lived grass does not make a great garden plant. Best used as a fast-growing "nurse crop" in wet prairie meadow seedings. Sow at a rate of 3–5 pounds per acre in conjunction with other wet prairie flowers and grasses to provide early soil cover. Birds eat the seeds.

Habitat: Wet prairies and meadows, streamsides, riverbanks, wet woods and thickets
Garden Uses: Informal gardens, meadows

USDA Hardiness Zones: 2–9
Light Requirements: Full sun to medium shade
Soil Types: Sand, loam, clay
Soil Moisture: Medium to moist
Bloom Time: July–August
Fall Color: Straw
Height: 4–5 feet
Life Expectancy: 3–5 years
Root Type: Fibrous, bunch former
Propagation: Dry stratify seed 60 days
Aggressiveness: Medium by seed

Attracts: Game birds
Deer Palatability: Low
Distinguishing Characteristics:
- Awns of mature seeds are straight (awns of Canada Wildrye are curved at maturity)
- Leaves often have a bluish cast
- Commonly grows in wet prairies, meadows, and riparian woodlands, and is more shade tolerant than Canada Wildrye (*E. canadensis*)

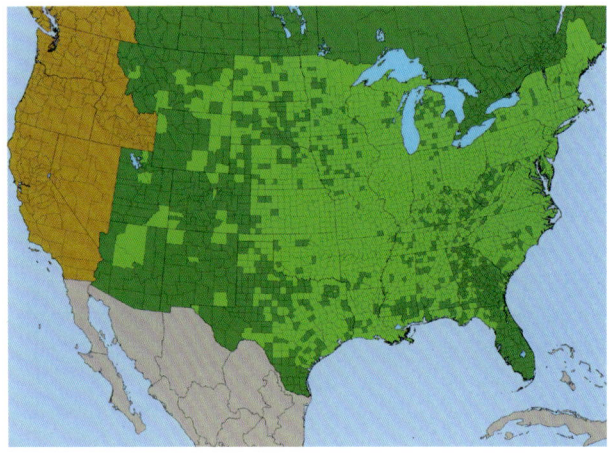

Seedling. Seedling virtually identical to *E. canadensis*.

Entire plant

Flower close-up

Flower

Emerging mature plant in spring

Seed. Individual seedheads less wide, at 1–1.5 in., than *E. canadensis*, at 1.5–2 in.

Ligule. Ligule essentially identical to *Elymus canadensis*.

LOOK-ALIKE PLANTS:

Natives: Virtually indistinguishable from Canada Wildrye (p. 312) until seeds are mature. Seeds of Canada Wildrye exhibit recurved awns, whereas Virginia Wildrye has straight awns. Hairy Wildrye (*E. villosus*) and Riverbank Wildrye (*E. riparius*) are woodland grasses of similar form and stature. Hairy Wildrye leaves are covered with fine white hairs. The seeds of Riverbank Wildrye have straight awns, making it difficult to distinguish from Virginia Wildrye.

Weeds: Leaves and ligules are very similar to Quackgrass (*Elytrigia repens*). Virginia Wildrye leaves are usually 50% wider at about 0.5 inches at their widest point. Wildryes are bunch-forming grasses, whereas Quackgrass has distinctive, widely ranging white rhizomes in the upper 4 inches of soil.

Eragrostis spectabilis
Purple Lovegrass, Tumblegrass, Petticoat Climber

FAMILY: GRASS (POACEAE, GRAMINEAE)

Demure and unobtrusive during spring and early summer, this denizen of dry soils explodes in a blaze of pink as the seeds ripen in August and September. Seedheads eventually break off and tumble in the wind as a dispersal mechanism. Mix clumps with low-growing prairie flowers for a stunning effect. Restricted to dry, sandy, and gravelly soils, a field of Purple Lovegrass resembles a pink cloud hovering over the ground. Petticoat Climber can also be a bit of a tribulation as its name indicates.

Habitat: Dry sand prairies, sand barrens, glades, dry roadsides
Garden Uses: Rock gardens, sand gardens, low-maintenance "wild lawns" on dry and sandy soil, seeded dry prairie meadows

USDA Hardiness Zones: 3–10
Light Requirements: Full sun
Soil Types: Sand, gravel
Soil Moisture: Dry
Bloom Time: July–September
Fall Color: Pink, red
Height: 1–2 feet
Life Expectancy: 5–10 years
Root Type: Fibrous, rhizomatous, bunch former
Propagation: Dry stratify seed 60 days; root division in fall or spring

Aggressiveness: Medium by seed
Attracts: Butterflies, songbirds
Deer Palatability: Low
Distinguishing Characteristics:
- Seedheads turn bright pink in fall
- Occurs as dense low-growing clumps, seldom more than 18 inches tall
- Wide, coarse leaves are covered with hairs at their bases and ligules

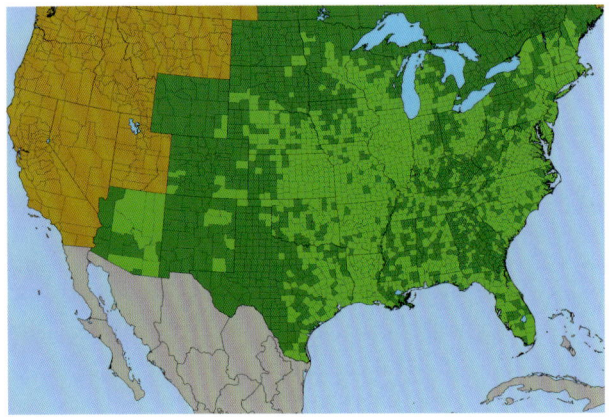

Seedling

Emerging mature plant in spring

Entire plant. Sprawling, distinctly clump-forming plants produce showy pink to purple flower heads in late summer.

Flower close-up

Flower

Seed. Pink or purple seedheads are open and airy, similar to some annual grasses, including Witchgrass (*Panicum capillare*) and *Agrostis* spp.

Ligule. Ligule has an abundance of long, upright hairs; a few long hairs extend sparsely up lower section of the leaf.

LOOK-ALIKE PLANTS:

Natives: Numerous other species of *Eragrostis* occur in North America, some similar in appearance. Purple Lovegrass has wider leaves (up to 0.5 inch) than most other *Eragrostis* species. Hairawn Muhly (*Muhlenbergia capillaris*), a southeastern native, has similar pink flowers and seedheads but does not have densely hairy leaves. It typically grows on clay soils and calcareous rocky outcrops in barrens and open woodlands, rather than on sandy soils in full sun.

Weeds: Witchgrass (*Panicum capillare*) is an annual grass that grows in dry habitats with a similar growth form, pinkish seedheads, and hairy stems and leaves.

Hesperostipa spartea (*Stipa spartea*)
Needlegrass, Porcupinegrass

FAMILY: GRASS (POACEAE, GRAMINEAE)

Aptly named, this grass is armed and dangerous! Its extremely sharp-pointed seeds turn a beautiful golden brown when ripe. As seeds ripen the stems arch gracefully downward and shimmer in the sun. The seed awns become like corkscrews when dry and literally drill the needlelike seeds into the soil.

Habitat: Dry to medium prairies, sand prairies, sand dunes, savannas, dry open woods, woodland edges
Garden Uses: Dry gardens, sand gardens, prairie meadows. Not a good candidate for children's gardens due to needle-sharp seeds.

USDA Hardiness Zones: 3–6
Light Requirements: Full sun
Soil Types: Sand, gravel, loam
Soil Moisture: Dry to medium
Bloom Time: April–June
Fall Color: Gold
Height: 2–4 feet
Life Expectancy: 10–20 years
Root Type: Fibrous, bunch former
Propagation: Moist stratify seed 30 days, or plant fresh for fall germination; root division in spring
Aggressiveness: Low by seed
Attracts: Butterflies, songbirds
Deer Palatability: Low

Distinguishing Characteristics:
- Leaf is tough and sinewy, with numerous prominent green veins running its length
- Large, sharply pointed seeds are golden brown with a ring of hairs near the base
- Seeds are surrounded with thin papery white husks
- Seedheads cause the stems to arc downward as they mature
- Long, corkscrew-shaped awns twist with humidity changes and drill seed into the soil

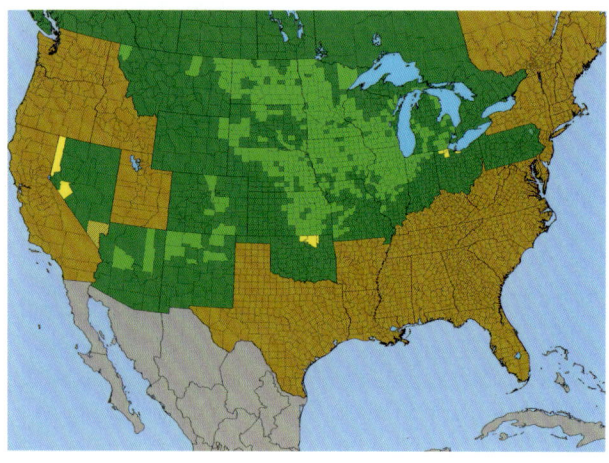

Seedling

Emerging mature plant in spring

Entire plant

Flower. Flowers are small and inconspicuous.

Flower close-up

Mature seed. Large golden seeds have a ring of hairs just below the needlelike tip; paperlike glumes surrounding the seed are readily visible, especially as the stems droop with the weight of mature seeds.

Ligule. Ligule, leaves, and stems are hairless; short extension of ligule above the base.

LOOK-ALIKE PLANTS:

Natives: Other species of *Stipa* have needlelike seeds. Most are smaller in stature and occur primarily in the northern Great Plains and Canadian prairies. Many species of *Aristida* occur in the Midwest, East, and South, and they are generally shorter with smaller, long-awned needlelike seeds.

Hierochloe hirta (*Hierochloe odorata*)
Northern Sweetgrass, Vanilla Sweetgrass

FAMILY: GRASS (POACEAE, GRAMINEAE)

Best known for its use when burned as a "smudge" in Native American purification ceremonies. Its shiny green leaves are harvested in summer, braided, and dried for later use. Widely distributed in northern and western latitudes in a variety of wetland habitats. Sweetgrass (*Hierochloe odorata*) is found in northeastern states and Canada.

Habitat: Moist to wet prairies and meadows, swales, lakeshores, edges of marshes, sedge meadows, moist woodlands
Garden Uses: Not recommended for most gardens due to its invasive nature but excellent as a groundcover in rain gardens

USDA Hardiness Zones: 3–9
Light Requirements: Full sun to partial shade
Soil Types: Sand, loam, clay
Soil Moisture: Medium to wet
Bloom Time: May–June
Fall Color: Straw
Height: 1–2 feet
Life Expectancy: 10–20 years
Root Type: Rhizome
Propagation: Moist stratify seed 30 days; root division in early spring
Aggressiveness: High by rhizome

Attracts: Songbirds
Deer Palatability: Low
Distinguishing Characteristics:
- Leaves emit a pleasant vanilla flavor when crushed
- Flowers and seedheads are produced on top of low-growing plants, one of our shortest native grasses, usually only a foot tall
- Ligule has a prominent diaphanous upward extension that wraps around the stem

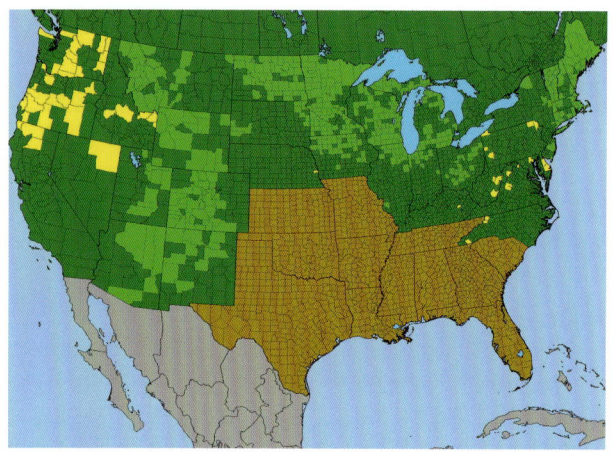

Seedling

Emerging mature plant in spring. Emerging plants exhibit short, wide, distinctly veined leaves.

Entire plant. Leaves hairless, creeps by rhizomes to form a loose patch, and the leaves and roots are highly aromatic.

Flower. Flower structures are smooth and shiny.

Mature seed

Ligule. Ligule hairless, with a thin papery extension rising 0.0625–0.125 in. above the leaf.

LOOK-ALIKE PLANTS:

Natives: Leaves are similar to Wildryes, but Northern Sweetgrass is readily distinguishable by its shorter stature and extended ligule.

Weeds: Quackgrass (*Elytrigia repens*) is similar in leaf and growth form but grows taller, has a distinctly different flower and seedhead, and does not have a long extended ligule.

Koeleria macrantha (Koeleria cristata)
Junegrass, Prairie Junegrass

FAMILY: GRASS (POACEAE, GRAMINEAE)

Great for rock gardens and very dry soils. Short and compact, Junegrass mixes well with low-growing, dry prairie flowers such as Prairie Smoke and Purple Prairie Clover. Its strong vertical element mixes well without dominating them. A cool-season grass, seed should be planted in early spring.

Habitat: Dry prairies, sand barrens, sand dunes, dry open woods
Garden Uses: Rock gardens, sand gardens, seeded dry prairie meadows

USDA Hardiness Zones: 3–8
Light Requirements: Full sun
Soil Types: Sand, gravel
Soil Moisture: Dry
Bloom Time: May–June
Fall Color: Straw
Height: 2–3 feet
Life Expectancy: 5–10 years
Root Type: Fibrous, bunch former
Propagation: Dry stratify seed 60 days; root division in early spring
Aggressiveness: Low by seed
Attracts: Butterflies, game birds, songbirds
Deer Palatability: Low

Distinguishing Characteristics:
- Leaves are thinner than most other prairie grasses at only 0.0625 inches, except Prairie Dropseed (*Sporobolus heterolepis*, p. 332)
- Flowering stems are 0.04 inches wide, half the diameter of most other prairie grasses, except Prairie Dropseed and Bluejoint (*Calamagrostis canadensis*, p. 310)
- Long compact upright seedheads are shiny silver green when young, turning white with age
- Grows primarily in dry, sandy soil

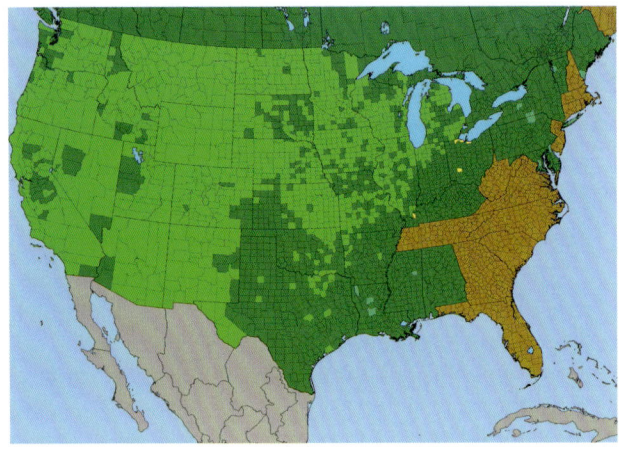

Seedling

Emerging mature plant in spring

Entire plant. Forms small clumps 3–6 in. across; leaves have numerous thin white veins and are covered in small, short hairs.

Flower close-up

Flower

Mature seed. Cylindrical seedheads have a silvery cast when young, turning straw color at maturity.

Ligule. Ligule hairless, with no extensions above the leaf at point of juncture with the sheath.

Panicum virgatum
Switchgrass, Prairie Switchgrass

FAMILY: GRASS (POACEAE, GRAMINEAE)

A popular prairie grass for gardens. Forms compact clumps and spreads slowly to form a sod. Plant with big, bold flowers and other grasses that can stand up to it. Creates high-quality bird habitat when planted thus in large meadows. Highest winter "standability" of all prairie grasses. Not recommended for most garden uses due to its tendency to self-sow and take over. Seed at low rates (0.25 pounds per acre or less) in prairie meadows. Sow at higher rates (up to 5 pounds per acre) to create wildlife cover. Easy from seed. Many selections are available.

Habitat: Dry to moist prairies, sand barrens, sand dunes, swales, riverbanks
Garden Uses: Groupings, informal gardens, prairie meadows

USDA Hardiness Zones: 3–10
Light Requirements: Full sun
Soil Types: Sand, loam, clay
Soil Moisture: Dry to moist
Bloom Time: July–September
Fall Color: Straw
Height: 3–6 feet
Life Expectancy: 20+ years
Root Type: Fibrous, sod former
Propagation: Dry stratify seed 60 days; root division in fall or spring
Aggressiveness: Medium by seed
Attracts: Butterflies, game birds, songbirds
Deer Palatability: Low

Distinguishing Characteristics:
- Seedhead is an open, conical panicle; seedheads of other tall prairie grasses are more compact and individual seeds are larger
- The upper membranous ligule is not pronounced as in Indiangrass (*Sorghastrum nutans*), but is densely hairy on the interior; Big Bluestem (*Andropogon gerardii*) has a small upper ligular membrane, with only a few hairs on the leaf blade

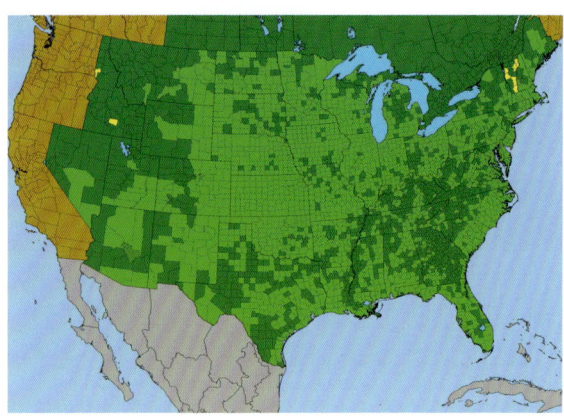

Seedling

Emerging mature plant in spring. New shoots are often reddish at base and hairless.

Entire plant. Forms distinct large clumps; leaves hairless, except for a small tuft at the ligule.

Flower close-up

Flower

Seed

Ligule. Small tufts of hairs occur at the ligule, with no fleshy extensions above the leaf-sheath junction. A fringe of hair extends about 0.125 in. above the leaf-sheath junction.

LOOK-ALIKE PLANTS:

Natives: Leaves resemble Big Bluestem (p. 302) and Indiangrass (p. 328) but can be distinguished prior to flowering by differences in ligules. Purpletop (*Tridens flavus*, p. 334) has a similar but narrower paniculate seedhead, and hairs around the *outside* of the ligule collar.

Weeds: Redtop (*Agrostis gigantea*) has somewhat similar reddish-tinged seedheads but only grows 2–3 feet tall. Its ligule is hairless, with a pronounced filamentous extension above the point of juncture of the leaf blade and grass stem.

Schizachyrium scoparium
(*Andropogon scoparius*)
Little Bluestem

FAMILY: GRASS (POACEAE, GRAMINEAE)

One of the showiest prairie grasses. Turns bright red in fall with shiny silver seedheads. Requires loose, well-drained soil. Easy from seed, can also be readily divided. One of the best prairie grasses for gardens due to its low stature, clumping form, and distinctive red fall color. Mixes well with flowers and other short grasses. Extremely effective when planted in drifts. Plant seed in spring or early summer. Many species of skipper butterflies utilize this plant as a larval food.

Habitat: Dry to medium prairies, savannas, barrens, glades, sand dunes, bluffs, occasionally in moist prairies on sandy soil
Garden Uses: Groupings, informal gardens, prairie meadows

USDA Hardiness Zones: 3–10
Light Requirements: Full sun
Soil Types: Sand, loam
Soil Moisture: Dry to medium
Bloom Time: August–October
Fall Color: Red
Height: 2–5 feet
Life Expectancy: 20+ years
Root Type: Fibrous bunch former; may produce rhizomes in damp soils
Propagation: Dry stratify seed 60 days; root division in fall or spring
Aggressiveness: Low by seed
Attracts: Butterflies, game birds, songbirds
Deer Palatability: Low

Distinguishing Characteristics:
- Foliage and stems turn bright red in fall, with silvery-white seedheads
- Upright, compact growth habit
- Leaves are flat, same as Broomsedge (*Andropogon virginicus*, p. 304), but brighter red in fall
- Seeds typically clustered more at the top of plant, are larger and less "fluffy" than Broomsedge, whose seeds are thinner and distributed along length of the plant
- Tends to grow on dry sandy soils, whereas Broomsedge prefers acid clay soils
- Does not generally tolerate moist soils, as will Broomsedge

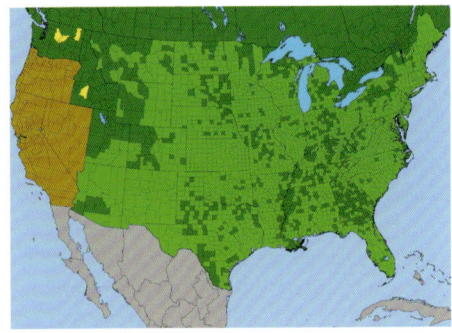

Seedling

Emerging mature plant in spring. New shoots are flattened with a few hairs; much less hirsute than Big Bluestem or Indiangrass.

Entire plant. Stems are flattened, not round; most other grasses have round or oval stems (Broomsedge also has flattened stems). Orchardgrass (*Dactylis glomerata*), a cool-season, nonnative forage grass, also has flattened stems and can be easily confused with Little Bluestem when small.

Flower close-up

Flower. Multiple flowers on flower stalk.

Mature seed. Distinctive white seedheads can be distinguished from Broomsedge by the seedstalk that extends above the point of attachment to the stem, whereas seeds of Broomsedge tend to be smaller and clustered nearer to the stem.

Ligule. Ligule is hairy, with little or no upward extension of leafy material.

LOOK-ALIKE PLANTS:

Natives: Broomsedge (p. 304) is virtually identical until seed forms in fall. Splitbeard Bluestem (*Andropogon ternarius*) is a southern native with similar leaves and growth form, but feathery, fluffy, silvery seedheads.

Weeds: Orchardgrass (*Dactylis glomerata*) is an imported, cool-season pasture grass that has flat stems and can appear similar to Little Bluestem in early spring.

Sorghastrum nutans
Indiangrass

FAMILY: GRASS (POACEAE, GRAMINEAE)

Beautiful golden seedheads make this a good choice for the back of garden or border. Has excellent color and form but tends to self-sow in the garden, so beware. Looks great when mixed with Little Bluestem (*Schizachyrium scoparium*, p. 326) and medium to tall flowers. Very adaptable, it makes fine bird habitat when seeded in larger meadows with other flowers and grasses. Easy from seed, it often volunteers on open soil.

Habitat: Dry to moist prairies, savannas, glades, dry open woods
Garden Uses: Back of borders, middle of island beds

USDA Hardiness Zones: 3–9
Light Requirements: Full sun
Soil Types: Sand, loam, clay
Soil Moisture: Dry to moist
Bloom Time: August–September
Fall Color: Gold
Height: 5–7 feet
Life Expectancy: 10–20 years
Root Type: Fibrous bunch former
Propagation: Dry stratify seed 60 days; root division in fall or spring
Aggressiveness: Medium by seed

Attracts: Butterflies, game birds, songbirds
Deer Palatability: Low
Distinguishing Characteristics:
- Bright-yellow, showy male flowers
- Golden-plumed seedheads
- Upper membranous ligule extends up the stem farther than other tall prairie grasses
- Leaf blade narrows dramatically at point of attachment to stem

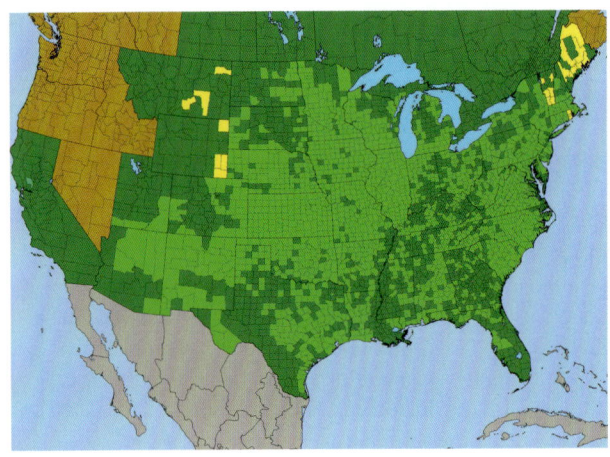

Seedling

Flower close-up

Emerging mature plant in spring. New shoots are reddish to purple and densely hairy (similar to Big Bluestem).

Entire plant

Flower. Flowers are among the showiest and most visible of all prairie grasses, with large, bright-yellow stamens.

Seed

Ligule. Hairless, with notable long notched extension (0.1875–0.375 in. long) above the leaf-sheath junction; leaf width narrows to approximately half of normal width as it approaches junction with the stem.

LOOK-ALIKE PLANTS:

Natives: Leaves resemble Big Bluestem (p. 302) and Switchgrass (p. 324) but can be distinguished prior to flowering by differences in ligules and narrowing leaf base.

Weeds: Plant may be mistaken for the nonnative ornamental grass Feather Reed Grass (*Calamagrostis acutiflora*) "Karl Foerster" when in flower and seed.

Spartina pectinata
Prairie Cordgrass, Ripgut

FAMILY: GRASS (POACEAE, GRAMINEAE)

The most robust wet prairie grass. Creeps rapidly by rhizomes, disqualifying it as a garden plant. Excellent for stabilizing sunny pond edges, stream banks, and earthen spillways. Plant plugs 2 feet on center to create a solid cover after three growing seasons. Plant in fall for best results. Seed is difficult to germinate.

Habitat: Moist to wet prairies, fens, sedge meadows, sandy lakeshores, swales, roadside ditches
Garden Uses: As hedgerow, along fenceline. Not recommended for gardens due to its tendency to spread rapidly by rhizomes.

USDA Hardiness Zones: 3–8
Light Requirements: Full sun
Soil Types: Sand, loam, clay
Soil Moisture: Moist to wet
Bloom Time: August–September
Fall Color: Gold
Height: 6–10 feet
Life Expectancy: 20+ years
Root Type: Rhizomatous or fibrous sod former; main roots are rhizomes, with minor fibrous feeder roots attached
Propagation: Moist stratify seed 90 days; root division in fall or spring
Aggressiveness: High by rhizome

Attracts: Game birds, songbirds
Deer Palatability: Low
Distinguishing Characteristics:
- Leaf blades are serrated and can cut skin; wear gloves when pulling this grass!
- Spreads by rhizomes to form large patches
- Thick rhizomes have sharp points similar to, but smaller than, Common Reed (*Phragmites australis*)
- Leaves are about half as wide (about 0.5 inch) as Common Reed (about 0.75–0.8 inch)

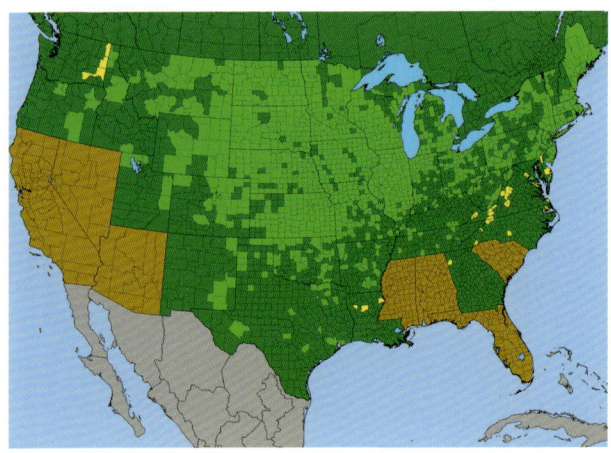

Seedling

Emerging mature plant in spring. Shoots emerge in matted groups due to rhizomatous nature of the species.

Entire plant. Seedstalks grow up to 10 ft. tall, with gracefully arching leaves up to 3 ft.; leaf edges serrated and sharp enough to cut skin.

Flower close-up. Stamens are reddish pink rather than the more common white or yellow of most grasses.

Flower

Mature seed. Individual seedheads are compact and tightly pressed together.

Ligule. Small tuft of hair at base of ligule opposite leaf; otherwise all leaves and stems are hairless.

LOOK-ALIKE PLANTS:

Natives: Eastern Gamagrass (*Tripsacum dactyloides*, p. 336) can form large sods that resemble Cordgrass clones; its leaves are also serrated but have a distinct white midrib, whereas Cordgrass has a green midrib.
Weeds: Common Reed (*Phragmites australis*) grows in similar wet habitats but has a large, fluffy seedhead and can reach up to 12 feet, forming extensive clones that eliminate most other plants.

Sporobolus heterolepis
Prairie Dropseed, Northern Dropseed

FAMILY: GRASS (POACEAE, GRAMINEAE)

Has the finest foliage of all prairie grasses. Grows in gorgeous emerald-green mounds, fabulous and well behaved in gardens and borders. Plant 2 feet on center in straight or curvilinear rows. Mixes well with many different prairie flowers, especially long-lived, taprooted species. Turns gold in fall. Plant seed in fall or early spring. Very slow from seed, requiring three to five years to bloom. Plants can be divided using a sharp spade or machete.

Habitat: Dry to moist prairies
Garden Uses: Excellent for front of borders, especially as a transition from lawn to prairie garden; informal gardens, prairie meadows

USDA Hardiness Zones: 3–8
Light Requirements: Full sun
Soil Types: Sand, loam, clay
Soil Moisture: Dry to moist
Bloom Time: August–September
Fall Color: Gold
Height: 2–4 feet
Life Expectancy: 20+ years
Root Type: Fibrous bunch former
Propagation: Dry stratify seed 60 days — plant seed in early spring or fresh in fall; root division in fall or spring
Aggressiveness: Low by seed
Attracts: Butterflies, game birds, songbirds
Deer Palatability: Low

Distinguishing Characteristics:
- Forms deep-green mounds of cascading leaves
- Seedstalks spray out from center of clump
- Leaves are thin compared to most other prairie grasses at 0.0625 inches wide
- Flowering stems are wiry and narrow, only 0.04 inch wide, similar to Bluejoint (*Calamagrostis canadensis*, p. 310) and Junegrass (*Koeleria macrantha*, p. 322)
- Flowers emit aroma of cilantro or popcorn

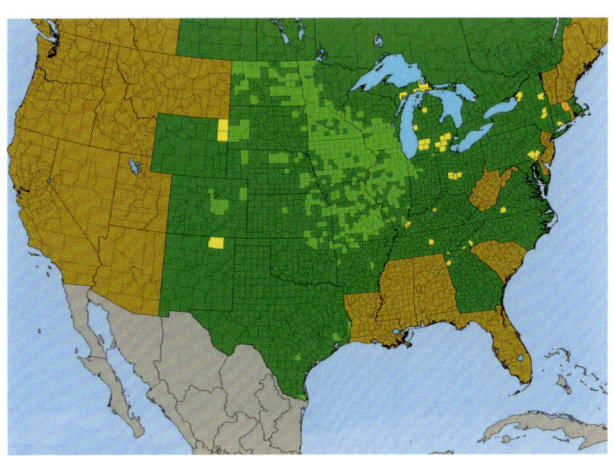

Seedling

Emerging mature plant in spring

Entire plant. Distinctive elegant clump form, with arching, narrow, emerald-green leaves.

Flower close-up

Flower. Flowers not showy or evident.

Seed. Seedstalks exude strong aroma sometimes likened to cilantro and buttered popcorn.

Ligule. A few small silky hairs occur at base or below the ligule; leaves are hairless.

LOOK-ALIKE PLANTS:

Weeds: Weeping Lovegrass (*Eragrostis curvula*), a South African native, has been planted for ornamental and erosion-control purposes in the southern United States. It has a similar leaf and growth form, but the seedheads are drooping panicles rather than erect.

Tridens flavus (Triodia flava)
Purpletop, Purpletop Tridens

FAMILY: GRASS (POACEAE, GRAMINEAE)

More common in eastern meadows than Midwestern prairies, this lovely grass regularly colonizes abandoned farm fields, often in association with Indiangrass (*Sorghastrum nutans*, p. 328). It is similar in appearance to Switchgrass (*Panicum virgatum*, p. 324) except has panicles of purple rather than golden-yellow seedheads. A nice grass for both gardens and seeded meadow plantings.

Habitat: Dry to medium meadows, open woods, old fields, roadsides, mostly in the eastern United States

Garden Uses: Specimens, small groups, back borders, prairie meadows. Tends to self-sow, so expect to do some weeding in garden situations.

USDA Hardiness Zones: 4–10
Light Requirements: Full sun to partial shade
Soil Types: Sand, loam, clay
Soil Moisture: Dry to medium
Bloom Time: July–September
Fall Color: Red
Height: 3–6 feet
Life Expectancy: 10–20 years
Root Type: Fibrous bunch former
Propagation: Dry stratify seed 60 days; root division in fall or spring

Aggressiveness: Medium by seed
Attracts: Butterflies, game birds, songbirds
Deer Palatability: Medium
Distinguishing Characteristics:
- Seedheads consist of bright purple panicles
- Ligule is encircled by hairs around the outside of the ligule collar
- Seed panicles droop

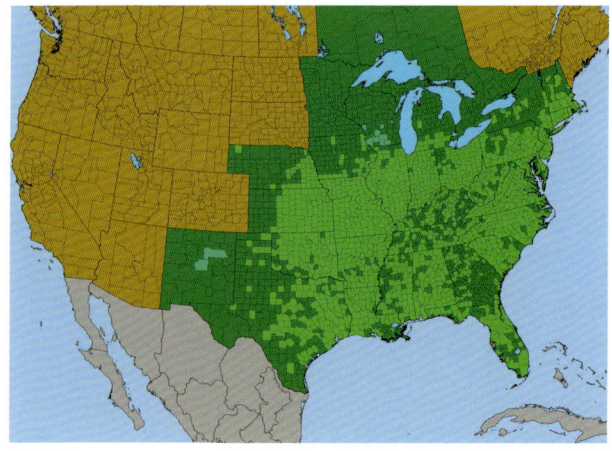

Seedling

Emerging mature plant in spring

Entire plant. Grows in small clumps with an open, airy habit.

Flower close-up   Flower

Early seed

Ligule. Ring of short hairs occurs at base of ligule, with hairs creeping upward on the upper side of leaf.

LOOK-ALIKE PLANTS:

Natives: Switchgrass (p. 324) is similar in size and form but has hairs inside the ligule collar. Redtop (*Agrostis gigantea*) has similar reddish seedheads but is typically shorter (2–3 feet); its ligule has a pronounced filamentous upward extension and is hairless, with no ring of short hairs at the base as with Purpletop.

Tripsacum dactyloides
Eastern Gamagrass

FAMILY: GRASS (POACEAE, GRAMINEAE)

This grand grass forms attractive, wide-spreading clumps. It is closely related to corn; their reproductive structures are similar, with large, distinctly separate male and female flowers. Although typically found in moist habitats, it grows well in a good garden soil. Mixes well with tall robust flowers such as False Indigo (*Baptisia* spp.), Joe Pye Weeds (*Eutrochium* spp.), and Silphium.

Habitat: Medium to wet prairies, swales, streamsides, river bottoms, seeps
Garden Uses: Informal gardens, prairie meadows, best used in standalone clumps or groups rather than in association with other species, due to its ability to spread by rhizomes

USDA Hardiness Zones: 5–10
Light Requirements: Full sun
Soil Types: Sand, loam, clay
Soil Moisture: Medium to wet
Bloom Time: July–September
Fall Color: Straw
Height: 4–8 feet
Life Expectancy: 20+ years
Root Type: Fibrous, rhizomatous sod former
Propagation: Moist stratify seed 30 days; root division in fall or spring
Aggressiveness: Medium by rhizome and seed
Attracts: Butterflies, game birds, songbirds
Deer Palatability: Low

Distinguishing Characteristics:
- Sharply serrated leaves have a distinct white line down the middle
- Bright orange male flowers are borne on long inflorescences up to 1.25 feet
- Female flowers are borne separately at the base of the flowering stalk
- Seeds are large, round, interconnected "cylinders" arranged in a line up the seedstalk
- Mature plants may form dense sods, covered in old grass leaves if left unburned or unmowed for many years

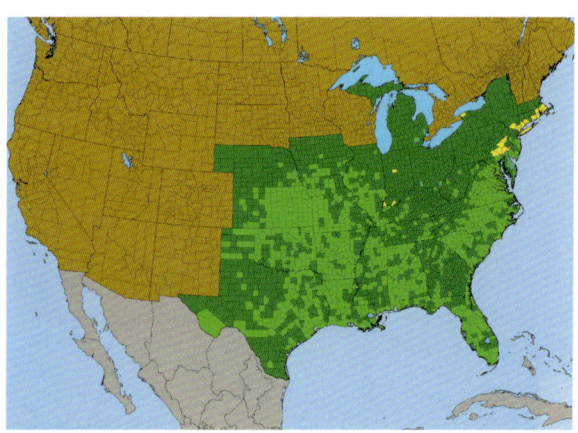

Seedling. Seedlings have unusually wide initial shoots, around 0.5 in. wide.

Female flower. Prominent reddish-purple pistils are borne on the lower portion of the flower stalk.

Emerging mature plant in spring

Entire plant. Forms large clumps up to 2 ft. wide, with thick seedstalks up to 10 ft. tall.

Seed. Large seeds are borne in tight clusters at the end of the seedstalks; each seed is pressed against the others and breaks apart as they dry and turn brown.

Male flower. Orange to purple stamens are borne on upper three-quarters of the long flower stalk and typically emerge after female pistils have bloomed on the plant.

Male flowers (*above*) with female flowers (*below*)

Leaf and ligule. Leaf (*left*) is widest of all prairie grasses at 0.5–1 in., with a distinctive white midrib. Ligule (*center*) has no hairs or other distinguishing structures.

LOOK-ALIKE PLANTS:

Natives: Prairie Cordgrass (*Spartina pectinata*, p. 330) forms a sod and has similar serrated leaves but does not form individual clumps. Prairie Cordgrass leaves have a green midrib rather than white.

Weeds: The leaves of Eastern Gamagrass resemble those of many species of Miscanthus, a genus of nonnative grasses planted for ornamental purposes. Miscanthus seedheads are fluffy plumes rather than hard cylindrical seeds.

6

Establishing a Successful Prairie Meadow

Establishing a prairie meadow by seed requires attention to specific procedures, outlined below.

Designing Prairie Meadows

Site Selection

Prairies and meadows require sunny, open sites with good air circulation. A minimum of a half day of full sun is necessary for most prairie plants to thrive and bloom. South-facing slopes receive more sun than level ground, are hotter and drier, and are well suited to prairie meadows. Similarly, west-facing slopes are subject to desiccation from hot afternoon sun and prevailing westerly winds. East-facing slopes are cooler than south-or west-facing slopes but receive sufficient sun to support prairie plants. Moderate north-facing slopes are OK, but steep north-facing slopes receive insufficient sun, stay cooler and moister in summer, and are not well suited to prairies.

Prairies can be seeded over septic fields and mound systems. Roots of herbaceous perennial flowers and grasses apparently do not grow into the pipes or pose a threat to their function. Deep-rooted prairie plants can utilize wastewater and nutrients it contains, preventing entry into groundwater — a great way to recycle wastewater using native plants.

Beware of seeding a prairie near adjacent fields and fencerows with aggressive, weedy plants, which can invade the site by underground rhizomes or seeds blown in on the wind. Problem neighbors include Quackgrass, Smooth Bromegrass, Tall Fescue, Johnsongrass, Canada Goldenrod, Tall Goldenrod, Canada Thistle, Crownvetch, Gray Dogwood, Sumac, Common and Glossy Buckthorn, Tatarian, Amur and Japanese Honeysuckles, and Multiflora Rose. Mow adjacent fields annually in early to midsummer before weeds go to seed. Make sure that any ground-nesting birds have fledged from their nests prior to mowing. A second mowing may be necessary if weeds rebloom and begin to produce new seed.

Weedy fields will require extensive site preparation to kill existing weeds and to reduce accumulated dormant weed seeds. This typically requires two to three growing seasons, using appropriate herbicides, smothering, or other options detailed in the section "Seeding Crop Fields," later in this chapter.

Tall and Short Prairies

Taller prairie plants tend to be more common on rich medium and moister soils. They are best planted on larger acreages or in background situations. Shorter plants predominate on poor, dry soils and are a good choice for small areas and around homes and buildings.

Short and tall prairies can be planted to create different landscape effects and habitat types. A short prairie mix can be seeded in the foreground, with taller species in the background to create a layered effect. For a prominent display of wildflowers, mix with shorter bunchgrasses, such as Little Bluestem, Prairie Dropseed, and Sideoats Grama. These low-growing, clump-forming grasses allow flowers to show off better, and their shallower roots are less competitive than those of deeper-rooted tall grasses. When tall prairie is planted on the windward side, seeds of tall plants will be blown by prevailing winds into the short prairie. Eventually the short prairie may become a tall prairie.

Two clump-forming grasses, Indiangrass and Little Bluestem, mixed with various flowers, make a good combination for taller prairies, as they allow room for flowers to develop. Larger, taller flowers are best able to compete with tall prairie grasses. Many low-growing taprooted flowers also do well with tall prairie grasses. They actually grow better and produce more flowers and seeds when grown with clump-forming grasses. Beware of seeding only one or two species of flowers in a given area, as there is no guarantee all will germinate and survive. If they do, most flowers do not have sufficiently dense root systems to discourage weeds by themselves. Taprooted flowers in particular do not occupy surrounding surface soil and are quickly invaded by weeds.

Tall prairie grasses tend to become dominant in prairie seedings over a period of many years. If a diverse display of forbs is desired in a prairie mix, avoid using Big Bluestem, Indiangrass, and Switchgrass.

The complementary root systems of prairie flowers and grasses work together to squeeze out weeds. By occupying different parts of the soil, they coexist as a tight-knit plant community. Seeding a wide variety of prairie flowers and grasses is the secret to creating low-maintenance prairie meadows. By understanding plant behavior and working with nature, the plants will do most of the work for you.

Early, Middle, and Late Successional Prairie Species

Revegetation of open soil by plants is referred to as *ecological succession*, in which first annual, then biennial, and eventually perennial herbaceous plants colonize the open space. In regions that receive 15 inches or more rainfall per year, shrubs and trees will eventually take over the landscape. Each plant has its own growth cycle and life span. Annuals, biennials, and short-lived perennials grow rapidly and are referred to as early successional species. There are relatively few annual and biennial prairie plants.

Biennials flower in the second year. It is important to include native biennials in a prairie-seed mix to provide early color and to compete with and suppress annual and biennial weeds.

Early successional perennial flowers and grasses usually reach flowering stage in the second or third year and tend to be dominant in years 3–6.

Mid-successional species include slower-growing perennials and grasses, which often require four to five years to reach maturity, with life spans of ten years or more.

Late successional species are slow-growing flowers and grasses that can live twenty years or more.

It is important to include a balance of early, middle, and late successional species in a prairie-seed mix. Biennials provide early color, usually fading by the fifth growing season. Short-lived perennial flowers often persist in the prairie for many years by reseeding. Middle and late successional flowers and grasses will become dominant as the prairie matures five to ten years after seeding.

The total number of species of a prairie seeding typically peaks between the eighth and twelfth year after sowing. This is also true of abandoned agricultural "old fields" that grow up to weeds and grasses, as well as other successional grasslands in the prairie region of North America. This is because biennial prairie species begin to fade after the first three to five years as perennial flowers and grasses create a sod, with little or no open soil for germination of biennials. Early successional perennials generally reach their peak in the fifth to tenth year. Long-lived prairie species begin to exert their dominance after about ten years, as many of them are slower growing and require that long to achieve full size. This is why Black-eyed Susans, other biennials, and short-lived perennials that are abundant in the first five years of a prairie seeding often become scarce after about ten years. The prairie will typically reach equilibrium among the long-lived flowers and grasses in about the fifteenth year.

TABLE 6.1.

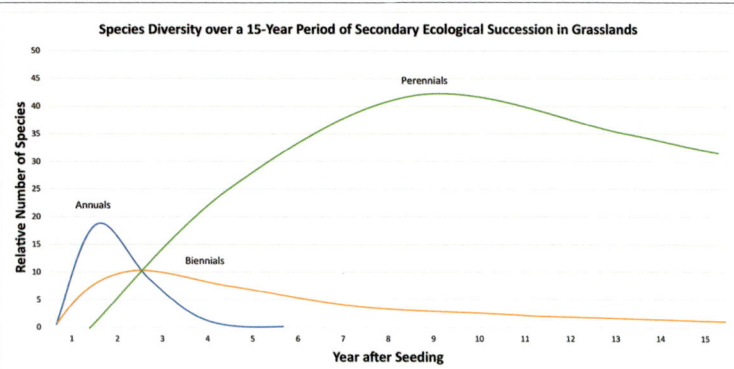

Spring versus Fall Seeding

Seeds can be sown in spring, early summer, or fall. The spring seeding window runs from early spring to mid-July. Most prairie grasses and many wildflowers are warm-season plants that germinate best after soil temperatures have warmed up, between midspring and early summer. This allows for additional weed control in spring prior to seeding, compared to early-spring seedings. July-seeded prairies typically have lower weed densities but will have less time to grow sufficient roots to survive winter. If seeding in July, include an annual nurse crop such as oats or Annual Rye to help stabilize the soil and reduce seedling mortality due to frost heaving over winter (see the section on nurse crops later in this chapter). Prairies sown after June 30 benefit from watering to stimulate rapid germination so seedlings can develop properly prior to winter.

Germination of most prairie species cease after July. Late seedings run the risk of not receiving rain for many weeks, and the germination window closes. This can be overcome by watering for the first four weeks, which is feasible for small areas but not large acreages. Many prairie flowers require exposure to cold, damp conditions in the soil over winter to break seed dormancy. These may not germinate the first season but will often overwinter in the soil and come up the following spring.

Fall "dormant seedings" will germinate the following spring. For many prairie flowers exposure to cold, moist conditions in the soil increases germination rates compared to spring seedings.. Many cool-season prairie grasses, such as Wildryes, Junegrass, Needlegrass, and Prairie Dropseed, do well with fall seeding. However, germination of warm-season prairie grasses such as Big Bluestem, Little Bluestem, Sideoats Grama, and Indiangrass is significantly reduced. To compensate, warm-season grass-seeding rates can be increased by 50%–100% in fall-dormant seedings.

If the primary goal is to establish a forb-rich prairie, seeding should be done in fall. For optimal germination of warm-season grasses, seedings in spring or early summer are best. Many prairie flowers also germinate well when seeded in spring and early summer.

Fall seeding pre-positions seeds to germinate in early spring and become established before the heat of summer. This is particularly beneficial on dry sandy and rocky soils, as well as heavy clay. Dry soils have more reliably available moisture in early spring and dry out later. Clay soils are often impossible to work in spring because of their high water-holding capacity. Clays are typically drier, more friable, and easier to work in fall. Heavy soils are often too wet to work in spring, and seeding may not occur until June, when the weather is often warmer and drier.

Fall seedings get a head start compared to late-spring or early-summer seedings, extending the effective growing season and allowing deeper root penetration into the soil. When clay dries out, it restricts downward root growth, causing stunted growth or death. Careful soil preparation and pre-planting weed control is essential with fall seedings, as there will be no opportunity to perform further weed control the following spring.

Fall seeding on erosion-prone sites requires addition of a nurse crop to the seed mix for soil stabilization. Seeding should be completed by the end of September, with the addition of nurse crops of oats or Annual Rye to provide sufficient protective cover. Nurse crops germinate and grow rapidly and will usually die over winter. Dead roots will continue to hold the soil until spring when prairie seeds begin to germinate.

TABLE 6.2.   SELECTED ANNUAL NURSE CROPS AND SEEDING RATES

Nurse crop	Spring plantings	Fall plantings
Oats	50–64 lbs./acre	128 lbs./acre
Annual Rye	5 lbs./acre	10 lbs./acre on level ground
		15 lbs./acre on slopes

Annual Rye should be used as a nurse crop only in USDA Zones 1–4. In zones that are warmer than that, Annual Rye will often survive winter and can become competitive with the germinating prairie seedlings the following spring. Annual Rye is mildly allelopathic and helps control weed growth, with limited effects on most prairie seedlings. However, it will have to be mowed regularly in the first growing season to prevent it from stunting growth of the small prairie seedlings if it survives winter. With climate change, Annual Rye is now surviving some winters in Zone 4. Oats reliably die over the winter and never come back the following spring. This makes it a safer choice as a fall nurse crop.

Steep slopes should be covered with a light-duty erosion-control blanket containing straw or excelsior (wood shavings) and staked into place. Jute netting can also be used to help hold seed and soil in place on slopes.

Fall seedings on sites not prone to erosion benefit from application of 1–2 inches of clean chopped and blown straw to retain moisture during spring germination. This is particularly beneficial when hand broadcasting seed directly onto an untilled soil surface in fall or spring.

Frost Seeding

A variation of fall-dormant seeding is frost seeding, in which prairie seeds are scattered on the surface of light snow or open soil in late winter. A phenomenon known as *frost heaving* occurs when daytime temperatures reach above freezing, the soil thaws, and refreezes at night when temperatures drop. This creates small openings in the soil into which seeds fall and then germinate in spring.

Frost seeding should not be done on crusted or deep snow. Seed is easily blown off crusted snow. Deep snow can melt rapidly during spring thaws, washing seed away. Frost seedings are best accomplished in early morning before the soil thaws, to avoid muddy working conditions. On large acreages, a broadcast seeder pulled by a tractor or all-terrain vehicle can cover large areas rapidly (see the section "Prairie-Seeding Methods").

An advantage of frost seedings is that warm-season grasses do not suffer the same loss of seed viability as with fall-dormant seedings. No increase in warm-season grass-seeding rates is needed. Many prairie flowers whose seeds require exposure to cold, damp conditions over winter to break dormancy show improved germination rates. Another advantage is that soil does not require tilling prior to seeding, saving a step while also preventing exposure of dormant weed seeds.

Site Preparation

Proper soil preparation is the single most important factor for success. The site must be completely free of weeds. If not eliminated, weeds can delay or prevent growth and maturation of the prairie.

Weeds and grasses can be killed using smothering, cultivating, smother cropping, herbiciding, or a combination of these techniques. On small areas of a few thousand square feet or less, smothering with black plastic, old rugs, pieces of old plywood, or cardboard is simple, effective, and requires no chemicals or special equipment. Lawn grasses can usually be killed in two to three months. If perennial weeds are present, a full year of smothering may be required to kill them. A few tough perennial weeds, including Canada Thistle, Carolina Horse Nettle, Field Bindweed, and Crownvetch require two consecutive seasons of smothering to kill.

If using herbicides, a broad-spectrum herbicide (glyphosate) will kill most grasses and weeds. If the above-mentioned tough perennial weeds, woody shrubs, or vines are present, it will be necessary to add the appropriate broadleaf herbicide to the mixture. The addition of a surfactant and water conditioner may be necessary to increase herbicide effectiveness. When using herbicides, *always read the label* and follow instructions.

The advantage of using herbicides as opposed to tillage is that every herbicide application kills existing weeds, plus weeds that germinate from dormant seeds. Tillage exposes new weed seeds that will germinate and reinfest the area.

Seeding Prairie into Former Lawns

Treat lawn with herbicide in early fall or midspring when actively growing, or smother it for 2–3 months during the growing season. After it is dead, prairie seed can be broadcast or machine-planted directly into dead sod in autumn, without tilling. This technique works well in fall, allowing seed to settle into the soil via frost heaving and then germinate the following spring. This method can also be used with seedings in spring and early summer if watered every morning for 60 days.

Prior to seeding in spring or early summer after killing turf, the dead sod will require thorough tilling to break up roots to create a good seedbed. Thatch can be reduced before tilling by burning when the grass turns brown after spraying. Tilling freshly killed sod is not easy. A good alternative is to spray the lawn two to three times every eight weeks from spring through late summer, or smothering for an entire growing season in preparation for a fall seeding. Dead grass roots will break down over summer, making tilling much easier. Alternatively, dead thatch can be removed using a sod cutter and the newly bare soil tilled to create a good seedbed.

To prepare a living lawn for seeding without herbicides, the upper 2–3 inches of sod can be removed using a sod cutter. Deep-rooted perennial weeds will not be removed by sod cutting and will need to be hand dug. After sod removal the area will be 2–3 inches lower than the surrounding lawn. It can either be tilled and sown immediately or filled with a weed-free compost or soil mixture prior to seeding.

An easier alternative to tilling is simply to scatter the seed over the dead sod in fall. The seed will work its way down into the soil over winter and germinate the following spring. Because the seed is not incorporated into the soil, 1–2 inches of clean, weed-free straw (Winter Wheat) should be applied over the seeded area to retain moisture and improve germination the next spring.

To kill existing turf using cultivation, rototill every week for three weeks when the lawn is actively growing. If rhizomatous perennial grasses such as Quackgrass, Smooth Brome, or Johnsongrass are present, a yearlong tilling program may be required, as for old agricultural fields (discussed below).

Seeding Crop Fields

Fields planted to corn, beans, and small grains in the previous year can be sprayed with glyphosate once in late spring or early summer after most weeds have emerged or germinated. If there are no perennial weeds on the site, it can be seeded one week after applying herbicide. If small grains were harvested in summer, the field can be sprayed once or twice in late summer and early fall, and then seeded that fall or the following spring, provided that all perennial weeds and grasses have been completely eliminated. If problem perennial weeds remain, further site preparation will be required.

If seeding many acres of prairie, the land can be rented to a local farmer and planted to Roundup Ready corn the first year, followed by Roundup Ready soybeans the second year. This cropping sequence will eliminate most, if not all, perennial weeds. If perennial weeds are still present after this two-year cropping sequence, a final application of glyphosate in mid- to late spring of the third year will be required prior to seeding. It should be noted that Roundup Ready crops are associated with the wholesale loss of plants in the Milkweed family and habitat for monarch butterflies and myriad other pollinators and insects. Neonicotinoid insecticides are also applied to most corn and soybean seedstock used in contemporary American agriculture, which can persist in the environment for many years. Although economical and effective, this method of site preparation has numerous ecological drawbacks.

Do *not* seed wildflowers or cool-season prairie grasses (e.g., Bromus, Calamagrostis, Elymus, Koeleria, Sporobolus) into cornfields treated with triazine herbicide within the previous two years. Although most warm-season prairie grasses can tolerate moderate levels of triazine, prairie flowers cannot. This herbicide breaks down in one to three years depending on soil type, precipitation, and the amount of herbicide originally applied. Warm-season grasses such as the Bluestems, Indiangrass, Sideoats Grama, and Switchgrass can all tolerate up to two pounds per acre and can be successfully seeded into soils with these levels.

Seeding Old Agricultural Fields

When establishing prairies on old fields, there is a choice: fast failure or slow success. Abandoned former agricultural fields contain a variety of perennial weeds and dormant weed seeds, which makes them difficult to prepare. An old field usually requires two full growing seasons to be properly prepared. This may seem like a long time, but a little patience at this stage is essential for success.

If perennial weeds are still alive at the end of the second growing season, leave the soil undisturbed over winter. An additional application of herbicide will be required the following spring, where vegetation has reached at least 12 inches. If only glyphosate is used in the spring application, seeding can occur one week after spraying.

When preparing old fields using herbicides only, without Roundup

Ready crops, begin by mowing in late July or early August. Allow vegetation to regrow to 12–16 inches (usually until four to six weeks after mowing) before applying herbicide. Add the appropriate broadleaf herbicide where necessary. Late summer and early fall can be cool, and air temperature must be 60 degrees Fahrenheit or higher for foliar herbicides to be effective. Dew must be allowed to evaporate prior to application, as it can dilute the solution and render it ineffective.

The following year apply herbicide three to four times. The first application should take place in midspring when vegetation is 12 inches tall, the second in early to midsummer, the third in late summer to early fall, with possibly a final midautumn application. The site can then be seeded, provided all perennial weeds have been eliminated.

To determine whether perennial weeds are dead, inspect the roots. If roots are white and firm, they are probably still alive. If yellow, brown, or black and soft, they are almost certainly dead. A 1-inch root segment of Quackgrass, Smooth Bromegrass, or Canada Goldenrod can resprout and rapidly become a 3-foot diameter patch in a single year.

Because old abandoned agricultural fields are subject to years or even decades of weed seed deposition, it is wise to prepare the site using a second full year of herbicide treatment to kill dormant perennial-weed seeds lurking in the soil. Follow the same procedure as outlined above for the second full year of treatment.

This herbicide method involves no tilling of the soil, which avoids exposing dormant weed seeds. Install prairie seeds by using a no-till prairie-seed drill or by broadcasting the seed on the bare soil in fall as a dormant seeding or in late winter as a frost seeding.

When using broadleaf herbicides, do not seed prairie flowers until one to five months after the final application, depending on the herbicide used. Broadleaf herbicides can remain active for many weeks or months. If prairie flowers are seeded too soon after spraying, seedlings may be stunted or killed by residual activity.

Using cultivation to prepare old fields is a labor and energy intensive method of site preparation, beginning in spring and continuing all summer through fall. Cultivate every two to three weeks at a depth of 4–8 inches using a spring-tooth harrow or rototiller to expose roots of weeds to the sun to kill them. A drawback is that new weed seeds are brought to the surface with every cultivation. If rhizomatous perennial weeds are present, waiting longer than three to four weeks between cultivations may allow them to reestablish. If hard-to-kill weeds are not completely eliminated after one year of tilling, a second year will be necessary.

Repeated herbiciding and cultivation are not recommended on slopes or other erosion-prone sites where soil loss and siltation may occur.

Smother crops, or "green manures," such as Buckwheat and Winter Wheat, can be used to prepare old fields for seeding without the use of herbicides. This is a two-year process. First, plow or till and level the field in late spring after all danger of frost has passed, as Buckwheat will be killed by frost. Sow seed of Buckwheat between June 1 and July 15 at a rate of 60 pounds per acre. Allow the crop to grow and develop a solid canopy. The thick leafy growth will smother most weeds by depriving them of light. When the Buckwheat has just finished flowering, usually about eight weeks after germinating, it should be plowed under or tilled into the soil. This will add organic matter to the soil while also helping to further control weeds.

After plowing under, allow two weeks for green material to break down. This prevents damage to the next crop from ammonia released during decomposition of dead vegetation. Plant Winter Wheat at a rate of 100 pounds per acre, at the recommended date for your climatic zone (usually sometime in September). Winter Wheat will germinate and grow in fall and come back the following spring. Allow it to grow, and then plow or till under in late spring when forming seedheads. Smooth the field to create a good seedbed, and seed a second crop of Buckwheat when danger of frost has passed. Let it grow and flower, mow down, and plow or till under in June. Wait two weeks, level the field, and the site is ready for a fall-dormant prairie seeding.

Do not allow Buckwheat or Winter Wheat to go to seed prior to tilling under, as seeds can germinate and compete with prairie plants. Planting smother crops is an excellent method of improving heavy clay soils and poor sandy soils. Adding organic plant matter helps break up clay soils and increases water and nutrient holding capacity in dry sandy soils.

Seeding Newly Disturbed Soils

Areas of bare soil resulting from recent construction may be relatively weed-free or full of hidden weed seeds and live roots waiting to sprout and grow. If all sod and roots have been removed, the exposed soil can be seeded immediately. Beware of subsoil sand and clay with little or no organic matter, as this makes a poor seedbed. The incorporation of compost or other non-woody organic matter into the subsoil will improve its nutrient- and water-holding capacity. Weed-free topsoil is another option.

If the weed status of recently disturbed soil is unknown, wait to see what comes up. When weeds emerge and grow to a height of 1 foot, the area should be sprayed with herbicide. If problem perennial weeds appear, the area will likely require a full season of treatment prior to seeding, as noted previously. An alternative to herbicides is cultivating or smothering the area for one full growing season.

Topsoil that has been scraped off and stockpiled to be respread on-site is often infested with weeds. A proactive option is to kill all vegetation prior to topsoil removal. The area to be scraped should be treated with appropriate herbicides every eight weeks during the growing season before moving soil. If topsoil is to be stockpiled for more than six weeks, it should be sprayed with glyphosate every eight weeks to prevent establishment of weeds that germinate from dormant seeds. If herbicides cannot be used, weedy vegetation can be eliminated using repeated tillage or cultivation, smother crops, or smothering on small areas.

Precautions need to be taken on erosion-prone sites. To avoid runoff and soil loss, the site should never be left bare for any length of time. Cultivation should be kept to a minimum. Repeated tillage is not recommended, as it can cause erosion. If the slope has a thick sod of weeds and grasses, it can be sprayed with herbicide for a full growing season with little danger of erosion. The dead roots will typically hold the soil in place for eight to twelve months before breaking down. If the process of using herbicide is started in spring and all weeds are completely dead by fall, prairie seed can be sown using a no-till prairie drill or broadcasting with a nurse crop no later than September 30. Seeds scattered into the dead sod will work their way down into the soil over winter with frost action. The nurse crop will germinate in fall to help hold the soil.

Following soil preparation on slopes, prairie seed should be sown immediately, with a nurse crop of oats or Annual Rye, and mulched (see section "Mulching" below). Erosion-prone sites that cannot be immediately seeded can be stabilized for weeks or months by seeding a cover crop of oats at 128 pounds per acre (3 pounds per 1,000 square feet). When ready to sow prairie seed, till the oats under, or spray with glyphosate to kill nurse crop and newly germinated weeds. Wait one week after spraying before seeding.

To ensure success in the battle against weeds, take no prisoners!

Prairie-Seeding Methods

Hand Broadcasting

Instead of using a lawn seeder, prairie seed is diluted with a larger volume of lightweight, inert carrier such as sawdust, peat moss, or vermiculite to ensure even coverage.

Mixing seed with a larger volume of an inert carrier such as sawdust ensures good coverage

The inert matter should be slightly damp, so seed sticks to it. For a 1,000-square-foot planting, 2.5 cubic feet (or four five-gallon buckets) of inert material is sufficient. A pickup truck full of sawdust from a sawmill will cover one acre. Sand is not recommended because of its excessive weight (see table 6.3 for quantities of inert matter to mix per area covered).

TABLE 6.3. **INERT MATTER VOLUMES FOR MIXING PRAIRIE SEED FOR HAND BROADCASTING**

Area	Cubic feet	Cubic yards	5-gallon buckets
1,000 sq. ft.	2.5	0.10	4
2,000 sq. ft.	5	0.20	8
3,000 sq. ft.	7.5	0.30	12
4,400 sq. ft. (0.10 acre)	10	0.40	16
11,000 sq. ft. (0.25 acre)	25	1.00	40
22,000 sq. ft. (0.50 acre)	50	1.80	80
33,000 sq. ft. (0.75 acre)	75	2.70	120
44,000 sq. ft. (1.00 acre)	100	3.60	160

Mix seed evenly into the inert material. Divide the mixed material into two equal piles, and spread each half across the entire area in two separate passes. After the first half of the seed mix is spread, broadcast the second half perpendicular to the first pass. When seeding in spring, rake or drag the area lightly to cover seed, 0.125 to 0.25 inches deep. Do not bury seed too deeply.

Broadcasting seed onto soil by hand   Raking seed into soil

Fall-dormant seedings do not require raking or dragging, as seeds will work their way into the soil over winter. Roll the seeded area with a roller, or drive across it with truck or tractor tires to firm seed into the soil. Never roll or drive on wet soil: seed and soil will stick to the roller or tires and can compromise the planting. Heavy soils are easily compacted if rolled when moist. Only sow seed when soils have dried out sufficiently to be easily worked.

Rolling seeded area to firm seed into soil   Scattering straw over seeded area

Drifting and Spot Seeding

All species in a prairie mix are typically spread uniformly across a planting area. Certain species may grow better in some locations than others given variations in soil type, moisture, and so on. To create variation or bursts of color, concentrations of one or more species can be seeded in drifts or patches in selected areas. This is most effective in high-visibility foreground areas and along walking trails.

A "general mix" that includes most or all species is first sown across the entire area. Additional flowers and/or grasses can be scattered in drifts, patches, or individual locations. When using a mechanical seeder, the species to be spot planted should be scattered onto the prepared seedbed in selected locations prior to machine seeding.

Sufficient grass seed must be sown uniformly to ensure that adequate sod develops to squeeze out weeds. Planting prairie flowers without grasses is not recommended, as flowers alone cannot discourage the growth of weeds and undesirable nonnative grasses.

Mechanical Seeders

For larger areas of one acre or more, seed can be sown more precisely and cost-effectively with specialized mechanical seeders pulled by a tractor. There are two basic types of mechanical seeders: no-till drills and broadcast seeders. A firm, level seedbed is required for good results with no-till seed drills.

Brillion Sure Stand double-box broadcast seeder, rear view of packer wheels

Close-up of Brillion Sure Stand's seed boxes for larger seeds (*right*) and small seeds (*left*)

Some small, pull-behind broadcast seeders can be calibrated to accommodate the low seeding rates required for prairie seedings. Models made by Agri-Fab can be pulled with a small tractor or all-terrain vehicle and can be calibrated for the low seeding rates typically used in prairie seedings. Although many small broadcast seeders can be calibrated to the low rates used for prairie seedings, coverage can be improved by diluting prairie seed with a larger volume of dry inert carrier such as pelletized lime, cracked corn, kitty litter, or Oil-Dri. The seeder should be calibrated using the carrier only prior to seeding. The prairie seed can then be mixed with the carrier and spread at the calibrated rate. The carrier is cheap, and pre-calibration ensures proper coverage.

A double-seed-box Brillion Sure Stand broadcast seeder works well for seeding worked-up soils. It has two boxes: a large box for fluffy grasses with awns and large-seeded flowers such as members of the genus *Silphium*, and a smaller box for flowers and small grass seeds without awns. It broadcasts seed rather than drilling it into the soil, creating a more natural effect (no rows). Two sets of heavy cast-iron rollers firm seed into the soil. For spring seedings, the Brillion requires a worked-up seedbed with loose surface soil. For fall-dormant seedings, it can be used on tilled or untilled, smooth, weed-free sites, such as soybean, small grain, or old fields previously prepared for planting. Cornfields need to be worked up and leveled prior to seeding with a Brillion.

No-till drills that successfully plant prairie-seed mixtures include Truax, Great Plains, and Tye. These drills use disklike "coulters" or wheel-like disk furrow openers to slice into the soil and create an opening for the seeds to be dropped into, and then packer wheels firm the seed into the ground. No-till drills can be set to various depths, with 0.25–0.5 inches the optimal depth for most prairie seeds. Soil does not require working up prior to seeding, which results in lower weed densities. A firm, smooth seedbed is required for good results with no-till seed drills.

Excessive plant debris, such as corn stubble or soybean stems, can foul the coulters so they do not cut into the soil to create a proper opening for the seed. Removal of plant debris may be necessary prior to seeding. A disadvantage of no-till drills is that, unlike when using broadcast seeders or hand-broadcasting, prairie plants are in rows that are noticeable for up to ten years after seeding.

Side view of Truax Flex II no-till drill

Rear view of Truax Flex II no-till drill with disc furrow openers (*right*) and packer wheels (*left*)

Truax Flex II no-till drill has three seed boxes: native grass and large forb center seed box (*left*), annual nurse crop seed box (*right*)

Truax Flex II no-till drill's seed box for wildflowers and small grass/sedge seeds, open on left

It is best to make two separate passes when seeding with a mechanical seeder. The machine should be calibrated to sow half the seed first time across the field and the second half on the second pass. This ensures good coverage and prevents running out of seed prematurely if the seeder has been improperly calibrated initially. In light, sandy soils, the heavy packing rollers of the Brillion provide better seed-soil contact than a no-till drill does. It is also beneficial to make a third pass on sandy soils with an empty Brillion seeder to further firm the soil and improve germination.

It is not necessary to mix prairie seed with inert material when using mechanical seeders, as doing so can clog the seeder. However, mixing prairie-grass seed with a small-seeded nurse crop such as Annual Rye helps prairie-grass seed flow smoothly through the machine. Do not exceed recommended seeding rates for Annual Rye nurse crops of 5 pounds per acre in spring and 15 pounds per acre in fall to prevent excessive competition from the nurse crop.

Hydroseeding does not ensure firm seed-soil contact and is generally not recommended for prairie seedings. Results with hydroseeded plantings tend to be variable, with numerous failures. An exception is fall-dormant seedings. Seed is applied in the hydroseeder slurry without mulch in a first pass, then covered with hydromulch in a second pass. This method is less effective with spring and early-summer seedings, as seeds will not work into the lower soil over winter. Minimal tackifier should be used when hydroseeding in spring or summer, as it binds the hydromulch tightly, obstructing emergence of seedlings of small-seeded prairie flowers.

Mulching

Between 1 and 2 inches of weed-free chopped and blown straw mulch helps retain soil moisture and increases germination. This is particularly important on dry sand and heavy clay. The straw should just cover the soil surface. On steep slopes and windy sites, erosion blankets are easier to use and more durable.

Apply straw at a rate of 3,000 pounds per acre (sixty 50-pound bales, or equivalent). On small areas use 75 pounds (1.5 50-pound bales) per 1,000 square feet. Never use hay for mulching a prairie planting, as it is invariably loaded with weed seeds. Winter Wheat typically has the lowest weed seed content of grain straw mulches. Avoid wood mulches such as shredded bark, bark nuggets, or sawdust, as they rob soil of nitrogen and can retard emergence and growth of prairie seedlings.

Protect slopes and swales from erosion with straw or excelsior-filled erosion blankets staked into the ground with 6-inch staples. A light-duty erosion-control blanket is best for normal applications on slopes and swales that receive modest runoff. Openings in the surface mesh should be 0.75 by 0.75 inches to allow successful emergence of prairie-flower seedlings. A heavier "high-velocity" erosion blanket may be required in swales and ditches subject to large volumes of fast-moving water. Although prairie-grass seedlings emerge readily through heavy blankets, some prairie-flower seedlings get caught in the thicker mulch blankets and might not emerge.

Watering

Watering requirements vary by season, soil type, and mulch cover. Sandy soils and unmulched clay soils require more frequent watering than loam or moist soils. Fall-dormant seedings require no watering, unless followed

Chopped and blown Winter Wheat straw mulch applied to newly seeded prairie

by a dry spring. Early to midspring plantings benefit from occasional watering, and late-spring and early-summer plantings from regular watering. Water when surface soil becomes dry. Mulched plantings require less watering. Water for fifteen to thirty minutes, or until soil surface is thoroughly moist. If water begins to run off, stop watering. Do not overwater! Overwatering can drown seedlings, especially on heavy clay soils.

Spring and summer prairie seedings watered regularly in the first eight weeks have higher germination and seedling survival. After eight weeks watering should not be necessary, except during hot, dry weather. Water only in the early morning. Watering during midday is often ineffective and wasteful because of evaporation. Afternoon and evening watering encourages high nighttime moisture levels that can lead to seedling loss by fungal attack.

Nurse Crops

Nurse crops such as Annual Rye and Oats can be planted with prairie seed to stabilize the soil and reduce weed growth. Nurse crops occupy the ecological niche that would otherwise be taken by annual weeds, thus reducing competition. Oats will not reseed. Annual Rye often reseeds in USDA Zones 5 and warmer and should be avoided in these climates.

Warning: Never use Winter Wheat, Winter Rye, or Perennial Rye as a nurse crop. Winter Rye and Wheat are allelopathic, producing chemicals in their roots that suppress germination of other plants. Perennial Rye is a fast-growing, cool-season turfgrass that can out-compete prairie seedlings and should never be used as a nurse crop. Annual Rye is mildly allelopathic and suppresses weeds without seriously affecting prairie-seed germination.

Successional Seeding of Grasses and Flowers

Competition from broadleaf weeds can be a significant problem in the first two to three years of a prairie seeding. A solution is to plant only prairie grasses in spring, followed by prairie flowers in fall. This method is easier and more cost-effective for smaller areas, given the logistics and expense involved with larger acreages. Warm-season prairie grasses germinate at higher rates when seeded in spring or early summer. Prairie grasses and weeds are allowed to grow the first summer. Mow at a height of 6 inches to prevent weeds from smothering slower-growing prairie-grass seedlings. In late summer, after prairie grasses have become established, the area is sprayed with broadleaf herbicide, as directed on the label. These herbicides kill only broadleaf weeds, leaving prairie grasses unharmed. If broadleaf weeds persist into early fall, they can be sprayed a second time while air temperatures remain above 60 degrees Fahrenheit.

In fall of the first growing season, prairie flowers can be overseeded into one-year-old grasses. On small areas of a few thousand square feet, flower seed can be scattered by hand, using sawdust or other inert material as a carrier. On larger areas of one acre or more, flower seed can be sown using a no-till drill or Brillion seeder, taking care not to disturb the soil or young prairie-grass seedlings. A no-till drill can be used but may disturb and uproot small one-year-old prairie-grass seedlings. Flower seed will germinate the following spring with less competition from broadleaf weeds. Once prairie flowers have germinated, broadleaf herbicides cannot be applied again, as they will kill them.

Post-Planting Management and Weed Control

Weed control in the first three to five years of a prairie planting is critical to success. Occasional well-timed mowing and burning (if possible) are essential. Once established, prairie flowers and grasses will suppress most weeds with minimum maintenance.

Care must be taken when mowing not to disturb grassland nesting birds between May 1 and July 15, earlier in southern latitudes. Occasionally, mowing may be necessary at these times. This may destroy some nests, but most birds will usually have time to renest and raise their brood.

First Year

Perennial wildflowers and grasses grow slowly, focusing on building their root systems rather than foliage. Annual and biennial weeds grow much faster in the first two years and must be controlled. When weeds reach a height of 12 inches, they should be mowed back to 6 inches (first year only). Most native prairie seedlings will not grow taller than 6 inches the first growing season. Mowing at this height will not damage them. Always mow when soil is dry. Mowing when soil is wet will compact it and can damage or kill seedlings. (See table 6.4 for some common annual weeds of the tallgrass prairie region.)

First-year prairie seeding being mowed at 6 inches to control annual weeds

Keeping weeds cut back in the first year prevents production of new weed seeds that can reinfest the planting. Mowing at 6 inches on a regular basis is essential. Left unmowed, weeds can grow 5–6 feet high and shade out slower-growing prairie seedlings. Prairie plants will be stunted, have shallower root systems, and be subject to mortality during their first winter.

A flail-type mower works best, as it chops up weeds and prevents clippings from smothering small prairie seedlings. Rotary mowers and sickle bar mowers are OK, but they often leave clumps of cut material which can smother seedlings. Rotary mowers can be problematic, as wet mowed material tends to accumulate on the mower housing and can be deposited in thick clumps, smothering seedlings. String trimmers, sometimes called weed eaters, are excellent for cutting back weeds on smaller plantings. These devices gently lay down cut material where it dries rapidly. Expect to mow weeds about once a month in the first year. Actual mowing frequency will depend on soil fertility, rainfall, and weed density and height.

Sometimes timely mowing may not be possible due to extended periods of rain. Weeds can grow well beyond the 12-inch height at which they should have been mowed. After the soil dries, mowing should be done in two separate phases. The first mowing should cut the weeds to half their height. Allow mowed material to dry out, and mow a second time at the 6-inch level. This two-step approach prevents seedlings from being smothered by excessive cut material that would have resulted from a single mowing.

TABLE 6.4.   COMMON ANNUAL WEEDS

	Scientific name	Common name	Notes
Broadleaf weeds	Abutilon theophrasti	Velvetleaf	Usually appears in first year only
	Amaranthus retroflexus	Pigweed	Usually appears in first year only
	Ambrosia artemisiifolia	Common Ragweed	Often persists in reduce form for years
	Ambrosia trifida	Giant Ragweed	Grows in rich, often damp soils
	Barbarea vulgaris	Yellow Rocket	Germinates in fall
	Chenopodium album	Lambsquarters	Usually appears in first year only
	Erigeron canadensis	Horseweed/Marestail	Germinates in fall
	Lactuca serriola	Prickly Wild Lettuce	Can sometimes be biennial
	Malva neglecta	Common Mallow	Can sometimes be biennial
	Medicago lupulina	Black Medic	Can be biennial or short-lived perennial
	Mollugo verticillata	Carpetweed	Sandy, dry soils; mat helps retain soil moisture
	Polygonum aviculare	Prostrate Knotweed	Beneficial; spreading mat retains soil moisture
	Polygonum convolvulus	Wild Buckwheat	Twining vine
	Polygonum pensylvanicum	Pennsylvania Smartweed	Prefers moist soils; usually appears first year only
	Portulaca oleracea	Purslane	Beneficial; spreading mat retains soil moisture
	Xanthium strumarium	Cocklebur	Native to both North America and Eurasia
Grassy weeds	Bromus tectorum	Downy Brome	Germinates in fall, as with other weedy Bromus
	Digitaria sanguinalis	Crabgrass	
	Echinocloa crus-galli	Barnyard Grass	
	Elusine indica	Goosegrass	
	Eragrostis cilianensis	Stinkgrass	
	Panicum capillare	Witchgrass	
	Panicum milliaceum	Wild Proso Millet	
	Setaria faberi	Giant Foxtail	
	Setaria pumila	Yellow Foxtail	
	Setaria viridis	Green Foxtail	

Note: For grassy weeds, all species except Bromus are warm season and germinate from midspring to early summer.

If a nurse crop was planted with prairie seed, it will be cut back along with the weeds. This will not jeopardize effectiveness of the nurse crop. Its purpose is to rapidly stabilize the soil, prevent erosion, and provide cover for newly germinated prairie seedlings. Once weeds reach 12 inches tall and need to be mowed, the nurse crop will have accomplished its job.

At the end of the first season, do not mow the seeding to the ground. Leave stubble in place to help protect young plants over the winter. Plant litter will catch and retain snow, which helps insulate the soil and reduce plant losses due to frost heaving.

Despite temptation, *do not pull weeds in the first year*. Prairie seedlings are small and easily pulled up along with weeds. If weed seedlings can be distinguished from prairie seedlings they can be delicately pulled, taking care not to disturb adjacent prairie seedlings. If a weed must be removed, it should be cut at the base with hand pruners.

Second Year

In midspring of the second year, the planting should be mowed close to the ground after cool-season weeds have initiated growth but before warm-season prairie plants have emerged (when buds of the Sugar Maple [*Acer saccharum*] are breaking open). This encourages soil warming, stimulating growth of predominantly warm-season prairie plants. Weeds will likely be dominant again in the second growing season as slower-growing prairie plants develop. Many prairie seeds germinate over a two- to three-year period and will emerge in the second spring. Burning at this time is not recommended, as it could kill newly germinated seedlings.

Second-year prairie dominated by biennial Blackeyed Susans

Second-year prairie seedings should be mowed at a height of 12 inches in late spring or early summer. Young prairie plants seldom exceed 1 foot high before summer of the second year and should not be harmed by mowing. Biennial weeds, such as Sweet Clover, Burdock, Wild Parsnip, Bull Thistle, and Queen Anne's Lace can be highly competitive. Just after biennials have completed blooming, cut or mow to a height of 12 inches. This kills many biennial weeds and sets back others significantly. If some rebloom later in summer, they should be mowed again just after second bloom is complete to prevent seed formation. If biennial weeds go to seed, they should be cut at ground level with pruning shears, carefully bagged, and removed from the site, taking care not to shatter the seedheads and reinfest the young prairie.

Two biennial weeds of particular concern are White Sweet Clover and Yellow Sweet Clover (*Melilotus alba* and *Melilotus officinale*). These extremely invasive weeds must be controlled. Their seeds remain viable in soil for decades and are stimulated to germinate by fire. In prairies to be

TABLE 6.5.   COMMON BIENNIAL BROADLEAF WEEDS

Scientific name	Common name	Notes
Arctium minus	Common Burdock	Large leaves can smother prairie seedlings
Carduus nutans	Musk Thistle	Usually not a long-term problem
Centaurea maculosa	Spotted Knapweed	Biennial to short-lived perennial; dry soils
Cirsium vulgare	Bull Thistle	Large leaves can smother prairie seedlings
Daucus carota	Queen Anne's Lace	Invasive; can persist for years once established
Dipsacus fullonum	Common Teasel	Can also act as a short-lived perennial
Melilotus alba	White Sweet Clover	Aggressive; seeds germinate stimulated by fire
Melilotus officinale	Yellow Sweet Clover	Aggressive; seeds germinate stimulated by fire
Pastinaca sativa	Wild Parsnip	Sap is phototoxic; can cause serious burns
Verbascum thapsus	Common Mullein	Large leaves can smother prairie seedlings

managed by controlled burning, this weed can become a long-term problem if not dealt with immediately in the second year and beyond.

If biennial weeds finish flowering and begin to form seed they will continue to ripen even after cutting. Mowing at this point will merely distribute seeds throughout the planting. Plants that have begun to form seed must be carefully cut (to prevent dispersal), bagged, and removed.

If either Sweet Clover reappears in the third or ensuing years, it will likely be on a limited basis and can be hand pulled or cut individually. Biennial weeds should never be allowed to go to seed. They can be hard to eliminate once a seed bank is established in the soil.

Third Year and Beyond

BURNING AND MOWING MANAGEMENT

Regular burning or mowing should begin in spring of the third growing season. Sufficient combustible plant material from the second year's growth is usually available to carry a fire. If fuel is insufficient to support a fire, mowing close to the ground and raking off debris should be substituted until the prairie produces more flammable plant material in ensuing years (see chapter 7 for details on burn timing and methods).

CONTROLLING PRAIRIE-GRASS DENSITY

As a prairie matures, some prairie grasses may become dominant and begin to crowd out some of the flowers. Prior to the prairie being settled by Europeans, bison and elk roamed the prairie, preferentially grazing grasses while leaving most flowers untouched. This reduced the vigor of prairie grasses and favored the flowers. Because most people do not stock bison and elk on their prairie plantings, tall grasses such as Big Bluestem, Switchgrass and Indiangrass may gradually become ascendant at the expense of shorter and less aggressive flowers.

The competitiveness of warm-season prairie grasses can be reduced by midsummer mowing when they are most actively growing and producing

Third-year prairie on dry sand on a hard-to-mow hillside in Verona, WI

Steve Hull's seventh-year prairie in Wisconsin

Ironweed blooming in the third year of fall-seeded detention basin/wet prairie mix in Sun Prairie, WI

seedstalks. Cutting them between ground level and 1 foot in height when they send up flower stalks in midsummer can set them back significantly, while favoring earlier-blooming prairie flowers and cool-season grasses. This is best done every other year, often for a number of years, to achieve a more balanced ratio of flowers to grasses. This will reduce the vigor of some late-blooming prairie flowers, as well as interrupt their growth at the height of summer activity in the year they are cut. They will typically recover and grow normally the next year.

Another option for controlling prairie grasses is to apply a grass-selective herbicide in early summer when these warm-season grasses are actively growing. These herbicides affect grasses only and will not damage prairie flowers. A single herbicide application usually does not kill most grasses but sets them back significantly and provides an opportunity for prairie flowers to flourish.

Some invasive perennial weeds may find their way into a prairie and will usually become evident in the second or third year. Although burning and/ or well-timed mowing will control most cool-season weeds and grasses, some perennial weeds may require other methods to obtain good control.

Some perennial weeds can be carefully pulled starting in the second growing season. The best time to pull is when soil is damp after a good rain. Roots of prairie plants should be sufficiently developed to resist being pulled up with the weeds. However, to prevent pulling up desirable prairie plants, place feet on both sides of the weed to be pulled to minimize soil and root disturbance in the adjacent area.

Taprooted weeds such as Canada Thistle, Crownvetch, and Carolina Horsenettle cannot be readily pulled, as their deep roots will break off and resprout. Densely rhizomatous weeds such as Quackgrass, Johnsongrass, Canada Goldenrod, Grass-Leaved Goldenrod, and Field Bindweed resist pulling, as their rhizomes are virtually impossible to fully remove. These weedy species can be carefully spot treated with specific herbicides. With the exception of grasses and goldenrods, these weeds are resistant to glyphosate, and specific broadleaf herbicides must be used to eliminate them. Please refer to table 6.6, "Problem and Invasive Perennials," for specific broadleaf herbicides for controlling glyphosate-resistant broadleaf weeds.

A time-saving and cost-effective method of controlling broadleaf perennial weeds in small areas is to hand treat with herbicides using "the glove of death." Wear thick rubber or neoprene herbicide (or chemical-resistant) gloves and place a large, absorbent cotton glove over one of the rubber gloves. Mix the recommended solution of appropriate herbicide in a small spray bottle or other non-spill container. Set the sprayer setting to "stream" rather than "mist" to avoid herbicide drift. Carefully saturate the cotton glove with the herbicide solution, but not so much that it drips onto adjacent plants. Grab leaves and stems of the offending weed and apply the herbicide by pulling gently upward along the stems. This will apply a concentrated dose of herbicide on the individual weed.

Do not touch other nearby desirable plants with the herbicide glove, as they can be harmed or killed. If desirable prairie plants are adjacent to or touching target weeds, push them aside using your feet or tie them up with string ahead of time to make sure they do not come in contact with treated plants. Use the glove-of-death method only on cool, windless days. Herbicides volatilize readily on warm days, and wind can blow mist onto adjacent plants, damaging or killing them. This is not a good choice for killing most weedy grasses, as they tend to have numerous stems intermingled with desirable plants and cannot be treated without risk of collateral damage to desirable plants growing among them. The method described below can be used to control unwanted tall grasses.

Gary Eldred, founder of the Prairie Enthusiasts, a nonprofit organization that preserves and manages remnant prairies in Wisconsin, Illinois, and

TABLE 6.6.   PROBLEM AND INVASIVE PERENNIALS

Scientific name	Common name	Notes
Problem broadleaf weeds		
Cirsium arvense	Canada Thistle	Highly invasive, deep-rooted rhizomatous perennial
		Cannot be pulled
		Resistant to glyphosate; use aminopyralid (Milestone), clopyralid (Stinger, Transline, Lontrel), or metsulfuron methyl (Escort)
Convolvulus arvensis	Field Bindweed	Highly invasive rhizomatous perennial
		Cannot be pulled
		Resistant to glyphosate; use quinclorac (Drive)
Plantago spp.	Plantain	Smothers plants at ground level
Securigera varia	Crownvetch	Highly invasive rhizomatous perennial
		Cannot be pulled
		Resistant to glyphosate; use aminopyralid (Milestone), clopyralid (Stinger, Transline, Lontrel), triclopyr (Garlon), or metsulfuron methyl (Escort)
Solanum caroliniense	Carolina Horsenettle	Resistant to glyphosate; use aminopyralid (Milestone)
Solidago canadensis	Canada Goldenrod	Highly invasive rhizomatous native
		Can be controlled by pulling rhizomes over a period of many years
Sonchus arvensis	Perennial Sowthistle	Invasive by seeds and rhizomes
		Can be controlled by pulling over a period of a few years
Trifolium repens	White Clover	Smothers plants with dense carpet
		Cannot be pulled out
Urtica dioica	Stinging Nettle	Native rhizomatous perennial
		Can be pulled up easily
Invasive rhizomatous grasses		
Bromus inermis	Smooth Brome	Cool season
		Not easily controlled by midspring burning
Elymus repens	Quackgrass	Cool season
		Readily controlled with midspring burning
Festuca arundinacea	Tall Fescue	Cool season
		Moderately controlled by midspring burning
Phalaris arundinacea	Reed Canary Grass	Cool season
		Best controlled with late-spring burning (June)
Poa pratensis	Kentucky Bluegrass	Cool season
		Readily controlled with midspring burning
Sorghum halepense	Johnsongrass	Warm season
		More common southward

Minnesota, uses an efficient and low-risk method for eliminating tall perennial weeds that suppress growth of prairie species. He knew he could not spray herbicides without harming nearby prairie plants but was frustrated with the invasion of his prairie by rhizomatous Canada Goldenrod (*Solidago canadensis*), so he bundled together five to ten adjacent goldenrod stems with binder twine at waist height when the plants were 5 feet tall in August. He then carefully applied herbicide to the upper leaves using a small spray bottle, with the nozzle set to "stream" rather than "mist," to prevent drift of herbicide onto adjacent desirable plants. The leaves are moistened with the herbicide so that no liquid drips onto prairie plants below.

This method can be used on Canada Thistle using specific broadleaf herbicides such as Clopyralid and Aminopyralid. Glyphosate is effective on most broadleaf weeds and grasses. It is also an excellent method for controlling invasive tall grasses such as Reed Canary Grass (*Phalaris arundinacea*) and Giant Reed Grass (*Phragmites australis*). It can be employed to reduce density of tall prairie grasses such as Big Bluestem, Indiangrass, and Switchgrass in situations where they have become dominant at the expense of lower-growing grasses and flowers.

A nonchemical method for controlling Canada Goldenrod was discovered by Dr. David Atwell when his prairie in Verona, WI, began to be invaded by this aggressive native species. He found that by mowing twice a year for two years, he was able to essentially eliminate Canada Goldenrod. To control the species, mow in late May to late June at a height of 6 inches, allow the prairie to regrow, then mow it a second time in late August to early September at a height of 12 inches. After two successive years of this mowing protocol, the Canada Goldenrod disappears. This method has been used by many others with equally effective results. It has also proved effective against Canada Thistle when practiced over a five year period. Perennial prairie flowers and grasses will be set back by the mowing but will survive and recover in the third year (sixth year for thistle), after the cessation of the mowing treatment.

CONTROLLING NONNATIVE COOL-SEASON GRASSES

Undesirable cool-season grasses can be killed or severely stunted by applying grass-selective herbicides such as sethoxydim, clethodim, and fluazifop-p-butyl after warm-season prairie grasses have gone dormant in fall or prior to their emergence in spring. Late in the season, after prairie grasses have turned their fall colors and their vegetation is dry, many nonnative cool-season grasses such as Quackgrass, Kentucky Bluegrass, and Smooth Bromegrass are still actively growing. They can be sprayed with grass-selective herbicides without danger of harming dormant warm-season prairie grasses. However, native cool-season prairie grasses which are still actively growing will be harmed or killed, so care must be taken when using this method. Air temperature should be between 60 degrees and 85 degrees Fahrenheit for foliar applied herbicides to be effective.

Spraying nonselective herbicides in your prairie when plants are actively growing is not recommended! Drift from spray can kill desirable native plants in a wide swath, leaving large dead patches that will grow up to weeds. One relatively safe method of spot-spraying problem perennial weeds is to install a shroud over a handheld sprayer nozzle. Cut the bottom off a one-liter soda bottle. Remove sprayer nozzle fittings from the sprayer wand,

and slide the bottle onto the wand, bottle neck up. Replace the nozzle and duct-tape the bottle neck onto the wand, just above the nozzle. To spray a weed, cover it with the soda bottle so that the bottle's base touches the ground to avoid drift. Spray the weed inside the bottle, taking care that no desirable plants are within the confines of the bottle. This method works best when weeds are short and not obscured by, or intertwined with, desirable prairie species. Using a properly selected broadleaf herbicide is effective on most dicotyledonous plants when applied to young spring growth that is 3–6 inches tall. Glyphosate is typically ineffective in killing weeds when applied in spring.

Dandelions seldom present a long-term threat to seeded prairies, despite appearances to the contrary in the first few years. As prairie plants develop, they will out-compete dandelions and most other common lawn weeds at both root and foliar level. By the fifth year after seeding, with proper post-planting management, few, if any, dandelions should be present in the prairie.

LOW-GROWING PERENNIAL RHIZOMATOUS WEEDS

Ground-hugging, creeping perennial weeds such as White Clover (*Trifolium repens*) and Creeping Charlie (*Glechoma hederacea*) can usually be controlled with regular midspring burning every other year. A hot prairie fire will burn the vegetation of these weeds to the ground, setting back their growth and favoring warm-season prairie plants. However, they cannot be effectively controlled by substituting close mowing and raking for burning. It is difficult to mow to ground level to remove all new spring growth, so these weeds recover more rapidly when mowed rather than burned.

Creeping Charlie will usually eventually be outcompeted by taller prairie flowers and grasses. White Clover is not so easily dispatched and can be a long-term problem. One strategy that has proved effective is to mow clover-infested prairies in mid- to late spring as close to the ground as possible at 1 inch. Apply a fertilizer which contains nitrogen but *no* phosphorus immediately after mowing, at about one-third to half the rate recommended for lawns on the label. This removes the nitrogen-fixing advantage of clover and effectively levels the playing field for prairie plants. The fertilizer can contain a combination of nitrogen and potassium, but phosphorus must be avoided. It stimulates the growth of leguminous plants such as clover, whose roots are capable of acquiring nitrogen from the atmosphere.

Within three to five years after seeding, the prairie should reach early maturity and the famous prairie sod will begin to develop. At this point prairie flowers and grasses should be well established, leaving little room for intrusion by weeds. Occasional weed removal or spot treatment with herbicides may be necessary.

Plateau-Resistant Prairie Seedings

The herbicide Plateau is sometimes used to control selected weeds in prairie seedings. The seed mix that is to be planted should consist only of those prairie flowers and grasses that are resistant to this herbicide. Most warm-season prairie grasses are quite tolerant of Plateau, with the notable exceptions of Switchgrass and Prairie Cordgrass. Most members of the Aster and Pea families are resistant, but this is not a universal trait and each species has different tolerances to the herbicide.

Many cool-season perennial grasses can be suppressed, but not eliminated, using a post-emergence foliar application of Plateau (see table 6.7).

TABLE 6.7.   **PERENNIAL COOL-SEASON GRASSES CONTROLLED BY POSTEMERGENT APPLICATION OF PLATEAU HERBICIDE**

Scientific name	Common name	Notes
Agrostis alba	Redtop	
Bromus inermis	Smooth Brome	Only suppressed at high application rates
Elymus repens	Quackgrass	
Festuca arundinacea	Tall Fescue	Readily controlled
Lolium perenne	Perennial Rye	
Phalaris arundinacea	Reed Canary Grass	
Poa pratensis	Kentucky Bluegrass	Only suppressed at high application rates

Better control of many cool-season grasses can be achieved by using grass-selective herbicides such as fluazifop (Fusilade), sethoxydim (Poast), and clethodim. These can be applied to control perennial cool-season grasses while warm-season prairie grasses are dormant in early spring or midautumn. If cool-season native prairie grasses are a component of the prairie, these herbicides should not be used unless you are willing to accept damage to native cool-season grasses.

Plateau is effective in controlling many annual and biennial weeds. However, some of the most problematic biennial and perennial weeds that afflict prairie seedings are not controlled by Plateau (see table 6.8).

TABLE 6.8.   **PROBLEM PERENNIAL WEEDS NOT CONTROLLED BY POSTEMERGENT APPLICATION OF PLATEAU HERBICIDE**

Scientific name	Common name
Cirsium arvense	Canada Thistle
Convolvulus arvensis	Field Bindweed
Securigera varia	Crownvetch
Melilotus spp.	Sweet Clover species
Solanum caroliniense	Horsenettle
Solidago canadensis	Canada Goldenrod

Prairie species that are proven to tolerate Plateau when used in established planting (one or more full years of plant growth) as a postemergent application include those listed in table 6.9.

TABLE 6.9. PRAIRIE PLANT TOLERANCE OF POSTEMERGENT APPLIED PLATEAU

Scientific name	Common name
Tolerant grasses	
Andropogon gerardii	Big Bluestem
Andropogon virginicus	Broomsedge
Schizachyrium scoparium	Little Bluestem
Bouteloua curtipendula	Sideoats Grama
Hesperostipa spartea	Needlegrass
Sorghastrum nutans	Indiangrass
Tripsacum dactyloides	Eastern Gamagrass
Intolerant grasses	
Elymus canadensis	Canada Wildrye
Elymus virginicus	Virginia Wildrye
Panicum virgatum	Switchgrass
Spartina pectinata	Prairie Cordgrass
Sporobolus heterolepis	Prairie Dropseed
Tolerant flowers	
Amorpha canescens	Leadplant
Baptisia alba	White False Indigo
Baptisia australis	Blue False Indigo
Coreopsis lanceolata	Lanceleaf Coreopsis
Dalea candida	White Prairie Clover
Dalea purpurea	Purple Prairie Clover
Desmanthus illinoensis	Illinois Bundleflower
Desmodium canadense	Canada Ticktrefoil
Desmodium illinoense	Illinois Ticktrefoil
Echinacea purpurea	Purple Coneflower
Lupinus perennis	Lupine
Oligoneuron rigidum	Stiff Goldenrod
Rudbeckia hirta	Blackeyed Susan
Symphyotrichum novae-angliae	New England Aster

Note: For Plateau-tolerant grasses and flowers, always consult the product label for recommended rates.

Plateau is best used in the early years of prairie establishment to control annual and biennial weeds, as well as selected cool-season grasses, provided that cool-season native grasses have not been included in the prairie-seed mix. A relatively small number of prairie forbs can be included in a seeding that is to be managed using Plateau herbicide, as many prairie flowers are not tolerant of its effects. Therefore, Plateau is not an option for managing diverse prairie seedings.

CONTROLLING WOODY PLANTS

Prairies established from seed normally require three to five years to develop a dense sod of native flowers and grasses. During these early years, the prairie is susceptible to invasion by woody plants from nearby trees and shrubs. Once established, they can be difficult to kill. Burning and mowing management will kill most conifers but will only set back deciduous woody plants to ground level. They will typically resprout and remain a component of the prairie. Left unburned or unmowed, deciduous woody plants will grow up and eventually shade out sun-loving prairie plants.

Once established, woody plants are hard to remove by pulling or digging. They can be controlled in three ways: (1) cutting and immediately treating the cut stumps either with concentrated glyphosate (40 percent) in fall or winter or with brush-killing herbicides any time except spring, (2) basal bark treatment during the dormant season using a silvicide, such as Garlon 4, mixed with crop oil or diesel fuel to penetrate the bark, or (3) mowing in midsummer (August). Cutting and treating individual stems with herbicides is a highly targeted and effective technique that does not disturb adjacent prairie plants. Basal bark treatment in winter can result in collateral damage to prairie plants at the base of the treated woody stem due to translocation of the herbicide into the soil. Summer mowing cuts back woody plants at the most vulnerable stage of their annual growth cycle but is nonselective. This will do temporary damage to prairie flowers and grasses, but they typically recover in the following year. Summer mowing seldom kills woody plants, but it weakens them and prevents them from taking over.

Controlling woody plants using herbicides is best accomplished in late fall or winter, when plants are dormant. It is not effective in spring or early summer, as sap in the plants is moving upward and will not translocate herbicides down into the roots to kill them. In fall the sap is moving downward, making this is an excellent time to treat woody invaders.

When using the stump treatment method, cut woody stems to the ground and immediately apply a concentrated solution of the appropriate herbicide. Concentrated glyphosate will kill almost all deciduous trees and shrubs when applied in fall or winter. A few species may require use of specific broadleaf herbicides designed for killing woody plants. Apply the herbicide drop by drop in a ring, a half inch inside the bark. This is the cambium layer, which contains the food and water transport systems of woody plants. It is not necessary to apply herbicide to the interior wood of the stump, as this region consists of dead wood and will not transport the herbicide down to the roots. Treat the cut stump immediately after cutting. Do not allow the stump to dry out after cutting, as it will develop a callus and be less likely to absorb the herbicide. The plant may not be killed as a result. Coniferous species will be killed simply by cutting them down and do not require herbicide treatment of their stumps.

Basal bark treatment works on young saplings that have not developed a thick outer bark layer. The mixture of silvicide and penetrating oil enters through the bark into the cambium, where it is translocated down to the roots to kill the woody plant. Be careful to avoid excess application of the herbicide mixture, as it can move through the soil and kill other desirable prairie species.

Never place herbicides in open containers, as they can spill and damage adjacent desirable vegetation and cause a safety problem. Pour herbicides carefully into a clean, plastic dish detergent bottle with a pull-top dispenser valve on top. The valve can be opened so one drop of herbicide can be dispensed at a time. This avoids spillage and ensures that the minimum necessary quantity of herbicide is used. Always read the herbicide label and follow all recommended safety procedures.

7

Burning Your Prairie Safely (and Having Fun, Too!)

The North American prairie evolved under the influence of regular wild-fires. When managing prairie meadows and gardens, a controlled burn is the magic bullet for discouraging many unwanted invasive plants. A good prairie fire is a special rite of spring. These tips will help ensure that your prairie burn is both inspirational and safe!

Burning removes accumulated plant litter from the previous year's growth, exposing soil to the sun so it warms more rapidly. Higher soil temperatures favor growth of warm-season prairie plants while discouraging cool-season weeds and grasses. This tips the balance in favor of warm-season natives. Numerous studies have shown that midspring burning of Midwestern prairies increases growth rates, total biomass, and flower and seed production.

To save time mowing firebreaks every time you want to burn, design your prairie to take advantage of as many existing firebreaks as possible, such as roads, driveways, ponds, and streams. These serve as built-in fire-breaks. You can also create mowed lawns or trails as boundaries.

Where firebreaks have to be mowed around the prairie to remove flammable material, make sure they are sufficiently wide to prevent fire from escaping. A general rule of thumb is to make your firebreak 15–20 feet wide. Firebreaks can also be disked or tilled on level ground during the growing season to expose soil and prevent plant growth. However, open soil encourages annual weeds and can lead to erosion on slopes.

Do not plant prairies near conifers or other trees easily damaged by fire. Fire-resistant native trees, such as Bur Oak (*Quercus macrocarpa*), White Oak (*Quercus alba*), and Shagbark Hickory (*Carya ovata*) can be scattered widely within your prairie. Including more than the occasional tree in a prairie is not recommended. Plant one specimen tree per acre, or a grouping of three to form a "destination grove" within the prairie. Most prairie plants grow best in full sun and do not compete well in shade. Because birds perch in trees, they drop seeds of invasive woody plants into your meadow.

Be sure to mow around less fire-tolerant trees, removing all flammable material before burning. Plant a low-maintenance fescue turf 10–20 feet in diameter around each tree. Lawn grass makes a fairly good firebreak that can be mowed monthly, or once in late fall, to reduce flammable thatch and control invasion by unwanted woody plants.

Divide large prairies into smaller, individual, burn-management units by including firebreak trails between each unit. Prairie fires on small areas are easier to manage than large ones. It is recommended that each unit be burned every two to three years to maintain plant and animal diversity. Burning every spring tends to favor warm-season prairie grasses over most flowers, reducing flowering density and pollinator habitat.

A number of butterflies and moths overwinter as pupae, adults, and even caterpillars. Undisturbed plots help preserve them, plus provide protective winter cover and spring nesting sites for grassland birds. Birds also forage for seeds and invertebrates in the prairie all winter long. A cleared-off, cleaned-up landscape may appear attractive but is a habitat wasteland similar to a lawn in terms of providing homes for garden visitors.

Each burn unit will respond differently to the management cycle, creating changing patterns of wildflowers and prairie grasses within the same planting. Creating different management zones makes burning more manageable and maximizes biodiversity.

Plant low-maintenance, cool-season grasses in your firebreaks, such as fine fescue grasses in northern climates and turf-type tall fescues in southern zones. Avoid Kentucky 31 and other unimproved tall fescues, as they can be invasive. Cool-season grasses green up in early spring, and their new green growth impedes the spread of fire. However, these are not impervious firebreaks, as dead thatch in the grass remains flammable.

Mow firebreaks closely in fall, leaving a minimum of potentially flammable stubble. Cut plant material will decompose over winter and essentially become nonflammable by spring.

Dry plant material mowed in spring must be raked off just prior to burning as it will carry fire. Fall-mowed material does not need to be raked unless it remains dry and flammable.

Raking firebreaks prior to a burn. Fine fescue grasses were seeded on the right to serve as partial firebreak; Indiangrass on left serves as fuel load. Photo courtesy of Dennis Healy.

Neil Diboll with drip torch, flapper, and backpack water-pump sprayer — tools of the trade for the prairie pyro! Photo courtesy of Dennis Healy.

Fire Equipment and Protective Clothing

Always wear appropriate clothing and have proper fire-control equipment on hand. Never wear flammable clothing or easily melted synthetics such as nylon or rayon. Molten fabrics stick to skin and can cause serious burns. Denim jeans and long-sleeved, cotton work shirts are relatively fire resistant and afford moderate protection. Fire-resistant Nomex suits provide optimal protection. Nomex gear is recommended when burning large prairies with heavy grass fuel loads and high-intensity heat is anticipated.

Water is an essential element for fire control. If conducting a small prairie burn in your yard, have a garden hose with a nozzle at the ready. Wet down firebreaks *before* lighting the first match. Pay careful attention to your rear firebreak when backfiring, as flames driven by headwinds can breach it. Be aware that during warm, dry, and windy weather, firebreaks will dry out quickly. The heat of the fire will also dry out vegetation in mowed firebreaks and can burn into residual dead thatch or other combustible material. Mowed firebreaks should be closely monitored and may need to be remoistened as the fire progresses.

Backpack water-pump sprayers are essential during large burns to prevent spot fires from jumping fire lines. These portable four-gallon tanks weigh 35–40 pounds fully loaded. Always test your equipment to ensure it is in proper working order before starting the fire.

On small areas, fires can be spread by raking and dragging flaming dry grass along the interior of the firebreak. A drip torch is an essential tool when burning larger areas. This handheld metal canister is filled with a mixture of 80% diesel fuel and 20% gasoline. A metal tube with a "fire

Laying down fire line with drip torch in prairie/oak savanna. Photo courtesy of Dennis Healy.

Controlling a backfire with a flapper, which is essentially a truck tire mudflap on a handle. Photo courtesy of Dennis Healy.

Controlling a backfire with a backpack water sprayer using a slide-action trombone handpump. For optimal control, aim the stream of water at the base of the fire. Photo courtesy of Dennis Healy.

pad" on the end extends 2–3 feet from the top of the canister. The tube is pointed at the ground to deliver fuel to the fire pad, which retains a flame to ignite the fuel. The flow of fuel is controlled with a pair of valves that regulate fuel output and air intake. This device can be used to rapidly lay down a fire line just inside the firebreak.

Fire Behavior

Four factors determine fire behavior: fuel load, air temperature, relative humidity, and wind speed. Together these influence the rate of spread and ease of control of a prairie burn.

Fuel Load

Mature prairies that have not been burned for two or more years may have a heavy fuel load composed of dry dead grass. They can burn rapidly because oxygen is readily available in the standing fuel. The rate of fire spread and height of flames can be significantly reduced by mowing the prairie with a heavy-duty, tractor-mounted mower just prior to burning. Flammable material will now be concentrated on the ground, and the resultant fire will spread more slowly, have lower flames, and be easier to control.

Air Temperature

The warmer the air temperature, the hotter the fire will be. Burning in the evening and at night results in cooler, slower-moving fires. Burning prairies with heavy fuel loads on a warm day can result in high flames, fast rates of spread, and potential loss of control.

Relative Humidity

Relative humidity is a measure of moisture in the air. Fire spreads more slowly under more humid conditions. Burning in the evening when humidity is higher can often be the most important factor in determining the rate of fire spread. Evening dew reduces flammability of dead plant material, even rendering it nonflammable. A prairie that would erupt into flames when burned in the afternoon may not burn at all when soaked in dew at night.

Wind Speed

Wind speed is a critical factor in determining fire intensity and rate of spread. Wind velocity is typically lower in the evening. Lower wind speeds and temperatures, combined with higher humidity, make evenings the safest time to burn. When wind speeds exceed five miles per hour, burning is not recommended, especially if fuel load is high and flammable structures are nearby. This is of particular concern when air temperature is high and relative humidity low. Attempting a prairie burn under such conditions often ensures a visit from the local fire department.

Many municipalities have strict regulations and specific hours of day when burning is allowed. Some mandate burning in evenings only. Always check your local regulations and burning permit requirements. Burning permits are issued primarily so people read and understand the conditions under which burning can be conducted safely.

Burning against the wind is almost always the best strategy. Never start a fire so it moves in the same direction as the wind, as it will move rapidly and can jump firebreaks. Burning into the wind, known as a *backfire*, ensures that the line of flames moves slowly along the ground as the wind pushes against its advance. This allows establishment of a secure rear firebreak, a critical first step in a controlled burn.

Continue to burn against the wind until the fire has consumed the fuel within the first 20–50 feet from the original firebreak. This burned area is called a *blackline*, and is the line of defense in case the wind reverses direction unexpectedly. New fire lines can now be established along one or both flanks, perpendicular to the blackline. After all three sides of a prairie have been burned in, the remaining portion can be ignited with the wind, referred to as a *headfire*. Flames advance rapidly, meeting the previously burned blacklines, at which point they will extinguish themselves for lack of fuel.

Burn downhill to maintain control. The heat of fire moves upward and away from the fuel below it. A downhill fire spreads more slowly than a fire on level land, making it more controllable. When a fire burns uphill, it preheats fuel ahead of it, making it burn faster and hotter. It can actually create its own wind, further increasing the rate of spread.

Burn when soil moisture is moderate to high. This is an important factor in determining the rate at which a fire moves, especially a backfire. Water in the soil absorbs heat energy and slows its pace compared to burning

Headfire burning rapidly uphill after establishing blackline with a backfire at the top of the hill

Backfire in early evening moving slowly with low flames — a safe burn

Tall prairie grasses burning hot with high flames due to heavy fuel load accumulated after five years of not burning

under dry soil conditions. Bunch grasses such as Little Bluestem and Prairie Dropseed may be damaged if burned under dry soil conditions. Fire can burn down into the crowns and kill new leaf buds.

Frequency and Timing of Burning

Burning frequency and seasonal timing depend on the plant species that compose the prairie, as well as on each prairie steward's goals. Midspring burns control cool-season weeds and grasses when actively growing and kill newly germinated weed seedlings. This knocks back new growth of problem invasive cool-season perennials, such as Quackgrass, Bluegrass, Tall Fescue, Bromegrass, and Clover.

Timing is essential to maximize damage inflicted on undesirable cool-season weeds and grasses. The optimal time to burn a typical mesic prairie is midspring, just as buds of the Sugar Maple (*Acer saccharum*) break open. Cool-season weeds will be greened up and growing. Fire will burn them back to ground level, setting them back. They now have to initiate new leaf

Seventy-five-foot-tall fire tornado at conclusion of ring fire in Switchgrass. Photo courtesy of Chris Klahn.

growth and in doing so will consume valuable root reserves. The primarily warm-season prairie plants are dormant during a midspring burn and are unharmed or only lightly singed. Cool-season prairie species will be leafed out and thus damaged, but they typically resprout and recover rapidly. Burning in midspring after unwanted woody trees, shrubs, and vines have leafed out also does proportionally more damage to them compared with burning in early spring before they initiate new growth, or in late fall after they have gone dormant.

After a burn the soil turns black and absorbs heat from the sun more readily than unburned prairies covered by old dead plant material. As soil temperatures rise in a burned prairie, it favors warm-season prairie plants while retarding growth of undesirable cool-season weeds.

Fall versus Spring Burning

Fall burns conducted after the growing season is complete and plants have gone dormant (after November 15 in the Midwest) result in little or no selective damage to most herbaceous plants, be they cool-season or warm-season species. Fall burns are therefore generally "neutral," in that they do not favor one group of plants over another.

Dry sandy prairies and goat prairie on steep limestone or dolomitic slopes typically have a number of early-blooming, cool-season flowers, such as Pasque Flower, Prairie Smoke, Puccoons, Downy Phlox, Lupine, Blue-Eyed Grass, Shooting Star, and Pussytoes. There are also cool-season sedges and grasses that grow in these dry prairies, including Junegrass

Post-burn landscape at a suburban home in Appleton, WI

Flowers in profusion ten weeks after burn

New green growth one month after prairie burn

Rattlesnake Master singed but undeterred after midspring burn

Prairie in bloom two months after burn

and Needlegrass. Midspring burns inflict serious damage to these desirable species, when most are in full bloom. To avoid this problem, dry prairies with early-blooming native species can be burned either in early spring before plants emerge from winter dormancy or in late fall when they are dormant.

Burning prairies every year in midspring is not recommended, as it tends to favor warm-season prairie grasses over many of the flowers. Burning every other year is generally recommended, as it significantly reduces invasion by woody plants. However, burning every three years allows unwanted trees and shrubs to develop sufficient root systems to survive fire.

Although midspring burning every two years provides good control of most woody and herbaceous weeds, populations of certain butterflies and other invertebrates have been shown to be negatively affected by this burning regime. Many restorationists interested in optimizing prairie invertebrates favor longer burning intervals of up to five years.

The impact of fire on vegetation is a function of intensity and duration. Slow-moving backfires do maximum damage to undesirable woody seedlings and cool-season weeds and grasses. Energy from a headfire escapes rapidly upward, with little time to damage vegetation at ground level. Although a headfire's roaring flames may look impressive, a backfire inflicts more damage. Backfires creep along the ground, so temperatures at soil level are higher than headfires. The slow pace of backfires increases the length of time a plant is exposed to damaging heat.

Prairie grasses, which provide the primary fuel for prairie fires, may not be sufficiently developed to carry a fire by the third growing season.

It is recommended that seeded prairies be first burned in their third spring. However, sometimes four to five years are required for prairie grasses to reach sufficient density to carry a fire. If nonnative cool-season grasses gain dominance, their new spring growth can reduce flammability. An earlier burn date may be necessary, before the ratio of flammable dead grass to new green foliage is reduced to the point at which there is insufficient fuel to carry a fire.

Prairies that will not burn in the evening due to excessive green-plant growth or limited fuel can be burned in the afternoon if local regulations allow it. Higher temperatures and lower humidity create conditions more conducive to carrying a fire. Never burn under windy conditions. High or shifting winds can create uncontrollable conditions that can lead to disaster.

Most states and some communities require individuals to obtain burning permits prior to beginning a controlled burn. The permit terms will specify the hours of the day when burning is allowed. Many states require the permit holder to check the permitting entity's website just prior to burning to confirm that burning is allowed that day.

Some states and nonprofit environmental organizations sponsor controlled-burn training classes for the public, to assist in teaching landowners safety techniques, precautionary measures, and fire-control methods. These classes are an excellent source of information and hands-on training for individuals who may have little or no experience in managing prairie fires.

When conducting a large burn, contact both the local fire department and sheriff's department. Inform them of your location and the time you expect to begin and end burning. They will appreciate this notice, even if you have a valid burning permit. Should someone call to report your fire, the authorities will be aware of the situation.

Inform your neighbors of your intent to burn your prairie. Make sure they know to close their windows and not hang out laundry when you will be burning. A little forewarning goes a long way to maintaining good neighborly relations.

This chapter is not intended as a comprehensive guide to fire manage-

Close mowing in midspring simulates a prairie burn

ment. For those who wish to become proficient in controlled burning, accredited courses in fire management are available through the Nature Conservancy (for more information, see the website https://www.nature.org/ourinitiatives/regions/northamerica/unitedstates/illinois/volunteer/il-vsn-fire-training.xml).

Always respect the energy and power of your prairie fire. Be careful out there!

Spring Mowing as an Alternative to Burning

In the event that burning management is not an option, you can substitute midspring mowing. Although not as effective as burning in controlling cool-season weeds and woody plants, mowing and removing cut material is a reasonable substitute. Dead vegetation and new green weedy growth should be mowed down to the soil surface, or at least to within an inch of the ground. This does maximum damage to cool-season nonnatives. Raking off cuttings exposes soil to the sun and encourages rapid soil warming, simulating the effects of burning.

Studies have shown that mowing and removing cut material in midspring is approximately 65% as effective as burning for controlling nonnative cool-season grasses. By July, a mowed and raked prairie will often be visually indistinguishable from a midspring burned prairie.[1]

Burning Small Prairie Gardens

Small prairie gardens are much easier to burn than large meadows because of the lower levels of fuel and reduced size of the area to be ignited. A prairie garden must have sufficient "fine fuel" in the form of prairie grasses to carry a fire. If no grasses are present, the fuel structure will usually not support a fire. One solution is to cut down all dead stems just prior to burning and raking them into small piles that can be more readily burned. Do not burn large piles of debris in the garden, as sustained heat can harm or kill plants in the soil below. Another option is to cut down the dead stalks and remove them from the garden. This exposes the soil to the sun's warming rays in spring to stimulate new growth of prairie plants.

As with cutting and removing vegetation from the garden, waiting until spring to burn allows for retention of dead plant material that serves as refugia for eggs, larvae, pupae, and adults of various insects and other invertebrates over winter. Birds will also find standing dead plants of interest as they forage for seed and insects over winter.

Be very careful when burning prairie gardens that have been mulched with shredded hardwood. Wood mulch can catch on fire and burn down into the soil, damaging or killing plants. Wood mulch fires can smolder undetected and then erupt into flame hours later when no one is around to tend them. Burning prairie gardens mulched with wood products is not generally recommended unless the area is thoroughly hosed down following the fire event.

Prairie gardens are often planted close to homes and buildings. This adds an extra level of difficulty to the burning process. Prairies can be safely ignited near brick, stone, stucco, masonry, and adobe buildings. Avoid burning anywhere near painted surfaces that can crack, craze, or catch fire. Burning prairies near structures with wood or vinyl siding is not recommended, as the wood can catch fire, and vinyl siding will deform with a mere hint of heat from burning vegetation. For the adventurous prairie pyro, burning can be carried out close to wooden structures as long as a steady stream of water from a hose is played on the wood wall. However, this is not a recommended procedure.

When in doubt, always err on the safe side. Cutting and raking dead vegetation in a prairie garden will yield similar benefits to the plants with a lower risk of uncontrolled conflagration. It's just not nearly as much fun!

8

Propagating Prairie Plants from Seed

Seed Harvesting, Cleaning, and Storage

With a little effort and experience, anyone can harvest and process prairie seed. The most important thing is to regularly observe each plant's life cycle, from flowering to seed formation to seed ripening. It is usually obvious when the seeds of most plants are ripe. In the case of grasses and fluffy-seeded flowers — such as Asters, Goldenrods, and *Liatris* spp. — seeds are easily stripped from the stalk or pulled from the inflorescence with little or no resistance. Ripe seeds will be firm and not milky inside when cut open or when nicked with a fingernail.

Many seedpods typically turn brown or black when ripe. Seeds will be firm and dry, and will fall readily from their point of attachment inside the pod. Many plants produce open seedheads, including Coneflowers (*Echinacea* spp.), Blackeyed Susans (*Rudbeckia* spp.), and Sunflowers (*Helianthus* spp.). Most seedheads are ripe when dry and seeds fall out easily when crushed. An exception is Lupine (*Lupinus perennis*), whose seeds are soft when pods are ripe.

A few species such as Roses and Great Solomon's Seal (*Polygonatum biflorum*) have fleshy seeds. Rose hips are ripe when they turn red and begin to shrivel. The seeds inside should be yellow to brownish and hard.

Seeds of Great Solomon's Seal have a fleshy outer covering that turns bluish black and begins to soften when mature. The actual seed inside should be white and hard.

Some have exploding seedpods, such as Lupine (*Lupinus perennis*) and Birdfoot Violet (*Viola pedata*). Lupine pods turn from green to yellow to black. Somewhere between yellow and black they dry out and twist in the heat of a warm sunny day, literally exploding and propelling their seeds up to 30 feet away. When inspecting Lupine pods for ripeness, seeds in the pods turn from green to white, and can be collected before all seeds have fully ripened. The best method to determine ripeness is to rap the pods with the back of the hand. Ripe pods rattle when rapped and can then be harvested. Yellow-colored pods are almost always ripe. Collected too early, Lupine, like all seeds, will show reduced or zero viability. Lupine seedpods dry and mature rapidly in late June and early July, and they must be monitored daily.

Violets produce small seedpods that droop downward as seeds are formed, elevate to a forty-five degree angle above horizontal, and when fully ripe . . . they explode! Violet seedpods can be collected when they are exactly horizontal to the ground or higher. If collected earlier, seed will not be fully formed and will germinate poorly, if at all.

Native Phloxes also have exploding seedpods, which are among the most difficult to collect. One must be present at the exact ripe moment. The seed matures rapidly from green to black inside paper-thin brown seedpods that always seem to have dispersed the day before! To ensure viability, seed should be dark brown or black before harvesting. Some people tie nylon stockings or similar breathable material over the pods or plants to capture exploding seeds.

Seeds of some species ripen indeterminately, one by one over an extended period. These cannot be harvested all at once, as only a small percentage of seed ripens at any given time. Seeds of Puccoons (*Lithospermum* spp.) are notoriously difficult to collect. Each plant produces only a few seeds, held deep within the axils of their leaves along the stem. They ripen individually and soon fall to the ground. The easiest way to collect Puccoon seed is to surround the base of the plant with a sheet of black plastic and collect the seeds that drop onto it. The whitish to brown seeds are easy to spot against the black, and they can be gathered together with a piece of thin cardboard and scooped up into a bag or other container.

A few plants defy the rules of seed collecting, notably Spiderworts (*Tradescantia* spp.) and Culver's Root (*Veronicastrum virginicum*). If one waits until the seedpods turn brown, seeds will be long gone, leaving only brown chaff. Spiderwort seeds ripen indeterminately and drop promptly when mature. They must be collected when young seedheads are green to greenish blue. Pick seedheads when still fleshy, with only one or no blooms remaining. Check in the morning, as flowers close up on sunny afternoons.

Culver's Root is a bit simpler to collect than Spiderwort, but it too must be harvested at the proper stage. Seedheads should be harvested when seedpods are yellow to yellowish brown, before they turn brown and eventually black. Seeds inside the pods will be red to light brown in color at this stage, and they will after-ripen to a nice auburn brown when dry.

Equipment for Hand Harvesting Seed

Most prairie seed can be harvested by hand using a few simple tools:

1 Leather or cotton gloves; rubber gloves optional for Spiderwort
2 Pruning shears for clipping seedheads and seedpods
3 Five-gallon plastic pails with metal handles
4 Old belt to hang a pail from the waist
5 Bags for stashing harvested seed: paper grocery bags for large-seeded species, large tight weave canvas bags for small-seeded species, and nylon feed bags for large quantities
6 Brimmed hat, sunscreen, and insect repellent

Although most prairie seeds ripen from late August through the end of October, early-spring bloomers begin to ripen in May: Birdfoot Violet (*Viola pedata*), Pasque Flower (*Pulsatilla patens*), Prairie Smoke (*Geum triflorum*), and Lupine. Spiderworts, Prairie Larkspur (*Delphinium virescens*), Downy Phlox, Shooting Star (*Dodecatheon meadia*), and Golden Alexanders (*Zizia* spp.) ripen between June and early August.

Harvesting Prairie-Grass Seed

Prairie grasses can easily be collected by hand, with the exception of Needlegrass (*Stipa spartea*). Aptly named, it requires special handling (see below). When seeds of prairie grasses are ripe, they dry out and turn a white, brown, red, or golden color. In addition, many will "fluff out." Grass seeds should strip off the stalk easily when ripe. If they do not come off readily, they are not ripe.

A few grass species have hard, grainlike rounded seeds, such as Purple Lovegrass (*Eragrostis spectabilis*), Switchgrass (*Panicum virgatum*), Prairie Dropseed (*Sporobolus heterolepis*), Purpletop (*Tridens flavus*), and Eastern Gamagrass (*Tripsacum dactyloides*). These should come off the stalk with little or no resistance when ripe. Cotton or leather gloves are recommended for harvesting prairie-grass seeds, as the stems can cut into fingers and hands.

Needlegrass has brutally sharp seeds, encased in a gorgeous golden seed coat. Leather gloves are a definite requirement when handling these dangerous little daggers. Seed is ripe when stems arch over, and seeds can be easily pulled from the white papery sheaths surrounding them. As they dry, the long seed awns twist and turn, literally drilling the sharply pointed seeds into the ground.

Immediately upon harvesting, the seeds become an unmanageable rat's nest as they dry. To overcome this, all points of the seeds should be placed facing one direction in a flat-bottomed bag or cardboard box. To prevent twisting, seeds should be sprayed with water every few minutes using a plant misting bottle. If possible, plant the seed in the desired location immediately. If seed is to be stored for planting later, it can be bundled while still moist and the awns are straight, using rubber bands or twist ties. Awns can also be cut off at this time to facilitate handling. Take a handful of Needlegrass seeds and gently tap the sharp, pointed ends against a table so that all seeds are even. Using pruning shears, clip off the awns just above the seed body. Beware: the seeds may be disarmed, but they are still dangerous!

Harvesting Prairie-Flower Seeds

Flower seedpods and seedheads can be stripped or broken off by hand (wear gloves) or clipped with pruning shears. Species with thick stems and prickly seedheads, such as Echinacea (*Echinacea* spp.) and Rattlesnake Master (*Eryngium yuccifolium*), are best harvested with pruners. Leather gloves are recommended when using pruning shears to avoid cutting off a finger!

Stripped or cut seed can be placed in a five-gallon plastic pail suspended from an old belt or rope tied around the waist. Small-seeded species should not be placed in bags with seams, such as paper grocery bags or nylon feedbags, because the seeds can leak out. Do not store freshly collected seed in plastic bags or containers that retain moisture. The seed may mold or rot, reducing viability. Exceptions include fleshy seeds such as roses and Great Solomon's Seal, which must remain moist in order to extract the seeds from the pulp.

Drying Prairie Seeds

Seeds should be laid out on a table to dry in cardboard boxes or on a smooth floor with no holes or cracks where seed can fall. Do not pile seed more than a few inches deep, as it will not dry properly. Good air circulation is essential to dry seed rapidly to prevent fungal attack. Seedheads that naturally contain high proportions of moisture, such as Lupine, Spiderworts, Flowering Spurge, and Purple Meadow-Rue, should be laid out to dry in a warm, well-ventilated location. Air circulation can be augmented using window fans to blow fresh air across drying seed. Collecting seed when damp is not recommended, as extensive drying may be necessary.

Hand-stripped seed often appears to be dry, but it commonly contains a surprisingly high percentage of moisture. It should be spread out to dry in a well-ventilated location, no more than a few inches deep to prevent molding.

Seed should be turned at least once a day, preferably two or three times. This is especially important in the first few days after harvesting when seed still contains significant moisture. After a few days, leaves and flower heads collected with the seed should begin to dry and turn brown. Seedheads of many species will open as they dry, and seeds will fall out. Others have seedheads that require breaking open to extract the seeds.

Pods become brittle as they dry. Seeds with aerial-dispersal "parachutes" will detach from the receptacles on the seedhead. After a number of days of good drying conditions, seed will feel dry to the touch. It is then ready to store in preparation for cleaning.

Exploding seeds such as Lupine, Violets, and Phlox must be placed in boxes or containers with window screen over the top to prevent seed from popping all over the place. Lupine can be effectively dried by placing it 6 inches deep in large cardboard boxes, covered with screens, in the sun. Violet and Phlox seedheads can be dried in deep, clear plastic jars (6–12 inches), placed in the sun with a layer of screen secured on top.

Seed Cleaning and Processing

To save the trouble of drying and cleaning seed, harvest and sow directly onto the ground in late summer or fall on sites which have been properly prepared and are ready to plant. Seeds will work their way into the soil with the freeze-thaw cycles of winter and germinate the following spring. A few species, such as Wildrye grasses (*Elymus* spp.), Blackeyed Susan (*Rudbeckia hirta*), Spiderworts (*Tradescantia* spp.), and some early-spring-flowering species will often germinate in autumn and grow sufficient root systems to overwinter successfully.

When sowing in spring, the seed should be cleaned, processed, and stored. Some species are easy to clean; others are tedious and challenging. Some basic equipment will be required:

Various-sized screens attached to wooden frames for sifting seed from chaff — For small seed quantities, frames can be made from 1×4s, 1–2 feet long on each side. Attach screens to the bottom of the frame using a staple gun. A separate box with a solid bottom should be made with 0.25- to 0.375-inch-thick smooth plywood instead of a screen. Place a screened frame with desired opening on top of the solid bottom box and rub the seed against the screen. The lower box will catch seed and chaff that falls through the screen.

Heavy-duty marble pastry rolling pin for breaking open seedheads and pods, such as *Baptisia* spp. — Place seed in a wooden box such as that described above. Crush seedheads and pods with the rolling pin to release seeds.

Small, electric motor-driven hammer mill for threshing larger quantities of seeds — This saves time hand-crushing seedheads and pods when working with large volumes. Blade speed can be adjusted using various pulleys and belts. More expensive hammer mills have a variable-speed motor control that allows for precise setting of the blade speed. Various sizes of heavy-duty metal screens are secured to the bottom of the hammer mill. Match screen to seed size so it will fall through. Blades grind up seedheads and pods so they can be separated by screening and/or blowing off chaff using an air source. It is important to select the correct speed for each species, as many a batch of valuable seed has been ground to pulp by using too fast a hammer mill speed. When in doubt, start by experimenting with small batches of seed, beginning slowly and increasing speed gradually. Upon sign of damage to seed, reduce speed to the point where no damage occurs.

A reversible canister-type vacuum cleaner such as a Shop-Vac for blowing off chaff — Place screened or hammer-milled seed and chaff mixture in a five-gallon plastic pail. When cleaning large, heavy seeds, the air stream can be aimed directly at the seed. The seed is too heavy to exit the container at air velocities of normal, household vacuum cleaners. When cleaning small, light seeds, aim the air stream onto the upper sides of the pail so it swirls around the inside walls. Aiming directly at small seeds can blow them out. As a precautionary measure, the pail can be placed inside a large cardboard or plywood box (4 feet by 4 feet) so any seed accidentally blown out can be recovered. Do not use this technique for cleaning extremely small, light seeds (e.g., Gentian, Lobelia, and Culver's Root) — as they will be scattered to the four winds. Small seeds should be processed by repeated screening, which will usually remove most chaff and other debris.

Small, old-fashioned "fanning mill" or "air screen cleaner" — In the old days, Grandpa cleaned seeds of grains and legumes using a hand-cranked fanning mill. These typically hold two or three different screens that slide into place, one above the other. Material to be cleaned is placed on the top screen as the crank is turned, shaking the screens back and forth. The upper screen size is selected so large debris is screened off, and seeds drop through to the second screen below. This lower screen is selected so it is smaller than the desired seed, and smaller debris and weed seeds fall through. Good seed travels across the lower screen. At the bottom of the screen, the seed falls off a ledge where a blower driven by the hand crank applies wind. This blows off most remaining chaff. Cleaned seed, being denser than chaff, falls to the bottom, where it is collected in a box. Shaking speed and air velocity are controlled by changing a leather drive belt, driven by variously sized wooden pulleys.

These old machines are easily converted to run on a small 0.5–1 horsepower electric motor by connecting the motor to the hand-crank shaft. The best old fanning mill is the venerable Clipper 2-B. This durable contraption is the right size for the home prairie-seed processor, as it fits in a garage or small outbuilding and is light enough to move. This machine can be easily repaired. They are still made today, using sheet metal instead of wood. Old Clipper fanning mills can be found at farm auctions. New models can also be purchased with the desired screens.

Storing Seed

The enemies of seed viability are heat and moisture. Seeds are living organisms and respire while dormant, consuming their energy reserves. They expend more energy at higher temperatures, which reduces seed longevity. Moisture can lead to growth of bacteria and fungi, damaging or destroying seed.

Seed can be stored over the winter in cool climates in an unheated building inside a sealed rodent-proof metal container. Dry seed can also be stored in a refrigerator inside two sealed plastic bags to keep out moisture. Totally dry seed can be stored in a freezer in plastic bags or other airtight containers at 0 degrees Fahrenheit.

Moist-stratified seed (see below) should not be stored in a refrigerator for extended periods (more than four months in most cases). Moisture will encourage growth of molds and bacteria, reducing viability. Do not place moist-stratified seed in a freezer, as this can damage it.

Seed Germination Techniques

Most prairie flowers, grasses, and shrubs can be germinated easily. A few exhibit complex seed dormancies and are more easily propagated by root division or stem cuttings. Most seeds require pretreatment to break dormancy. There are four basic types of treatments to overcome dormancy:

1 Dry stratification — Seed is exposed to near-freezing or freezing temperatures for 30 or more days.
2 Moist stratification — Seed is mixed with a damp, inert substrate and stored in a refrigerated environment at 34–36 degrees Fahrenheit. Seed should not be frozen, to prevent damage.
3 Scarification — Hard-coated seeds are scratched with sandpaper to score the outer layer, allowing moisture to penetrate and initiate the germination process.
4 Hot water — Seeds stimulated to germinate by the heat of wildfires are treated with near-boiling water.

Most native seeds require exposure to cold temperatures to prevent fall germination and the death of tiny seedlings over winter. The term *seed stratification* originated years ago when seeds were pretreated in layers of damp, clean sand and refrigerated to mimic the effects of winter. Many species require exposure only to cold temperatures without addition of moisture. This process is referred to as *dry stratification*.

Dry Stratification

Most prairie grasses and many prairie flowers require simple dry stratification. Seed can be treated by placing in a refrigerator or freezer for 30–90 days prior to sowing. Large quantities of seed can be stored in an unheated building over winter in rodent-proof metal containers.

Moist Stratification

Many prairie-flower seeds require moist stratification to break dormancy and increase germination rates. For example, Shooting Star (*Dodecatheon meadia*) yields no germination when dry stratified but nearly 100% after 30 days of moist stratification.

Different species require varying lengths of moist stratification. Lupine (*Lupinus perennis*) requires only 7 days. It will often begin germinating after a week in the refrigerator. Members of the genus *Iris* require at least 90 days of moist stratification to yield good germination. Dormancy in most species can be broken within 30 days. Some species germinate at higher rates with this treatment.

Seed can be moist stratified by mixing with an equal or greater volume of slightly damp, inert material. Oak or pine sawdust works well. Sawdust readily absorbs and transfers moisture to seed, and its high acidity limits bacterial growth. Clean builder's sand, vermiculite, perlite, and peat moss may also be used. Soil is *not* recommended, as it contains fungi and bacteria that can attack the seed.

Inert matter should be lightly dampened prior to mixing with seed. If water can be wrung out when squeezed, it is too wet. Vermiculite and perlite should be moistened in a bowl or colander and excess water drained off. Mix seed and inert matter together thoroughly, place in a zip-top plastic bag labeled with species and date, and place in refrigerator for the specified amount of time.

Another method is sowing directly into flats in winter or spring. Cover with plastic wrap to retain moisture, and place in the refrigerator or a larger cooler. Alternatively, flats can be sown in fall and stored over winter in an unheated building or greenhouse. Make sure flats are protected from damage by mice and other animals during winter storage.

TIMING OF MOIST STRATIFICATION PRETREATMENT

Seed should be moist stratified on a schedule that allows for removal from refrigeration at the optimal time. Cool-season plants germinate best in early spring when temperatures are cool. Warm-season species are best started when air temperatures reach the high 70s or low 80s Fahrenheit in mid- to late spring. Seed should be sown and refrigerated in advance of the desired germination date.

Scarification

Scarification is accomplished by placing a single layer of seed in the bottom of a wooden box and rubbing it lightly with sandpaper wrapped around a wooden block. Rub the seed just hard enough to scratch the outer surface, being careful not to grind it into flour. Light pressure is usually sufficient to scarify all but the most resistant seeds.

Some seeds, such as irises and roses, require scarification followed by a period of moist stratification. Roses should be treated as specified in the section "Double-Dormant Seeds."

Hot Water

The effects of fire on seeds can be mimicked by treatment with hot water. The prairie shrub New Jersey Tea (*Ceanothus americanus*) exhibits higher germination when soaked in hot water, followed by 30 days of moist stratification.

Place seed in a bowl. Boil water and allow to cool for a minute. Pour the still-hot water over seed. Allow to cool, pour off water, and sow the seed. New Jersey Tea exhibits increased germination when heat-treated seed is moist stratified for 30 days.

Some growers have reported good results using hot-water treatment with the genus *Baptisia*, followed by freezing the seed briefly until ice crystals begin to form on the wet seed (about an hour or less). Repeat treatments three times in succession (hot water followed by near freezing) have resulted in improved germination.

Other Considerations in Native Seed Propagation

FLESHY-FRUITED SEEDS

Some seeds are encased in fleshy pulp. Pulp often possesses compounds that retard seed germination and must be removed prior to sowing. Collected when ripe, pulp is usually soft and readily removed. Wash the seed while rubbing it carefully across a screen with openings smaller than the seed (a 0.25-inch screen works for most species). The flesh will go through the screen, and the seeds will remain on top, where they can be collected. Large quantities of fleshy-fruited seed can be carefully run through a hand-cranked meat grinder to loosen the pulpy material, which can then be washed off. If the flesh is hard when collected, allow the seed to soften for a week or longer by storing in a cool, damp place.

Most plants with fleshy seeds are woodland species. The only prairie species in this book with fleshy fruits are Roses (*Rosa* spp.) and Great Solomon's Seal (*Polygonatum biflorum* var. *commutatum*).

DOUBLE-DORMANT SEEDS

Some species, especially members of the Lily and Rose families, exhibit a phenomenon known as *double dormancy*. These seeds require exposure to two consecutive winters in the soil before they will germinate. Some species germinate in the first year, but their development occurs underground, and no visible leaves are produced. Seedlings will emerge in spring after the second winter.

Double-dormant seed is typically sown fresh directly in beds in the ground, or in flats stored over their first winter in a cooler or greenhouse. They are allowed to experience a second summer and winter in flats, stored over winter, or in an unheated greenhouse. Seeds will germinate the following spring.

This process can be accelerated by tricking the seed into "thinking" it has experienced two winters in an eight month period. Freshly collected seeds are planted in flats and refrigerated in early fall. After three months, remove flats from refrigeration, cover with plastic wrap, and store at room temperature for two months during midwinter. Refrigerate a second time for three months, and remove in spring to initiate germination.

Timing of Seed Sowing

Different species germinate best at different times of year. Most summer-blooming prairie flowers and grasses are warm-season plants, so germinate best at or above 75 degrees Fahrenheit. Warm-season prairie grasses are best sown in mid- to late spring or early summer. Cool-season native grasses do best seeded in fall or early spring. Most spring-blooming prairie flowers germinate in late summer, early fall, or early spring when temperatures are between 55 and 70 degrees Fahrenheit. See table 8.1 for prairie grasses that germinate well at warm temperatures.

TABLE 8.1.　　PRAIRIE GRASSES THAT GERMINATE BEST WHEN SOWN IN LATE SPRING AND EARLY SUMMER

Scientific name	Common name
Andropogon species	Big Bluestem, Broomsedge
Bouteloua curtipendula	Sideoats Grama
Elymus spp.	Wildrye (also germinate well in early fall)
Eragrostis spectabilis	Purple Lovegrass
Panicum virgatum	Switchgrass
Schizachyrium scoparium	Little Bluestem
Sorghastrum nutans	Indiangrass
Tridens flavus	Purpletop
Tripsacum dactyloides	Eastern Gamagrass

Cool-season prairie grasses germinate best in early to midspring. They can also be "dormant-seeded" in fall and will germinate the following spring when conditions are optimal. Although Prairie Cordgrass is a warm-season grass, it requires an extended period of moist stratification to break dormancy. It germinates best when sown in fall, with minimal germination when sown in spring or early summer. See table 8.2 for prairie grasses that germinate best at cool temperatures.

PLANTING FRESHLY COLLECTED SEED OF SPRING-BLOOMING PRAIRIE FLOWERS

Seeds of many spring bloomers ripen in late spring to early summer and some germinate in summer or fall. They are best sown immediately after harvesting. Seedlings will develop during cool fall weather, in preparation for active growth the following spring. See table 8.3 for species that often germinate in summer or early fall.

TABLE 8.2.

TABLE 8.2. **COOL-SEASON PRAIRIE GRASSES THAT GERMINATE BEST WHEN SOWN IN EARLY SPRING OR FALL**

Scientific name	Common name
Anthoxanthum hirtum	Northern Sweetgrass
Bromus kalmii	Prairie Bromegrass
Calamagrostis canadensis	Bluejoint Grass
Elymus species	Wildryes
Hesperostipa spartea	Needlegrass
Koeleria macrantha	Junegrass
Sporobolus heterolepis	Prairie Dropseed

TABLE 8.3. **PRAIRIE FLOWERS AND GRASSES THAT GERMINATE WELL WHEN SOWN FRESH AFTER HARVESTING**

Scientific name	Common name
Delphinium virescens	Prairie Larkspur
Geum triflorum	Prairie Smoke
Lupinus perennis	Lupine
Pulsatilla patens	Pasque Flower
Tradescantia spp.	Spiderworts
Viola spp.	Birdfoot Violet, Prairie Violet
Hesperostipa spartea	Needlegrass

Starting Prairie Seeds Indoors

This brief segment is by no means intended as a guide to greenhouse design, setup, and plant propagation. It will, however, assist home gardeners in starting prairie plants from seed on a small scale.

Use the following materials to grow prairie plants in a south-facing window, small greenhouse, or growing chamber using plant lights:

1   Small pots, plug trays, or standard 10-inch-by-20-inch open flats — preferably new and sterile, or thoroughly washed, rinsed, and disinfected with a 10% bleach solution
2   Self-adhesive labels and permanent marking pen to label pots and flats
3   Sterile seed-starting soil medium (*not* potting soil)
4   Vermiculite for lightly covering seeded flats
5   Saltshaker for sowing small seeds
6   Rainwater collected from a roof gutter — it's better than well water or tap water.

Seeding Pots, Flats, and Trays

Prairie seeds vary in size from species to species   A variety of containers can be used for seed sowing

Label pots, trays, or flats. Use a permanent marker on a self-adhesive label to mark species and date on each pot or tray. If growing individual rows of different species in a plug tray or flat, mark each row with the species so you can tell who's who.

Fill pots and flats evenly with sterile seed-starting medium. Firm soil by placing another pot or flat on top and pushing down firmly. Add more soil and continue to firm until soil level is half an inch below the top. When seeding plug trays, fill each cell to just below the top.

If growing dry prairie species that require well-drained soil, it is recommended that the sterile soil medium be mixed with sterilized sand for a volume ratio of one-to-one. You can make your own sterilized sand by placing it in a 250 degree Fahrenheit oven for one hour. Let the sand cool and mix it thoroughly with seed-starting soil medium.

Sterile seed-starting medium soaking before seeding. Pots filled with soil are placed in a pan with water that is absorbed from below by capillary action, rather than by watering from above.

Use pots or flats to compress seed-starting medium to half an inch below rims

Water pots or trays thoroughly to moisten and settle the soil before sowing. Let stand for one hour or longer to drain.

For growing small numbers of plants, pots and flats can be hand seeded. A general rule of thumb is to place three to five seeds in each pot or cell. Germination can vary depending on seed viability, growing conditions, time of seeding, and seed pretreatment. When seeding rows in a 10-inch-by-20-inch flat, place larger seeds about 0.125 inches apart. Smaller seeds can be planted more closely together, but be careful not to seed too densely. They will come up too thickly, will grow weakly, and will require thinning in order to develop into strong individuals.

To sow very small seeds, put seed in a saltshaker. Tilt the shaker so seed just starts to come out, tapping gently as you go. This provides better seed spacing and plant density.

Small-seeded species can be sown evenly and efficiently using a saltshaker with appropriately sized openings for various seed sizes. Metal or plastic shaker tops can be drilled to the desired diameter to accommodate seeds of various sizes.

After pots or plugs have been sown, sprinkle a thin layer of vermiculite over each pot, flat, or plug cell, just enough to cover the seed. Water once more, using a fine mist to avoid disturbing the seed. Place flats in a greenhouse or a window that receives direct light but not full sun. If moist stratifying seed prior to germination, place in refrigerator or cooler for the appropriate length of time.

When growing small quantities of seedlings, it is easier to moist stratify seed prior to sowing into pots or trays. Because the seed mixture will be moist when removed from refrigeration, it may clump when handled. Spread seed mixture thinly on a tray and allow to dry overnight. The dried mixture can be evenly distributed in pots and flats.

When growing large numbers of seedlings, it is easier to sow flats and place them in a cooler to break seed dormancy. The tops of seeded flats should be completely covered with plastic wrap to retain moisture. Place a piece of cardboard or similar material, cut to 10 inches by 20 inches, between flats to prevent damage when stacking.

If a large refrigerator is not available, allow nature to treat the seed for you. Sow flats in late fall or early winter, cover each one with plastic wrap, and place in plastic garbage bags in an unheated shed or garage. Although many prairie flowers require only 30 days of treatment, most can be stratified for longer periods without harm (the exception being Lupine, which often germinates after only 7–10 days).

Germinating Seed

Butterflyweed seedlings germinating in a 72-cell plug tray

Many prairie plants germinate better when the day's length is increasing. Seed sown from November through January often does not germinate well, unless natural sunlight is augmented with high-intensity growing lights to artificially extend daylength. If not using lights, seed should be sown or removed from refrigeration in March or April as days grow longer. Many species germinate well in July and even August in a climate-controlled greenhouse. However, these seedlings may have insufficient time to develop and become winter hardy when transplanted outdoors in fall.

As seedlings grow, artificial lights should be raised above the tops of the seedlings. If the light source is placed too high above plants, they grow long and leggy. Seedlings grown under lights should be moved into natural light as soon as possible, where they will grow faster and stronger.

Do not allow seedlings to get taller than 4 inches when growing indoors. Their stems will be weak and easily blown over or broken by wind when planted outdoors. If this is not possible, a fan can be set up to mimic windy conditions and encourage strong stems.

Watering

One of the biggest mistakes is overwatering. Too much water keeps the soil cool, reduces germination, and encourages growth of fungi and other pathogens which attack seedlings. Excessive watering can actually drown seedlings by preventing proper air exchange to their roots. Dry prairie species that are not adapted to high moisture conditions are particularly susceptible to overwatering.

Never use chlorinated tap water. Rainwater is best. Collect it from roof gutters in spring, summer and fall, and store in large jugs or barrels in a shady location to prevent algal growth. If rainwater is unavailable, well water is usually OK unless it is very hard. Calcium and magnesium in hard water can tie up phosphorus and other micronutrients. Fill open containers with hard water and allow to stand for forty-eight hours to settle out calcium and magnesium carbonates before using.

Water seedlings from early morning into early afternoon, never late afternoon or night. Watering late in the day encourages high moisture and humidity conditions, which can lead to "damping off," a fungal disease that wipes out seedlings in a flash. Morning watering allows soil to dry out during the day, so fungi have less favorable growing conditions when temperatures drop and humidity rises at night.

When placing plug trays inside 10-inch-by-20-inch tray bottoms, make sure the bottom tray has holes for drainage. Do not allow standing water to accumulate in the bottom of the tray, as this can cause saturated soil conditions that may lead to plant death. Water only as needed. More seedlings have suffered premature death due to overwatering than underwatering.

CAPILLARY MATTING

Capillary matting is an excellent alternative to overhead watering for open flats, pots, and individual plugs (not full plug trays). It does not disturb the seeds and keeps soil more evenly moist from the bottom up. In the greenhouse, place matting along a bench, with one end submerged in a clean, deep plastic container (6–12 inches) such as a wash basin or cat-litter box filled with water. Add desired liquid fertilizer in the recommended quantity. Arrange flats, pots, or plugs on the matting as close to one another as possible. Water and fertilizer will be wicked up into the flats and pots by capillary action. Add water and fertilizer when the reservoir runs low. This method prevents both over- and underwatering, reducing the chance of plant mortality.

Capillary matting waters from below, preventing over- or underwatering and helping prevent diseases, such as "damping off." Apply fertilizer in the water reservoir.

Fertilizing

Seedling development is increased with a proper fertilization regime. Begin fertilizing when the first true leaves appear. For grasses and monocotyledonous flowers, which do not have two seedling leaves, wait until plants produce two to three "true" leaves before fertilizing. Use a balanced, soluble fertilizer with relatively equal amounts of nitrogen, phosphorus, and potassium (listed on the label as N, P, and K respectively). Mix soluble fertilizer in water at the recommended rate and apply using a spray mister or capillary matting to avoid seedling damage.

Seedlings are ready to transplant into large pots or outdoors when soil in the pot or plug cell is filled with roots. Check root development in a plug using a clean knife to carefully pry it out of the tray. If it pops out easily in one piece with no loose soil and is completely rooted, it is ready to be transplanted.

Hardening Off and Transplanting

A week to 10 days after transplanting seedlings into pots, it is time to start the hardening off process. This allows plants to acclimate gradually to drying winds and seasonal weather, including driving rain and bright sunshine. Place plants outdoors in an area protected from high winds during the day and bring them in at night. After one week, they can be left outdoors day and night. They can then be planted in the ground.

Cold frames can be used to begin the hardening off process. Rather than bringing plants inside at night, they can be left in cold frames with the lids closed. Don't forget to open them in the morning, or you could fry your plants!

After transplanting small plugs into the ground, monitor daily and water regularly for the first few weeks. If desired, mulch with 1–2 inches of clean straw or shredded hardwood to help retain soil moisture and reduce weed germination. Do not overmulch and bury seedlings. Overmulching encourages high moisture levels around the plant and encourages fungal diseases. Mulch creates excellent slug habitat, so be judicious in application. Do not mulch with bark nuggets, as they can lead to high plant mortality.

If deer, rabbits, groundhogs, or other plant eating critters are a problem, newly transplanted seedlings should be protected using fencing or liquid spray animal repellents. For more information on controlling animal damage and selecting deer-resistant plants, see the University of Colorado Extension's website (http://www.ext.colostate.edu/pubs/natres/06520.html).

Propagating Plants by Seeding Directly into Soil

Site Preparation and Seeding

Ensure that the site is completely free of perennial weeds before sowing. Weeds grow much faster than prairie plants and must be controlled. An existing well-maintained vegetable garden is an ideal location. Long-established, well-maintained lawns are also good, as they usually have low weed densities.

For sowings in spring or early summer, pretreat seeds of each species according to directions to break dormancy. If not pretreated, they may not germinate at sufficient rates. Fall-dormant seedings require no pretreatment, with the exception of double-dormant species.

Work up soil to create a good seedbed and then rake level. Mark rows with stakes on both ends, labeling at least one end with the species using an indelible, sun-resistant marker (nursery marking pen). For maximum space efficiency, place rows 18 inches apart. This makes for tight working conditions but consumes a minimum of space and leaves less open ground for weed growth.

Use a string row marker between stakes to define rows. Make a small trench down the row using a trowel or fingers. The depth of the trench depends on the size of seed to be sown. A general rule of thumb is seed should be planted two to three times deeper than its diameter. For extremely small, fine seed, no trench is necessary. A common mistake when sowing very small seed is placing it too deeply. Seedlings may germinate but have difficulty emerging. Cover small seeds with a light layer of well-rotted, screened compost or vermiculite instead of soil to hold in moisture without burying the seeds.

Sow seeds as you would in a flat, about 0.125 inches apart, except for small seeds, which can be planted more closely. The saltshaker method described earlier can be used for small seeds. Cover seeds with an appropriate amount of soil so they are at the proper depth. Firm soil lightly by pressing down with your hand or by rolling with the tire of an empty wheelbarrow down the row.

Watering and Fertilizing

Water seeded areas every morning, unless surface soil is damp. Never water late in the day or night, as this encourages damping off. Seeds such as Lupine will emerge in a day or two, whereas slower-germinating species may require many weeks. Many species of prairie flowers do not require moist stratification for germination but will germinate earlier and more uniformly when so treated.

Seedlings can be fertilized with a balanced fertilizer when true leaves appear. Use a soluble fertilizer mixed in water, applied with a watering can.

Once plants reach a height of 2–3 inches, a slow-release granular fertilizer can be applied, if desired. Use a low nitrogen formulation with a high proportion of phosphorus and moderate amounts of potassium. Phosphorus stimulates root growth and builds healthy transplants. Small monthly applications are better than one large one early on.

Weed Control

As prairie seedlings appear, weeds must be controlled. Weeds between rows can be uprooted using a "Ro Ho" push tiller, or regular hoe, early on a sunny day. Weeds in the rows must be carefully pulled by hand or tweezed out. Use a regular table fork to work around the seedlings and remove weeds. Weeds must be removed before they are an inch tall and firmly entrenched.

Digging and Transplanting Seedlings

Most prairie flowers and grasses seeded in spring or early summer will develop sufficient roots for transplanting by early fall of the first year. Once leaves die back and plants go dormant, they can be dug and transplanted to their new home. Use a long, narrow spade to dig deep-rooted prairie plants. A pitchfork works well for shallow-rooted species such as prairie grasses and non-taprooted flowers.

If preferred, plants can be left in the ground over winter and transplanted before they emerge in spring. To protect seedlings over winter, cover with 4–6 inches of clean Winter Wheat straw after plants have gone dormant in fall. This prevents losses due to extreme cold and frost heaving, which can literally push roots up out of the ground and kill them.

9

Propagating Prairie Plants Vegetatively

A few species can be difficult or expensive to propagate using seeds. It is usually much faster to produce slow-growing plants vegetatively, from divisions or cuttings, rather than waiting for seedlings to grow into mature plants. Some slow-growing species such as Shooting Star (*Dodecatheon meadia*), Gentians (*Gentiana* spp.), Michigan Lily (*Lilium michiganense*), and Great Solomon's Seal (*Polygonatum biflorum* var. *commutatum*) may take four to five years to mature from seed. Other species have exploding seedpods that are challenging to collect or expensive to purchase, such as some species of Phlox (*Phlox* spp.) and Violets (*Viola* spp.). Some seeds require extensive effort to break dormancy, including double-dormant Roses (*Rosa* spp.) and many members of the Lily family.

There are three basic methods of propagating herbaceous native plants: root division when plants are dormant; stem cuttings when plants are actively growing, usually in early or midspring; and tissue culture in a specially designed, climate-controlled, aseptic lab. Because most people do not have access to tissue-culture labs, this chapter focuses on root division and stem cuttings, which can be done easily and at little cost by the average gardener or nursery owner.

Root Division

There are five basic root types you'll encounter with prairie plants:

1   Fibrous — Numerous individual spreading roots, usually with many associated root "clumps" that possess their own buds, without a central taproot, rhizome, bulb, or corm.
2   Rhizome — A root that spreads horizontally, usually in the upper 6–12 inches of soil. Rhizomes are actually underground stems that have evolved to function as roots.
3   Bulb — A structure composed of many smaller plantlets or scales that surround a central stem. The core of the bulb is a modified stem, and the individual scales, as in lilies, are modified underground "leaves" that have evolved into storage organs. Each scale can be removed to form a new plant.
4   Corm — A bulblike structure that is solid when cut in cross section and does not possess individual scales. Corms are modified stems that store nutrients and water. Most corms are not easily divided.
5   Taproot — One or more thick roots that grow straight down into the lower soil with little or no branching in the upper soil. Some taproots can be cut into long segments and propagated, but resulting plants are often weak and short-lived. Dividing taproots is generally not recommended and seldom successful.

Most plants with fibrous roots, rhizomes, or bulbs can be readily propagated by root division. Plants with taproots (including most legumes) and corms (e.g., *Liatris* spp.) are not easily divided. They have a single, distinct root rather than many individual roots, rhizomes, or scales that can be separated from one another.

Most grasses are easily propagated using root division but not stem cuttings. The growing points of most grass leaves (and most monocot leaves) are located at or below ground level. They do not produce leaf nodes along the stem, where new roots can be stimulated to develop. The growing points of most dicotyledonous (broadleaf) plants are located at their stem tips, and there are usually sufficient leaf nodes along the stems where roots can be induced to grow.

Timing of Root Division

It is usually best to divide roots when plants are dormant. This minimizes stress, as the roots do not need to support demands for water and nutrients of actively growing leaves, and allows time to acclimate to the new environment. Most plants can be safely divided in early spring before they initiate new growth or in fall after they have gone dormant. Some plants are best divided only in spring, especially species with fine, thin roots, such as Sideoats Grama (*Bouteloua curtipendula*), Vanilla Sweetgrass (*Hierochloe odorata*), Blue-Eyed Grasses (*Sisyrinchium* spp.), and Phlox. The fine roots of these plants are more susceptible to winter damage compared to plants with more robust root systems.

A few species do best when divided in fall rather than spring. This is particularly true of rhizomatous members of the Lily family, such as Michigan Lily and Great Solomon's Seal.

Some early-blooming spring flowers exhibit a period of near dormancy in midsummer and can be divided after setting seed. These include Prairie Smoke (*Geum triflorum*), Blue-Eyed Grass (*Sisyrinchium* spp.), and Violets (*Viola* spp.). Species in a few genera can be divided almost any time of year, including Iris, Aster, Goldenrod, and most grasses. When dividing plants during the growing season, cut back at least 50% of the top leafy growth to reduce stress on root systems. To ensure survival, water regularly for the first two to three weeks after dividing.

Dividing Fibrous-Rooted Plants

Plants with fibrous root systems are generally easy to divide, although there can be variation in root structure from species to species. Roots of most nonrhizomatous grasses are very dense and intertwined. They are usually best divided by simply slicing clumps into quarters or eighths using a sharp narrow shovel, such as a tile spade. It is easiest to slice into grass clumps while still in the ground before digging. Individual precut pieces will be more cleanly cut and are then easier to remove than an entire clump, which can weigh up to 100 pounds. When removed prior to cutting, large clumps tend to flop around as one attempts to cut them up.

If the goal is to rapidly increase numbers of a certain grass plant or selection, most fibrous-rooted species can be divided down to small, individual plants consisting of a bud and a root system. These can be planted directly in the ground in an irrigated nursery or into small pots in a greenhouse under mist. Once pots are well rooted, they can be moved into large containers or planted out in the field.

Beware of dividing Prairie Dropseed (*Sporobolus heterolepis*) too aggressively (pieces with a diameter less than 2 inches). Its roots are easily torn and damaged, and high mortality can result. It is easiest to divide Prairie Dropseed with a sharp knife, machete, or hatchet on a firm wooden surface.

Mature root systems of most Aster, Goldenrod, Rattlesnake Master (*Eryngium yuccifolium*), Rosinweed and Cup Plant (*Silphium* spp.), *Eupatorium*, *Helenium*, *Heliopsis*, *Lobelia*, and *Penstemon* occur as masses that can be divided. Lobelia generally separates easily, whereas Penstemon usually requires some forceful tugging. The thick roots of Aster, Goldenrod, and *Eupatorium* can be carefully ripped apart or cut with a sharp knife or pruning shears. Rosinweed and Cup Plant require major elbow grease to divide. Use of a chain saw on these two bad boys is not out of the question. Be sure to rinse sand and soil off roots before firing up the saw.

A helpful hint: most fibrous-rooted species retain a significant amount of soil in their roots, making some heavy to handle and difficult to divide. Soaking plants in a five-gallon pail or large tub of water for a few minutes, and then hosing them down to remove most of the soil, makes the work much easier.

Swamp Milkweed (*Asclepias incarnata* spp.), Shooting Star (*Dodecatheon meadia* spp.), Gentian (*Gentianaceae*), Spiderwort (*Tradescantia* spp.), and Phlox have closely intertwined roots and should be carefully teased apart. Soaking in water to remove extraneous soil makes this process much easier.

Never divide Shooting Star plants once they have emerged in spring or when actively growing. This can severely harm or kill them. When dormant in late summer and fall, Shooting Star is nearly impervious to damage. Individual roots can be cut from the central core and planted so the top of each root is an inch below soil level. Roots will develop new buds and emerge from the ground as small plants the following spring.

Dividing Rhizomatous Plants

Rhizomatous plants are among the easiest to divide. Rhizomes contain ample amounts of energy reserves, and most rhizomes will develop new shoots from nodes that occur at intervals along their length. Studies at Prairie Nursery have shown best results with most rhizomes are attained when using sections at least 6 inches long. Shorter segments exhibited lower survival rates and less vigor. Some species require a terminal leaf bud be present on the rhizome. These include Queen of the Prairie and Prairie Cordgrass. Heath Aster and Stiff Coreopsis will emerge sooner and grow more rapidly if a terminal leaf bud is present.

Rhizomes should be placed horizontally about 1 inch below soil surface, and the soil packed firmly. If terminal leaf buds are present, they should be positioned just below the soil surface.

PROTECTING FALL DIVISIONS

Roots and bulbs divided in fall should be covered with 3–6 inches of loose clean straw mulch to protect them over winter. The mulch will compress down to 1–3 inches by spring, and the new growth of most species should be able to come up through it. Mark your fall transplants with stakes or flags at time of transplanting. If some plants do not emerge as expected the next spring, scrape away mulch where the crown of the plant should be. Leave surrounding mulch in place to retain soil moisture and help prevent weed germination and growth.

Dividing Bulbs

There is some variability in the structure of bulbs, and techniques for dividing them will differ. For instance, Nodding Pink Onion and Prairie Onion (*Allium* spp.) produce numerous new "sets" around a central plant, each with its own root system and leaves. These are easily separated into new plants at almost any time of the year, except when plants are flowering.

Michigan Lily (*Lilium michiganense*) produces numerous small scales attached to a central stem. An individual scale can be separated from the main bulb in fall after the plant is completely dormant and, placed 2 inches deep in soil, will emerge the following spring. Cover with three inches of clean straw mulch for winter protection and weed control. New bulbs will develop at a depth of about 4 inches. Dividing lily scales in early spring is not recommended. They typically remain dormant for a year and will not emerge until the following spring.

Nondivisible Taprooted Plants

Very few taprooted plants can be successfully divided. Not only are their deep roots difficult to extract from the soil, but a lack of buds on their roots prevents them from sprouting. Taprooted prairie flowers that do not divide well include the following:

Blue Sage
Butterfly Milkweed
Canada Milkvetch
Compassplant
False Indigo
Goatsrue
Illinois Bundleflower
Leadplant
Marbleseed
New Jersey Tea
Ozark Coneflower
Pale Purple Coneflower
Poppymallow
Prairie Clover
Prairie Dock
Prairie Turnip
Puccoons
Roundhead Bush Clover
Royal Catchfly
Ticktrefoils
Wild Lupine
Wild Petunia
Wild Quinine

Stem Cuttings

Many dicotyledonous plants are easily propagated by taking cuttings of fresh, actively growing stems and "sticking" them in sterilized potting-soil medium. A typical cutting consists of the upper three to four leaf nodes at the top of the stem, usually 4–6 inches in length. The upper two leaf nodes will remain above soil. Leaves on the lower one or two nodes are carefully removed. The prepared stem is placed in the potting-soil mixture in plug tray cells or small pots. Soil must be kept consistently moist, and within a few days new roots will develop at the buried nodes.

Taking Cuttings

Use a sharp knife or pruners to cut stems. To prevent transmission of disease to new cuttings, dip tools periodically in rubbing alcohol or 10% bleach solution.

Cuttings taken from terminal shoots of new growth generally perform best. Lateral side shoots often root better than top shoots taken from the upper portion of the plant. Only healthy plants should be used as stock plants. Plants that are diseased, under moisture stress, suffering from nutrient shortages, or otherwise low in vitality should not be used. Plants that have received excessive nitrogen fertilizer often will not root properly.

Take cuttings from stems that do not have flower buds, as flowers will consume energy that would otherwise be used to develop new roots. If cuttings can be taken only from stems with flower buds, remove the buds before sticking them.

Take cuttings early in the morning before plants begin to experience moisture stress. Cuttings should be kept cool and damp until they are stuck. When working in the field or warm greenhouse, place fresh cuttings in damp paper towels inside a small cooler. Process and stick them immediately. If cuttings have to be stored prior to sticking, wrap their lower half in damp paper towels, place them in a plastic bag, and store in the refrigerator.

Sticking Cuttings

Cuttings should be inserted into a low-nutrient, sterilized soil mix that is well drained but also holds moisture. Do not use potting soil that has nutrients added to it. Large bags of seed-starting mix work well. You can also make your own soil mix for growing cuttings. Two options are as follows: (1) 50% clean, coarse builders' sand and 50% peat by volume and (2) 50% perlite and 50% peat by volume.

The soil mixture should be thoroughly homogenized and then moistened prior to sticking cuttings. Add just enough water so the mix is damp, not sopping wet. Some people prefer to fill their plug trays and pots first, and then water repeatedly with a gentle spray wand to saturate the soil mix. If watering this way, allow mix to drain for an hour or so before sticking your cuttings.

Research has indicated that the development of roots can be enhanced by dipping lower stem nodes into a rooting hormone solution or powder consisting of gibberellic acid (GA3). Gibberellins are a class of plant hormones that promote growth and cell elongation. They stimulate rapid cell division, and hence faster root development. When using powdered GA3, be sure to tap each cutting after dipping to knock off any excess powder that could result in an excessive dose of the hormone.

When using rooting hormone, always prepare a small amount in a separate bowl that will be used for dipping the plants. Never dip plants directly into main container, as this can contaminate it with plant disease organisms. Once all cuttings have been treated, throw away any remaining rooting hormone solution or powder that has been used.

Place cuttings in soil mix so one-third to one-half of the stem is covered, with the leaves on top and the nodes without leaves in the soil. Leaves of cuttings should not touch one another. If this occurs, use a larger plug or pot size. Water cuttings with a fine mist or light spray after each tray or group of pots has been stuck.

Care of New Cuttings

Place newly stuck cuttings in a location that receives indirect sun. Avoid direct light, as it can dry out cuttings that have yet to form roots. If an automatic misting system is available, set the interval and duration of misting so that soil moisture is maintained but leaves have sufficient time to dry out between mistings. Typical misting intervals range from every fifteen minutes during hot weather to every hour or so under cooler, more humid conditions. Discontinue misting by late afternoon and during the night to prevent damp conditions that can lead to fungal diseases and plant losses.

If a misting system is not available, flats or pots should be covered with a plastic "tent," similar to a miniature hoop-type greenhouse. Use metal wire or old clothes hangers bent into semicircles to serve as hoops along the length of the flat. Anchor hoops in holes drilled in wooden 1×2s, or similar material that will hold the wire rigid. Stretch plastic such as Saran wrap, window treatment material, or Visqueen over the hanger "hoops." Leave the ends open for good air circulation. Clearance needs to be reasonably high so a misting head can be inserted into the center of the structure from either end to ensure good coverage when watering. Water regularly to maintain soil moisture and high humidity around the new cuttings.

Good air circulation is essential to prevent the development of fungal diseases. Make sure that leaves of the cuttings dry off quickly after each misting event. Consistent, gentle air circulation is all that is usually needed.

Rooting time varies from species to species. Most herbaceous plants root readily under warm, moist conditions and will be ready to transplant in a matter of a few weeks. Newly rooted cuttings should not be placed outdoors unless in a maintained nursery bed that receives regular watering and care. Most rooted cuttings are potted up and grown to a larger size for a month or longer prior to planting outside. After potting, a regular fertilizing program can be implemented to increase growth rates.[1]

10

The Prairie Food Web

The North American prairie is recognized by ecologists as the second most productive regional ecosystem after freshwater estuaries. The total annual production of plant and animal biomass (living material) exceeds that of forests, inland wetlands, and farm fields. The key to its productivity is an underground organic milieu that supports extraordinary levels of both plant and animal life. Its deep fertile soil provides prairie plants and associated soil organisms with a huge reservoir of nutrients. There are literally thousands of species of soil organisms per square foot of prairie soil. Studies have shown that when one or more groups of soil life are eliminated in grasslands, plant productivity is significantly reduced.

A tremendous diversity of other creatures is present in the prairie ecosystem, such as insects, spiders, birds, rodents, reptiles, amphibians, and mammals of all kinds. Some of the more readily observed forms of wildlife that utilize each prairie plant species are listed on the plant description pages as well as in the tables at the end of this book.

Insects: The Foundation of the Prairie Food Web

The number one "grazing group" in the tallgrass prairie is insects. They consume more vegetable material than bison, elk, antelope, and all other ungulates combined. Insects and other invertebrates (organisms with external skeletons) form the foundation of the prairie food web upon which most other animals depend. As the eminent entomologist E. O. Wilson of Harvard University has stated, "Insects are the little things that run the world."

Insects are perhaps most important in maintaining the balance of nature among their fellow invertebrates. For instance, predaceous and parasitic wasps help regulate populations of other insects, mites, spiders, ticks, and small invertebrates. Without these essential but still consistently underappreciated miniature workhorses, the balance of nature would soon be compromised in our gardens and the global agricultural economy.

Some people might be wary of attracting grasshoppers, leafhoppers, leaf miners, mosquitoes, bees, wasps, and beetles to their gardens and landscapes. However, these creatures play an essential role in the prairie food web and are fascinating in their own right. The prairie's amazing productivity is directly tied to its ability to support a diversity of invertebrates that serve as food for birds, reptiles, amphibians, and small mammals.

Mosquitoes on Common Milkweed; males especially can be found on flowers

Many insects are pollinators extraordinaire; some are "specialists" (oligoleges) that pollinate only one genus of plants or sometimes a single species within the genus. Honeybees and bumblebees are generalists and pollinate a wide variety of wild plants as well as many economically important crops. Nearly one-third of the food we eat is pollinated by insects. It was estimated in 2018 that honeybees alone accounted for more than $20 billion in crop value annually in the United States (Modern Agriculture, April 27, 2018). Wild bees are estimated to provide $4 billion in crop value annually in the United States.

Another $40 billion is accounted for by secondary agricultural products such as milk and meat, which also depend upon insect pollination to grow seeds for feed crops such as alfalfa and clover.

Without insects, our diet would be far less diverse than we enjoy today.

The diversity of bees and wasps is significant worldwide. For instance, a gardener in Leicester, England, Jennifer Owen, trapped insects in her garden over a period of ten years. She found 46 species of bees, 40 species of predatory wasps, and 553 species of parasitic wasps.[1]

Native Bees

Bumblebees and other native bees are indefatigable pollinators. Compared with nonnative honeybees, many native bees come into more direct contact with the pollen on flowers' anthers which increases the chance of cross-pollination.

Carpenter bee on Phlox

Bees are by nature docile and will only sting when provoked or their nests are threatened. Bumblebees can pollinate during cool, wet weather when honeybees and other pollinators are not active. They have been documented to pollinate hundreds of native plants, as well as a wide variety of economically important vegetables, berries, and fruit trees. Bumblebees are particularly important because they can pollinate certain plant species that other insects cannot. The power and frequency of their buzzing loosens recalcitrant pollen grains. They collect pollen from plants that do not produce nectar, thus pollinating species that honeybees find unattractive, including tomatoes, peppers, eggplants, cucumbers, and other important food crops. The importance of bumblebees as pollinators of native plants can hardly be overstated.

Bumblebee on Bergamot

Bumblebee on an aster

Numerous other native bees are vital to pollination. For example, sweat bees are solitary ground-nesting bees half the size of honeybees and are important generalist pollinators. They are attracted to the salt in human sweat and often land on bare limbs. They usually will not sting unless swatted, and the sting is mild and of short duration.

Sweat bee on an aster

Leaf-cutter bees are solitary bees similar in size to honeybees. They are important pollinators of many native flowers, especially legumes, which play a critical role in prairie ecosystems by adding nitrogen to the soil and supporting a wide variety of animal life, by producing seeds, forage, pollen, and nectar. Leaf-cutter bees are docile and have a mild sting which they use only when provoked. They cut plant leaves to line their nests in soft, rotten wood or in large, pithy plant stems.

Mason bees are dark, fuzzy, and slightly smaller than honeybees. They are solitary and build nests using mud, sometimes mixed with small sand grains. They are among the hardest-working and most effective pollinators. Two to three hundred mason bees are capable of doing the work of 10,000–20,000 honeybees. Like bumblebees, mason bees are active during cool, wet weather, even in drizzle, when many other pollinators are confined to their nests. This is particularly important for pollination of early-season wildflowers and orchard crops.

Mason bee houses, with predrilled holes of a diameter preferred by these bees, are available for sale.

Wasps

Wasps are divided into two groups: *predaceous* and *parasitic*. Predaceous wasps prey directly on other insects, spiders, and invertebrates, whereas

Wasp pollinating Rattlesnake Master

Tiny wasp pollinating Boneset

parasitic wasps lay eggs in these organisms that serve as hosts for their larvae. Nearly every pest species on Earth is preyed upon by a wasp species. The vast majority of wasps are never encountered by humans, many do not sting, and the group as a whole is instrumental in pollination as well as in controlling plant pests in our gardens and farm fields.

Wasps are also an important food source for other creatures. Small wasps are eaten by dragonflies, robber flies, and hoverflies. Large wasps are eaten by many different birds, as well as bats, bears, badgers, skunks, and mice.

Predaceous Wasps

Most predaceous wasps focus on one type of prey, such as moths, butterflies, beetles, spiders, other wasps, weevils, locusts, or even cockroaches. Some are generalists and prey on a wide variety of flies, bugs, and wasps.

Thread-waisted wasp

Left, grass-carrying wasp; *right*, digger wasp

The only truly aggressive social wasps are yellow jackets. They seek sustenance from any available source — which is why they are all too often associated with picnics and soda cans. Hornets and paper wasps, though well-armed and highly protective of their nests, will only sting when provoked. People can coexist in close quarters with these wasps as long as their territories are respected. Commonly reviled, social wasps perform a huge service in maintaining the balance of nature. They consume massive amounts of destructive insects that might otherwise reduce crop yields if not controlled by these natural predators.

Parasitic Wasps and Sawflies

The diverse group of parasitic wasps and sawflies helps control garden pests and other insects and arachnids by parasitizing their eggs, larvae, pupae, and adults. Most parasitic wasps do not sting. Organic gardeners value parasitic wasps as a front-line defense against pests.

Stinging parasitic wasps are not considered "true" parasitoid wasps, but their life cycles still involve the parasitism of a host organism. Interestingly, their most common prey is their close relatives: bees and predaceous wasps.

Milkweed tussock moth caterpillar being parasitized by wasp

Wasp laying eggs in milkweed tussock moth caterpillar

Potter (or mason) wasp

Parasitic cuckoo wasp maneuvering prey into an underground hole for its young

Some larger wasps specialize in parasitizing ground-nesting grubs, such as Japanese beetles and rose chafers. Others attack the larval, nymph, and adult stages of true bugs, including planthoppers and leafhoppers. Some internal feeders are unique in that their larvae shed successive skins until they break through the abdominal wall. Fully formed larvae then leave the sac to find a protected location to pupate.

Beautiful metallic green or blue cuckoo wasps are a nonstinging parasitic wasp. Their ovipositor has evolved so it can grasp objects and inject eggs into the larvae of bees, predatory wasps, sawflies, and walking sticks.

Spider wasps are mostly predaceous spider killers. The female stings and paralyzes a spider and lays an egg on it. The larva hatches and consumes the spider. Bethylid wasps specialize in parasitizing moth larvae, including grain moths, clothes moths, and plant-feeding moths such as leafminers, leafrollers, and fruit and shoot borers.

Sawflies are the least common parasitic wasps, and the most likely to be encountered by people. Sawflies look more like flies than wasps and do not sting. Their larvae feed on plants rather than other creatures and can cause problems in gardens and orchards. Most adult sawflies subsist on nectar and pollen except for a few species that apparently feed on small insects, and are known to be effective pollinators.

Plant-Feeding "Parasitic" Wasps

Although classified as parasitic wasps, a few species feed on plants. These include the gall-forming wasps, most commonly observed as round galls on goldenrod stems and the undersides of oak leaves. Adults lay eggs in plant tissue, and the larvae induce plants to construct a home for them in the form of protective galls. Larvae rely on food supplied by plants for their nourishment.

The velvet ant, sometimes known as the cow-killer ant, is actually a wasp

Velvet Ants

These denizens of dry sandy and rocky prairies look like ants but are actually parasitic, solitary, stinging wasps covered with dense hairs. Females are wingless and often seen roaming on the ground. Males are wingless, stingless, and appear very different from the females. The sting of the female is extremely painful. There are many questions surrounding the habits of velvet ants, but it appears that most of them parasitize mature larvae and pupae of ground-nesting solitary bees and solitary predaceous wasps. The most commonly encountered velvet ant in eastern North America is the cow killer, easily recognizable by its fuzzy bright orange and black body. It preys on large ground-nesting parasitic and predaceous wasps, laying eggs on the cocoon of the developing pupa. The ant larva devours the pupa over winter and emerges next spring as an adult.

Ants

Ants are an often overlooked component of the prairie. They can account for up to 33% of insect biomass in prairie ecosystems, especially in dry prairies and shortgrass prairies of the Great Plains.

Ants incorporate plant and animal organic matter into the lower soil, enriching the lower rooting zone. Their tunnels encourage rapid infiltration of rainfall into the soil, reducing runoff and facilitating subsoil moisture recharge. Because most of their work is accomplished underground, it is easy to discount the impact ants have on a given ecosystem.

Anthill in a prairie

Most ants of the North American prairie are generalists when it comes to diet, although there are some species that specialize in carnivory, seed eating, recycling detritus, aphid farming (see below), and even enslaving other ants. Larger ants actively hunt other insects and spiders, especially soft-bodied insects and larvae, although grasshoppers, crickets, and other large prey are taken as well. Smaller ants tend to be scavengers, foraging for dead invertebrates, worms, and fruit.

Flies

Tachinid flies often look like houseflies. Most are parasites that develop inside their hosts, killing them. A few are strictly parasitic and do not kill their hosts. Favorite targets include moth and butterfly caterpillars, sawflies, beetles, true bugs, grasshoppers, and even centipedes.

The tachinid fly's larval stage is parasitic. Photo courtesy of Doug Tallamy.

Robber flies are powerfully built, with stout, spiny legs for grasping prey. Favorite foods include grasshoppers, dragonflies, and other robber flies. Some species attack wasps and bees.

Hoverflies

Hoverflies are the helicopters of the fly world. Adults of many species feed on flower pollen and nectar. Stingless, they mimic small wasps and sweat bees as a defense mechanism. The larvae of many species eat aphids, thrips, and other plant-sucking insects, making them desirable members of the garden. Others subsist on decaying plant and animal matter in their larval stage.

The hoverfly, also known as a syrphid fly or flower fly, is a wasp look-alike, on Zizia (*left*) and Hypericum (*right*)

Bee flies belong to a large family of parasitic flies, with species ranging in size from tiny (0.08 inch) to quite large (3 inches). Most have long, swept-back wings longer than their bodies and appear similar to bees. Most adults feed on flower nectar and pollen, and some are important pollinators. Bee flies typically possess a very long proboscis and can pollinate long, tubular flowers that other pollinators cannot reach.

Bee fly, a bee mimic

Dragonflies

Dragonflies are predaceous, and their nymphs (larvae) develop in aquatic ecosystems. Adults lay eggs on floating leaves or other materials in or near water. Nymphs typically climb up an emergent plant stem and exit the larval stage as adults. Nymphs consume small invertebrates, and mosquito larvae are among their favorites. Adult dragonflies subsist on mosquitoes and small flies, bees, ants, wasps, and rarely butterflies. In turn, they are eaten by fish, aquatic bugs, birds, frogs, lizards, and spiders. Strong fliers, they have been clocked at over 30 miles per hour. Dragonflies and their close relatives, damselflies, are commonly found in wet prairies, sloughs, prairie ponds and potholes, and along rivers and streams.

Sixteen of North America's 326 dragonfly species migrate seasonally. The common green darner travels from Canada and the northern United States to southern states, Mexico, and the Caribbean. A resident population of green darners breeds in the north without migrating. A separate population migrates south every year in September to breed in Mexico and central America in winter, and their offspring return north to breed in spring. While the resident northern populations are believed to remain in their home waters, migratory populations return to different bodies of water each year. Migrating green darners are an important food source for migrating kestrels.

Dragonflies and damselflies hunt their prey over prairies and lay their eggs in prairie pools. Some dragonfly species migrate farther than monarch butterflies.

Damselflies

Similar to dragonflies in appearance, damselflies typically have smaller, narrower bodies. The most commonly encountered damselflies in eastern North America are blue. They hold their wings folded next to their bodies when at rest, while dragonflies extend their wings outward. Damselflies' life cycle and food habits are similar to those of dragonflies.

Damselfly and dragonfly naiads (nymphs) are a vital part of the aquatic food chain in wetlands and wet prairies and a voracious predator of mosquitoes

Beetles

Beetles are the most numerous group of creatures in the world. It is estimated that only about 25% of the planet's beetles have been described to date, and they are likely to account for some 40% of all life forms. They come in all shapes and sizes and typically consume plant leaves, flowers, fruit, and pollen. Beetles as a group perform two very important functions in ecosystems: they decompose a wide variety of plant and animal waste, and they serve as food for a host of other insects, spiders, birds, reptiles, and small mammals.

Cerambycid (longhorn) beetles on Common Milkweed

Many beetles are associated with the pollination process. Others are active in surface soil and the upper root zone, and they recycle feces and dead plant and animal material. Some are predators of other insects and invertebrates, such as the well-known aphid-eating ladybug. Lightning bugs, common in prairies and wet meadows, are vicious predators. Their larvae hunt down earthworms, snails, and slugs, inject them with toxins and digestive enzymes, and suck them dry. As with bees and wasps, there are also parasitic beetles that live on other insects as well as small rodents and beavers. A few species even consume fungi.

Although some beetles can be serious garden pests, the vast majority go about their business of pollinating, recycling, and inadvertently feeding other creatures.

Goldenrod soldier beetle on Yarrow flowers (*Achillea millefolium*) common to prairies and meadows throughout the northern hemisphere

Weevils

Weevils are among the most destructive of beetles, attacking crops and stored grain. The Baptisia seedpod weevil commonly infests seedpods of the prairie legumes Cream False Indigo (*Baptisia bracteata*) and White False Indigo (*Baptisia* alba). In spring, overwintering adults lay eggs in young seedpods after flowering. The larvae consume developing seeds, pupate in midsummer, and adults emerge as the seedpods ripen and open in late summer. *B. bracteata* is almost always infested annually, but infestations appear to be cyclic with *Baptisia alba*.

Plants in the genera *Helianthus* (Sunflowers), *Echinacea* (Coneflowers), and *Silphium* (Compassplant, Prairie Dock, Cup Plant, Rosinweed) may be attacked by the sunflower head-clipping weevil. It partially severs stems a few inches below the flowers so they droop. Eggs are laid in the cut, drooping stems. Eggs hatch, and larvae consume the remnants of the flower. The heads drop to the ground, and larvae overwinter in the soil. They pupate in spring, and adults emerge in midsummer in time to start the cycle again.

The goldenrod soldier beetle is common on Goldenrods (*Solidago* spp.), *Silphium*, and Blackeyed Susans (*Rudbeckia* spp.). The metallic green dogbane beetle occurs on Milkweeds (*Asclepias* spp.) and Dogbanes (*Apocynum* spp.).

Bugs

"True bugs" have two sets of wings, with the outer pair being armored for protection. They possess piercing mouthparts for sucking fluids from plants and animals, and they undergo incomplete metamorphosis without a larval or pupal stage. Two of the most commonly encountered types of true bugs include assassin bugs and ambush bugs.

Assassin bugs are usually brown and gray in color, and they possess a short, thin proboscis for sucking fluids from their victims. Most live outdoors and attack a wide variety of insects, spiders, and mammals. Some frequent flowers, attacking bees, flies, and other pollinators. The spined assassin bug feeds on a wide variety of destructive insects and is considered a beneficial garden predator. The wheel bug gets its name from the wheel-like structure on its back. This large bug (up to 2 inches) is a ferocious predator. It injects venom and enzymes into its prey and slowly extracts their fluids. Its bite is extremely painful to humans and can take months to heal, often leaving a lifelong scar. Wheel bugs are particularly fond of Japanese beetles.

Ambush bugs are more brightly colored, often orange, red, and yellow. They have bigger, stronger front legs for holding their quarry. While most assassin bugs actively pursue their prey, ambush bugs are content to lie in wait for their victims. Most assassin bugs primarily frequent vegetation, but ambush bugs commonly hunt for pollinators on flowers.

Nymphs of ambush bugs and assassin bugs are eaten by birds, praying mantises, spiders, and rodents, as well as other ambush and assassin bugs. Both can fly, but not very well. Adults are also subject to predation

The wheel bug, an assassin bug, is a predator of Japanese beetles. Photo courtesy of Sam Bahr.

but have harder shells than nymphs and are thus better armored against attack.

Leafhoppers, Spittlebugs, Aphids, Scale, and Cicadas

Aphids, leafhoppers, leaf miners, plant lice, and other sucking insects abound in the prairie. These are consumed by a variety of predaceous invertebrates, including spiders, wasps, and ladybird beetles. Eggs of leafhoppers are attacked by certain parasitic wasps. Although leafhoppers and other sucking insects may cause discoloration and localized damage, they are usually kept in check by other invertebrates.

Leafhoppers suck sap from a variety of trees, shrubs, flowers, and grasses. Some species specialize on prairie grasses, such as Sideoats Grama (*Bouteloua curtipendula*), Needlegrass (*Stipa spartea*), Prairie Cordgrass (*Spartina pectinata*), and Little Bluestem (*Schizachyrium scoparium*). Prairie Dropseed (*Sporobolus heterolepis*) is an obligate host for a couple of these, too. A few leafhoppers are known to eat aphids, but the vast majority of leafhopper species are vegetarian.

Leafhoppers have strong rear legs covered with hairs that are used to apply secretions to the body; the hairs repel water and emit pheromones. The sister or six-spotted leafhopper carries the aster yellows virus, which affects members of the Aster family, as well as other widely used ornamental garden plants. Birds, lizards, spiders, assassin bugs, robber flies, and predaceous wasps are primary predators of leafhoppers.

Spittlebug larvae (nymphs) surround themselves with a white foamy liquid to obscure and protect them from predators. The foam helps moderate temperature and moisture, plus it has an acrid flavor that helps repel predators. Nevertheless, they are eaten by birds, mantids, spiders, and wasps.

Many prairie forbs are targets of spittlebugs, which are readily visible after long periods without rain. Most gardeners have seen these gooey masses attached to their plants, which often result in deformed leaves.

Aphids and Scale

Aphids are one of the most common and successful groups of plant-sucking insects. Widely distributed in temperate climates, they attack a variety of plants. They are known to migrate long distances on wind currents. Most aphids are soft-bodied insects, ranging in color from bright red to black, brown, green, and nearly colorless. Many aphid species are specialists and feed on only one or a few related species of plant. Others are generalists and feed on dozens or hundreds of different, unrelated plants.

Aphids are small sap-sucking insects; their predators include ladybird beetles, lacewings, and syrphid flies

Gardeners commonly encounter aphids on plant stems, often in large numbers, where they suck copious quantities of plant fluids. Some species of aphids live in the ground and feed on roots. Because aphids are essentially defenseless, living underground affords protection from many aboveground predators.

Aphids produce highly nutritious honeydew on their posteriors, attracting ants, which collect it for food. Certain species of ants and aphids have a symbiotic relationship in which the ants actively "farm" the aphids. The ants will "milk" the aphids by stroking their abdomens with their antennae. Some "farming ants" go so far as to carry aphid eggs to their nest to protect them over the winter, and then they deliver the freshly hatched aphids back to plant stems or roots in spring.

Ants "farming" aphids   An ant "milking" an aphid

Aphids are particularly fond of milkweeds. It is not uncommon to see aphids literally coating the stems of members of this genus. Ladybird beetles, both larvae and adults, are the best known predators of aphids.

Lacewing and syrphid fly larvae also prey upon them. Aphids are parasitized by a number of parasitic wasps.

Milkweed aphids on Butterflyweed

Ladybird beetle larvae and adults are the best-known predators of aphids

Lacewings

Lacewings are characterized by their diaphanous wings, often with slightly iridescent green, white, and sometimes light-brownish coloration. Adults are nocturnal and can most commonly be observed at dawn or dusk. Some species feed on plant pollen, nectar, and honeydew. Certain larvae feed on aphids, mites, and tiny arthropods, and they are valued for controlling aphids in gardens. Names for lacewings include "aphid lion" and "aphid wolf" because of their appetite for aphids. They are capable of wiping out entire aphid colonies, hence their value to agriculture.

Green lacewings are another predator of aphids in both the larval and adult stages

Green lacewings are the best known of the group, with some species commonly sold for controlling aphids, red spider mites, whitefly, and thrips.

Adult lacewings typically feed on pollen and nectar. They overwinter in leaf litter, emerging in spring to lay eggs for the next generation. Larvae are voracious, consuming not only aphids but also larger caterpillars and other insects. When these food sources are scarce, they can turn to cannibalism and consume their own kind. Under favorable warm-weather conditions, a new generation can be produced every month.

Mantids

Three species of mantises are typically encountered in the prairie region of North America. Of these only the Carolina mantis is native. The praying mantis was introduced from Europe, and the Chinese mantis from the Far East.

Mantises are efficient predators that eat other insects and invertebrates. Their powerful front legs possess sharp barbs with which they impale and hold their prey in place as they consume them. Mantises experience "incomplete" metamorphosis, wherein nymphs molt throughout the summer until they reach adulthood. The female famously consumes her mate following, or even during, copulation, which provides her with additional nutrition for egg production.

The female Carolina mantis consumes part of the male during and after mating for extra protein to produce eggs

She lays her eggs in late summer or fall, inside a protective substance that allows them to overwinter and hatch the following spring. It is not unusual for emerging young mantises to eat their brothers and sisters as they hatch.

Grasshoppers, Crickets, and Katydids

GRASSHOPPERS

The grasshoppers include both grasshoppers and locusts. They feed primarily on prairie grasses and are an important component of the diet of birds, snakes, frogs, small mammals, and even hungry coyotes. In years when they are abundant, the various species of grasshoppers represent one of the most significant grazers of the prairie, rivaling bison in terms of total biomass consumed. Highly nutritious, grasshoppers are an essential source of protein in the prairie food web.

CRICKETS

Represented by nine different groups known to occur in North America, crickets are nearly ubiquitous in our grassland ecosystems.

The prairie mole cricket was once common to the tallgrass prairie. Mole crickets can grow up to 2 inches long, and have broad front feet that resemble those of moles. They live in burrows and typically have a two- to three-year life cycle, from egg to adult. Once thought to be extinct, mole crickets are rare across their former range, sensitive to plowing and grazing. They are now found primarily in hayed meadows and undisturbed grasslands, and they do no economic damage to vegetation. Their diet includes insect larvae, worms, roots, and grass leaves. They are eaten by various beetles, assassin bugs, wolf spiders, birds, skunks, snakes, frogs, raccoons, and foxes.

Field crickets are among the most common crickets in North American grasslands. They are omnivorous, eating plant material, seeds, small fruits, grasshopper eggs, butterfly and moth larvae, and flies. Unlike the prairie mole cricket, field crickets do not overwinter as adults or nymphs, and rely upon eggs laid in fall to regenerate the following spring.

KATYDIDS

Both male and female katydids typically rub their forewings together to communicate, unlike grasshoppers and crickets where only the male rubs his hind legs against the thorax to "sing."

Most katydids reside in trees, shrubs, and brushy fields. Katydids are weak fliers and prefer to climb or walk on plants. They eat the leaves of deciduous trees and shrubs, especially oaks. Katydids are eaten by birds, bats, frogs, snakes, and spiders.

Early-instar field katydid on Fringed Puccoon

Many species of the cricket-like meadow katydids reside in prairies, meadows, and fields. Like crickets, only the males sing to attract female mates. Nymphs feed on grasses, with adults augmenting their diet with aphids, caterpillars, and bugs. Some species of meadow katydids are considered pests of the farm and garden, as they are known to eat certain agricultural crops.

Butterflies and Moths

Many prairie flowers are well known as butterfly magnets. The genera *Asclepias, Aster, Echinacea, Eupatorium* and *Eutrochium, Liatris,* and *Vernonia* produce nectar that is particularly attractive to these lovely garden visitors.

Many prairie plants also serve as hosts for caterpillars, such as the Milkweed, which support the monarch butterfly. Even prairie grasses are hosts for

Monarch on Blazing Star flower

Monarch on Swamp Milkweed

Common wood-nymph on Joe Pye Weed; their larvae eat bluestem grasses

certain butterfly larvae, notably skippers. Grass stems are utilized by some butterflies for suspending their chrysalids, including swallowtail species.

Swallowtails hang from grass stems during metamorphosis

The Karner blue butterfly (*Plebejus melissa samuelis*) is federally endangered. Its larvae depend exclusively on the leaves of *Lupinus perennis* (p. 214).

A seasonal succession of nectar-producing prairie flowers will attract a wide variety of butterflies. In fall allow dead plant material to stand to preserve overwintering eggs (coral and Edwards' hairstreaks, coppers, and Karner blue), larvae (eastern tailed blue, fritillaries, checkerspots, and skippers), and chrysalids (swallowtails, hairstreaks) that will produce the next generation in spring.

Moths are an often overlooked but fascinating part of native woodlands and prairies. There are 800 species of butterflies, but 11,000 species of moths known to occur in North America. The importance of moths as pollinators is greatly undervalued. Most of the highly visible "sexy" moths (e.g., *Cecropia, Polyphemus, Luna*) are active at night, with larvae that feed on native trees. Hummingbird moths or sphinx moths are diurnal (active during the day) and nectar on a variety of prairie flowers. Commonly mistaken for hummingbirds, they are also prodigious pollinators. They insert their long tongues into tubular flowers, picking up pollen in the process. Preferred flowers include Evening Primrose (*Oenothera* spp.), *Liatris* spp., Lily (*Lilium* spp.), Milkweed (*Asclepias* spp.), *Monarda* spp., *Penstemon* spp., *Phlox* spp., and *Verbena* spp. Thousands of less-noticeable moths perform valuable pollination services on the night shift.

A hummingbird "mimic," hummingbird moths are a diurnal species (active during the day), unlike nocturnal moths. There are four species in the United States: hummingbird clearwing, snowberry clearwing, slender clearwing, and white-lined sphinx moths.

The Clymene moth is active day and night and feeds on Joe Pye species as well as on oaks, willows, and other plants

Moths are in decline due to the disruption of their stellar navigation systems by streetlights, dusk-to-dawn lights, and vehicle headlights. Backyard bug zappers kill millions of moths annually. The indiscriminate and unnecessary use of insecticides has decimated populations of butterflies, moths, and other pollinators. The encroachment of civilization into their habitats makes it ever more critical that home gardeners establish sanctuaries of native trees, shrubs, and prairie flowers and grasses. These complementary plant communities provide multiple habitat requirements for a wide variety of birds, butterflies, pollinators, and other desirable wildlife.

The tables at the end of this book provide specific information on which prairie plants serve as hosts for various butterfly and moth larvae.

Arachnids

SPIDERS

Spiders are well represented in the prairie, from nearly microscopic species to the large and glorious, web-weaving black-and-yellow garden spider. Their large orbicular webs with Z-shaped patterns that glisten with morning dew are characteristic of the garden spider's work.

An orb-weaving garden spider familiar to gardeners

The garden spider's close relative, the banded garden spider, is also a common prairie resident, along with the smaller shamrock orb weaver. Interestingly, females, with their large, rounded bodies, are generally three to four times larger than the less conspicuous males.

Not all spiders on the prairie are web weavers. Acrobatic jumping spiders are members of the largest group of spiders, many occurring in grasslands. Rather than weave webs, they stalk their prey and pounce by jumping, using an internal hydraulic system rather than muscles to propel themselves. When they jump they produce a thin filament of silk as a tether to prevent falling to the ground.

The bright colors of some jumping spiders, in combination with their shiny eyes, can give them an almost jewel-like appearance. Although almost all spiders are carnivorous some jumping spiders utilize plant nectar and pollen. Agile wolf spiders pounce on and sometimes chase after their small insect prey. Most do not weave webs although some lurk in funnel-like webs they construct at ground level in grass. Intrepid gardeners may come upon the secret lairs of wolf spiders, hidden at ground level among the prairie grasses.

An amazing diversity of spiders lurks in prairie flowers, some nearly invisible, lying in wait for unsuspecting visitors. The goldenrod crab spider, for example, inhabits yellow and white flowers and can change its color

Crab spider on a Seedbox flower (*Ludwigia alternifolia*), native to the eastern half of the United States

Big Bluestem flowers with crab spider and its prey

from yellow to white in about a week, and from white to yellow in about a month. They are strictly hunters and do not spin webs to catch prey, but they will create webs to hold their eggs. Their favorite prey is wasps, bees, and flies. They also consume ants, stinkbugs, and beetles, including the spotted cucumber beetle, a common garden pest.

DADDY LONGLEGS

Also known as harvestmen, daddy longlegs differ from spiders in that they lack venom glands, have only one body segment instead of two, and have two eyes instead of the usual eight in most spiders. Among their favorite foods are aphids. They also eat other small insects and spiders, dead invertebrates, and vegetable matter.

Daddy longlegs can emit a foul-smelling liquid from scent glands located above the hip joints of their front pairs of legs. They also have legs that are readily detached when grasped by a predator, allowing them to make an escape on their seven remaining appendages.

Birds

Hummingbirds are frequent visitors to a wide variety of nectar-producing prairie flowers. They require trees for nesting sites and roosting, and are commonly found in backyard prairie meadows and gardens with nearby trees and woodlands. Hummingbirds utilize tubular-shaped flowers of many colors, not just red flowers, as is commonly believed.

Ruby-throated hummingbird taking nectar from royal catchfly flowers. Photo courtesy of Jim McCormac.

Hummingbirds are predaceous, with small insects constituting a critical protein source in their diet. The prairie supports a huge diversity of small invertebrates, thus providing protein in addition to nectar for hummingbirds. Hummingbird feeders provide only nectar, but your prairie will supply them with a more balanced and complete diet.

Numerous songbirds, game birds, and raptors frequent prairies, savannas, and woodland edges. Many require large acreages of open grassland, but a number of them will visit small prairie gardens and meadows.

Barn owls need large areas of open land over which to hunt

Great-horned owls forage over grasslands; a baby great-horned owl nests here with an adult

Sandhill cranes in flight

Sandhill cranes foraging in a prairie in early spring

Young red-tailed hawk in an Indiana garden. Red-tailed hawks usually hunt in open prairies and grasslands.

The prairies' main attraction for songbirds and game birds is the smorgasbord of insects and invertebrates sustained by prairie plants. Although seeds are an important element of some birds' diets, the high protein diet afforded by insects is critical to chick survival, nesting success, and overall avian welfare.

Birds are credited with billions of dollars' worth of insect and pest control annually. They serve the same purpose in the home landscape. Healthy, diverse native landscapes are inherently sustainable and require little or no pesticides when a diversity of insects, birds, and other life is encouraged.

The inclusion of woody plants provides additional insect food sources,

The American kestrel is a wide-ranging small falcon that inhabits grasslands and woodland edges. Its diet consists of large insects, spiders, and small rodents, as well as amphibians, reptiles, and small birds. Photo courtesy of Joel Trick.

Bluebirds nest on the edges of prairies and woodlands. They thrive on the abundance of insects that utilize prairie flowers. Berries are also an important part of their diet, as well as occasional larger creatures such as lizards and tree frogs.

Red-headed woodpeckers nest in savannas and forests and hunt insects in the air and on the ground in prairies and meadows

as well as nesting sites and seeds for birds. Caterpillars of many butterflies and moths also feed on leaves of native trees and shrubs. A combination of prairie plants with native woody species will attract the largest variety of birds, moths, and butterflies.

Polyphemus moth caterpillars feed on trees associated with prairie edges. Photo courtesy of Doug Tallamy.

Small Mammals

Rodents

Rodents and other small mammals are an important part of the prairie eco-system. Although considered "vermin" by some people, mice, voles, and shrews are preferred foods for many hawks, owls, foxes, and coyotes. Were it not for small mammals, there would be precious few soaring raptors gracing our prairie landscapes.

Red-tailed hawk soaring high. Photo courtesy of Joel Trick.

One of the most entertaining creatures of the prairie is the meadow jumping mouse, which has a tail three times as long as its body and can jump five feet in the air. This seldom-seen almost comical creature is a fa-vorite with those who have the good fortune to encounter it. This mouse dwells strictly outdoors and will not leave its calling card in your silverware drawer, unlike the house mouse.

Moles

The eastern mole is the most common mole found in North America and ranges across all but the northernmost reaches of the tallgrass prairie. It prefers well-drained sandy and loamy soils in prairies, meadows, fields, and open woodlands. It digs burrows up to 6 feet deep. Moles help to aerate soil while also controlling grubs, slugs, earthworms, and other creatures that feed on plant roots. Foxes and coyotes are common predators of the eastern mole, as are dogs and cats.

The star-nosed mole is found in wet lowland prairies in the northeastern portion of the eastern tallgrass prairie. It is strictly carnivorous, consuming worms, aquatic insects, mollusks, grubs and other invertebrates, and even small amphibians and fish. This mole is active both day and night, and it re-mains active in winter. It has been known to tunnel through snow and has been seen swimming in icy streams. It is eaten by skunks, weasels, otters, red-tailed hawks, owls, and even large fish.

Groundhogs (Woodchucks)

The solitary vegetarian groundhog prefers grasslands and forest edges. Groundhogs' primary diet consists of grasses, but they will also consume various flowers. This true hibernator is most active in the mornings and evenings of summer and fall. An unwelcome garden visitor, it is a favorite repast of coyotes, bobcats, large raptors, and snakes. The best defense against marauding groundhogs is a large dog.

Lagomorphs (Rabbits)

Eastern cottontail rabbits occur throughout the eastern half of the United States and the entire prairie region. They thrive in open grasslands, savannas, shrubby thickets, and forest clearings. Although their nests are constructed of grass and other herbaceous plant material, cottontails require brushy cover to escape from predators. The eastern cottontail digs shallow nesting holes 4–6 inches deep but does not create its own burrows. For winter cover it utilizes abandoned dens originally created by burrowing animals such as groundhogs and badgers.

The cottontail is primarily vegetarian, consuming grasses, sedges, leaves, flowers, fruit, seeds, twigs, buds, and bark. It is known to eat arthropods, such as insects, spiders, and small crustaceans.

Cottontails make a fine repast for many predators, including hawks (they are a red-tailed hawk favorite), great horned owls, barred owls, coyotes, foxes, bobcats, weasels, raccoons, minks, snakes, and even crows. Young rabbits are taken by badgers, skunks, and possums.

The prairie hare, or white-tailed jackrabbit, is actually a hare not a rabbit. Hares are born with fur, their eyes open, and are able to move on their own, whereas rabbits are born hairless and blind, and they are helpless for the first few days of their life. The white-tailed jackrabbit is the largest of the North American rabbits and hares, weighing up to 10 pounds (4.5 kg). They are common in the northern Great Plains but also occur on the western edge of the eastern tallgrass prairie in Wisconsin, Minnesota, Iowa, northwestern Missouri, northeastern Nebraska, and eastern North and South Dakota.

Like eastern cottontails, white-tailed jackrabbits depend on a variety of herbaceous and woody vegetation for their sustenance. Common in open grasslands, they are not as dependent as cottontails on woody vegetation for protective cover. They are preyed upon by foxes, coyotes, great horned owls, and large hawks.

Weasel Family

The American badger is a true creature of prairies and meadows. It prefers open grasslands and is seldom found in wooded areas. It has few natural enemies by virtue of its fierce disposition and powerful build. Its strong legs and sharp claws make the badger a digging machine so efficient at eliminating burrowing rodents that it must pursue a nomadic life in search of new prey. It feeds on rodents such as ground squirrels, chipmunks, gophers, mice, and voles. Badgers are also known to eat ground-nesting birds and their eggs, as well as snakes, lizards, insects, and beetle grubs. On the Great Plains, badgers often subsist on prairie dogs. They dig deep burrows, usually 15–30 feet in length. Abandoned dens and hunting excavations make fine homes for tortoises, coyotes, snakes, skunks, and burrowing owls.

Other members of the Weasel family found on the eastern tallgrass prairie include the North American river otter and American mink, which inhabit rivers and streams in grasslands and woodlands. Otters subsist primarily on fish and crustaceans, especially crayfish. They have been known to catch and eat ducks and other small birds. Other food sources include reptiles, amphibians, small mammals, aquatic insects, and mollusks.

One might encounter various weasels on the prairie, including the long-tailed weasel, short-tailed weasel (or stoat), and the least weasel. They eat primarily small rodents such as the white-footed mouse, deer mouse, meadow vole, and North American least shrew.

The wolverine is a former resident of the eastern tallgrass prairie. It competes directly with humans for the same top-of-the-food-chain ecological niche and is known to track down people who violate its territory. As a result the species has been pushed close to extirpation in its former range.

Black-footed ferrets were once common on the Great Plains, preying at night upon sleeping prairie dogs. The last known population of black-footed ferrets in North America was taken into captivity in 1985, and animals were bred and released into the wild. They have since made a modest comeback but are still considered an endangered species. As populations of prairie dogs have declined due to hunting and poisoning by ranchers, populations of ferrets have also plummeted.

The black-footed ferret is a member of the weasel family and roughly the same size as a mink. Its diet consists primarily of prairie dogs, augmented by mice, ground squirrels, rabbits, and the occasional reptile, bird, or large insect. Photo courtesy of USGS National Wildlife Health Center.

The prairie dog's range includes the shortgrass, mixed-grass, and the western edge of the tallgrass prairie regions. They are considered a keystone species.

Reptiles and Amphibians

Various frogs, lizards, turtles, and snakes are prairie residents. Chorus frogs, wood frogs, northern leopard frogs, and American toads are common prairie amphibians.

Green frogs are commonly found in most permanent aquatic habitats, including prairie ponds and potholes. Photo courtesy of Joel Trick.

Northern leopard frogs can be found in grassy meadows a fair distance from water during summer

Gray tree frogs are nocturnal; this one was in a burr oak in a barrens in Kentucky

Northern cricket frogs get their name from their calls, which sound like crickets. Their preferred habitats are sunny shallow ponds with cover afforded by shoreline vegetation and slow-moving streams with sunny banks.

Chorus frogs and wood frogs are known for their spring concerts as the ponds in which they hibernate thaw, and mating season begins.

Eastern tiger salamanders inhabit ponds, vernal pools, and banks of slow-moving streams in wet prairies, marshes, and woodlands. They are most common in areas with sandy or friable soils in which they can burrow to maintain skin moisture and to hibernate. Like many reptiles and amphibians, salamanders are becoming increasingly rare.

The ornate box turtle and the legless slender glass lizard are two rare species that prefer dry sandy prairies. They feed primarily on grasshoppers and other insects and invertebrates. Glass lizards have the ability to disconnect their tails to escape from predators.

Snakes play an important role in prairies, subsisting on insects and small mammals. Among the most familiar is the common garter snake.

The American toad, a master of camouflage

Eastern tiger salamanders tend to be associated with grasslands, savannas, and woodland edges and less so with forests. Photo courtesy of Matt Jeppson, Shutterstock number 85160425.

The ornate box turtle is one of only two terrestrial species of turtles native to the Great Plains

Slender glass lizards resemble a snake but are legless lizards. Photo courtesy of Joel Trick.

Smooth green snakes are found in moist, grassy areas, usually in prairies, pastures, meadows, and marshes

The rough green snake appears very similar to the smooth green snake but prefers woodland edges and open fields

Garter snakes can be found in many types of habitat, including lowland and upland grasslands

One of the gardener's best friends, the garter snake's diet consists of slugs, leeches, and rodents, as well as earthworms, lizards, toads, and other amphibians. The entertaining and harmless eastern hog-nosed snake mimics a cobra when threatened, complete with arched neck, extended head, and loud hissing. If the cobra imitation proves unsuccessful, the eastern hog-nosed snake will play dead and emit a foul-smelling substance in the hope that the predator loses interest.

Eastern hognose snakes are mostly found in areas with dry, sandy, or mixed sandy soils and in grassy fields, crop fields, and along woodland edges. Photo courtesy of Ryan M. Bolton, Shutterstock number 13526554(3).

The bull snake can grow to an impressive eight feet in length. Similar in appearance to the western diamondback rattlesnake, it does an excellent impression when threatened by rearing up its head, hissing, and rattling its tail against leaves or grass.

The bull snake is a prairie-dwelling species that eats mice, squirrels, rabbits, and birds. Photo courtesy of Matt Jeppson, Shutterstock number 82495150(1).

The eastern racer is another common grassland dweller that feeds on insects and small mammals. True to its name, it is extremely fast and prefers to "run" away from humans rather than confront them.

Poisonous snakes are infrequent on eastern prairies, limited to the rare massasauga or pygmy massasauga rattlesnake. This small rattlesnake prefers marshy areas and stream banks but will also venture into upland

Eastern racers prefer dry open fields, meadows, forest clearings, and prairies. Photo courtesy of Todd W. Pierson.

Due to habitat loss, the Massasauga pygmy rattlesnake is listed as a threatened species. Photo courtesy of James Horton and the Western Pennsylvania Conservancy.

grasslands. Chances of encountering a poisonous snake in the eastern tallgrass prairie are slim, and far less likely than in woodlands and wetlands.

Canines

Wolves can thrive in a diversity of habitats, from the tundra to woodlands, forests, grasslands, and deserts

The original prairies were home to the gray wolf and the eastern timber wolf. Not much is known about the original ranges of these species, as early American settlers hunted them relentlessly. What is known is that wolves were the top predator in the eastern tallgrass prairie food chain, along with the occasional wolverine. They subsisted on large mammals such as deer and elk, removing weak animals from the population and helping to maintain strong herds. Wolves are also known to eat rodents such as mice, voles, and shrews when larger prey are not available. Wolves are larger than coyotes, and their pack size is typically bigger, which makes them a dominant predator.

With the reduction of bison, deer, and elk populations by settlers, wolves became a threat to livestock and, as mentioned, were hunted relentlessly. Until recently, wolves remained only in western Canada, Alaska, and northern Minnesota. With protection under the Endangered Species Act, reintroduction efforts, and natural migration in the Upper Midwest, remnant wolf populations have recently reoccupied significant portions of their historic range.

Until the early twentieth century, the coyote was a predominantly western species, rarely found east of the Mississippi River. Where wolves have been displaced, leaving an open ecological niche for a top predator, coyotes have proved adept at colonizing territory. Coyotes are more tolerant of people than wolves, and so adaptable that they can thrive in urban and suburban settings wolves would carefully avoid.

Coyotes, a member of the dog family, once lived primarily in open prairies and deserts. Along with other predators, they play critically important roles in grassland ecosystems.

Coyotes are opportunistic omnivores, preferring small mammals such as rabbits, mice, voles, and ground squirrels. They will take down weak deer, elk, and farm livestock, and they are known to eat snakes, lizards, birds, and large invertebrates such as grasshoppers and grubs. Suburban coyotes frequent parks, cemeteries, and green spaces, subsisting primarily on small rodents, rabbits, deer, and fruit. Studies have shown that less than 2% of suburban coyotes' diets include garbage, cats, and small dogs. Urban coyotes often subsist on a diet of rats and mice.

The primarily nocturnal and omnivorous red fox has established populations in both urban and rural situations worldwide. It subsists on mice, squirrels, rabbits, birds, crickets, grasshoppers, beetles, and fruit, among

Red foxes are commonly associated with grasslands, boreal forests, coniferous forests, deciduous forests, and tundra

others. Red foxes prefer open spaces within moderately populated suburban areas, and coyotes reside in less developed wilder areas, often on the edge of human settlement. These two canids generally coexist in and around urban environments and help control vermin.

Ungulates

The eastern tallgrass prairie previously supported a wide variety of ungulates, or hoofed quadrupeds, particularly American bison (commonly referred to as buffalo), elk, and the extinct eastern woodland bison. Although elk are now found primarily in the mountain west, they were once common in grasslands and savannas and roamed the eastern United States. In 1828, the explorer Henry Schoolcraft noted elk, moose, and caribou at the interface of northern prairie and woodland in the area of Green Bay, Wisconsin.

Prairies that are grazed and trampled by bison and other ungulates have higher plant diversity

The hooves of these animals created small disturbances in the prairie sod, encouraging growth of early successional species, which are otherwise squeezed out by longer-lived plants over time. Bison also created "buffalo wallows" in low-lying depressions where they rolled in mud and dirt, ostensibly to control external parasites. Prairies that are grazed and trampled by bison and other ungulates have higher plant diversity due to the creation of microsites of soil disturbance where annuals, biennials, and short-lived perennials can germinate from dormant seeds in the soil. Wallows created similar but larger areas where ecological succession could begin anew.

The diet of bison and elk consists mainly of grasses. Early European accounts of prairies noted an abundance of flowers. Today's prairie restorations often become dominated by grasses over time as a result of a lack of selective grazing pressure on grasses by ungulates. To compensate for the lack of grazers in contemporary prairie restorations, the ratio of flowers to grasses in seed mixes should be high, and seeding rates for tall prairie grasses should be minimal. More detailed information is available in chapter 6, "Successful Prairie Meadow Establishment."

Large ungulates no longer roam freely across the American prairie. However, one can discover the incredible diversity of creatures that depend on prairie plants in one's own backyard. While birds, butterflies, and bees will flock to prairie gardens, a magnifying glass can reveal an undiscovered world of tiny creatures that are ecologically essential and make these flowers and grasses their homes. These seemingly unimportant organisms are the true building blocks of the prairie food web upon which so many others depend — hidden in plain sight for all to see, if we only look.

Although the North American prairie was nearly extirpated as a plant community by the early twentieth century, it now holds the key to the future for modern gardeners. These plants survive and thrive through heat and cold, flood and drought, with little or no fertilizer, pesticide, or irrigation. Properly designed and managed native landscapes can save significant time and money so that we can devote resources to more pressing priorities.

Native plants support a plethora of pollinators, upon which one-third of our food depends. The future of our world depends on wise stewardship of these resources, and respect for all forms of life, no matter how small or seemingly insignificant. Our quality of life depends directly upon fostering diversity instead of destroying it. The very survival of the human race depends upon the survival of all Earth's inhabitants.

Go native!

11

Prairie-Seed Mixes

TABLE 11.1.   NORTHERN SHORTGRASS PRAIRIE MIX FOR MEDIUM SOILS (60% GRASSES, 40% FORBS BY WEIGHT)

		Area covered & recommended seeding rate		
		1,000 sq. ft.	4,400 sq. ft. 1/10 acre	11,000 sq. ft. 1/4 acre
		1/4 lb.	1 lb.	2.5 lbs.
Scientific name	Common name	Seed weight in grams for each mix quantity above		
Forbs				
Agastache foeniculum	Lavender Hyssop	0.71	2.84	7.09
Allium cernuum	Nodding Pink Onion	1.42	5.67	14.18
Asclepias tuberosa	Butterflyweed	2.84	11.34	28.35
Ceanothus americanus	New Jersey Tea	0.71	2.84	7.09
Coreopsis lanceolata	Lanceleaf Coreopsis	1.42	5.67	14.18
Dodecatheon meadia	Shootingstar	0.71	2.84	7.09
Echinacea pallida	Pale Purple Coneflower	5.67	22.68	56.70
Echinacea purpurea	Purple Coneflower	4.25	17.01	42.53
Eryngium yuccifolium	Rattlesnake Master	2.84	11.34	28.35
Liatris ligulistylis	Meadow Blazing Star	0.71	2.84	7.09
Liatris pycnostachya	Prairie Blazing Star	3.54	14.18	35.44
Oligoneuron rigidum	Stiff Goldenrod	0.71	2.84	7.09
Parthenium integrifolium	Wild Quinine	1.42	5.67	14.18
Penstemon digitalis	Smooth Penstemon	1.06	4.25	10.63
Rudbeckia hirta	Blackeyed Susan	1.42	5.67	14.18
Rudbeckia triloba	Browneyed Susan	0.71	2.84	7.09
Symphyotrichum laeve	Smooth Aster	0.71	2.84	7.09
Symphyotrichum oolentangiense	Skyblue Aster	0.71	2.84	7.09
Tradescantia ohiensis	Ohio Spiderwort	2.13	8.51	21.26
Verbena stricta	Hoary Vervain	1.06	4.25	10.63
Zizia aptera	Heartleaf Golden Alexander	2.84	11.34	28.35
Total forbs		37.56	150.26	375.64
Legumes				
Astragalus canadensis	Canada Milkvetch	0.71	2.84	7.09
Dalea candida	White Prairie Clover	2.13	8.51	21.26
Dalea purpurea	Purple Prairie Clover	3.54	14.18	35.44
Lespedeza capitata	Roundhead Bush Clover	1.42	5.67	14.18
Total legumes		7.80	31.19	77.96
Total forbs and legumes		45.36	181.44	453.6
Grasses				
Bouteloua curtipendula	Sideoats Grama	34.02	136.08	340.20
Schizachyrium scoparium	Little Bluestem	25.52	102.06	255.15
Sporobolus heterolepis	Prairie Dropseed	8.51	34.02	85.05
Total grasses		68.04	272.16	680.40
Total prairie seed		113.4	453.6	1134

Area covered & recommended seeding rate

22,000 sq. ft.	43,560 sq. ft.
1/2 acre	1 acre
5 lbs.	10 lbs.

Seed weight in grams for each mix quantity above		Seed wt./acre (10 PLS lbs.) (oz.)	Seeds/ oz.	Seeds/ sq. ft.	Mix by weight (%)	Mix by Seeds/sq. ft. (%)
14.18	28.35	1.00	65,000	1.49	0.63	3.07
28.35	56.70	2.00	7,700	0.35	1.25	0.73
56.70	113.40	4.00	3,500	0.32	2.50	0.66
14.18	28.35	1.00	7,000	0.16	0.63	0.33
28.35	56.70	2.00	12,500	0.57	1.25	1.18
14.18	28.35	1.00	75,000	1.72	0.63	3.54
113.40	226.80	8.00	5,000	0.92	5.00	1.89
85.05	170.10	6.00	6,600	0.91	3.75	1.87
56.70	113.40	4.00	8,000	0.73	2.50	1.51
14.18	28.35	1.00	13,000	0.30	0.63	0.61
70.88	141.75	5.00	12,000	1.38	3.13	2.84
14.18	28.35	1.00	46,000	1.06	0.63	2.17
28.35	56.70	2.00	6,800	0.31	1.25	0.64
21.26	42.53	1.50	100,000	3.44	0.94	7.09
28.35	56.70	2.00	100,000	4.59	1.25	9.45
14.18	28.35	1.00	33,000	0.76	0.63	1.56
14.18	28.35	1.00	48,000	1.10	0.63	2.27
14.18	28.35	1.00	82,000	1.88	0.63	3.87
42.53	85.05	3.00	8,000	0.55	1.88	1.13
21.26	42.53	1.50	32,000	1.10	0.94	2.27
56.70	113.40	4.00	9,000	0.83	2.50	1.70
751.28	1502.55	53.00		24.49	33.13	50.40
14.18	28.35	1.00	16,000	0.37	0.63	0.76
42.53	85.05	3.00	15,000	1.03	1.88	2.13
70.88	141.75	5.00	20,000	2.30	3.13	4.73
28.35	56.70	2.00	10,000	0.46	1.25	0.95
155.93	311.85	11.00		4.16	6.88	8.55
907.2	1814.4	64.00		28.64	40.00	58.95
		4.00 lbs.				
680.40	1360.80	48.00	8,000	8.82	30.00	18.14
510.30	1020.60	36.00	8,800	7.27	22.50	14.97
170.10	340.20	12.00	14,000	3.86	7.50	7.94
1360.80	2721.60	96.00		19.94	60.00	41.05
		6.00 lbs.				
2268	4536	160.00		48.59	100.00	100.00
		10.00 lbs.				

TABLE 11.2. **NORTHERN SHORTGRASS PRAIRIE MIX FOR DRY SOILS (60% GRASSES, 40% FORBS BY WEIGHT)**

Scientific name	Common name	Area covered & recommended seeding rate		
		1,000 sq. ft. 1/4 lb.	4,400 sq. ft. 1/10 acre 1 lb.	11,000 sq. ft. 1/4 acre 2.5 lbs.
		Seed weight in grams for each mix quantity above		
Forbs				
Coreopsis lanceolata	Lanceleaf Coreopsis	2.13	8.51	21.26
Echinacea pallida	Pale Purple Coneflower	4.96	19.85	49.61
Helianthus occidentalis	Western Sunflower	0.71	2.84	7.09
Liatris aspera	Rough Blazing Star	2.13	8.51	21.26
Monarda punctata	Dotted Mint	0.71	2.84	7.09
Oligoneuron rigidum	Stiff Goldenrod	0.35	1.42	3.54
Penstemon grandiflorus	Beardtongue	2.13	8.51	21.26
Rudbeckia hirta	Blackeyed Susan	2.13	8.51	21.26
Ruellia humilis	Wild Petunia	1.42	5.67	14.18
Solidago ptarmicoides	White Aster	0.71	2.84	7.09
Solidago speciosa	Showy Goldenrod	0.35	1.42	3.54
Symphyotrichum laeve	Smooth Aster	0.71	2.84	7.09
Tradescantia ohiensis	Ohio Spiderwort	2.84	11.34	28.35
Verbena stricta	Hoary Vervain	1.42	5.67	14.18
Total forbs		27.64	110.57	276.41
Legumes				
Amorpha canescens	Leadplant	0.71	2.84	7.09
Astragalus canadensis	Canada Milkvetch	0.71	2.84	7.09
Dalea candida	White Prairie Clover	2.13	8.51	21.26
Dalea purpurea	Purple Prairie Clover	4.25	17.01	42.53
Lespedeza capitata	Roundhead Bush Clover	1.42	5.67	14.18
Lupinus perennis	Lupine	8.51	34.02	85.05
Total legumes		17.72	70.88	177.19
Total forbs and legumes		45.36	181.44	453.60
Grasses				
Bouteloua curtipendula	Sideoats Grama	34.02	136.08	340.20
Schizachyrium scoparium	Little Bluestem	25.52	102.06	255.15
Sporobolus heterolepis	Prairie Dropseed	8.51	34.02	85.05
Total grasses		68.04	272.16	680.40
Total prairie seed		113.40	453.60	1134.00

Area covered & recommended seeding rate		Seed wt./acre (10 PLS lbs.) (oz.)	Seeds/ oz.	Seeds/ sq. ft.	Mix by weight (%)	Mix by Seeds/sq. ft. (%)
22,000 sq. ft.	43,560 sq. ft.					
1/2 acre	1 acre					
5 lbs.	10 lbs.					
Seed weight in grams for each mix quantity above						
42.53	85.05	3.00	12,500	0.86	1.88	1.77
99.23	198.45	7.00	5,000	0.80	4.38	1.66
14.18	28.35	1.00	13,000	0.30	0.63	0.62
42.53	85.05	3.00	13,500	0.93	1.88	1.92
14.18	28.35	1.00	94,000	2.16	0.63	4.45
7.09	14.18	0.50	46,000	0.53	0.31	1.09
42.53	85.05	3.00	11,000	0.76	1.88	1.56
42.53	85.05	3.00	100,000	6.89	1.88	14.19
28.35	56.70	2.00	4,000	0.18	1.25	0.38
14.18	28.35	1.00	70,000	1.61	0.63	3.31
7.09	14.18	0.50	105,000	1.21	0.31	2.48
14.18	28.35	1.00	48,000	1.10	0.63	2.27
56.70	113.40	4.00	8,000	0.73	2.50	1.51
28.35	56.70	2.00	32,000	1.47	1.25	3.03
552.83	1105.65	39.00		23.30	24.38	48.02
14.18	28.35	1.00	17,000	0.39	0.63	0.80
14.18	28.35	1.00	16,000	0.37	0.63	0.76
42.53	85.05	3.00	15,000	1.03	1.88	2.13
85.05	170.10	6.00	20,000	2.75	3.75	5.68
28.35	56.70	2.00	10,000	0.46	1.25	0.95
170.10	340.20	12.00	1,000	0.28	7.50	0.57
354.38	708.75	25.00		5.28	15.63	10.88
907.20	1814.40	64.00		28.58	40.00	58.90
		4.00 lbs.				
680.40	1360.80	48.00	8,000	8.82	30.00	18.17
510.30	1020.60	36.00	8,800	7.27	22.50	14.99
170.10	340.20	12.00	14,000	3.86	7.50	7.95
1360.80	2721.60	96.00		19.94	60.00	41.10
		6.00 lbs.				
2268.00	4536.00	160.00		48.53	100.00	100.00
		10.00 lbs.				

TABLE 11.3. NORTHERN DIVERSE PRAIRIE MIX FOR MEDIUM SOILS (50% GRASS, 50% FORBS BY WEIGHT)

Scientific name	Common name	Area covered & recommended seeding rate		
		1,000 sq. ft. 1/4 lb.	4,400 sq. ft. 1/10 acre 1 lb.	11,000 sq. ft. 1/4 acre 2.5 lbs.
		Seed weight in grams for each mix quantity above		
Forbs				
Allium cernuum	Nodding Pink Onion	2.13	8.51	21.26
Asclepias tuberosa	Butterflyweed	4.25	17.01	42.53
Cacalia atriplicifolia	Pale Indian Plantain	0.71	2.84	7.09
Ceanothus americanus	New Jersey Tea	0.71	2.84	7.09
Coreopsis lanceolata	Lanceleaf Coreopsis	1.42	5.67	14.18
Dodecatheon meadia	Shooting Star	0.71	2.84	7.09
Echinacea pallida	Pale Purple Coneflower	6.38	25.52	63.79
Echinacea purpurea	Purple Coneflower	4.96	19.85	49.61
Eryngium yuccifolium	Rattlesnake Master	2.13	8.51	21.26
Helianthus laetiflorus	Showy Sunflower	0.35	1.42	3.54
Heliopsis helianthoides	Oxeye Sunflower	0.71	2.84	7.09
Liatris ligulistylis	Meadow Blazing Star	0.71	2.84	7.09
Liatris pycnostachya	Prairie Blazing Star	2.84	11.34	28.35
Monarda fistulosa	Bergamot	0.35	1.42	3.54
Oligoneuron ohioense	Ohio Goldenrod	0.35	1.42	3.54
Oligoneuron rigidum	Stiff Goldenrod	0.35	1.42	3.54
Parthenium integrifolium	Wild Quinine	2.13	8.51	21.26
Penstemon digitalis	Smooth Penstemon	0.71	2.84	7.09
Ratibida pinnata	Yellow Coneflower	1.42	5.67	14.18
Rudbeckia hirta	Blackeyed Susan	1.42	5.67	14.18
Rudbeckia subtomentosa	Sweet Blackeyed Susan	0.35	1.42	3.54
Rudbeckia triloba	Browneyed Susan	0.35	1.42	3.54
Silphium integrifolium	Rosinweed	0.35	1.42	3.54
Silphium laciniatum	Compassplant	1.77	7.09	17.72
Silphium terebinthinaceum	Prairie Dock	1.06	4.25	10.63
Symphyotrichum laeve	Smooth Aster	1.06	4.25	10.63
Symphyotrichum novae-angliae	New England Aster	0.71	2.84	7.09
Symphyotrichum oolentangiense	Skyblue Aster	0.71	2.84	7.09
Tradescantia ohiensis	Ohio Spiderwort	2.13	8.51	21.26
Verbena hastata	Blue Vervain	0.71	2.84	7.09
Veronicastrum virginicum	Culver's Root	0.71	2.84	7.09
Zizia aurea	Golden Alexander	1.42	5.67	14.18
Total forbs		46.07	184.28	460.69

Area covered & recommended seeding rate

22,000 sq. ft.	43,560 sq. ft.
1/2 acre	1 acre
5 lbs.	10 lbs.

Seed weight in grams for each mix quantity above		Seed wt./acre (10 PLS lbs.) (oz.)	Seeds/ oz.	Seeds/ sq. ft.	Mix by weight (%)	Mix by Seeds/sq. ft. (%)
42.53	85.05	3.00	7,700	0.53	1.88	0.82
85.05	170.10	6.00	3,500	0.48	3.75	0.75
14.18	28.35	1.00	6,500	0.15	0.63	0.23
14.18	28.35	1.00	7,000	0.16	0.63	0.25
28.35	56.70	2.00	12,500	0.57	1.25	0.89
14.18	28.35	1.00	75,000	1.72	0.63	2.67
127.58	255.15	9.00	5,000	1.03	5.63	1.60
99.23	198.45	7.00	6,600	1.06	4.38	1.64
42.53	85.05	3.00	8,000	0.55	1.88	0.85
7.09	14.18	0.50	4,600	0.05	0.31	0.08
14.18	28.35	1.00	6,500	0.15	0.63	0.23
14.18	28.35	1.00	13,000	0.30	0.63	0.46
56.70	113.40	4.00	12,000	1.10	2.50	1.71
7.09	14.18	0.50	78,000	0.90	0.31	1.39
7.09	14.18	0.50	90,000	1.03	0.31	1.60
7.09	14.18	0.50	46,000	0.53	0.31	0.82
42.53	85.05	3.00	6,800	0.47	1.88	0.73
14.18	28.35	1.00	100,000	2.30	0.63	3.56
28.35	56.70	2.00	27,000	1.24	1.25	1.92
28.35	56.70	2.00	100,000	4.59	1.25	7.11
7.09	14.18	0.50	46,000	0.53	0.31	0.82
7.09	14.18	0.50	33,000	0.38	0.31	0.59
7.09	14.18	0.50	4,000	0.05	0.31	0.07
35.44	70.88	2.50	650	0.04	1.56	0.06
21.26	42.53	1.50	1,100	0.04	0.94	0.06
21.26	42.53	1.50	48,000	1.65	0.94	2.56
14.18	28.35	1.00	70,000	1.61	0.63	2.49
14.18	28.35	1.00	82,000	1.88	0.63	2.92
42.53	85.05	3.00	8,000	0.55	1.88	0.85
14.18	28.35	1.00	100,000	2.30	0.63	3.56
14.18	28.35	1.00	750,000	17.22	0.63	26.67
28.35	56.70	2.00	12,000	0.55	1.25	0.85
921.38	1842.75	65.00		45.70	40.63	70.80

(continued)

TABLE 11.3.   (*continued*)

Scientific name	Common name	Area covered & recommended seeding rate		
		1,000 sq. ft. 1/4 lb.	4,400 sq. ft. 1/10 acre 1 lb.	11,000 sq. ft. 1/4 acre 2.5 lbs.
		Seed weight in grams for each mix quantity above		
Legumes				
Astragalus canadensis	Canada Milkvetch	0.71	2.84	7.09
Baptisia australis	Blue False Indigo	1.06	4.25	10.63
Baptisia lactea	White False Indigo	1.06	4.25	10.63
Lespedeza capitata	Roundhead Bush Clover	0.71	2.84	7.09
Dalea candida	White Prairie Clover	2.13	8.51	21.26
Dalea purpurea	Purple Prairie Clover	4.25	17.01	42.53
Senna hebecarpa	Wild Senna	0.71	2.84	7.09
Total legumes		10.63	42.53	106.31
Total forbs and legumes		56.70	226.80	567.00
Grasses				
Andropogon gerardii	Big Bluestem	1.42	5.67	14.18
Bouteloua curtipendula	Sideoats Grama	14.18	56.70	141.75
Elymus canadensis	Canada Wildrye	14.18	56.70	141.75
Schizachyrium scoparium	Little Bluestem	19.85	79.38	198.45
Sorghastrum nutans	Indiangrass	2.84	11.34	28.35
Sporobolus heterolepis	Prairie Dropseed	4.25	17.01	42.53
Total grasses		56.70	226.80	567.00
Total prairie seed		113.40	453.60	1134.00

22,000 sq. ft.	43,560 sq. ft.
1/2 acre	1 acre
5 lbs.	10 lbs.

Seed weight in grams for each mix quantity above		Seed wt./acre (10 PLS lbs.) (oz.)	Seeds/ oz.	Seeds/ sq. ft.	Mix by weight (%)	Mix by Seeds/sq. ft. (%)
14.18	28.35	1.00	16,000	0.37	0.63	0.57
21.26	42.53	1.50	1,600	0.06	0.94	0.09
21.26	42.53	1.50	1,600	0.06	0.94	0.09
14.18	28.35	1.00	10,000	0.23	0.63	0.36
42.53	85.05	3.00	15,000	1.03	1.88	1.60
85.05	170.10	6.00	20,000	2.75	3.75	4.27
14.18	28.35	1.00	1,400	0.03	0.63	0.05
212.63	425.25	15.00		4.53	9.38	7.01
1134.00	2268.00	80.00 *5.00 lbs.*		50.23	50.00	77.81
28.35	56.70	2.00	8,200	0.38	1.25	0.58
283.50	567.00	20.00	8,000	3.67	12.50	5.69
283.50	567.00	20.00	4,200	1.93	12.50	2.99
396.90	793.80	28.00	8,800	5.66	17.50	8.76
56.70	113.40	4.00	8,300	0.76	2.50	1.18
85.05	170.10	6.00	14,000	1.93	3.75	2.99
1134.00	2268.00	80.00 *5.00 lbs.*		14.33	50.00	22.19
2268.00	4536.00	160.00 *10.00 lbs.*		64.55	100.00	100.00

TABLE 11.4. DIVERSE PRAIRIE MIX FOR DRY SOILS (60% GRASS, 40% FORBS BY WEIGHT)

		Area covered & recommended seeding rate		
		1,000 sq. ft.	4,400 sq. ft. 1/10 acre	11,000 sq. ft. 1/4 acre
		1/4 lb.	1 lb.	2.5 lbs.
Scientific name	Common name	Seed weight in grams for each mix quantity above		
Forbs				
Agastache foeniculum	Lavender Hyssop	0.71	2.84	7.09
Asclepias tuberosa	Butterflyweed	4.25	17.01	42.53
Coreopsis lanceolata	Lanceleaf Coreopsis	1.42	5.67	14.18
Echinacea pallida	Pale Purple Coneflower	4.25	17.01	42.53
Helianthus laetiflorus	Showy Sunflower	0.35	1.42	3.54
Helianthus mollis	Downy Sunflower	0.35	1.42	3.54
Helianthus occidentalis	Western Sunflower	0.35	1.42	3.54
Liatris aspera	Rough Blazing Star	1.42	5.67	14.18
Monarda fistulosa	Bergamot	0.71	2.84	7.09
Monarda punctata	Dotted Mint	0.71	2.84	7.09
Oligoneuron rigidum	Stiff Goldenrod	0.71	2.84	7.09
Penstemon grandiflorus	Beardtongue	2.13	8.51	21.26
Ratibida pinnata	Yellow Coneflower	1.42	5.67	14.18
Rudbeckia hirta	Blackeyed Susan	2.13	8.51	21.26
Ruellia humilis	Wild Petunia	1.42	5.67	14.18
Silphium integrifolium	Rosinweed	0.71	2.84	7.09
Silphium laciniatum	Compassplant	2.13	8.51	21.26
Solidago ptarmicoides	White Aster	0.71	2.84	7.09
Solidago speciosa	Showy Goldenrod	0.71	2.84	7.09
Symphyotrichum laeve	Smooth Aster	0.71	2.84	7.09
Symphyotrichum oolentangiense	Skyblue Aster	0.71	2.84	7.09
Tradescantia ohiensis	Ohio Spiderwort	2.48	9.92	24.81
Verbena stricta	Hoary Vervain	1.42	5.67	14.18
Total forbs		31.89	127.58	318.94
Legumes				
Amorpha canescens	Leadplant	0.71	2.84	7.09
Astragalus canadensis	Canada Milkvetch	0.71	2.84	7.09
Dalea candida	White Prairie Clover	2.13	8.51	21.26
Dalea purpurea	Purple Prairie Clover	2.84	11.34	28.35
Lespedeza capitata	Roundhead Bush Clover	1.42	5.67	14.18
Lupinus perennis	Lupine	5.67	22.68	56.70
Total legumes		13.47	53.87	134.66
Total forbs and legumes		45.36	181.44	453.60

Seed weight in grams for each mix quantity above		Seed wt./acre (10 PLS lbs.) (oz.)	Seeds/ oz.	Seeds/ sq. ft.	Mix by weight (%)	Mix by Seeds/sq. ft. (%)
14.18	28.35	1.00	65,000	1.49	0.63	2.58
85.05	170.10	6.00	3,500	0.48	3.75	0.83
28.35	56.70	2.00	12,500	0.57	1.25	0.99
85.05	170.10	6.00	5,000	0.69	3.75	1.19
7.09	14.18	0.50	4,600	0.05	0.31	0.09
7.09	14.18	0.50	7,700	0.09	0.31	0.15
7.09	14.18	0.50	13,000	0.15	0.31	0.26
28.35	56.70	2.00	13,000	0.60	1.25	1.03
14.18	28.35	1.00	78,000	1.79	0.63	3.10
14.18	28.35	1.00	94,000	2.16	0.63	3.73
14.18	28.35	1.00	46,000	1.06	0.63	1.83
42.53	85.05	3.00	11,000	0.76	1.88	1.31
28.35	56.70	2.00	27,000	1.24	1.25	2.14
42.53	85.05	3.00	100,000	6.89	1.88	11.91
28.35	56.70	2.00	4,000	0.18	1.25	0.32
14.18	28.35	1.00	4,000	0.09	0.63	0.16
42.53	85.05	3.00	650	0.04	1.88	0.08
14.18	28.35	1.00	70,000	1.61	0.63	2.78
14.18	28.35	1.00	105,000	2.41	0.63	4.17
14.18	28.35	1.00	48,000	1.10	0.63	1.90
14.18	28.35	1.00	82,000	1.88	0.63	3.25
49.61	99.23	3.50	8,000	0.64	2.19	1.11
28.35	56.70	2.00	32,000	1.47	1.25	2.54
637.88	1275.75	45.00		27.45		
14.18	28.35	1.00	17,000	0.39	0.63	0.67
14.18	28.35	1.00	16,000	0.37	0.63	0.63
42.53	85.05	3.00	15,000	1.03	1.88	1.79
56.70	113.40	4.00	20,000	1.84	2.50	3.17
28.35	56.70	2.00	10,000	0.46	1.25	0.79
113.40	226.80	8.00	1,000	0.18	5.00	0.32
269.33	538.65	19.00		4.27		
907.20	1814.40	64.00		31.72		
		4.00 lbs.				

(continued)

TABLE 11.4. **(*continued*)**

		Area covered & recommended seeding rate		
		1,000 sq. ft.	4,400 sq. ft. 1/10 acre	11,000 sq. ft. 1/4 acre
		1/4 lb.	1 lb.	2.5 lbs.
Scientific name	**Common name**	Seed weight in grams for each mix quantity above		
Grasses				
Andropogon gerardii	Big Bluestem	1	4	12
Bouteloua curtipendula	Sideoats Grama	18	72	216
Elymus canadensis	Canada Wildrye	18	72	216
Koeleria macrantha	Junegrass	4	16	48
Schizachyrium scoparium	Little Bluestem	19	76	228
Sorghastrum nutans	Indiangrass	4	16	48
Sporobolus heterolepis	Prairie Dropseed	4	16	48
Total grasses		68	272	816
		68.04	*272.16*	*816.48*
Total prairie seed		113.36	453.44	1269.60

Area covered & recommended seeding rate		Seed wt./acre (10 PLS lbs.) (oz.)	Seeds/ oz.	Seeds/ sq. ft.	Mix by weight (%)	Mix by Seeds/sq. ft. (%)
22,000 sq. ft.	43,560 sq. ft.					
1/2 acre	1 acre					
5 lbs.	10 lbs.					
Seed weight in grams for each mix quantity above						
20	40	2.00	8,200	0.38	1.25	0.65
360	720	24.00	8,000	4.41	15.00	7.62
360	720	24.00	4,200	2.31	15.00	4.00
80	160	4.00	100,000	9.18	2.50	15.87
380	760	24.00	8,800	4.85	15.00	8.38
80	160	6.00	8,300	1.14	3.75	1.98
80	160	12.00	14,000	3.86	7.50	6.67
1360	2720	96.00		26.13		
1360.8	2721.6	6.00 lbs.				
2267.20	4534.40	160.00		57.85		
		10.00 lbs.				

TABLE 11.5. NORTHERN BUTTERFLY PRAIRIE MIX FOR MEDIUM SOILS (50% GRASS, 50% FORBS BY WEIGHT)

Scientific name	Common name	Area covered & recommended seeding rate		
		1,000 sq. ft. 1/4 lb.	4,400 sq. ft. 1/10 acre 1 lb.	11,000 sq. ft. 1/4 acre 2.5 lbs.
		Seed weight in grams for each mix quantity above		
Forbs				
Agastache foeniculum	Lavender Hyssop	1.06	4.25	10.63
Allium cernuum	Nodding Pink Onion	2.13	8.51	21.26
Asclepias tuberosa	Butterflyweed	6.38	25.52	63.79
Coreopsis lanceolata	Lanceleaf Coreopsis	2.84	11.34	28.35
Echinacea pallida	Pale Purple Coneflower	7.80	31.19	77.96
Echinacea purpurea	Purple Coneflower	5.67	22.68	56.70
Liatris pycnostachya	Prairie Blazing Star	4.96	19.85	49.61
Liatris spicata	Dense Blazing Star	4.96	19.85	49.61
Monarda fistulosa	Bergamot	0.71	2.84	7.09
Oligoneuron ohioense	Ohio Goldenrod	0.71	2.84	7.09
Oligoneuron rigidum	Stiff Goldenrod	0.71	2.84	7.09
Ratibida pinnata	Yellow Coneflower	1.42	5.67	14.18
Rudbeckia fulgida	Orange Coneflower	0.71	2.84	7.09
Rudbeckia hirta	Blackeyed Susan	2.13	8.51	21.26
Rudbeckia subtomentosa	Sweet Blackeyed Susan	0.71	2.84	7.09
Rudbeckia triloba	Browneyed Susan	0.71	2.84	7.09
Symphyotrichum laeve	Smooth Aster	1.06	4.25	10.63
Symphyotrichum novae-angliae	New England Aster	1.06	4.25	10.63
Symphyotrichum oolentangiense	Skyblue Aster	0.71	2.84	7.09
Verbena hastata	Blue Vervain	0.71	2.84	7.09
Vernonia fasciculata	Ironweed	1.77	7.09	17.72
Zizia aurea	Golden Alexander	3.54	14.18	35.44
Total forbs		52.45	209.79	524.48
Legumes				
Dalea purpurea	Purple Prairie Clover	4.25	17.01	42.53
Total legumes		4.25	17.01	42.53
Total forbs and legumes		56.70	226.80	567.00
Grasses				
Bouteloua curtipendula	Sideoats Grama	17.01	68.04	170.10
Elymus canadensis	Canada Wildrye	17.01	68.04	170.10
Schizachyrium scoparium	Little Bluestem	17.01	68.04	170.10
Sporobolus heterolepis	Prairie Dropseed	5.67	22.68	56.70
Total grasses		56.70	226.80	567.00
Total prairie seed		113.40	453.60	1134.00

Area covered & recommended seeding rate

22,000 sq. ft.	43,560 sq. ft.
1/2 acre	1 acre
5 lbs.	10 lbs.

Seed weight in grams for each mix quantity above		Seed wt./acre (10 PLS lbs.) (oz.)	Seeds/ oz.	Seeds/ sq. ft.	Mix by weight (%)	Mix by Seeds/sq. ft. (%)
21.26	42.53	1.50	65,000	2.24	0.94	4.14
42.53	85.05	3.00	7,700	0.53	1.88	0.98
127.58	255.15	9.00	3,500	0.72	5.63	1.34
56.70	113.40	4.00	12,500	1.15	2.50	2.12
155.93	311.85	11.00	5,000	1.26	6.88	2.34
113.40	226.80	8.00	6,600	1.21	5.00	2.24
99.23	198.45	7.00	12,000	1.93	4.38	3.57
99.23	198.45	7.00	12,000	1.93	4.38	3.57
14.18	28.35	1.00	78,000	1.79	0.63	3.31
14.18	28.35	1.00	90,000	2.07	0.63	3.82
14.18	28.35	1.00	46,000	1.06	0.63	1.95
28.35	56.70	2.00	27,000	1.24	1.25	2.29
14.18	28.35	1.00	25,000	0.57	0.63	1.06
42.53	85.05	3.00	100,000	6.89	1.88	12.74
14.18	28.35	1.00	46,000	1.06	0.63	1.95
14.18	28.35	1.00	33,000	0.76	0.63	1.40
21.26	42.53	1.50	48,000	1.65	0.94	3.06
21.26	42.53	1.50	70,000	2.41	0.94	4.46
14.18	28.35	1.00	82,000	1.88	0.63	3.48
14.18	28.35	1.00	100,000	2.30	0.63	4.25
35.44	70.88	2.50	20,000	1.15	1.56	2.12
70.88	141.75	5.00	12,000	1.38	3.13	2.55
1048.95	2097.90	74.00		37.16	46.25	68.75
85.05	170.10	6.00	20,000	2.75	3.75	5.10
85.05	170.10	6.00		2.75	3.75	5.10
1134.00	2268.00	80.00 5.00 lbs.		39.92	50.00	73.84
340.20	680.40	24.00	8,000	4.41	15.00	8.15
340.20	680.40	24.00	4,200	2.31	15.00	4.28
340.20	680.40	24.00	8,800	4.85	15.00	8.97
113.40	226.80	8.00	14,000	2.57	5.00	4.76
1134.00	2268.00	80.00 5.00 lbs.		14.14	50.00	26.16
2268.00	4536.00	160.00 10.00 lbs.		54.06	100.00	100.00

TABLE 11.6. NORTHERN BUTTERFLY MIX FOR DRY SOILS (50% GRASS, 50% FORBS BY WEIGHT)

		Area covered & recommended seeding rate		
		1,000 sq. ft. 1/4 lb.	4,400 sq. ft. 1/10 acre 1 lb.	11,000 sq. ft. 1/4 acre 2.5 lbs.
Scientific name	Common name	Seed weight in grams for each mix quantity above		
Forbs				
Agastache foeniculum	Lavender Hyssop	0.71	2.84	7.09
Asclepias tuberosa	Butterflyweed	7.09	28.35	70.88
Coreopsis lanceolata	Lanceleaf Coreopsis	2.84	11.34	28.35
Echinacea pallida	Pale Purple Coneflower	8.51	34.02	85.05
Liatris aspera	Rough Blazing Star	2.84	11.34	28.35
Monarda fistulosa	Bergamot	0.71	2.84	7.09
Monarda punctata	Dotted Mint	0.71	2.84	7.09
Oligoneuron rigidum	Stiff Goldenrod	0.71	2.84	7.09
Penstemon grandiflorus	Beardtongue	2.84	11.34	28.35
Ratibida pinnata	Yellow Coneflower	2.13	8.51	21.26
Rudbeckia hirta	Blackeyed Susan	2.13	8.51	21.26
Ruellia humilis	Wild Petunia	2.84	11.34	28.35
Solidago ptarmicoides	White Aster	0.71	2.84	7.09
Solidago speciosa	Showy Goldenrod	0.71	2.84	7.09
Symphyotrichum laeve	Smooth Aster	0.71	2.84	7.09
Symphyotrichum oolentangiense	Skyblue Aster	0.71	2.84	7.09
Verbena stricta	Hoary Vervain	2.13	8.51	21.26
Total forbs		38.98	155.93	389.81
Legumes				
Amorpha canescens	Leadplant	0.71	2.84	7.09
Dalea purpurea	Purple Prairie Clover	5.67	22.68	56.70
Lupinus perennis	Lupine	11.34	45.36	113.40
Total legumes		17.72	70.88	177.19
Total forbs and legumes		56.70	226.80	567.00
Grasses				
Bouteloua curtipendula	Sideoats Grama	17.01	68.04	170.10
Elymus canadensis	Canada Wildrye	14.18	56.70	141.75
Koeleria macrantha	Junegrass	2.84	11.34	28.35
Schizachyrium scoparium	Little Bluestem	17.01	68.04	170.10
Sporobolus heterolepis	Prairie Dropseed	5.67	22.68	56.70
Total grasses		56.70	226.80	567.00
		56.70	*226.80*	*680.40*
Total prairie seed		113.40	453.60	1134.00

Area covered & recommended seeding rate						
22,000 sq. ft.	43,560 sq. ft.					
1/2 acre	1 acre					
5 lbs.	10 lbs.	Seed wt./acre (10 PLS lbs.) (oz.)	Seeds/ oz.	Seeds/ sq. ft.	Mix by weight (%)	Mix by Seeds/sq. ft. (%)
Seed weight in grams for each mix quantity above						

14.18	28.35	1.00	65,000	1.49	0.63	2.39
141.75	283.50	10.00	3,500	0.80	6.25	1.29
56.70	113.40	4.00	12,500	1.15	2.50	1.84
170.10	340.20	12.00	5,000	1.38	7.50	2.21
56.70	113.40	4.00	13,000	1.19	2.50	1.92
14.18	28.35	1.00	78,000	1.79	0.63	2.87
14.18	28.35	1.00	94,000	2.16	0.63	3.46
14.18	28.35	1.00	46,000	1.06	0.63	1.69
56.70	113.40	4.00	11,000	1.01	2.50	1.62
42.53	85.05	3.00	27,000	1.86	1.88	2.98
42.53	85.05	3.00	100,000	6.89	1.88	11.05
56.70	113.40	4.00	4,000	0.37	2.50	0.59
14.18	28.35	1.00	70,000	1.61	0.63	2.58
14.18	28.35	1.00	105,000	2.41	0.63	3.87
14.18	28.35	1.00	48,000	1.10	0.63	1.77
14.18	28.35	1.00	82,000	1.88	0.63	3.02
42.53	85.05	3.00	32,000	2.20	1.88	3.54
779.63	1559.25	55.00		30.35	34.38	48.71
14.18	28.35	1.00	17,000	0.39	0.63	0.63
113.40	226.80	8.00	20,000	3.67	5.00	5.89
226.80	453.60	16.00	1,000	0.37	10.00	0.59
354.38	708.75	25.00		4.43	15.63	7.11
1134.00	2268.00	80.00		34.78	50.00	55.82
		5.00 lbs.				
340.20	680.40	24.00	8,000	4.41	15.00	7.07
283.50	567.00	20.00	4,200	1.93	12.50	3.09
56.70	113.40	4.00	150,000	13.77	2.50	22.11
340.20	680.40	24.00	8,800	4.85	15.00	7.78
113.40	226.80	8.00	14,000	2.57	5.00	4.13
1134.00	2268.00	80.00		27.53	50.00	44.18
1134.00	2268.00	5.00 lbs.				
2268.00	4536.00	160.00		62.31	100.00	100.00
		10.00 lbs.				

TABLE 11.7.

NORTHERN DEER-RESISTANT SHORT PRAIRIE MIX FOR MEDIUM SOIL (60% GRASS, 40% FORBS BY WEIGHT)

		Area covered & recommended seeding rate		
		1,000 sq. ft.	4,400 sq. ft.	11,000 sq. ft.
			1/10 acre	1/4 acre
		1/4 lb.	1 lb.	2.5 lbs.
Scientific name	Common name	Seed weight in grams for each mix quantity above		
Forbs				
Agastache foeniculum	Lavender Hyssop	1.00	4.00	12.00
Allium cernuum	Nodding Pink Onion	4.00	16.00	48.00
Asclepias tuberosa	Butterflyweed	4.00	16.00	48.00
Coreopsis lanceolata	Lanceleaf Coreopsis	2.00	8.00	24.00
Dodecatheon meadia	Shooting Star	1.50	6.00	18.00
Echinacea pallida	Pale Purple Coneflower	7.00	28.00	84.00
Echinacea purpurea	Purple Coneflower	5.00	20.00	60.00
Eryngium yuccifolium	Rattlesnake Master	3.00	12.00	36.00
Liatris pycnostachya	Prairie Blazing Star	5.00	20.00	60.00
Oligoneuron ohioensis	Ohio Goldenrod	0.50	2.00	6.00
Oligoneuron rigidum	Stiff Goldenrod	0.50	2.00	6.00
Parthenium integrifolium	Wild Quinine	3.00	12.00	36.00
Penstemon digitalis	Smooth Penstemon	2.00	8.00	24.00
Rudbeckia hirta	Blackeyed Susan	2.00	8.00	24.00
Zizia aurea	Golden Alexander	2.00	8.00	24.00
Total forbs		42.50	170.00	510.00
Legumes				
Astragalus canadensis	Canada Milkvetch	1.00	4.00	12.00
Lespedeza capitata	Roundhead Bush Clover	2.00	8.00	24.00
Total legumes		3.00	12.00	36.00
Total forbs and legumes		45.50	182.00	546.00
Grasses				
Bouteloua curtipendula	Sideoats Grama	31.00	124.00	372.00
Schizachyrium scoparium	Little Bluestem	28.00	112.00	336.00
Sporobolus heterolepis	Prairie Dropseed	9.00	36.00	108.00
Total grasses		68.00	272.00	816.00
		68.05	*272.20*	*816.60*
Total prairie seed		113.50	454.00	1362.00

Area covered & recommended seeding rate						
22,000 sq. ft.	43,560 sq. ft.					
1/2 acre	1 acre					
5 lbs.	10 lbs.					
Seed weight in grams for each mix quantity above		Seed wt./acre (10 PLS lbs.) (oz.)	Seeds/ oz.	Seeds/ sq. ft.	Mix by weight (%)	Mix by Seeds/sq. ft. (%)
20.00	40.00	1.00	65,000	1.49	0.63	2.98
80.00	160.00	6.00	7,700	1.06	3.75	2.12
80.00	160.00	6.00	3,500	0.48	3.75	0.96
40.00	80.00	3.00	12,500	0.86	1.88	1.72
30.00	60.00	2.00	75,000	3.44	1.25	6.88
140.00	280.00	10.00	5,000	1.15	6.25	2.29
100.00	200.00	7.00	6,600	1.06	4.38	2.12
60.00	120.00	4.00	8,000	0.73	2.50	1.47
100.00	200.00	7.00	12,000	1.93	4.38	3.85
10.00	20.00	0.50	90,000	1.03	0.31	2.06
10.00	20.00	0.50	46,000	0.53	0.31	1.05
60.00	120.00	4.00	6,800	0.62	2.50	1.25
40.00	80.00	3.00	100,000	6.89	1.88	13.75
40.00	80.00	3.00	100,000	6.89	1.88	13.75
40.00	80.00	3.00	12,000	0.83	1.88	1.65
850.00	1700.00	60.00		29.00	37.50	57.91
20.00	40.00	1.00	16,000	0.37	0.63	0.73
40.00	80.00	3.00	10,000	0.69	1.88	1.38
60.00	120.00	4.00		1.06	2.50	2.11
910.00	1820.00	64.00		30.05	40.00	60.02
		4.00 lbs.				
620.00	1240.00	44.00	8,000	8.08	27.50	16.14
560.00	1120.00	40.00	8,800	8.08	25.00	16.14
180.00	360.00	12.00	14,000	3.86	7.50	7.70
1360.00	2720.00	96.00		20.02	60.00	39.98
1361.00	2722.00	6.00lb.				
2270.00	4540.00	160.00		50.07	100.00	100.00
		10.00 lbs.				

TABLE 11.8. NORTHERN DEER-RESISTANT SHORT PRAIRIE MIX FOR DRY SOIL (60% GRASS, 40% FORBS BY WEIGHT)

Scientific name	Common name	Area covered & recommended seeding rate		
		1,000 sq. ft. 1/4 lb.	4,400 sq. ft. 1/10 acre 1 lb.	11,000 sq. ft. 1/4 acre 2.5 lbs.
		Seed weight in grams for each mix quantity above		
Forbs				
Agastache foeniculum	Lavender Hyssop	0.71	2.84	7.09
Asclepias tuberosa	Butterflyweed	4.96	19.85	49.61
Coreopsis lanceolata	Lanceleaf Coreopsis	2.13	8.51	21.26
Echinacea pallida	Pale Purple Coneflower	7.09	28.35	70.88
Helianthus occidentalis	Western Sunflower	0.71	2.84	7.09
Liatris aspera	Rough Blazing Star	2.13	8.51	21.26
Monarda punctata	Dotted Mint	0.71	2.84	7.09
Oligoneuron rigidum	Stiff Goldenrod	0.71	2.84	7.09
Penstemon grandiflorus	Beardtongue	2.84	11.34	28.35
Rudbeckia hirta	Blackeyed Susan	2.13	8.51	21.26
Ruellia humilis	Wild Petunia	2.13	8.51	21.26
Solidago ptarmicoides	White Aster	0.71	2.84	7.09
Solidago speciosa	Showy Goldenrod	0.71	2.84	7.09
Symphyotrichum oolentangiense	Skyblue Aster	0.71	2.84	7.09
Verbena stricta	Hoary Vervain	2.13	8.51	21.26
Total forbs		30.48	121.91	304.76
Legumes				
Amorpha canescens	Leadplant	0.71	2.84	7.09
Astragalus canadensis	Canada Milkvetch	1.42	5.67	14.18
Lespedeza capitata	Roundhead Bush Clover	4.25	17.01	42.53
Lupinus perennis	Lupine	8.51	34.02	85.05
Total legumes		14.88	59.54	148.84
Total forbs and legumes		45.36	181.44	453.60
Grasses				
Bouteloua curtipendula	Sideoats Grama	42.53	170.10	425.25
Schizachyrium scoparium	Little Bluestem	17.01	68.04	170.10
Sporobolus heterolepis	Prairie Dropseed	8.51	34.02	85.05
Total grasses		68.04	272.16	680.40
		68.05	*272.20*	*816.60*
Total prairie seed		113.40	453.60	1134.00

Seed weight in grams for each mix quantity above		Seed wt./acre (10 PLS lbs.) (oz.)	Seeds/ oz.	Seeds/ sq. ft.	Mix by weight (%)	Mix by Seeds/sq. ft. (%)
14.18	28.35	1.00	65,000	1.49	0.63	3.28
99.23	198.45	7.00	3,500	0.56	4.38	1.24
42.53	85.05	3.00	12,500	0.86	1.88	1.89
141.75	283.50	10.00	5,000	1.15	6.25	2.52
14.18	28.35	1.00	13,000	0.30	0.63	0.66
42.53	85.05	3.00	13,500	0.93	1.88	2.04
14.18	28.35	1.00	94,000	2.16	0.63	4.74
14.18	28.35	1.00	46,000	1.06	0.63	2.32
56.70	113.40	4.00	11,000	1.01	2.50	2.22
42.53	85.05	3.00	100,000	6.89	1.88	15.12
42.53	85.05	3.00	4,000	0.28	1.88	0.60
14.18	28.35	1.00	48,000	1.10	0.63	2.42
14.18	28.35	1.00	105,000	2.41	0.63	5.29
14.18	28.35	1.00	82,000	1.88	0.63	4.13
42.53	85.05	3.00	32,000	2.20	1.88	4.84
609.53	1219.05	43.00		24.28	26.88	53.31
14.18	28.35	1.00	17,000	0.39	0.63	0.86
28.35	56.70	2.00	16,000	0.73	1.25	1.61
85.05	170.10	6.00	1,000	0.14	3.75	0.30
170.10	340.20	12.00	1,000	0.28	7.50	0.60
297.68	595.35	21.00		1.54	13.13	3.38
907.20	1814.40	64.00		25.81	40.00	56.69
		4.00 lbs.				
850.50	1701.00	60.00	8,000	11.02	37.50	24.20
340.20	680.40	24.00	8,800	4.85	15.00	10.65
170.10	340.20	12.00	14,000	3.86	7.50	8.47
1360.80	2721.60	96.00		19.72	60.00	43.31
1361.00	2722.00	6.00 lbs.				
2268.00	4536.00	160.00		45.54	100.00	100.00
		10.00 lbs.				

TABLE 11.9. **NORTHERN WET PRAIRIE, MOIST MEADOW, RAIN GARDEN MIX (60% GRAMINOIDS, 40% FORBS BY WEIGHT)**

		Area covered & recommended seeding rate		
		1,000 sq. ft.	4,400 sq. ft. 1/10 acre	11,000 sq. ft. 1/4 acre
		1/4 lb.	1 lb.	2.5 lbs.
Scientific name	**Common name**	Seed weight in grams for each mix quantity above		
Forbs				
Allium cernuum	Nodding Pink Onion	2.84	11.34	28.35
Asclepias incarnata	Red Milkweed	2.84	11.34	28.35
Baptisia lactea	White False Indigo	3.54	14.18	35.44
Eupatorium perfoliatum	Boneset	0.71	2.84	7.09
Eutrochium maculatum	Joe Pye Weed	0.71	2.84	7.09
Helenium autumnale	Sneezeweed	0.71	2.84	7.09
Heliopsis helianthoides	Oxeye Sunflower	1.42	5.67	14.18
Iris versicolor	Blue Flag Iris	1.42	5.67	14.18
Iris virginica var. *shrevei*	Wild Iris	1.42	5.67	14.18
Liatris pycnostachya	Prairie Blazing Star	2.84	11.34	28.35
Liatris spicata	Dense Blazing Star	2.13	8.51	21.26
Lobelia siphilitica	Great Blue Lobelia	0.71	2.84	7.09
Mimulus ringens	Monkeyflower	0.71	2.84	7.09
Monarda fistulosa	Bergamot	0.71	2.84	7.09
Oligoneuron ohioense	Ohio Goldenrod	0.71	2.84	7.09
Oligoneuron rigidum	Stiff Goldenrod	0.71	2.84	7.09
Parthenium integrifolium	Wild Quinine	2.13	8.51	21.26
Penstemon digitalis	Smooth Penstemon	0.71	2.84	7.09
Ratibida pinnata	Yellow Coneflower	2.13	8.51	21.26
Rudbeckia hirta	Blackeyed Susan	2.13	8.51	21.26
Rudbeckia laciniata	Green-Headed Coneflower	1.06	4.25	10.63
Rudbeckia subtomentosa	Sweet Blackeyed Susan	0.71	2.84	7.09
Rudbeckia triloba	Browneyed Susan	0.71	2.84	7.09
Senna hebecarpa	Wild Senna	3.54	14.18	35.44
Symphyotrichum novae-angliae	New England Aster	1.06	4.25	10.63
Thalictrum dasycarpum	Tall Meadow-Rue	2.13	8.51	21.26
Verbena hastata	Blue Vervain	0.71	2.84	7.09
Vernonia fasciculata	Ironweed	1.42	5.67	14.18
Veronicastrum virginicum	Culver's Root	0.71	2.84	7.09
Zizia aurea	Golden Alexander	2.13	8.51	21.26
Total forbs		45.36	181.44	453.6

Area covered & recommended seeding rate		Seed wt./acre (10 PLS lbs.) (oz.)	Seeds/ oz.	Seeds/ sq. ft.	Mix by weight (%)	Mix by Seeds/sq. ft. (%)
22,000 sq. ft.	43,560 sq. ft.					
1/2 acre	1 acre					
5 lbs.	10 lbs.					
Seed weight in grams for each mix quantity above						
56.70	113.40	4.00	6,600	0.61	2.50	0.31
56.70	113.40	4.00	4,500	0.41	2.50	0.21
70.88	141.75	5.00	1,600	0.18	3.13	0.09
14.18	28.35	1.00	200,000	4.59	0.63	2.35
14.18	28.35	1.00	85,000	1.95	0.63	1.00
14.18	28.35	1.00	100,000	2.30	0.63	1.17
28.35	56.70	2.00	6,500	0.30	1.25	0.15
28.35	56.70	2.00	1,500	0.07	1.25	0.04
28.35	56.70	2.00	1,400	0.06	1.25	0.03
56.70	113.40	4.00	12,000	1.10	2.50	0.56
42.53	85.05	3.00	12,000	0.83	1.88	0.42
14.18	28.35	1.00	470,000	10.79	0.63	5.52
14.18	28.35	1.00	2,500,000	57.39	0.63	29.34
14.18	28.35	1.00	78,000	1.79	0.63	0.92
14.18	28.35	1.00	90,000	2.07	0.63	1.06
14.18	28.35	1.00	46,000	1.06	0.63	0.54
42.53	85.05	3.00	6,800	0.47	1.88	0.24
14.18	28.35	1.00	100,000	2.30	0.63	1.17
42.53	85.05	3.00	27,000	1.86	1.88	0.95
42.53	85.05	3.00	100,000	6.89	1.88	3.52
21.26	42.53	1.50	15,000	0.52	0.94	0.26
14.18	28.35	1.00	46,000	1.06	0.63	0.54
14.18	28.35	1.00	33,000	0.76	0.63	0.39
70.88	141.75	5.00	1,400	0.16	3.13	0.08
21.26	42.53	1.50	70,000	2.41	0.94	1.23
42.53	85.05	3.00	13,000	0.90	1.88	0.46
14.18	28.35	1.00	100,000	2.30	0.63	1.17
28.35	56.70	2.00	20,000	0.92	1.25	0.47
14.18	28.35	1.00	750,000	17.22	0.63	8.80
42.53	85.05	3.00	12,000	0.83	1.88	0.42
907.2	1814.4	64.00 _4.00 lbs._		124.06	40.00	63.42

(continued)

TABLE 11.9.   (*continued*)

Scientific name	Common name	Area covered & recommended seeding rate		
		1,000 sq. ft.	4,400 sq. ft.	11,000 sq. ft.
			1/10 acre	1/4 acre
		1/4 lb.	1 lb.	2.5 lbs.
		Seed weight in grams for each mix quantity above		
Grasses, sedges, and rushes				
Carex bebbii	Bebb's Sedge	2.84	11.34	28.35
Carex comosa	Bottlebrush Sedge	4.25	17.01	42.53
Carex hystericina	Porcupine Sedge	2.84	11.34	28.35
Carex stipata	Awl Fruited Sedge	2.84	11.34	28.35
Carex vulpinoidea	Fox Sedge	2.13	8.51	21.26
Elymus canadensis	Canada Wildrye	22.68	90.72	226.80
Elymus virginicus	Virginia Wildrye	22.68	90.72	226.80
Scirpus atrovirens	Dark Green Bulrush	4.25	17.01	42.53
Scirpus cyperinus	Woolgrass	3.54	14.18	35.44
Total grasses, sedges, and rushes		68.04	272.16	680.4
Total native seed		113.4	453.6	1134

Area covered & recommended seeding rate		Seed wt./acre (10 PLS lbs.) (oz.)	Seeds/ oz.	Seeds/ sq. ft.	Mix by weight (%)	Mix by Seeds/sq. ft. (%)
22,000 sq. ft.	43,560 sq. ft.					
1/2 acre	1 acre					
5 lbs.	10 lbs.					
Seed weight in grams for each mix quantity above						
56.70	113.40	4.00	100,000	9.18	2.50	4.69
85.05	170.10	6.00	24,000	3.31	3.75	1.69
56.70	113.40	4.00	36,000	3.31	2.50	1.69
56.70	113.40	4.00	40,000	3.67	2.50	1.88
42.53	85.05	3.00	90,000	6.20	1.88	3.17
453.60	907.20	32.00	4,200	3.09	20.00	1.58
453.60	907.20	32.00	3,900	2.87	20.00	1.46
85.05	170.10	6.00	140,000	19.28	3.75	9.86
70.88	141.75	5.00	180,000	20.66	3.13	10.56
1360.8	2721.6	96.00 *6.00 lbs.*		71.56	60.00	36.58
2268	4536	160 *10.00 lbs.*		195.62	100.00	100.00

TABLE 11.10. NORTHERN TALLGRASS PRAIRIE MIX FOR MEDIUM SOILS (60% GRASS, 40% FORBS BY WEIGHT)

		Area covered & recommended seeding rates		
		1,000 sq. ft.	4,400 sq. ft. 1/10 acre	11,000 sq. ft. 1/4 acre
		1/4 lb.	1 lb.	2.5 lbs.
		Seed weight in grams for each mix		
Scientific name	**Common name**		quantity above	
Forbs				
Allium cernuum	Nodding Pink Onion	1.42	5.67	14.18
Coreopsis lanceolata	Lanceleaf Coreopsis	2.13	8.51	21.26
Dodecatheon meadia	Shooting Star	1.06	4.25	10.63
Echinacea pallida	Pale Purple Coneflower	5.67	22.68	56.70
Echinacea purpurea	Purple Coneflower	4.25	17.01	42.53
Eryngium yuccifolium	Rattlesnake Master	2.13	8.51	21.26
Heliopsis helianthoides	Oxeye Sunflower	1.42	5.67	14.18
Liatris pycnostachya	Prairie Blazing Star	3.54	14.18	35.44
Monarda fistulosa	Bergamot	0.71	2.84	7.09
Oligoneuron rigidum	Stiff Goldenrod	0.71	2.84	7.09
Parthenium integrifolium	Wild Quinine	1.42	5.67	14.18
Penstemon digitalis	Smooth Penstemon	0.71	2.84	7.09
Ratibida pinnata	Yellow Coneflower	1.42	5.67	14.18
Rudbeckia hirta	Black Eyed Susan	1.77	7.09	17.72
Rudbeckia subtomentosa	Sweet Black Eyed Susan	0.71	2.84	7.09
Rudbeckia triloba	Brown Eyed Susan	0.71	2.84	7.09
Silphium laciniatum	Compassplant	2.13	8.51	21.26
Silphium terebinthinaceum	Prairie Dock	1.42	5.67	14.18
Symphyotrichum laeve	Smooth Aster	0.71	2.84	7.09
Symphyotrichum novae-angliae	New England Aster	0.71	2.84	7.09
Veronicastrum virginicum	Culver's Root	0.71	2.84	7.09
Zizia aurea	Golden Alexander	2.13	8.51	21.26
Total forbs		37.56	150.26	375.64
Legumes				
Astragalus canadensis	Canada Milkvetch	0.71	2.84	7.09
Baptisia australis	Blue False Indigo	1.42	5.67	14.18
Baptisia lactea	White False Indigo	1.42	5.67	14.18
Dalea purpurea	Purple Prairie Clover	2.84	11.34	28.35
Senna hebecarpa	Wild Senna	1.42	5.67	14.18
Total legumes		7.80	31.19	77.96
Total forbs and legumes		45.36	181.44	453.60

Area covered & recommended seeding rates

22,000 sq. ft.	43,560 sq. ft.
1/2 acre	1 acre
5 lbs.	10 lbs.

Seed weight in grams for each mix quantity above		Seed wt./acre (10 PLS lbs.) (oz.)	Seeds/ oz.	Seeds/ sq. ft.	Mix by weight (%)	Mix by Seeds/sq. ft. (%)
28.35	56.70	2.00	7,700	0.35	1.25	0.57
42.53	85.05	3.00	12,500	0.86	1.88	1.38
21.26	42.53	1.50	75,000	2.58	0.94	4.14
113.40	226.80	8.00	5,000	0.92	5.00	1.47
85.05	170.10	6.00	6,600	0.91	3.75	1.46
42.53	85.05	3.00	8,000	0.55	1.88	0.88
28.35	56.70	2.00	6,500	0.30	1.25	0.48
70.88	141.75	5.00	12,000	1.38	3.13	2.21
14.18	28.35	1.00	78,000	1.79	0.63	2.87
14.18	28.35	1.00	46,000	1.06	0.63	1.69
28.35	56.70	2.00	6,800	0.31	1.25	0.50
14.18	28.35	1.00	100,000	2.30	0.63	3.68
28.35	56.70	2.00	27,000	1.24	1.25	1.99
35.44	70.88	2.50	100,000	5.74	1.56	9.20
14.18	28.35	1.00	46,000	1.06	0.63	1.69
14.18	28.35	1.00	33,000	0.76	0.63	1.21
42.53	85.05	3.00	650	0.04	1.88	0.07
28.35	56.70	2.00	1,100	0.05	1.25	0.08
14.18	28.35	1.00	48,000	1.10	0.63	1.77
14.18	28.35	1.00	70,000	1.61	0.63	2.58
14.18	28.35	1.00	750,000	17.22	0.63	27.59
42.53	85.05	3.00	12,000	0.83	1.88	1.32
751.28	1502.55	53.00		42.95	33.13	68.82
14.18	28.35	1.00	16,000	0.37	0.63	0.59
28.35	56.70	2.00	1,600	0.07	1.25	0.12
28.35	56.70	2.00	1,600	0.07	1.25	0.12
56.70	113.40	4.00	20,000	1.84	2.50	2.94
28.35	56.70	2.00	1,400	0.06	1.25	0.10
155.93	311.85	11.00		2.42	6.88	3.87
907.20	1814.40	64.00		45.36	40.00	72.69
		4.00 lbs.				

(continued)

TABLE 11.10.   (*continued*)

		Area covered & recommended seeding rates		
		1,000 sq. ft.	4,400 sq. ft. 1/10 acre	11,000 sq. ft. 1/4 acre
		1/4 lb.	1 lb.	2.5 lbs.
Scientific name	**Common name**		Seed weight in grams for each mix quantity above	
Grasses				
Andropogon gerardii	Big Bluestem	2.84	11.34	28.35
Elymus canadensis	Canada Wildrye	17.01	68.04	170.10
Panicum virgatum	Switchgrass	1.42	5.67	14.18
Schizachyrium scoparium	Little Bluestem	35.44	141.75	354.38
Sorghastrum nutans	Indiangrass	11.34	45.36	113.40
Sporobolus heterolepis	Prairie Dropseed	8.51	34.02	85.05
Total grasses		68.04	272.16	680.40
Total prairie seed		113.40	453.60	1134.00

22,000 sq. ft.	43,560 sq. ft.
1/2 acre	1 acre
5 lbs.	10 lbs.

Seed weight in grams for each mix quantity above		Seed wt./acre (10 PLS lbs.) (oz.)	Seeds/ oz.	Seeds/ sq. ft.	Mix by weight (%)	Mix by Seeds/sq. ft. (%)
56.70	113.40	4.00	8,200	0.75	2.50	1.21
340.20	680.40	24.00	4,200	2.31	15.00	3.71
28.35	56.70	2.00	18,000	0.83	1.25	1.32
708.75	1417.50	50.00	8,800	10.10	31.25	16.19
226.80	453.60	16.00	8,300	3.05	10.00	4.89
170.10	340.20	12.00	14,000	3.86	7.50	6.19
1360.80	2721.60	96.00		17.04	60.00	27.31
		6.00 lbs.				
2268.00	4536.00	160.00		62.40	100.00	100.00
		10.00 lbs.				

TABLE 11.11. XERCES POLLINATOR PRAIRIE FOR MEDIUM SOILS (50% GRASS, 50% FORBS BY WEIGHT)

		Area covered & recommended seeding rate		
		1,000 sq. ft. 1/4 lb.	4,400 sq. ft. 1/10 acre 1 lb.	11,000 sq. ft. 1/4 acre 2.5 lbs.
Scientific name	Common name	Seed weight in grams for each mix quantity above		
Forbs				
Agastache foeniculum	Lavender Hyssop	0.71	2.84	7.09
Allium cernuum	Nodding Pink Onion	2.13	8.51	21.26
Asclepias incarnata	Marsh Milkweed	2.84	11.34	28.35
Asclepias tuberosa	Butterflyweed	1.42	5.67	14.18
Coreopsis lanceolata	Lanceleaf Coreopsis	2.13	8.51	21.26
Echinacea pallida	Pale Purple Coneflower	5.67	22.68	56.70
Echinacea purpurea	Purple Coneflower	4.25	17.01	42.53
Eryngium yuccifolium	Rattlesnake Master	2.13	8.51	21.26
Eupatorium purpureum	Sweet Joe Pye Weed	0.71	2.84	7.09
Liatris pycnostachya	Prairie Blazing Star	2.84	11.34	28.35
Liatris spicata	Dense Blazing Star	2.84	11.34	28.35
Monarda fistulosa	Bergamot	0.71	2.84	7.09
Oligoneuron ohioense	Ohio Goldenrod	0.35	1.42	3.54
Oligoneuron rigidum	Stiff Goldenrod	0.35	1.42	3.54
Penstemon digitalis	Smooth Penstemon	0.71	2.84	7.09
Ratibida pinnata	Yellow Coneflower	1.42	5.67	14.18
Rudbeckia hirta	Blackeyed Susan	1.42	5.67	14.18
Rudbeckia subtomentosa	Sweet Blackeyed Susan	0.71	2.84	7.09
Rudbeckia triloba	Browneyed Susan	0.71	2.84	7.09
Symphyotrichum laeve	Smooth Aster	0.71	2.84	7.09
Symphyotrichum novae-angliae	New England Aster	0.71	2.84	7.09
Symphyotrichum oolentangiense	Skyblue Aster	0.71	2.84	7.09
Tradescantia ohiensis	Spiderwort	2.13	8.51	21.26
Verbena hastata	Blue Vervain	0.71	2.84	7.09
Vernonia fasciculata	Ironweed	1.42	5.67	14.18
Veronicastrum virginicum	Culver's Root	0.71	2.84	7.09
Zizia aurea	Golden Alexander	2.84	11.34	28.35
Total forbs		43.94	175.77	439.43
Legumes				
Chamaechrista fasciculata	Partridge Pea	5.67	22.68	56.70
Dalea purpurea	Purple Prairie Clover	4.25	17.01	42.53
Senna hebecarpa	Wild Senna	2.84	11.34	28.35
Total legumes		12.76	51.03	127.58
Total forbs and legumes		56.70	226.80	567.00

Area covered & recommended seeding rate		
22,000 sq. ft.	43,560 sq. ft.	
1/2 acre	1 acre	
5 lbs.	10 lbs.	

Seed weight in grams for each mix quantity above		Seed wt./acre (10 PLS lbs.) (oz.)	Seeds/ oz.	Seeds/ sq. ft.	Mix by weight (%)	Mix by Seeds/sq. ft. (%)
14.18	28.35	1.00	65,000	1.49	0.63	2.10
42.53	85.05	3.00	7,700	0.53	1.88	0.74
56.70	113.40	4.00	4,500	0.41	2.50	0.58
28.35	56.70	2.00	3,500	0.16	1.25	0.23
42.53	85.05	3.00	12,500	0.86	1.88	1.21
113.40	226.80	8.00	5,000	0.92	5.00	1.29
85.05	170.10	6.00	6,600	0.91	3.75	1.28
42.53	85.05	3.00	8,000	0.55	1.88	0.77
14.18	28.35	1.00	200,000	4.59	0.63	6.45
56.70	113.40	4.00	12,000	1.10	2.50	1.55
56.70	113.40	4.00	12,000	1.10	2.50	1.55
14.18	28.35	1.00	78,000	1.79	0.63	2.51
7.09	14.18	0.50	90,000	1.03	0.31	1.45
7.09	14.18	0.50	46,000	0.53	0.31	0.74
14.18	28.35	1.00	100,000	2.30	0.63	3.22
28.35	56.70	2.00	27,000	1.24	1.25	1.74
28.35	56.70	2.00	100,000	4.59	1.25	6.45
14.18	28.35	1.00	46,000	1.06	0.63	1.48
14.18	28.35	1.00	33,000	0.76	0.63	1.06
14.18	28.35	1.00	48,000	1.10	0.63	1.55
14.18	28.35	1.00	70,000	1.61	0.63	2.26
14.18	28.35	1.00	82,000	1.88	0.63	2.64
42.53	85.05	3.00	8,000	0.55	1.88	0.77
14.18	28.35	1.00	100,000	2.30	0.63	3.22
28.35	56.70	2.00	20,000	0.92	1.25	1.29
14.18	28.35	1.00	750,000	17.22	0.63	24.17
56.70	113.40	4.00	12,000	1.10	2.50	1.55
878.85	1757.70	62.00		52.60	38.75	73.85
113.40	226.80	8.00	3,800	0.70	5.00	0.98
85.05	170.10	6.00	20,000	2.75	3.75	3.87
56.70	113.40	4.00	1,400	0.13	2.50	0.18
255.15	510.30	18.00		3.58	11.25	5.03
1134.00	2268.00	80.00		56.18	50.00	78.88

(continued)

TABLE 11.11. *(continued)*

Scientific name	Common name	1,000 sq. ft. 1/10 acre 1/4 lb.	4,400 sq. ft. 1/10 acre 1 lb.	11,000 sq. ft. 1/4 acre 2.5 lbs.
			Seed weight in grams for each mix quantity above	
Grasses				
Bouteloua curtipendula	Sideoats Grama	17.01	68.04	170.10
Elymus canadensis	Canada Wildrye	14.18	56.70	141.75
Schizachyrium scoparium	Little Bluestem	17.01	68.04	170.10
Sporobolus heterolepis	Prairie Dropseed	8.51	34.02	85.05
Total grasses		56.70	226.80	567.00
Total prairie seed		113.40	453.60	1134.00

Header: Area covered & recommended seeding rate

Area covered & recommended seeding rate		Seed wt./acre (10 PLS lbs.) (oz.)	Seeds/ oz.	Seeds/ sq. ft.	Mix by weight (%)	Mix by Seeds/sq. ft. (%)
22,000 sq. ft.	43,560 sq. ft.					
1/2 acre	1 acre					
5 lbs.	10 lbs.					
Seed weight in grams for each mix quantity above						
340.20	680.40	24.00	8,000	4.41	15.00	6.19
283.50	567.00	20.00	4,200	1.93	12.50	2.71
340.20	680.40	24.00	8,800	4.85	15.00	6.81
170.10	340.20	12.00	14,000	3.86	7.50	5.42
1134.00	2268.00	80.00		15.04	50.00	21.12
2268.00	4536.00	160.00		71.22	100.00	100.00

TABLE 11.12. XERCES POLLINATOR PRAIRIE FOR DRY SOILS (50% GRASS, 50% FORBS BY WEIGHT)

		Area covered & recommended seeding rate		
		1,000 sq. ft.	4,400 sq. ft. 1/10 acre	11,000 sq. ft. 1/4 acre
		1/4 lb.	1 lb.	2.5 lbs.
Scientific name	Common name	Seed weight in grams for each mix quantity above		
Forbs				
Agastache foeniculum	Lavender Hyssop	0.71	2.84	7.09
Asclepias syriaca	Common Milkweed	1.42	5.67	14.18
Asclepias tuberosa	Butterflyweed	4.25	17.01	42.53
Coreopsis lanceolata	Lanceleaf Coreopsis	2.13	8.51	21.26
Echinacea pallida	Pale Purple Coneflower	5.67	22.68	56.70
Liatris aspera	Rough Blazing Star	1.42	5.67	14.18
Monarda fistulosa	Bergamot	0.71	2.84	7.09
Monarda punctata	Dotted Mint	0.71	2.84	7.09
Oligoneuron rigidum	Stiff Goldenrod	0.71	2.84	7.09
Penstemon grandiflorus	Beardtongue	2.13	8.51	21.26
Ratibida pinnata	Yellow Coneflower	2.13	8.51	21.26
Rudbeckia hirta	Blackeyed Susan	2.13	8.51	21.26
Ruellia humilis	Wild Petunia	2.13	8.51	21.26
Solidago ptarmicoides	White Aster	0.71	2.84	7.09
Solidago speciosa	Showy Goldenrod	0.71	2.84	7.09
Symphyotrichum laeve	Smooth Aster	0.71	2.84	7.09
Symphyotrichum oolentangiense	Skyblue Aster	0.71	2.84	7.09
Tradescantia ohiensis	Ohio Spiderwort	2.84	11.34	28.35
Verbena stricta	Hoary Vervain	1.42	5.67	14.18
Total forbs		33.31	133.25	333.11
Legumes				
Amorpha canescens	Leadplant	0.71	2.84	7.09
Chamaechrista fasciculata	Partridge Pea	5.67	22.68	56.70
Dalea candida	White Prairie Clover	2.84	11.34	28.35
Dalea purpurea	Purple Prairie Clover	5.67	22.68	56.70
Lupinus perennis	Lupine	8.51	34.02	85.05
Total legumes		23.39	93.56	233.89
Total forbs and legumes		56.70	226.80	567.00
Grasses				
Bouteloua curtipendula	Sideoats Grama	17.01	68.04	170.10
Elymus canadensis	Canada Wildrye	11.34	45.36	113.40
Koeleria macrantha	Junegrass	2.84	11.34	28.35
Schizachyrium scoparium	Little Bluestem	17.01	68.04	170.10
Sporobolus heterolepis	Prairie Dropseed	8.51	34.02	85.05
Total grasses		56.70	226.80	567.00
Total prairie seed		113.40	453.60	1134.00

Area covered & recommended seeding rate

22,000 sq. ft.	43,560 sq. ft.
1/2 acre	1 acre
5 lbs.	10 lbs.

Seed weight in grams for each mix quantity above		Seed wt./acre (10 PLS lbs.) (oz.)	Seeds/ oz.	Seeds/ sq. ft.	Mix by weight (%)	Mix by Seeds/sq. ft. (%)
14.18	28.35	1.00	65,000	1.49	0.63	2.35
28.35	56.70	2.00	4,000	0.18	1.25	0.29
85.05	170.10	6.00	3,500	0.48	3.75	0.76
42.53	85.05	3.00	12,500	0.86	1.88	1.36
113.40	226.80	8.00	5,000	0.92	5.00	1.45
28.35	56.70	2.00	13,500	0.62	1.25	0.98
14.18	28.35	1.00	78,000	1.79	0.63	2.82
14.18	28.35	1.00	94,000	2.16	0.63	3.40
14.18	28.35	1.00	46,000	1.06	0.63	1.67
42.53	85.05	3.00	11,000	0.76	1.88	1.20
42.53	85.05	3.00	27,000	1.86	1.88	2.93
42.53	85.05	3.00	100,000	6.89	1.88	10.86
42.53	85.05	3.00	4,000	0.28	1.88	0.43
14.18	28.35	1.00	70,000	1.61	0.63	2.54
14.18	28.35	1.00	105,000	2.41	0.63	3.80
14.18	28.35	1.00	48,000	1.10	0.63	1.74
14.18	28.35	1.00	82,000	1.88	0.63	2.97
56.70	113.40	4.00	8,000	0.73	2.50	1.16
28.35	56.70	2.00	32,000	1.47	1.25	2.32
666.23	1332.45	47.00		28.55	29.38	45.03
14.18	28.35	1.00	17,000	0.39	0.63	0.62
113.40	226.80	8.00	3,800	0.70	5.00	1.10
56.70	113.40	4.00	15,000	1.38	2.50	2.17
113.40	226.80	8.00	20,000	3.67	5.00	5.79
170.10	340.20	12.00	1,000	0.28	7.50	0.43
467.78	935.55	33.00		6.41	20.63	10.12
1134.00	2268.00	80.00		34.96	50.00	55.15
340.20	680.40	24.00	8,000	4.41	15.00	6.95
226.80	453.60	16.00	4,200	1.54	10.00	2.43
56.70	113.40	4.00	150,000	13.77	2.50	21.73
340.20	680.40	24.00	8,800	4.85	15.00	7.65
170.10	340.20	12.00	14,000	3.86	7.50	6.08
1134.00	2268.00	80.00		28.43	50.00	44.85
2268.00	4536.00	160.00		63.39	100.00	100.00

12

Tables

TABLE 12.1. PLANT HABITATS AND CHARACTERISTICS

Scientific name (4/1/2020)	Common name	Height	Flower color	Bloom time	USDA zones	Sun	Soil moisture
Flowers and shrubs							
Agastache foeniculum	Lavender Hyssop	1'–3'	Purple	July–Sept.	2–6	F, P	D, M
Allium cernuum	Nodding Pink Onion	1'–2'	White/pink	June–Aug.	3–8	F, P	M, Mo
Allium stellatum	Prairie Onion	1'–2'	Lavender	July–Aug.	3–8	F, P	D, M
Amorpha canescens	Leadplant	2'–3'	Purple	June–July	3–8	F, P	D, M
Amsonia tabernaemontana	Eastern Bluestar	2'–3'	Blue/white	May–June	4–8	F, P	M, Mo
Anemone canadensis	Canada Anemone	1'–2'	White	May–July	3–6	F, P	M, Mo
Anemone cylindrica	Thimbleweed	1'–3'	White	May–June	3–6	F	D, M
Antennaria neglecta	Pussytoes	6"–1'	White	Apr.–May	3–7	F	D
Arnoglossum atriplicifolium	Pale Indian Plantain	5'–10'	White	July–Sept.	4–8	F, P	D, M, Mo
Asclepias incarnata	Marsh Milkweed	3'–5'	Red/pink	June–July	3–9	F	Mo, W
Asclepias sullivantii	Sullivant's Milkweed	3'–5'	Pink/yellow	June–Aug.	4–7	F	M, Mo
Asclepias syriaca	Common Milkweed	2'–4'	Lavender	June–Aug.	3–8	F, P	D, M
Asclepias tuberosa	Butterflyweed	2'–3'	Orange	June–Aug.	3–10	F	D, M
Astragalus canadensis	Canada Milkvetch	2'–3'	Yellow	July–Aug.	3–8	F	D, M
Baptisia alba	White False Indigo	3'–5'	White	June–July	4–9	F, P	M, Mo
Baptisia australis	Blue False Indigo	3'–5'	Blue	June–July	3–10	F, P	M
Baptisia bracteata	Cream False Indigo	1'–2'	Cream	May–June	4–9	F, P	D, M
Blephilia ciliata	Downy Wood Mint	1'–2'	White/purple/violet	May–July	4–8	F, P	D, M
Callirhoe involucrata	Winecups	1'	Magenta	May–June	4–9	F, P	D, M
Callirhoe triangulata	Poppymallow	1'–2'	Magenta	July–Aug.	4–8	F	D
Ceanothus americanus	New Jersey Tea	2'–3'	White	July–Aug.	3–9	F, P	D, M
Chelone glabra	White Turtlehead	2'–4'	White	Aug.–Sept.	3–8	F, P	Mo, W
Conoclinium coelestinum	Blue Mistflower	1'–3'	Blue/purple	July–Oct.	4–10	F, P	M, Mo
Coreopsis lanceolata	Lanceleaf Coreopsis	1'–2'	Yellow	June–July	3–9	F	D, M
Coreopsis palmata	Stiff Coreopsis	2'–3'	Yellow	June–Aug.	3–8	F	D, M
Coreopsis tripteris	Tall Coreopsis	3'–7'	Yellow	July–Sept.	3–8	F, P	M, Mo, W
Dalea candida	White Prairie Clover	1'–2'	White	July–Aug.	3–9	F	D, M
Dalea purpurea	Purple Prairie Clover	1'–2'	Purple/yellow	July–Aug.	3–8	F	D, M
Delphinium carolinianum subsp. *virescens*	Prairie Larkspur	1'–4'	White	June–July	3–8	F	D, M
Desmanthus illinoensis	Illinois Bundleflower	2'–6'	White	June–Aug.	5–8	F	D, M, Mo
Desmodium canadense	Canada Ticktrefoil	3'–5'	Purple	July–Aug.	3–7	F	M, Mo
Dodecatheon meadia	Shooting Star	1'–2'	White/pink	May–June	4–8	F, P	M, Mo
Echinacea pallida	Pale Purple Coneflower	3'–5'	Purple	June–July	4–8	F	D, M
Echinacea paradoxa	Ozark Coneflower	3'–5'	Yellow	June–July	4–7	F, P	D, M
Echinacea purpurea	Purple Coneflower	3'–4'	Purple	July–Sept.	4–8	F, P	D, M
Eryngium yuccifolium	Rattlesnake Master	3'–5'	White	June–Aug.	4–9	F	D, M
Eupatorium perfoliatum	Boneset	3'–4'	White	Aug.–Sept.	3–8	F	M, Mo, W
Euphorbia corollata	Flowering Spurge	2'–4'	White	July–Aug.	3–9	F	D, M
Eutrochium fistulosum	Tall Joe Pye Weed	5'–8'	Purple/pink	Aug.–Sept.	4–9	F, P	M, Mo, W

Soil texture	pH range	Plant spacing in prairie gardens	Plant spacing in flower beds	Root type	Plant life expectancy in years	Aggressiveness and method	Cut flowers	Dried arrangements	Deer palatability
S, L	?–8.0	1'	1'–2'	FIB	3–5	M–S	x	FL/SH	L
S, L, C	6.5–8.0	6"–1'	1'	B	10–20	L–S	x	FL/SH	L
S, L	5.5–8.0	6"–1'	1'	B	10–20	L–S	x	FL/SH	L
S, L	5.5–8.0	2'	3'	TAP	20+	L–S		Leaves	L
S, L, C	?–8.0	2'–3'	3'	FIB	20+	L–S		SP	L
S, L, C	?	NR	NR	R	10–20	H–R			L
S, L	?	1'	1'	FIB	10–20	L–S		SD	L
S, G	5.5–7.5	1'	1'	R	5–10	L–R	x	FL	L
S, L, C	?–8.0	2'	3'–4'	FIB	3–20	H–S (north); L–S (south)	x	FL	L
S, L, C	5.0–8.0	1'–2'	3'	FIB	5–10	L–S	x	SP	L
L, C	?	1'–2'	3'	R	10–20	L–R		SP	L
S, L, C	5.0–8.0	NR	NR	R	10–20	H–R		SP	L
S, L	5.0–7.5	2'	3'	TAP	10–20	L–S	x	SP	L
S, L	6.0–8.0	1'–2'	3'	TAP	10–20	M–S		SP	M
S, L, C	?–7.5	3'	3'	TAP	20+	L–S	x	SP	L
S, L, C	?–8.0	3'	3'	TAP	20+	L–S	x	SP	L
S, L	?	2'	3'	TAP	20+	L–S	x	SP	L
G, L, C	?–8.00	1'	1'–2'	TAP/ short R	3–5	L–S		FL/SH	L
S, L	5.5–7.5	1'–2'	2'	TAP	10–20	L–S			L
S	?	1'–2'	2'	TAP	10–20	L–S			L
S, L	4.5–7.0	3'	3'	TAP	20+	L–S		SD	L
S, L, C	?	1'	1'–2'	FIB	5–10	L–S			M
S, L, C	5.0–8.0	1'–2'	2'	R	3–5	M–R			L
S, L	5.0–7.5	1'	1'–2'	FIB	3–5	M–S	x		M
S, L	?	1'–2'	2'	R	20+	M–R			L
S, L, C	?	1'–2'	NR	FIB	10–20	L–S	x		L
S, L	?	6"–1'	1'–2'	TAP	10–20	L–S			H
S, L, C	?	6"–1'	1'–2'	TAP	10–20	L–S	x		H
S, L	?	1'	1.5'	FIB	5–10	L–S	x	SP	L
S, L, C	6.0–8.0	1'	1.5'	TAP	5–10	L–S		SP	M
S, L, C	?–7.5	NR	NR	TAP	10–20	M–S			H
S, L, C	4.5–8.0	6"–1'	1'	FIB	20+	L–S		SP	L
S, L, C	6.0–7.5	1'	2'–3'	TAP	10–20	L–S	x	SH	M
S, L, C	?–7.5	1'	2'–3'	TAP	10–20	L–S	x	SH	M
S, L, C	5.5–8.0	1'–2'	3'	FIB	5–10	M–S	x	SH	M
S, L, C	6.0–8.0	1'	2'	FIB	5–10	M–S	x	SH	L
S, L, C	?–7.5	1'	NR	FIB	5–10	L–S			L
S, L	?	1'–2'	1'–2'	R	10–20	M–R			L
S, L, C	4.5–7.5	2'	3'–4'	FIB	5–10	L–S			H

(continued)

TABLE 12.1. *(continued)*

Scientific name (4/1/2020)	Common name	Height	Flower color	Bloom time	USDA zones	Sun	Soil moisture
Eutrochium maculatum	Joe Pye Weed	4'–6'	Pink	Aug.–Sept.	3–6	F, P	Mo, W
Filipendula rubra	Queen of the Prairie	4'–5'	Pink	June–July	3–6	F	M, Mo
Gentiana alba	Cream Gentian	1'–2'	White	July–Aug.	3–7	F, P	M
Gentiana andrewsii	Bottle Gentian	1'–2'	Blue	Aug.–Oct.	3–6	F, P	Mo, W
Geum triflorum	Prairie Smoke	6"	Pink	May–June	3–6	F	D, M
Helenium autumnale	Sneezeweed	4'–5'	Yellow	Aug.–Sept.	3–9	F	Mo, W
Helianthus mollis	Downy Sunflower	4'–6'	Yellow	Aug.–Sept.	4–9	F	D, M
Helianthus occidentalis	Western Sunflower	2'–3'	Yellow	July–Aug.	3–9	F	D, M
Helianthus ×laetiflorus	Showy Sunflower	3'–6'	Yellow	Aug.–Sept.	3–8	F	D, M
Heliopsis helianthoides	Oxeye Sunflower	3'–6'	Yellow	June–Sept.	3–8	F	D, M, Mo
Hibiscus moscheutos	Rosemallow	3'–7'	White/pink/red	July–Sept.	5–9	F, P	M, Mo, W
Hypericum ascyron	Great St. Johnswort	4'–6'	Yellow	July–Aug.	3–5	F, P	Mo, W
Iris virginica var. *shrevei*	Wild Iris	2'–3'	Blue	June–July	3–8	F, P	W
Lespedeza capitata	Roundhead Bush Clover	3'–5'	White	Aug.–Sept.	3–8	F, P	D, M
Liatris aspera	Rough Blazing Star	2'–5'	Purple/pink	Aug.–Sept.	3–8	F	D, M
Liatris ligulistylis	Meadow Blazing Star	3'–5'	Purple/pink	Aug.–Sept.	3–6	F	M, Mo
Liatris punctata	Dotted Blazing Star	1'–2'	Purple/pink	Aug.–Oct.	3–9	F	D
Liatris pycnostachya	Prairie Blazing Star	3'–5'	Purple/pink	July–Aug.	3–9	F	M, Mo
Liatris spicata	Dense Blazing Star	3'–6'	Purple/pink	Aug.–Sept.	4–10	F	M, Mo
Liatris squarrosa	Scaly Blazing Star	1'–2'	Purple/pink	Aug.–Sept.	4–8	F	D, M
Lilium michiganense	Michigan Lily	4'–6'	Orange	July–Aug.	3–7	F, P	Mo
Lithospermum canescens	Hoary Puccoon	1'	Orange	May–June	3–7	F, P	D, M
Lithospermum caroliniense	Hairy Puccoon	1'–2'	Yellow	June–July	3–8	F	D
Lithospermum incisum	Fringed Puccoon	1'–2'	Yellow	June–July	3–9	F	D
Lobelia cardinalis	Cardinal Flower	2'–5'	Red	July–Sept.	3–9	F, P	Mo, W
Lobelia siphilitica	Great Blue Lobelia	1'–4'	Blue	July–Sept.	3–9	F, P	M, Mo
Lupinus perennis	Wild Lupine	1'–2'	Blue	May–June	3–8	F, P	D
Monarda fistulosa	Bergamot	2'–5'	Lavender	July–Sept.	3–9	F, P	D, M, Mo
Monarda punctata	Dotted Mint	1'–2'	Lavender	July–Sept.	3–10	F	D
Oenothera gaura	Biennial Beeblossom	3'–6'	White/pink	June–Sept.	4–8	F, P	D, M, Mo
Oligoneuron ohioense	Ohio Goldenrod	3'–4'	Yellow	Aug.–Sept.	4–5	F, P	M, Mo
Oligoneuron rigidum	Stiff Goldenrod	3'–5'	Yellow	Aug.–Sept.	3–9	F	D, M, Mo
Onosmodium bejariense var. *occidentale*	Marbleseed (Western Marbleseed)	1'–3'	White	May–June	3–8	F	D, M
Parthenium integrifolium	Wild Quinine	3'–5'	White	June–Sept.	4–8	F	M, Mo
Pediomelum esculentum	Prairie Turnip	1'–2'	Blue	June–July	3–7	F	D
Penstemon cobaea	Cobaea Beardtongue	2'–3'	Lavender	Apr.–June	5–8	F	D
Penstemon digitalis	Foxglove Beardtongue	2'–3'	White	June–July	3–8	F, P	M, Mo
Penstemon gracilis	Slender Beardtongue	1'–2'	Lavender	May–June	3–6	F, P	D
Penstemon grandiflorus	Large-Flowered Beardtongue	2'–4'	Lavender	May–June	3–7	F	D

Soil texture	pH range	Plant spacing in prairie gardens	Plant spacing in flower beds	Root type	Plant life expectancy in years	Aggressiveness and method	Cut flowers	Dried arrangements	Deer palatability
S, L, C	5.5–7.5	1'–2'	3'–4'	FIB	5–10	L–S			L
S, L, C	?–8.0	2'	3'	R	20+	M–R	x		L
S, L, C	?–7.5	1'	1.5'	FIB	10–20	L–S	x		L
S, L, C	5.8–7.2	1'	1.5'	FIB	5–10	L–S	x		L
S, L	6.0–7.5	6"	1'	R	20+	L–R		SH	L
S, L, C	4.0–7.5	1'–2'	2'–3'	FIB	5–10	L–S	x	SH	L
S, L, C	6.0–8.0	2'	3'	R	20+	M–R	x	SH	L
S, L	?	2'	3'	R	20+	M–R	x		L
S, G, L, C	?	NR	NR	R	20+	H–R	x	SH	L
S, L, C	6.0–8.0	1'–2'	3'	FIB	5–10	H–S	x		L
S, L, C	4.0–8.0	2'	4'	R	10–20	L–R/S			L
S, L, C	5.7–7.1	1'–2'	3'	FIB	5–10	L–S	x	SP	L
S, L, C	4.8–7.3	1'	2'	R/CO	10–20	L–R	x	SP	L
S, G, L	5.5–8.0	1'	2'	TAP	10–20	L–S		SH	M
S, L	?	6"–1'	1'	CO	10–20	L–S	x	FL/SH	L
L	?	6"–1'	1'	CO	3–5	L–S	x	FL/SH	L
S, L	6.0–8.0	1'	1.5'–2'	TAPCO	10–20	L–S			L
S, L, C	6.0–8.0	6"–1'	1'	CO	10–20	L–S	x	FL/SH	L
S, L, C	5.5–7.5	6"–1'	1'	CO	10–20	L–S	x	FL/SH	L
S, L	?	1'	1.5'–2'	CO	10–20	L–S	x	FL/SH	L
S, L, C	?–6.8	2'	2'	B	20+	L–S	x	SP	H
S, L	?	1'	1'	FIB	5–10	L–S			L
S	?	1'	1'	TAP	10–20	L–S			L
S, G	?	1'	1'	TAP	5–10	L–S			L
S, L	5.8–7.8	6"–1'	2'	FIB	3–5	L–S	x		M
S, L, C	?–8.0	6"–1'	2'	FIB	5–10	M–S	x		L
S	5.0–?	1'	1.5'	TAP	10–20	M–S			L
S, L, C	6.0–8.0	1'–2'	3'	R	10–20	M–R	x	SH	L
S, G	5.0–?	1'	1'–2'	FIB	2	M–S	x	SH	L
S, L, C	?	2'	3'	FIB	2	M–S	x		L
S, L, C	?–8.0	1'–2'	3'	FIB	5–10	L–S	x	SH	L
S, L, C	5.5–8.0	1'–1.5'	3'	FIB	5–10	M–S	x	SH	L
S, L	?–7.5	1'–1.5'	2'	TAP	10–20	L–S			L
S, L, C	?–8.0	1'–2'	3'	TAP	10–20	L–S	x		M
S, G	?–8.5	1'–1.5'	2'	TAP	20+	L–S			L
S, G	?–7.5	1'	1'	FIB	3–5	L–S	x	SP	L
S, L, C	5.5–7.5	1'	2'	FIB	5–10	L–S	x		L
S, G	?–7.5	6"–1'	1'	FIB	5–10	L–S	x	SP	L
S, G	?	1'	2'–3'	FIB	3–5	L–S		SP	L

(continued)

TABLE 12.1. *(continued)*

Scientific name (4/1/2020)	Common name	Height	Flower color	Bloom time	USDA zones	Sun	Soil moisture
Penstemon hirsutus	Hairy Penstemon	1'–2'	Lavender	May–June	3–6	F, P	D
Phlox pilosa	Downy Phlox	1'–2'	Pink	May–June	3–9	F	D, M, Mo
Physostegia virginiana	Obedient Plant	1'–2'	Pink	Aug.–Sept.	3–9	F	M, Mo
Polygonatum biflorum var. commutatum	Great Solomon's Seal	2'–4'	Cream	May–June	3–9	F, P	D, M
Pulsatilla patens	Pasque Flower	6"–1'	White	Apr.–May	3–6	F	D
Pycnanthemum tenuifolium	Narrowleaf Mountainmint	1'–3'	White	July–Sept.	4–8	F, P	D, M
Pycnanthemum virginianum	Virginia Mountainmint	1'–3'	White	July–Sept.	3–7	F, P	Mo, W
Ratibida pinnata	Yellow Coneflower	3'–6'	Yellow	July–Sept.	3–9	F	D, M, Mo
Rosa carolina	Carolina Rose	1'–2'	Pink	June–July	3–8	F	D, M
Rosa setigera	Climbing Prairie Rose	6'–15'	Pink/white	May–July	4–9	F, P	M, Mo
Rudbeckia fulgida	Orange Coneflower	2'–4'	Yellow	July–Sept.	4–8	F, P	M, Mo
Rudbeckia hirta	Blackeyed Susan	1'–3'	Yellow	June–Sept.	3–10	F, P	D, M, Mo
Rudbeckia laciniata	Green-Headed Coneflower	5'–8'	Yellow	Aug.–Sept.	3–8	F, P	Mo, W
Rudbeckia subtomentosa	Sweet Blackeyed Susan	4'–6'	Yellow	Aug.–Oct.	4–8	F, P	M, Mo
Rudbeckia triloba	Browneyed Susan	2'–5'	Yellow	July–Oct.	4–9	F, P	M, Mo
Ruellia humilis	Wild Petunia	1'–2'	Violet	June–Aug.	4–9	F	D, M
Salvia azurea	Blue Sage	3'–5'	Blue	June–Aug.	4–9	F, P	D, M
Senna hebecarpa	Wild Senna	4'–6'	Yellow	July–Aug.	4–7	F	M, Mo, W
Silene regia	Royal Catchfly	2'–4'	Red	July–Aug.	4–9	F, P	M
Silphium integrifolium	Rosinweed	2'–6'	Yellow	July–Sept.	4–8	F	M, Mo
Silphium laciniatum	Compassplant	3'–10'	Yellow	June–Sept.	4–9	F	D, M
Silphium perfoliatum	Cup Plant	3'–10'	Yellow	July–Sept.	3–8	F, P	M, Mo, W
Silphium terebinthinaceum	Prairie Dock	3'–10'	Yellow	July–Sept.	4–7	F	M, Mo
Sisyrinchium albidum	White Blue-Eyed Grass	6"–1'	White	Apr.–May	4–9	F	D, M
Sisyrinchium angustifolium	Narrow-Leaved Blue-Eyed Grass	6"–2'	Blue/purple	Apr.–June	3–10	F, P	M, Mo
Sisyrinchium campestre	Prairie Blue-Eyed Grass	6"–1'	Blue/white	Apr.–June	3–8	F, P	D, M
Solidago ptarmicoides	White Aster	1'–2'	White	Aug.–Sept.	3–7	F	D
Solidago speciosa	Showy Goldenrod	1'–3'	Yellow	Aug.–Sept.	3–8	F	D, M
Spiraea alba	Meadowsweet	2'–4'	White/pink	June–Sept.	3–7	F, P	Mo, W
Spiraea tomentosa	Steeplebush	2'–4'	Pink	Aug.–Sept.	3–8	F, P	Mo, W
Symphyotrichum ericoides	Heath Aster	1'–3'	White	Aug.–Oct.	3–9	F	D, M
Symphyotrichum laeve	Smooth Aster	2'–4'	Blue	Aug.–Oct.	4–8	F	D, M
Symphyotrichum novae-angliae	New England Aster	3'–6'	Pink/purple/blue	Aug.–Oct.	3–7	F, P	M, Mo
Symphyotrichum oblongifolium	Aromatic Aster	1'–2'	Purple/violet	Sept.–Oct.	3–8	F	D, M
Symphyotrichum oolentangiense	Skyblue Aster	2'–3'	Blue	Aug.–Oct.	3–8	F, P	D, M
Tephrosia virginiana	Goatsrue	1'–2'	Pink/yellow	June–Aug.	4–9	F, P	D

Soil texture	pH range	Plant spacing in prairie gardens	Plant spacing in flower beds	Root type	Plant life expectancy in years	Aggressiveness and method	Cut flowers	Dried arrangements	Deer palatability
S, G	5.0–8.0	1'	1.5'	FIB	3–5	L–S	x	SP	L
S, L	?	6"–1'	1.5'	FIB	3–5	L–S	x		L
S, L, C	6.0–8.0	1'–2'	NR	R	10–20	H–R	x	SH	M
S, L, C	?	1'–2'	3'	R	10–20	L–R/S			H
S, G	?–8.0	6"–1'	1'	FIB	5–10	L–S		SD	H
S, G, L, C	?	1'	1.5'	TAP/R	5–10	L–S		SH	M
S, L, C	?–8.0	1'–2'	3'	R	5–10	L–R		SH	L
S, L, C	5.5–8.0	1'–2'	2'	FIB	10–20	M–S	x	SH	L
S, L	4.0–7.0	2'	3'	R	10–20	L–R/S	x	RH	L
S, L, C	5.0–8.0	2'	5'	FIB	10–20	L–S	x	RH	L
L, C	?–8.0	1'–2'	3'	FIB	5–10	L–S	x	SH	L
S, L, C	5.0–8.0	1'	2'	FIB	2	M–S	x	SH	L
S, L, C	4.5–7.5	2'	2'	FIB	3–5	M–S	x	SH	L
S, L, C	?–8.0	1'–2'	2'	FIB	10–20	L–S	x	SH	L
S, L	?–8.0	1'–2'	3'	FIB	2	M–S	x	SH	L
S, L	4.5–7.5	1'	2'	FIB	5–10	L–S			L
S, G, L	?	1'	1.5'	FIB	5–10	L–S	x	SH	L
S, L, C	?	1.5'–2'	3'–4'	FIB/TAP	10–20	L–S	x	FL/SH	M
L	?	1'	1.5'	TAP	5–10	L–S	x	SH	L
S, L, C	?–8.0	2'	3'	TAP	20+	M–S	x		L
S, L, C	?–8.0	2'	4'	TAP	20+	L–S	x		L
S, L, C	4.5–8.0	2'–3'	NR	FIB	20+	H–S	x		L
S, L, C	?–8.0	2'	4'	TAP	20+	L–S	x		L
S, L	?–8.0	6"	6"	FIB	3–5	L–S			L
S, L, C	?–8.0	6"	9"	FIB	3–5	L–S			L
S, L	?–7.5	6"	6"	FIB	3–5	L–S			L
S, G	6.5–8.0	1'	2'	FIB	5–10	L–S		Empty SH	L
S, L	?	1'	3'	FIB	5–10	L–S	x	SH	L
S, L, C	4.3–7.2	2'	3'	FIB	10–20	L–S	x		L
S, L, C	4.5–7.0	2'	3'	R	10–20	L–R/S	x		L
S, G, L	?	1'–2'	2'	R	10–20	M–R		Empty SH	L
S, L	?	1'	3'	FIB	5–10	M–S	x	Empty SH	H
S, L, C	5.0–8.0	1'	3'	FIB	5–10	L–S	x	Empty SH	H
S, L, C	?–8.0	1'	3'	R	5–10	M–R?		Empty SH	L
S, L	5.0–8.0	1'	2'	FIB	5–10	L–S	x	Empty SH	M
S, G	4.5–?	2'	3'	TAP	10–20	L–S			L

(continued)

TABLE 12.1. *(continued)*

Scientific name (4/1/2020)	Common name	Height	Flower color	Bloom time	USDA zones	Sun	Soil moisture
Thalictrum dasycarpum	Tall Meadow-Rue	3'–6'	White	June–July	3–9	F, P	Mo, W
Tradescantia bracteata	Prairie Spiderwort	1'–2'	Blue	June–July	3–7	F, P	D, M
Tradescantia occidentalis	Western Spiderwort	1'	Pink	June–July	3–9	F	D, M
Tradescantia ohiensis	Ohio Spiderwort	2'–4'	Blue	June–July	3–9	F, P	D, M, Mo
Verbena hastata	Blue Vervain	3'–6'	Blue	July–Sept.	3–8	F, P	M, Mo, W
Verbena stricta	Hoary Vervain	2'–4'	Blue	July–Sept.	3–8	F	D, M
Vernonia fasciculata	Ironweed	4'–6'	Red/pink	July–Sept.	3–7	F	Mo
Vernonia gigantea	Tall Ironweed	5'–8'	Red/pink	Aug.–Sept.	4–9	F, P	Mo, W
Veronicastrum virginicum	Culver's Root	3'–6'	White	July–Aug.	3–8	F, P, S	M, Mo
Viola pedata	Birdfoot Violet	6"	Blue/purple	Apr.–June	3–9	F, P	D
Zizia aptera	Heartleaf Golden Alexanders	1'–2'	Yellow	May–June	3–8	F	M, Mo
Zizia aurea	Golden Alexanders	1'–2'	Yellow	May–July	3–8	F, P, S	M, Mo, W

Scientific name (4/1/2020)	Common name	Height	Fall leaf color	Bloom time	USDA zones	Sun	Soil moisture
Grasses and sedges							
Andropogon gerardii	Big Bluestem	5'–8'	Bronze/red	Aug.–Oct.	3–9	F	D, M, Mo
Andropogon virginicus	Broomsedge	2'–5'	Reddish brown	Aug.–Oct.	5–10	F	D, M, Mo
Bouteloua curtipendula	Sideoats Grama	2'–3'	Straw	Aug.–Sept.	3–9	F	D, M
Bromus kalmii	Prairie Bromegrass	2'–4'	Straw	June–July	3–6	F	D
Calamagrostis canadensis	Bluejoint Grass	4'–6'	Straw	June–July	3–6	F	Mo, W
Carex brevior	Shortbeak Sedge	1'–4'	Straw	May–June	3–7	F, P	D, M, Mo
Elymus canadensis	Canada Wildrye	4'–5'	Straw	July–Aug.	2–9	F	D, M, Mo
Elymus virginicus	Virginia Wildrye	4'–5'	Straw	July–Aug.	2–9	F, P, S	M, Mo
Eragrostis spectabilis	Purple Lovegrass	1'–2'	Purple/pink	July–Sept.	3–10	F, P	D
Hesperostipa spartea	Needlegrass	2'–4'	Gold	Apr.–June	3–6	F	D, M
Hierochloe hirta	Vanilla Sweetgrass	1'–2'	Straw	May–June	3–9	F, P	M, Mo, W
Koeleria macrantha	Junegrass	2'–3'	Gold	May–June	3–8	F	D
Panicum virgatum	Switchgrass	3'–6'	Gold	Aug.–Sept.	3–10	F	D, M, Mo
Schizachyrium scoparium	Little Bluestem	2'–5'	Crimson red	Aug.–Oct.	3–10	F	D, M
Sorghastrum nutans	Indiangrass	5'–7'	Gold	Aug.–Sept.	3–9	F	D, M, Mo
Spartina pectinata	Prairie Cordgrass	6'–9'	Gold	Aug.–Sept.	3–8	F	Mo, W
Sporobolus heterolepis	Prairie Dropseed	2'–4'	Gold	Aug.–Sept.	3–8	F	D, M
Tridens flavus	Purpletop	3'–6'	Straw	July–Sept.	5–10	F	D, M
Tripsacum dactyloides	Eastern Gamagrass	4'–8'	Gold	July–Sept.	4–10	F, P	M, Mo, W

Notes: For soil moisture, D = dry, M = medium, Mo = moist, and W = wet. For soil texture, S = sand, L = loam, C = clay, and G = gravel. For sun, F = full sun, or at least a half day of full sun; P = partial sun, or a quarter to a half day of sun; and S = shade, or less than a quarter day of sun. For plant spacing, NR = not recommended. For root type, B = bulb, CO = corm, FIB = fibrous, R = rhizome, TAP =

Soil texture	pH range	Plant spacing in prairie gardens	Plant spacing in flower beds	Root type	Plant life expectancy in years	Aggressiveness and method	Cut flowers	Dried arrangements	Deer palatability
S, L, C	?	1'–2'	2'	FIB	5–10	L–S			L
S, L, C	?	1'	2'	R	20+	M–R			M
S, L	?	1'	NR	R	20+	H–R			M
S, L, C	?–7.5	1'	2'	FIB	20+	M–S			H
S, L, C	?	1'	2'	FIB	2	M–S	x	SH	L
S, L	?	1'	2'	FIB	3–5	M–S	x	SH	L
S, L, C	?	1'	3'	FIB	5–10	L–S	x		L
S, L, C	?–8.0	1'–2'	3'	FIB	10–20	L–S	x		L
S, L, C	?–8.0	1'	3'	FIB	10–20	L–S	x	SH	L
S, G	?	6"	1'	R	5–10	L–S			L
S, L	?–7.5	1'	2'	FIB	5–10	L–S			L
S, L, C	?–8.0	1'	3'	FIB	5–10	L–S			L

Soil texture	pH range	Plant spacing in prairie gardens	Plant spacing in flower beds	Root type	Plant life expectancy in years	Aggressiveness and method	Cut flower stalks	Dried seed arrangements	Deer palatability
S, L, C	5.0–7.5	2'	NR	FIB	20+	M–S		x	L
S, L, C	4.0–7.0	1'–1.5'	NR	FIB	20+	H–S		x	L
S, L	5.5–8.5	1'	2'	FIB	10–20	L–S		x	L
S, G	5.7–7.0	1'	2'	FIB	3–5	L–S		x	L
S, L, C	4.5–8.0	1'	3'	FIB	10–20	M–R/S		x	L
S, L, C	4.5–7.5	1'	1'	FIB	5–10	L–S			L
S, L, C	5.0–8.0	1'	3'	FIB	3–5	M–S		x	L
S, L, C	5.0–7.5	1'	3'	FIB	3–5	M–S		x	L
S, G	4.0–7.5	1'–1.5'	2'	FIB	5–10	M–S	x	x	L
S, G, L	5.0–7.5	1'	NR	FIB	10–20	L–S		x	L
S, L, C	5.7–8.0	1'	NR	R	10–20	H–R			L
S, G	6.0–8.0	6"–1'	1'	FIB	5–10	L–S		x	L
S, L, C	4.5–8.0	1.5'–2'	NR	FIB	20+	M–S		x	L
S, L	5.0–8.5	1'–1.5'	3'	FIB	20+	L–S		x	L
S, L, C	5.0–8.0	1'–2'	3'	FIB	10–20	M–S	x	x	L
S, L, C	6.0–8.5	2'	NR	R	20+	H–R		x	L
S, L	6.0–8.0	2'	2'	FIB	20+	L–S	x	x	L
S, L, C	5.0–7.5	1'	1.5'	FIB	5–10	M–S		x	L
S, L, C	4.5–7.5	2'	3'	FIB	20+	H–S		x	M

taproot, and TAPCO = taprooted corm. For aggressiveness and method, L = low, M = medium, H = high, R = by rhizome, S = by seed, and R/S = rhizome and seed. For dried arrangements, FL = flowers, SH = seedheads, SP = seedpods, SD = seed, and RH = rose hips. For deer palatability, L = low, M = medium, and H = high.

TABLE 12.2. WILDLIFE ATTRACTED

Scientific name (4/1/2020)	Common name	Attracts hummingbirds	Attracts songbirds	Attracts upland game birds	Attracts bees and bumblebees	Flowers attract butterflies
Flowers and shrubs						
Agastache foeniculum	Lavender Hyssop	x	x	x	x	x
Allium cernuum	Nodding Pink Onion	x	x	x	x	x
Allium stellatum	Prairie Onion				x	
Amorpha canescens	Leadplant		x	x	x	x
Amsonia tabernaemontana	Eastern Bluestar	x			x	x
Anemone canadensis	Canada Anemone				x	x
Anemone cylindrica	Thimbleweed				x	
Antennaria neglecta	Pussytoes				x	x
Arnoglossum atriplicifolium	Pale Indian Plantain				x	x
Asclepias incarnata	Marsh Milkweed	x			x	x
Asclepias sullivantii	Sullivant's Milkweed	x			x	x
Asclepias syriaca	Common Milkweed				x	x
Asclepias tuberosa	Butterflyweed	x			x	x
Astragalus canadensis	Canada Milkvetch	x	x	x	x	x
Baptisia alba	White False Indigo		x	x	x	
Baptisia australis	Blue False Indigo		x	x	x	
Baptisia bracteata	Cream False Indigo		x	x	x	
Blephilia ciliata	Downy Wood Mint				x	x
Callirhoe involucrata	Winecups				x	x
Callirhoe triangulata	Poppymallow				x	x
Ceanothus americanus	New Jersey Tea	x		x	x	x
Chelone glabra	White Turtlehead	x			x	x
Conoclinium coelestinum	Blue Mistflower		x		x	x
Coreopsis lanceolata	Lanceleaf Coreopsis	x			x	x
Coreopsis palmata	Stiff Coreopsis				x	x
Coreopsis tripteris	Tall Coreopsis		x		x	x
Dalea candida	White Prairie Clover		x	x	x	x
Dalea purpurea	Purple Prairie Clover		x	x	x	x
Delphinium carolinianum subsp. virescens	Prairie Larkspur	x			x	x
Desmanthus illinoensis	Illinois Bundleflower		x	x	x	
Desmodium canadense	Canada Ticktrefoil	x	x	x	x	x
Dodecatheon meadia	Shooting Star				x	
Echinacea pallida	Pale Purple Coneflower	x	x	x	x	x
Echinacea paradoxa	Ozark Coneflower		x	x	x	x
Echinacea purpurea	Purple Coneflower	x	x	x	x	x
Eryngium yuccifolium	Rattlesnake Master		x		x	x
Eupatorium perfoliatum	Boneset		x		x	x
Euphorbia corollata	Flowering Spurge				x	x

Attracts native moths	Attracts predaceous wasps	Attracts parasitic wasps	Attracts pollinating flies	Attracts beetles	Butterfly larval host
x			x	x	
			x	x	
			x	x	
x	x	x	x	x	Dogface Butterfly/Underwings
x					Coral Hairstreak
			x	x	
			x		
x			x		American Painted Lady
x	x	x	x		
x	x	x	x	x	Monarch; Queen
	x		x	x	Monarch
x	x		x		Monarch
	x			x	Monarch; Queen; Gray Hairstreak
x				x	Wild Indigo Dusky Wing Skipper; Clouded Sulphur; Southern Dogface; Orange Sulphur; Hoary Edge
x				x	Wild Indigo Dusky Wing Skipper; Hoary Edge; Frosted Elfin; Marin Blue; Orange Sulphur
x				x	Wild Indigo Dusky Wing Skipper; Hoary Edge; Frosted Elfin; Marin Blue; Orange Sulphur
			x		
x					Gray Hairstreak; Checkered Skipper
x					
	x	x	x	x	Spring Azure; Summer Azure; Mottled Duskywing
x			x	x	Baltimore Checkerspot
x			x	x	
x	x	x	x	x	
x	x	x	x	x	
x	x	x	x	x	
					Dogface Butterfly; Dog Sulphur; Reakirt's Blue
	x		x	x	Dogface Butterfly; Dog Sulphur; Reakirt's Blue
x			x		
			x		
x			x	x	
x					Silvery Checkerspot
x					
x					Silvery Checkerspot
	x	x	x	x	Black Swallowtail
x	x	x	x	x	
	x	x	x	x	

(continued)

TABLE 12.2. *(continued)*

Scientific name (4/1/2020)	Common name	Attracts hummingbirds	Attracts songbirds	Attracts upland game birds	Attracts bees and bumblebees	Flowers attract butterflies
Eutrochium fistulosum	Tall Joe Pye Weed		x		x	x
Eutrochium maculatum	Joe Pye Weed		x		x	x
Filipendula rubra	Queen of the Prairie				x	x
Gentiana alba	Cream Gentian				x	
Gentiana andrewsii	Bottle Gentian				x	
Geum triflorum	Prairie Smoke				x	x
Helenium autumnale	Sneezeweed				x	x
Helianthus mollis	Downy Sunflower		x	x	x	x
Helianthus occidentalis	Western Sunflower		x	x	x	x
Helianthus ×laetiflorus	Showy Sunflower		x	x	x	x
Heliopsis helianthoides	Oxeye Sunflower	x	x	x	x	x
Hibiscus moscheutos	Rosemallow	x			x	x
Hypericum ascyron	Great St. Johnswort				x	
Iris virginica var. *shrevei*	Wild Iris	x	x		x	x
Lespedeza capitata	Roundhead Bush Clover		x	x	x	x
Liatris aspera	Rough Blazing Star	x	x		x	x
Liatris ligulistylis	Meadow Blazing Star	x	x		x	x
Liatris punctata	Dotted Blazing Star	x	x		x	x
Liatris pycnostachya	Prairie Blazing Star	x	x		x	x
Liatris spicata	Dense Blazing Star	x	x		x	x
Liatris squarrosa	Scaly Blazing Star	x	x		x	x
Lilium michiganense	Michigan Lily	x			x	x
Lithospermum canescens	Hoary Puccoon			x	x	x
Lithospermum caroliniense	Hairy Puccoon			x	x	x
Lithospermum incisum	Fringed Puccoon			x		x
Lobelia cardinalis	Cardinal Flower	x			x	x
Lobelia siphilitica	Great Blue Lobelia	x			x	x
Lupinus perennis	Wild Lupine	x	x	x	x	x
Monarda fistulosa	Bergamot	x	x		x	x
Monarda punctata	Dotted Mint	x			x	x
Oenothera gaura	Biennial Beeblossom				x	x
Oligoneuron ohioense	Ohio Goldenrod		x	x	x	x
Oligoneuron rigidum	Stiff Goldenrod		x	x	x	x
Onosmodium bejariense var. occidentale	Marbleseed (Western Marbleseed)				x	x
Parthenium integrifolium	Wild Quinine				x	x
Pediomelum esculentum	Prairie Turnip			x	x	x
Penstemon cobaea	Cobaea Beardtongue	x			x	x
Penstemon digitalis	Foxglove Beardtongue	x			x	
Penstemon gracilis	Slender Beardtongue	x			x	

Attracts native moths	Attracts predaceous wasps	Attracts parasitic wasps	Attracts pollinating flies	Attracts beetles	Butterfly larval host
x	x		x	x	
x	x		x	x	
			x	x	
				x	
				x	
x	x		x	x	Dainty Sulphur
x			x		Silvery Checkerspot; Gorgone Checkerspot
x			x	x	Silvery Checkerspot; Gorgone Checkerspot; Painted Lady
x			x	x	Silvery Checkerspot; Gorgone Checkerspot; Painted Lady
x	x	x	x	x	Silvery Checkerspot
x					Painted Lady; Common Checkered Skipper; Gray Hairstreak
x	x		x	x	Gray Hairstreak
x			x		
x				x	Northern and Southern Cloudywings; Silver Spotted Skipper; Eastern Tailed Blue
x			x		
x			x		
x					
x			x		
x					
x			x		
x				x	
x			x	x	
x			x	x	
x			x	x	
x					
x					
x			x		Karner Blue; Frosted Elfin; Wild Indigo Duskywing; Persius Duskywing
x	x		x	x	
x	x			x	
x					
		x			
x	x	x	x	x	
x				x	Painted Lady
x	x	x	x	x	
x					Dotted Checkerspot
x			x		

(continued)

TABLE 12.2. *(continued)*

Scientific name (4/1/2020)	Common name	Attracts hummingbirds	Attracts songbirds	Attracts upland game birds	Attracts bees and bumblebees	Flowers attract butterflies
Penstemon grandiflorus	Large-Flowered Beardtongue	x			x	
Penstemon hirsutus	Hairy Penstemon	x			x	x
Phlox pilosa	Downy Phlox	x			x	x
Physostegia virginiana	Obedient Plant	x			x	x
Polygonatum biflorum var. *commutatum*	Great Solomon's Seal	x	x	x	x	
Pulsatilla patens	Pasque Flower				x	
Pycnanthemum tenuifolium	Narrowleaf Mountainmint		x		x	x
Pycnanthemum virginianum	Virginia Mountainmint				x	x
Ratibida pinnata	Yellow Coneflower		x		x	x
Rosa carolina	Carolina Rose		x	x	x	x
Rosa setigera	Climbing Prairie Rose		x	x	x	x
Rudbeckia fulgida	Orange Coneflower		x		x	x
Rudbeckia hirta	Blackeyed Susan		x		x	x
Rudbeckia laciniata	Green-Headed Coneflower		x		x	x
Rudbeckia subtomentosa	Sweet Blackeyed Susan		x		x	x
Rudbeckia triloba	Browneyed Susan		x		x	x
Ruellia humilis	Wild Petunia	x			x	x
Salvia azurea	Blue Sage	x			x	x
Senna hebecarpa	Wild Senna	x	x	x	x	x
Silene regia	Royal Catchfly	x				x
Silphium integrifolium	Rosinweed		x	x	x	x
Silphium laciniatum	Compassplant		x	x	x	x
Silphium perfoliatum	Cup Plant	x	x	x	x	x
Silphium terebinthinaceum	Prairie Dock	x	x	x	x	x
Sisyrinchium albidum	White Blue-Eyed Grass				x	
Sisyrinchium angustifolium	Narrow-Leaved Blue-Eyed Grass			x	x	x
Sisyrinchium campestre	Prairie Blue-Eyed Grass			x	x	x
Solidago ptarmicoides	White Aster		x	x	x	x
Solidago speciosa	Showy Goldenrod		x	x	x	x
Spiraea alba	Meadowsweet		x	x	x	x
Spiraea tomentosa	Steeplebush		x		x	x
Symphyotrichum ericoides	Heath Aster			x	x	x
Symphyotrichum laeve	Smooth Aster		x	x	x	x
Symphyotrichum novae-angliae	New England Aster			x	x	x
Symphyotrichum oblongifolium	Aromatic Aster			x	x	x
Symphyotrichum oolentangiense	Skyblue Aster		x	x	x	x
Tephrosia virginiana	Goatsrue		x	x		x

Attracts native moths	Attracts predaceous wasps	Attracts parasitic wasps	Attracts pollinating flies	Attracts beetles	Butterfly larval host
x					
x	x				Baltimore Checkerspot
x			x		
			x		
		x	x	x	
	x	x	x	x	
x	x	x	x	x	Silvery Checkerspot
x	x		x	x	
x	x		x	x	
x			x	x	
	x		x	x	Silvery Checkerspot
x	x		x	x	Silvery Checkerspot
x	x		x	x	Silvery Checkerspot
	x		x	x	
					Buckeye
x					
x			x	x	Cloudless Sulphur; Sleepy Orange; Orange Barred Sulphur
x	x	x	x	x	
		x	x	x	
	x	x	x	x	
		x	x	x	
			x	x	
			x		
			x		
				x	
x	x	x		x	
	x		x	x	Spring Azure
x			x	x	Spring Azure; Columbia Silkmoth
	x	x	x	x	Pearl Crescent; Silvery Checkerspot
x	x		x	x	Pearl Crescent
x					Pearl Crescent; Silvery Checkerspot
x					Silvery Checkerspot
x			x	x	Pearl Crescent; Silvery Checkerspot
x				x	Southern Cloudywing

(continued)

TABLE 12.2. *(continued)*

Scientific name (4/1/2020)	Common name	Attracts hummingbirds	Attracts songbirds	Attracts upland game birds	Attracts bees and bumblebees	Flowers attract butterflies
Thalictrum dasycarpum	Tall Meadow-Rue				x	
Tradescantia bracteata	Prairie Spiderwort			x	x	
Tradescantia occidentalis	Western Spiderwort			x	x	
Tradescantia ohiensis	Ohio Spiderwort			x	x	x
Verbena hastata	Blue Vervain	x	x		x	x
Verbena stricta	Hoary Vervain	x	x	x	x	x
Vernonia fasciculata	Ironweed		x		x	x
Vernonia gigantea	Tall Ironweed		x		x	x
Veronicastrum virginicum	Culver's Root				x	x
Viola pedata	Birdfoot Violet		x		x	x
Zizia aptera	Heartleaf Golden Alexanders				x	x
Zizia aurea	Golden Alexanders				x	x

Scientific name (4/1/2020)	Common name	Attracts songbirds	Attracts upland game birds	Flowers attract butterflies
Grasses and sedges				
Andropogon gerardii	Big Bluestem	x	x	
Andropogon virginicus	Broomsedge	x	x	
Bouteloua curtipendula	Sideoats Grama	x	x	
Bromus kalmii	Prairie Bromegrass	x		
Calamagrostis canadensis	Bluejoint Grass	x	x	
Carex brevior	Shortbeak Sedge			
Elymus canadensis	Canada Wildrye	x	x	
Elymus virginicus	Virginia Wildrye	x	x	
Eragrostis spectabilis	Purple Lovegrass	x		
Hesperostipa spartea	Needlegrass			
Hierochloe hirta	Vanilla Sweetgrass	x		
Koeleria macrantha	Junegrass			
Panicum virgatum	Switchgrass	x	x	
Schizachyrium scoparium	Little Bluestem	x	x	
Sorghastrum nutans	Indiangrass	x	x	
Spartina pectinata	Prairie Cordgrass	x	x	
Sporobolus heterolepis	Prairie Dropseed	x	x	
Tridens flavus	Purpletop	x		Leaves
Tripsacum dactyloides	Eastern Gamagrass	x	x	Leaves

Attracts native moths	Attracts predaceous wasps	Attracts parasitic wasps	Attracts pollinating flies	Attracts beetles	Butterfly larval host
x					
			x	x	
			x	x	
x	x		x	x	Common Buckeye
x	x		x	x	Common Buckeye
x			x		American Painted Lady
x			x		
x			x		
			x	x	Black Swallowtail; Missouri Woodland Swallowtail
x	x	x	x	x	Black Swallowtail; Missouri Woodland Swallowtail

Larval host

Delaware Skipper; Dusted Skipper; Arogos Skipper; Cobweb Skipper; Ottoe Skipper; Byssus Skipper; Common Wood Nymph
Zabulon Skipper; The Thinker
Green Skipper; Dotted Skipper; Sheep Skipper; Bronze Roadside Skipper; Orange Skipperling

Zabulon Skipper

Delaware Skipper; Dotted Skipper; Dion Skipper; Dun Skipper
Ottoe Skipper; Indian Skipper; Crossline Skipper; Dusted Skipper; Dixie Skipper; Cobweb Butterfly; Delaware Skipper

Crossline Skipper; Broad-Winged Skipper; Little Glassywing Skipper; Large Wood Nymph
Bunchgrass Skipper; Byssus Skipper

TABLE 12.3. PLANTS FOR DRY SOILS, BY BLOOM TIME

Bloom time	Scientific name (4/1/2020)	Common name	Height	Flower color	USDA zones	Sun	Soil moisture
Flowers and shrubs							
Apr.–May	*Antennaria neglecta*	Pussytoes	6"–1'	White	3–7	F	D
Apr.–June	*Penstemon cobaea*	Cobaea Beardtongue	2'–3'	Lavender	5–8	F	D
Apr.–May	*Pulsatilla patens*	Pasque Flower	6"–1'	White	3–6	F	D
Apr.–May	*Sisyrinchium albidum*	White Blue-Eyed Grass	6"–1'	White	4–9	F	D, M
Apr.–June	*Sisyrinchium campestre*	Prairie Blue-Eyed Grass	6"–1'	Blue/white	3–8	F, P	D, M
Apr.–June	*Viola pedata*	Birdfoot Violet	6"	Blue/purple	3–9	F, P	D
May–June	*Anemone cylindrica*	Thimbleweed	1'–3'	White	3–6	F	D, M
May–June	*Baptisia bracteata*	Cream False Indigo	1'–2'	Cream	4–9	F, P	D, M
May–July	*Blephilia ciliata*	Downy Wood Mint	1'–2'	White/purple/violet	4–8	F, P	D, M
May–June	*Callirhoe involucrata*	Winecups	1'	Magenta	4–9	F, P	D, M
May–June	*Geum triflorum*	Prairie Smoke	6"	Pink	3–6	F	D, M
May–June	*Lithospermum canescens*	Hoary Puccoon	1'	Orange	3–7	F, P	D, M
May–June	*Lupinus perennis*	Wild Lupine	1'–2'	Blue	3–8	F, P	D
May–June	*Onosmodium bejariense* var. *occidentale*	Marbleseed (Western Marbleseed)	1'–3'	White	3–8	F	D, M
May–June	*Penstemon gracilis*	Slender Beardtongue	1'–2'	Lavender	3–6	F, P	D
May–June	*Penstemon grandiflorus*	Large-Flowered Beardtongue	2'–4'	Lavender	3–7	F	D
May–June	*Penstemon hirsutus*	Hairy Penstemon	1'–2'	Lavender	3–6	F, P	D
May–June	*Phlox pilosa*	Downy Phlox	1'–2'	Pink	3–9	F	D, M, Mo
May–June	*Polygonatum biflorum* var. *commutatum*	Great Solomon's Seal	2'–4'	Cream	3–9	F, P	D, M
June–July	*Amorpha canescens*	Leadplant	2'–3'	Purple	3–8	F, P	D, M
June–Aug.	*Asclepias syriaca*	Common Milkweed	2'–4'	Lavender	3–8	F, P	D, M
June–Aug.	*Asclepias tuberosa*	Butterflyweed	2'–3'	Orange	3–10	F	D, M
June–July	*Coreopsis lanceolata*	Lanceleaf Coreopsis	1'–2'	Yellow	3–9	F	D, M
June–Aug.	*Coreopsis palmata*	Stiff Coreopsis	2'–3'	Yellow	3–8	F	D, M
June–July	*Delphinium carolinianum* subsp. *virescens*	Prairie Larkspur	1'–4'	White	3–8	F	D, M
June–Aug.	*Desmanthus illinoensis*	Illinois Bundleflower	2'–6'	White	5–8	F	D, M, Mo
June–July	*Echinacea pallida*	Pale Purple Coneflower	3'–5'	Purple	4–8	F	D, M
June–July	*Echinacea paradoxa*	Ozark Coneflower	3'–5'	Yellow	4–7	F, P	D, M
June–Aug.	*Eryngium yuccifolium*	Rattlesnake Master	3'–5'	White	4–9	F	D, M
June–Sept.	*Heliopsis helianthoides*	Oxeye Sunflower	3'–6'	Yellow	3–8	F	D, M, Mo
June–July	*Lithospermum caroliniense*	Hairy Puccoon	1'–2'	Yellow	3–8	F	D
June–July	*Lithospermum incisum*	Fringed Puccoon	1'–2'	Yellow	3–9	F	D
June–Sept.	*Oenothera gaura*	Biennial Beeblossom	3'–6'	White/pink	4–8	F, P	D, M, Mo
June–July	*Pediomelum esculentum*	Prairie Turnip	1'–2'	Blue	3–7	F	D
June–July	*Rosa carolina*	Carolina Rose	1'–2'	Pink	3–8	F	D, M
June–Sept.	*Rudbeckia hirta*	Blackeyed Susan	1'–3'	Yellow	3–10	F, P	D, M, Mo

Soil texture	pH range	Plant spacing in prairie gardens	Plant spacing in flower beds	Root type	Plant life expectancy in years	Aggressiveness and method	Cut flowers	Dried arrangements	Deer palatability
S, G	5.5–7.5	1'	1'	R	5–10	L–R	x	FL	L
S, G	?–7.5	1'	1'	FIB	3–5	L–S	x	SP	L
S, G	?–8.0	6"–1'	1'	FIB	5–10	L–S		SD	H
S, L	?–8.0	6"	6"	FIB	3–5	L–S			L
S, L	?–7.5	6"	6"	FIB	3–5	L–S			L
S, G	?	6"	1'	R	5–10	L–S			L
S, L	?	1'	1'	FIB	10–20	L–S		SD	L
S, L	?	2'	3'	TAP	20+	L–S	x	SP	L
G, L, C	?–8.00	1'	1'–2'	TAP/short Rs	3–5	L–S		FL/SH	L
S, L	5.5–7.5	1'–2'	2'	TAP	10–20	L–S			L
S, L	6.0–7.5	6"	1'	R	20+	L–R		SH	L
S, L	?	1'	1'	FIB	5–10	L–S			L
S	5.0–?	1'	1.5'	TAP	10–20	M–S			L
S, L	?–7.5	1'–1.5'	2'	TAP	10–20	L–S			L
S, G	?–7.5	6"–1'	1'	FIB	5–10	L–S	x	SP	L
S, G	?	1'	2'–3'	FIB	3–5	L–S		SP	L
S, G	5.0–8.0	1'	1.5'	FIB	3–5	L–S	x	SP	L
S, L	?	6"–1'	1.5'	FIB	3–5	L–S	x		L
S, L, C	?	1'–2'	3'	R	10–20	L–R/S			H
S, L	5.5–8.0	2'	3'	TAP	20+	L–S		Leaves	L
S, L, C	5.0–8.0	NR	NR	R	10–20	H–R		SP	L
S, L	5.0–7.5	2'	3'	TAP	10–20	L–S	x	SP	L
S, L	5.0–7.5	1'	1'–2'	FIB	3–5	M–S	x		M
S, L	?	1'–2'	2'	R	20+	M–R			L
S, L	?	1'	1.5'	FIB	5–10	L–S	x	SP	L
S, L, C	6.0–8.0	1'	1.5'	TAP	5–10	L–S		SP	M
S, L, C	6.0–7.5	1'	2'–3'	TAP	10–20	L–S	x	SH	M
S, L, C	?–7.5	1'	2'–3'	TAP	10–20	L–S	x	SH	M
S, L, C	6.0–8.0	1'	2'	FIB	5–10	M–S	x	SH	L
S, L, C	6.0–8.0	1'–2'	3'	FIB	5–10	H–S	x		L
S	?	1'	1'	TAP	10–20	L–S			L
S, G	?	1'	1'	TAP	5–10	L–S			L
S, L, C	?	2'	3'	FIB	2	M–S	x		L
S, G	?–8.5	1'–1.5'	2'	TAP	20+	L–S			L
S, L	4.0–7.0	2'	3'	R	10–20	L–R/S	x	RH	L
S, L, C	5.0–8.0	1'	2'	FIB	2	M–S	x	SH	L

(continued)

TABLE 12.3. (continued)

Bloom time	Scientific name (4/1/2020)	Common name	Height	Flower color	USDA zones	Sun	Soil moisture
June–Aug.	*Ruellia humilis*	Wild Petunia	1'–2'	Violet	4–9	F	D, M
June–Aug.	*Salvia azurea*	Blue Sage	3'–5'	Blue	4–9	F, P	D, M
June–Sept.	*Silphium laciniatum*	Compassplant	3'–10'	Yellow	4–9	F	D, M
June–Aug.	*Tephrosia virginiana*	Goatsrue	1'–2'	Pink/yellow	4–9	F, P	D
June–July	*Tradescantia bracteata*	Prairie Spiderwort	1'–2'	Blue	3–7	F, P	D, M
June–July	*Tradescantia occidentalis*	Western Spiderwort	1'	Pink	3–9	F	D, M
June–July	*Tradescantia ohiensis*	Ohio Spiderwort	2'–4'	Blue	3–9	F, P	D, M, Mo
July–Sept.	*Agastache foeniculum*	Lavender Hyssop	1'–3'	Purple	2–6	F, P	D, M
July–Aug.	*Allium stellatum*	Prairie Onion	1'–2'	Lavender	3–8	F, P	D, M
July–Sept.	*Arnoglossum atriplicifolium*	Pale Indian Plantain	5'–10'	White	4–8	F, P	D, M, Mo
July–Aug.	*Astragalus canadensis*	Canada Milkvetch	2'–3'	Yellow	3–8	F	D, M
July–Aug.	*Callirhoe triangulata*	Poppymallow	1'–2'	Magenta	4–8	F	D
July–Aug.	*Ceanothus americanus*	New Jersey Tea	2'–3'	White	3–9	F, P	D, M
July–Aug.	*Dalea candida*	White Prairie Clover	1'–2'	White	3–9	F	D, M
July–Aug.	*Dalea purpurea*	Purple Prairie Clover	1'–2'	Purple/yellow	3–8	F	D, M
July–Sept.	*Echinacea purpurea*	Purple Coneflower	3'–4'	Purple	4–8	F, P	D, M
July–Aug.	*Euphorbia corollata*	Flowering Spurge	2'–4'	White	3–9	F	D, M
July–Aug.	*Helianthus occidentalis*	Western Sunflower	2'–3'	Yellow	3–9	F	D, M
July–Sept.	*Monarda fistulosa*	Bergamot	2'–5'	Lavender	3–9	F, P	D, M, Mo
July–Sept.	*Monarda punctata*	Dotted Mint	1'–2'	Lavender	3–10	F	D
July–Sept.	*Pycnanthemum tenuifolium*	Narrowleaf Mountainmint	1'–3'	White	4–8	F, P	D, M
July–Sept.	*Ratibida pinnata*	Yellow Coneflower	3'–6'	Yellow	3–9	F	D, M, Mo
July–Sept.	*Verbena stricta*	Hoary Vervain	2'–4'	Blue	3–8	F	D, M
Aug.–Sept.	*Helianthus mollis*	Downy Sunflower	4'–6'	Yellow	4–9	F	D, M
Aug.–Sept.	*Helianthus ×laetiflorus*	Showy Sunflower	3'–6'	Yellow	3–8	F	D, M
Aug.–Sept.	*Lespedeza capitata*	Roundhead Bush Clover	3'–5'	White	3–8	F, P	D, M
Aug.–Sept.	*Liatris aspera*	Rough Blazing Star	2'–5'	Purple/pink	3–8	F	D, M
Aug.–Oct.	*Liatris punctata*	Dotted Blazing Star	1'–2'	Purple/pink	3–9	F	D
Aug.–Sept.	*Liatris squarrosa*	Scaly Blazing Star	1'–2'	Purple/pink	4–8	F	D, M
Aug.–Sept.	*Oligoneuron rigidum*	Stiff Goldenrod	3'–5'	Yellow	3–9	F	D, M, Mo
Aug.–Sept.	*Solidago ptarmicoides*	White Aster	1'–2'	White	3–7	F	D
Aug.–Sept.	*Solidago speciosa*	Showy Goldenrod	1'–3'	Yellow	3–8	F	D, M
Aug.–Oct.	*Symphyotrichum ericoides*	Heath Aster	1'–3'	White	3–9	F	D, M
Aug.–Oct.	*Symphyotrichum laeve*	Smooth Aster	2'–4'	Blue	4–8	F	D, M
Aug.–Oct.	*Symphyotrichum oolentangiense*	Skyblue Aster	2'–3'	Blue	3–8	F, P	D, M
Sept.–Oct.	*Symphyotrichum oblongifolium*	Aromatic Aster	1'–2'	Purple/violet	3–8	F	D, M

Soil texture	pH range	Plant spacing in prairie gardens	Plant spacing in flower beds	Root type	Plant life expectancy in years	Aggressiveness and method	Cut flowers	Dried arrangements	Deer palatability
S, L	4.5–7.5	1'	2'	FIB	5–10	L–S			L
S, G, L	?	1'	1.5'	FIB	5–10	L–S	x	SH	L
S, L, C	?–8.0	2'	4'	TAP	20+	L–S	x		L
S, G	4.5–?	2'	3'	TAP	10–20	L–S			L
S, L, C	?	1'	2'	R	20+	M–R			M
S, L	?	1'	NR	R	20+	H–R			M
S, L, C	?–7.5	1'	2'	FIB	20+	M–S			H
S, L	?–8.0	1'	1'–2'	FIB	3–5	M–S	x	FL/SH	L
S, L	5.5–8.0	6"–1'	1'	B	10–20	L–S	x	FL/SH	L
S, L, C	?–8.0	2'	3'–4'	FIB	3–20	H–S (north); L–S (south)	x	FL	L
S, L	6.0–8.0	1'–2'	3'	TAP	10–20	M–S		SP	M
S	?	1'–2'	2'	TAP	10–20	L–S			L
S, L	4.5–7.0	3'	3'	TAP	20+	L–S		SD	L
S, L	?	6"–1'	1'–2'	TAP	10–20	L–S			H
S, L, C	?	6"–1'	1'–2'	TAP	10–20	L–S	x		H
S, L, C	5.5–8.0	1'–2'	3'	FIB	5–10	M–S	x	SH	M
S, L	?	1'–2'	1'–2'	R	10–20	M–R			L
S, L	?	2'	3'	R	20+	M–R	x		L
S, L, C	6.0–8.0	1'–2'	3'	R	10–20	M–R	x	SH	L
S, G	5.0–?	1'	1'–2'	FIB	2	M–S	x	SH	L
S, G, L, C	?	1'	1.5'	TAP/R	5–10	L–S		SH	M
S, L, C	5.5–8.0	1'–2'	2'	FIB	10–20	M–S	x	SH	L
S, L	?	1'	2'	FIB	3–5	M–S	x	SH	L
S, L, C	6.0–8.0	2'	3'	R	20+	M–R	x	SH	L
S, L, C, G	?	NR	NR	R	20+	H–R	x	SH	L
S, G, L	5.5–8.0	1'	2'	TAP	10–20	L–S		SH	M
S, L	?	6"–1'	1'	CO	10–20	L–S	x	FL/SH	L
S, L	6.0–8.0	1'	1.5'–2'	TAPCO	10–20	L–S			L
S, G	6.5–8.0	1'	2'	FIB	5–10	L–S		Empty SH	L
S, L	?	1'	1.5'–2'	CO	10–20	L–S	x	FL/SH	L
S, L, C	5.5–8.0	1'–1.5'	3'	FIB	5–10	M–S	x	SH	L
S, L	?	1'	3'	FIB	5–10	L–S	x	SH	L
S, G, L	?	1'–2'	2'	R	10–20	M–R		Empty SH	L
S, L	?	1'	3'	FIB	5–10	M–S	x	Empty SH	H
S, L	5.0–8.0	1'	2'	FIB	5–10	L–S	x	Empty SH	M
S, L, C	?–8.0	1'	3'	R	5–10	M–R?		Empty SH	L

(continued)

TABLE 12.3.   *(continued)*

Bloom time	Scientific name (4/1/2020)	Common name	Height	Fall leaf color	USDA zones	Sun	Soil moisture
Grasses and sedges							
Apr.–June	*Hesperostipa spartea*	Needlegrass	2'–4'	Gold	3–6	F	D, M
May–June	*Carex brevior*	Shortbeak Sedge	1'–4'	Straw	3–7	F, P	D, M, Mo
May–June	*Koeleria macrantha*	Junegrass	2'–3'	Gold	3–8	F	D
June–July	*Bromus kalmii*	Prairie Bromegrass	2'–4'	Straw	3–6	F	D
July–Aug.	*Elymus canadensis*	Canada Wildrye	4'–5'	Straw	2–9	F	D, M, Mo
July–Sept.	*Eragrostis spectabilis*	Purple Lovegrass	1'–2'	Purple/pink	3–10	F, P	D
July–Sept.	*Tridens flavus*	Purpletop	3'–6'	Straw	5–10	F	D, M
Aug.–Oct.	*Andropogon gerardii*	Big Bluestem	5'–8'	Bronze/red	3–9	F	D, M, Mo
Aug.–Oct.	*Andropogon virginicus*	Broomsedge	2'–5'	Reddish brown	5–10	F	D, M, Mo
Aug.–Sept.	*Bouteloua curtipendula*	Sideoats Grama	2'–3'	Straw	3–9	F	D, M
Aug.–Sept.	*Panicum virgatum*	Switchgrass	3'–6'	Gold	3–10	F	D, M, Mo
Aug.–Oct.	*Schizachyrium scoparium*	Little Bluestem	2'–5'	Crimson red	3–10	F	D, M
Aug.–Sept.	*Sorghastrum nutans*	Indiangrass	5'–7'	Gold	3–9	F	D, M, Mo
Aug.–Sept.	*Sporobolus heterolepis*	Prairie Dropseed	2'–4'	Gold	3–8	F	D, M

Notes: For soil moisture, D = dry, M = medium, Mo = moist, and W = wet. For soil texture, S = sand, L = loam, C = clay, and G = gravel. For sun, F = full sun, or at least a half day of full sun; P = partial sun, or a quarter to a half day of sun; and S = shade, or less than a quarter day of sun. For plant spacing, NR = not recommended. For root type, B = bulb, CO = corm, FIB = fibrous, R = rhizome, TAP =

Soil texture	pH range	Plant spacing in prairie gardens	Plant spacing in flower beds	Root type	Plant life expectancy in years	Aggressiveness and method	Cut flower stalks	Dried seed arrangements	Deer palatability
S, G, L	5.0–7.5	1′	NR	FIB	10–20	L–S		x	L
S, L, C	4.5–7.5	1′	1′	FIB	5–10	L–S			L
S, G	6.0–8.0	6″–1′	1′	FIB	5–10	L–S		x	L
S, G	5.7–7.0	1′	2′	FIB	3–5	L–S		x	L
S, L, C	5.0–8.0	1′	3′	FIB	3–5	M–S		x	L
S, G	4.0–7.5	1′–1.5′	2′	FIB	5–10	M–S	x	x	L
S, L, C	5.0–7.5	1′	1.5′	FIB	5–10	M–S		x	L
S, L, C	5.0–7.5	2′	NR	FIB	20+	M–S		x	L
S, L, C	4.0–7.0	1′–1.5′	NR	FIB	20+	H–S		x	L
S, L	5.5–8.5	1′	2′	FIB	10–20	L–S		x	L
S, L, C	4.5–8.0	1.5′–2′	NR	FIB	20+	M–S		x	L
S, L	5.0–8.5	1′–1.5′	3′	FIB	20+	L–S		x	L
S, L, C	5.0–8.0	1′–2′	3′	FIB	10–20	M–S	x	x	L
S, L	6.0–8.0	2′	2′	FIB	20+	L–S	x	x	L

taproot, and TAPCO = taprooted corm. For aggressiveness and method, L = low, M = medium, H = high, R = by rhizome, S = by seed, and R/S = rhizome and seed. For dried arrangements, FL = flowers, SH = seedheads, SP = seedpods, SD = seed, and RH = rose hips. For deer palatability, L = low, M = medium, and H = high.

TABLE 12.4. PLANTS FOR MEDIUM SOILS, BY BLOOM TIME

Bloom time	Scientific name (4/1/2020)	Common name	Height	Flower color	USDA zones	Sun	Soil moisture
Flowers and shrubs							
Apr.–May	Sisyrinchium albidum	White Blue-Eyed Grass	6"–1'	White	4–9	F	D, M
Apr.–June	Sisyrinchium angustifolium	Narrow-Leaved Blue-Eyed Grass	6"–2'	Blue/purple	3–10	F, P	M, Mo
Apr.–June	Sisyrinchium campestre	Prairie Blue-Eyed Grass	6"–1'	Blue/white	3–8	F, P	D, M
May–June	Amsonia tabernaemontana	Eastern Bluestar	2'–3'	Blue/white	4–8	F, P	M, Mo
May–July	Anemone canadensis	Canada Anemone	1'–2'	White	3–6	F, P	M, Mo
May–June	Anemone cylindrica	Thimbleweed	1'–3'	White	3–6	F	D, M
May–June	Baptisia bracteata	Cream False Indigo	1'–2'	Cream	4–9	F, P	D, M
May–July	Blephilia ciliata	Downy Wood Mint	1'–2'	White/purple/violet	4–8	F, P	D, M
May–June	Callirhoe involucrata	Winecups	1'	Magenta	4–9	F, P	D, M
May–June	Dodecatheon meadia	Shooting Star	1'–2'	White/pink	4–8	F, P	M, Mo
May–June	Geum triflorum	Prairie Smoke	6"	Pink	3–6	F	D, M
May–June	Lithospermum canescens	Hoary Puccoon	1'	Orange	3–7	F, P	D, M
May–June	Onosmodium bejariense var. occidentale	Marbleseed (Western Marbleseed)	1'–3'	White	3–8	F	D, M
May–June	Phlox pilosa	Downy Phlox	1'–2'	Pink	3–9	F	D, M, Mo
May–June	Polygonatum biflorum var. commutatum	Great Solomon's Seal	2'–4'	Cream	3–9	F, P	D, M
May–July	Rosa setigera	Climbing Prairie Rose	6'–15'	Pink/white	4–9	F, P	M, Mo
May–June	Zizia aptera	Heartleaf Golden Alexanders	1'–2'	Yellow	3–8	F	M, Mo
May–July	Zizia aurea	Golden Alexanders	1'–2'	Yellow	3–8	F, P, S	M, Mo, W
June–Aug.	Allium cernuum	Nodding Pink Onion	1'–2'	White/pink	3–8	F, P	M, Mo
June–July	Amorpha canescens	Leadplant	2'–3'	Purple	3–8	F, P	D, M
June–Aug.	Asclepias sullivantii	Sullivant's Milkweed	3'–5'	Pink/yellow	4–7	F	M, Mo
June–Aug.	Asclepias syriaca	Common Milkweed	2'–4'	Lavender	3–8	F, P	D, M
June–Aug.	Asclepias tuberosa	Butterflyweed	2'–3'	Orange	3–10	F	D, M
June–July	Baptisia alba	White False Indigo	3'–5'	White	4–9	F, P	M, Mo
June–July	Baptisia australis	Blue False Indigo	3'–5'	Blue	3–10	F, P	M
June–July	Coreopsis lanceolata	Lanceleaf Coreopsis	1'–2'	Yellow	3–9	F	D, M
June–Aug.	Coreopsis palmata	Stiff Coreopsis	2'–3'	Yellow	3–8	F	D, M
June–July	Delphinium carolinianum subsp. virescens	Prairie Larkspur	1'–4'	White	3–8	F	D, M
June–Aug.	Desmanthus illinoensis	Illinois Bundleflower	2'–6'	White	5–8	F	D, M, Mo
June–July	Echinacea pallida	Pale Purple Coneflower	3'–5'	Purple	4–8	F	D, M
June–July	Echinacea paradoxa	Ozark Coneflower	3'–5'	Yellow	4–7	F, P	D, M
June–Aug.	Eryngium yuccifolium	Rattlesnake Master	3'–5'	White	4–9	F	D, M
June–July	Filipendula rubra	Queen of the Prairie	4'–5'	Pink	3–6	F	M, Mo
June–Sept.	Heliopsis helianthoides	Oxeye Sunflower	3'–6'	Yellow	3–8	F	D, M, Mo
June–Sept.	Oenothera gaura	Biennial Beeblossom	3'–6'	White/pink	4–8	F, P	D, M, Mo

Soil texture	pH range	Plant spacing in prairie gardens	Plant spacing in flower beds	Root type	Plant life expectancy in years	Aggressiveness and method	Cut flowers	Dried arrangements	Deer palatability
S, L	?–8.0	6"	6"	FIB	?	L–S			?
S, L, C	?–8.0	6"	9"	FIB	3–5	L–S			L
S, L	?–7.5	6"	6"	FIB	3–5	L–S			L
S, L, C	?–8.0	2'–3'	3'	FIB	20+	L–S		SP	L
S, L, C	?	NR	NR	R	10–20	H–R			L
S, L	?	1'	1'	FIB	10–20	L–S		SD	L
S, L	?	2'	3'	TAP	20+	L–S	x	SP	L
L, C, G	?–8.00	1'	1'–2'	TAP/short Rs	3–5	L–S		FL/SH	L
S, L	5.5–7.5	1'–2'	2'	TAP	10–20	L–S			L
S, L, C	4.5–8.0	6"–1'	1'	FIB	20+	L–S		SP	L
S, L	6.0–7.5	6"	1'	R	20+	L–R		SH	L
S, L	?	1'	1'	FIB	5–10	L–S			L
S, L	?–7.5	1'–1.5'	2'	TAP	10–20	L–S			L
S, L	?	6"–1'	1.5'	FIB	3–5	L–S	x		L
S, L, C	?	1'–2'	3'	R	10–20	L–R/S			H
S, L, C	5.0–8.0	2'	5'	FIB	10–20	L–S	x	RH	L
S, L	?–7.5	1'	2'	FIB	5–10	L–S			L
S, L, C	?–8.0	1'	3'	FIB	5–10	L–S			L
S, L, C	6.5–8.0	6"–1'	1'	B	10–20	L–S	x	FL/SH	L
S, L	5.5–8.0	2'	3'	TAP	20+	L–S		Leaves	L
L, C	?	1'–2'	3'	R	10–20	L–R		SP	L
S, L, C	5.0–8.0	NR	NR	R	10–20	H–R		SP	L
S, L	5.0–7.5	2'	3'	TAP	10–20	L–S	x	SP	L
S, L, C	?–7.5	3'	3'	TAP	20+	L–S	x	SP	L
S, L, C	?–8.0	3'	3'	TAP	20+	L–S	x	SP	L
S, L	5.0–7.5	1'	1'–2'	FIB	3–5	M–S	x		M
S, L	?	1'–2'	2'	R	20+	M–R			L
S, L	?	1'	1.5'	FIB	5–10	L–S	x	SP	L
S, L, C	6.0–8.0	1'	1.5'	TAP	5–10	L–S		SP	M
S, L, C	6.0–7.5	1'	2'–3'	TAP	10–20	L–S	x	SH	M
S, L, C	?–7.5	1'	2'–3'	TAP	10–20	L–S	x	SH	M
S, L, C	6.0–8.0	1'	2'	FIB	5–10	M–S	x	SH	L
S, L, C	?–8.0	2'	3'	R	20+	M–R	x		L
S, L, C	6.0–8.0	1'–2'	3'	FIB	5–10	H–S	x		L
S, L, C	?	2'	3'	FIB	2	M–S	x		L

(continued)

TABLE 12.4. *(continued)*

Bloom time	Scientific name (4/1/2020)	Common name	Height	Flower color	USDA zones	Sun	Soil moisture
June–Sept.	*Parthenium integrifolium*	Wild Quinine	3'–5'	White	4–8	F	M, Mo
June–July	*Penstemon digitalis*	Foxglove Beardtongue	2'–3'	White	3–8	F, P	M, Mo
June–July	*Rosa carolina*	Carolina Rose	1'–2'	Pink	3–8	F	D, M
June–Sept.	*Rudbeckia hirta*	Blackeyed Susan	1'–3'	Yellow	3–10	F, P	D, M, Mo
June–Aug.	*Ruellia humilis*	Wild Petunia	1'–2'	Violet	4–9	F	D, M
June–Aug.	*Salvia azurea*	Blue Sage	3'–5'	Blue	4–9	F, P	D, M
June–Sept.	*Silphium laciniatum*	Compassplant	3'–10'	Yellow	4–9	F	D, M
June–July	*Tradescantia bracteata*	Prairie Spiderwort	1'–2'	Blue	3–7	F, P	D, M
June–July	*Tradescantia occidentalis*	Western Spiderwort	1'	Pink	3–9	F	D, M
June–July	*Tradescantia ohiensis*	Ohio Spiderwort	2'–4'	Blue	3–9	F, P	D, M, Mo
July–Sept.	*Agastache foeniculum*	Lavender Hyssop	1'–3'	Purple	2–6	F, P	D, M
July–Aug.	*Allium stellatum*	Prairie Onion	1'–2'	Lavender	3–8	F, P	D, M
July–Sept.	*Arnoglossum atriplicifolium*	Pale Indian Plantain	5'–10'	White	4–8	F, P	D, M, Mo
July–Aug.	*Astragalus canadensis*	Canada Milkvetch	2'–3'	Yellow	3–8	F	D, M
July–Aug.	*Ceanothus americanus*	New Jersey Tea	2'–3'	White	3–9	F, P	D, M
July–Oct.	*Conoclinium coelestinum*	Blue Mistflower	1'–3'	Blue/purple	4–10	F, P	M, Mo
July–Sept.	*Coreopsis tripteris*	Tall Coreopsis	3'–7'	Yellow	3–8	F, P	M, Mo, W
July–Aug.	*Dalea candida*	White Prairie Clover	1'–2'	White	3–9	F	D, M
July–Aug.	*Dalea purpurea*	Purple Prairie Clover	1'–2'	Purple/yellow	3–8	F	D, M
July–Aug.	*Desmodium canadense*	Canada Ticktrefoil	3'–5'	Purple	3–7	F	M, Mo
July–Sept.	*Echinacea purpurea*	Purple Coneflower	3'–4'	Purple	4–8	F, P	D, M
July–Aug.	*Euphorbia corollata*	Flowering Spurge	2'–4'	White	3–9	F	D, M
July–Aug.	*Gentiana alba*	Cream Gentian	1'–2'	White	3–7	F, P	M
July–Aug.	*Helianthus occidentalis*	Western Sunflower	2'–3'	Yellow	3–9	F	D, M
July–Sept.	*Hibiscus moscheutos*	Rosemallow	3'–7'	White/pink/red	5–9	F, P	M, Mo, W
July–Aug.	*Liatris pycnostachya*	Prairie Blazing Star	3'–5'	Purple/pink	3–9	F	M, Mo
July–Sept.	*Lobelia siphilitica*	Great Blue Lobelia	1'–4'	Blue	3–9	F, P	M, Mo
July–Sept.	*Monarda fistulosa*	Bergamot	2'–5'	Lavender	3–9	F, P	D, M, Mo
July–Sept.	*Pycnanthemum tenuifolium*	Narrowleaf Mountainmint	1'–3'	White	4–8	F, P	D, M
July–Sept.	*Ratibida pinnata*	Yellow Coneflower	3'–6'	Yellow	3–9	F	D, M, Mo
July–Sept.	*Rudbeckia fulgida*	Orange Coneflower	2'–4'	Yellow	4–8	F, P	M, Mo
July–Oct.	*Rudbeckia triloba*	Browneyed Susan	2'–5'	Yellow	4–9	F, P	M, Mo
July–Aug.	*Senna hebecarpa*	Wild Senna	4'–6'	Yellow	4–7	F	M, Mo, W
July–Aug.	*Silene regia*	Royal Catchfly	2'–4'	Red	4–9	F, P	M
July–Sept.	*Silphium integrifolium*	Rosinweed	2'–6'	Yellow	4–8	F	M, Mo
July–Sept.	*Silphium perfoliatum*	Cup Plant	3'–10'	Yellow	3–8	F, P	M, Mo, W
July–Sept.	*Silphium terebinthinaceum*	Prairie Dock	3'–10'	Yellow	4–7	F	M, Mo
July–Sept.	*Verbena hastata*	Blue Vervain	3'–6'	Blue	3–8	F, P	M, Mo, W
July–Sept.	*Verbena stricta*	Hoary Vervain	2'–4'	Blue	3–8	F	D, M
July–Aug.	*Veronicastrum virginicum*	Culver's Root	3'–6'	White	3–8	F, P, S	M, Mo
Aug.–Sept.	*Eupatorium perfoliatum*	Boneset	3'–4'	White	3–8	F	M, Mo, W
Aug.–Sept.	*Eutrochium fistulosum*	Tall Joe Pye Weed	5'–8'	Purple/pink	4–9	F, P	M, Mo, W

Soil texture	pH range	Plant spacing in prairie gardens	Plant spacing in flower beds	Root type	Plant life expectancy in years	Aggressiveness and method	Cut flowers	Dried arrangements	Deer palatability
S, L, C	?–8.0	1'–2'	3'	TAP	10–20	L–S	x		M
S, L, C	5.5–7.5	1'	2'	FIB	5–10	L–S	x		L
S, L	4.0–7.0	2'	3'	R	10–20	L–R/S	x	RH	L
S, L, C	5.0–8.0	1'	2'	FIB	2	M–S	x	SH	L
S, L	4.5–7.5	1'	2'	FIB	5–10	L–S			L
S, L, G	?	1'	1.5'	FIB	5–10	L–S	x	SH	L
S, L, C	?–8.0	2'	4'	TAP	20+	L–S	x		L
S, L, C	?	1'	2'	R	20+	M–R			M
S, L	?	1'	NR	R	20+	H–R			M
S, L, C	?–7.5	1'	2'	FIB	20+	M–S			H
S, L	?–8.0	1'	1'–2'	FIB	3–5	M–S	x	FL/SH	L
S, L	5.5–8.0	6"–1'	1'	B	10–20	L–S	x	FL/SH	L
S, L, C	?–8.0	2'	3'–4'	FIB	3–20	H–S (north); L–S (south)	x	FL	L
S, L	6.0–8.0	1'–2'	3'	TAP	10–20	M–S		SP	M
S, L	4.5–7.0	3'	3'	TAP	20+	L–S		SD	L
S, L, C	5.0–8.0	1'–2'	2'	R	3–5	M–R			L
S, L, C	?	1'–2'	NR	FIB	10–20	L–S	x		L
S, L	?	6"–1'	1'–2'	TAP	10–20	L–S			H
S, L, C	?	6"–1'	1'–2'	TAP	10–20	L–S	x		H
S, L, C	?–7.5	NR	NR	TAP	10–20	M–S			H
S, L, C	5.5–8.0	1'–2'	3'	FIB	5–10	M–S	x	SH	M
S, L	?	1'–2'	1'–2'	R	10–20	M–R			L
S, L, C	?–7.5	1'	1.5'	FIB	10–20	L–S	x		L
S, L	?	2'	3'	R	20+	M–R	x		L
S, L, C	4.0–8.0	2'	4'	R	10–20	L–R/S			L
S, L, C	6.0–8.0	6"–1'	1'	CO	10–20	L–S	x	FL/SH	L
S, L, C	?–8.0	6"–1'	2'	FIB	5–10	M–S	x		L
S, L, C	6.0–8.0	1'–2'	3'	R	10–20	M–R	x	SH	L
S, L, C, G	?	1'	1.5'	TAP/R	5–10	L–S		SH	M
S, L, C	5.5–8.0	1'–2'	2'	FIB	10–20	M–S	x	SH	L
L, C	?–8.0	1'–2'	3'	FIB	5–10	L–S	x	SH	L
S, L	?–8.0	1'–2'	3'	FIB	2	M–S	x	SH	L
S, L, C	?	1.5'–2'	3'–4'	FIB/TAP	10–20	L–S	x	FL/SH	M
L	?	1'	1.5'	TAP	5–10	L–S	x	SH	L
S, L, C	?–8.0	2'	3'	TAP	20+	M–S	x		L
S, L, C	4.5–8.0	2'–3'	NR	FIB	20+	H–S	x		L
S, L, C	?–8.0	2'	4'	TAP	20+	L–S	x		L
S, L, C	?	1'	2'	FIB	2	M–S	x	SH	L
S, L	?	1'	2'	FIB	3–5	M–S	x	SH	L
S, L, C	?–8.0	1'	3'	FIB	10–20	L–S	x	SH	L
S, L, C	?–7.5	1'	NR	FIB	5–10	L–S			L
S, L, C	4.5–7.5	2'	3'–4'	FIB	5–10	L–S			H

(continued)

TABLE 12.4. *(continued)*

Bloom time	Scientific name (4/1/2020)	Common name	Height	Flower color	USDA zones	Sun	Soil moisture
Aug.–Sept.	*Helianthus mollis*	Downy Sunflower	4'–6'	Yellow	4–9	F	D, M
Aug.–Sept.	*Helianthus ×laetiflorus*	Showy Sunflower	3'–6'	Yellow	3–8	F	D, M
Aug.–Sept.	*Lespedeza capitata*	Roundhead Bush Clover	3'–5'	White	3–8	F, P	D, M
Aug.–Sept.	*Liatris aspera*	Rough Blazing Star	2'–5'	Purple/pink	3–8	F	D, M
Aug.–Sept.	*Liatris ligulistylis*	Meadow Blazing Star	3'–5'	Purple/pink	3–6	F	M, Mo
Aug.–Sept.	*Liatris spicata*	Dense Blazing Star	3'–6'	Purple/pink	4–10	F	M, Mo
Aug.–Sept.	*Liatris squarrosa*	Scaly Blazing Star	1'–2'	Purple/pink	4–8	F	D, M
Aug.–Sept.	*Oligoneuron ohioense*	Ohio Goldenrod	3'–4'	Yellow	4–5	F, P	M, Mo
Aug.–Sept.	*Oligoneuron rigidum*	Stiff Goldenrod	3'–5'	Yellow	3–9	F	D, M, Mo
Aug.–Sept.	*Physostegia virginiana*	Obedient Plant	1'–2'	Pink	3–9	F	M, Mo
Aug.–Oct.	*Rudbeckia subtomentosa*	Sweet Blackeyed Susan	4'–6'	Yellow	4–8	F, P	M, Mo
Aug.–Sept.	*Solidago speciosa*	Showy Goldenrod	1'–3'	Yellow	3–8	F	D, M
Aug.–Oct.	*Symphyotrichum ericoides*	Heath Aster	1'–3'	White	3–9	F	D, M
Aug.–Oct.	*Symphyotrichum laeve*	Smooth Aster	2'–4'	Blue	4–8	F	D, M
Aug.–Oct.	*Symphyotrichum novae-angliae*	New England Aster	3'–6'	Pink/purple/blue	3–7	F, P	M, Mo
Aug.–Oct.	*Symphyotrichum oolentangiense*	Skyblue Aster	2'–3'	Blue	3–8	F, P	D, M
Sept.–Oct.	*Symphyotrichum oblongifolium*	Aromatic Aster	1'–2'	Purple/violet	3–8	F	D, M

Bloom time	Scientific name (4/1/2020)	Common name	Height	Fall leaf color	USDA zones	Sun	Soil moisture
Grasses and sedges							
Apr.–June	*Hesperostipa spartea*	Needlegrass	2'–4'	Gold	3–6	F	D, M
May–June	*Carex brevior*	Shortbeak Sedge	1'–4'	Straw	3–7	F, P	D, M, Mo
May–June	*Hierochloe hirta*	Vanilla Sweetgrass	1'–2'	Straw	3–9	F, P	M, Mo, W
July–Aug.	*Elymus canadensis*	Canada Wildrye	4'–5'	Straw	2–9	F	D, M, Mo
July–Aug.	*Elymus virginicus*	Virginia Wildrye	4'–5'	Straw	2–9	F, P, S	M, Mo
July–Sept.	*Tridens flavus*	Purpletop	3'–6'	Straw	5–10	F	D, M
July–Sept.	*Tripsacum dactyloides*	Eastern Gamagrass	4'–8'	Gold	4–10	F, P	M, Mo, W
Aug.–Oct.	*Andropogon gerardii*	Big Bluestem	5'–8'	Bronze/red	3–9	F	D, M, Mo
Aug.–Oct.	*Andropogon virginicus*	Broomsedge	2'–5'	Reddish brown	5–10	F	D, M, Mo
Aug.–Sept.	*Bouteloua curtipendula*	Sideoats Grama	2'–3'	Straw	3–9	F	D, M
Aug.–Sept.	*Panicum virgatum*	Switchgrass	3'–6'	Gold	3–10	F	D, M, Mo
Aug.–Oct.	*Schizachyrium scoparium*	Little Bluestem	2'–5'	Crimson red	3–10	F	D, M
Aug.–Sept.	*Sorghastrum nutans*	Indiangrass	5'–7'	Gold	3–9	F	D, M, Mo
Aug.–Sept.	*Sporobolus heterolepis*	Prairie Dropseed	2'–4'	Gold	3–8	F	D, M

Notes: For soil moisture, D = dry, M = medium, Mo = moist, and W = wet. For soil texture, S = sand, L = loam, C = clay, and G = gravel. For sun, F = full sun, or at least a half day of full sun; P = partial sun, or a quarter to a half day of sun; and S = shade, or less than a quarter day of sun. For plant spacing, NR = not recommended. For root type, B = bulb, CO = corm, FIB = fibrous, R = rhizome, TAP =

Soil texture	pH range	Plant spacing in prairie gardens	Plant spacing in flower beds	Root type	Plant life expectancy in years	Aggressiveness and method	Cut flowers	Dried arrangements	Deer palatability
S, L, C	6.0–8.0	2'	3'	R	20+	M–R	x	SH	L
S, G, L, C	?	NR	NR	R	20+	H–R	x	SH	L
S, L, G	5.5–8.0	1'	2'	TAP	10–20	L–S		SH	M
S, L	?	6"–1'	1'	CO	10–20	L–S	x	FL/SH	L
L	?	6"–1'	1'	CO	3–5	L–S	x	FL/SH	L
S, L, C	5.5–7.5	6"–1'	1'	CO	10–20	L–S	x	FL/SH	L
S, L	?	1'	1.5'–2'	CO	10–20	L–S	x	FL/SH	L
S, L, C	?–8.0	1'–2'	3'	FIB	5–10	L–S	x	SH	L
S, L, C	5.5–8.0	1'–1.5'	3'	FIB	5–10	M–S	x	SH	L
S, L, C	6.0–8.0	1'–2'	NR	R	10–20	H–R	x	SH	M
S, L, C	?–8.0	1'–2'	2'	FIB	10–20	L–S	x	SH	L
S, L	?	1'	3'	FIB	5–10	L–S	x	SH	L
S, G, L	?	1'–2'	2'	R	10–20	M–R		Empty SH	L
S, L	?	1'	3'	FIB	5–10	M–S	x	Empty SH	H
S, L, C	5.0–8.0	1'	3'	FIB	5–10	L–S	x	Empty SH	H
S, L	5.0–8.0	1'	2'	FIB	5–10	L–S	x	Empty SH	M
S, L, C	?–8.0	1'	3'	R	5–10	M–R?		Empty SH	L

Soil texture	pH range	Plant spacing in prairie gardens	Plant spacing in flower beds	Root type	Plant life expectancy in years	Aggressiveness and method	Cut flower stalks	Dried seed arrangements	Deer palatability
S, G, L	5.0–7.5	1'	NR	FIB	10–20	L–S		x	L
S, L, C	4.5–7.5	1'	1'	FIB	5–10	L–S			L
S, L, C	5.7–8.0	1'	NR	R	10–20	H–R			L
S, L, C	5.0–8.0	1'	3'	FIB	3–5	M–S		x	L
S, L, C	5.0–7.5	1'	3'	FIB	3–5	M–S		x	L
S, L, C	5.0–7.5	1'	1.5'	FIB	5–10	M–S		x	L
S, L, C	4.5–7.5	2'	3'	FIB	20+	H–S		x	M
S, L, C	5.0–7.5	2'	NR	FIB	20+	M–S		x	L
S, L, C	4.0–7.0	1'–1.5'	NR	FIB	20+	H–S		x	L
S, L	5.5–8.5	1'	2'	FIB	10–20	L–S		x	L
S, L, C	4.5–8.0	1.5'–2'	NR	FIB	20+	M–S		x	L
S, L	5.0–8.5	1'–1.5'	3'	FIB	20+	L–S		x	L
S, L, C	5.0–8.0	1'–2'	3'	FIB	10–20	M–S	x	x	L
S, L	6.0–8.0	2'	2'	FIB	20+	L–S	x	x	L

taproot, and TAPCO = taprooted corm. For aggressiveness and method, L = low, M = medium, H = high, R = by rhizome, S = by seed, and R/S = rhizome and seed. For dried arrangements, FL = flowers, SH = seedheads, SP = seedpods, SD = seed, and RH = rose hips. For deer palatability, L = low, M = medium, and H = high.

TABLE 12.5. PLANTS FOR MOIST TO WET SOILS, BY BLOOM TIME

Bloom time	Scientific name (4/1/2020)	Common name	Height	Flower color	USDA zones	Sun	Soil moisture
Flowers and shrubs							
Apr.–June	*Sisyrinchium angustifolium*	Narrow-Leaved Blue-Eyed Grass	6"–2'	Blue/purple	3–10	F, P	M, Mo
May–June	*Amsonia tabernaemontana*	Eastern Bluestar	2'–3'	Blue/white	4–8	F, P	M, Mo
May–July	*Anemone canadensis*	Canada Anemone	1'–2'	White	3–6	F, P	M, Mo
May–June	*Dodecatheon meadia*	Shooting Star	1'–2'	White/pink	4–8	F, P	M, Mo
May–June	*Phlox pilosa*	Downy Phlox	1'–2'	Pink	3–9	F	D, M, Mo
May–July	*Rosa setigera*	Climbing Prairie Rose	6'–15'	Pink/white	4–9	F, P	M, Mo
May–June	*Zizia aptera*	Heartleaf Golden Alexanders	1'–2'	Yellow	3–8	F	M, Mo
May–July	*Zizia aurea*	Golden Alexanders	1'–2'	Yellow	3–8	F, P, S	M, Mo, W
June–Aug.	*Allium cernuum*	Nodding Pink Onion	1'–2'	White/pink	3–8	F, P	M, Mo
June–July	*Asclepias incarnata*	Marsh Milkweed	3'–5'	Red/pink	3–9	F	Mo, W
June–Aug.	*Asclepias sullivantii*	Sullivant's Milkweed	3'–5'	Pink/yellow	4–7	F	M, Mo
June–July	*Baptisia alba*	White False Indigo	3'–5'	White	4–9	F, P	M, Mo
June–Aug.	*Desmanthus illinoensis*	Illinois Bundleflower	2'–6'	White	5–8	F	D, M, Mo
June–July	*Filipendula rubra*	Queen of the Prairie	4'–5'	Pink	3–6	F	M, Mo
June–Sept.	*Heliopsis helianthoides*	Oxeye Sunflower	3'–6'	Yellow	3–8	F	D, M, Mo
June–July	*Iris virginica* var. *shrevei*	Wild Iris	2'–3'	Blue	3–8	F, P	W
June–Sept.	*Oenothera gaura*	Biennial Beeblossom	3'–6'	White/pink	4–8	F, P	D, M, Mo
June–Sept.	*Parthenium integrifolium*	Wild Quinine	3'–5'	White	4–8	F	M, Mo
June–July	*Penstemon digitalis*	Foxglove Beardtongue	2'–3'	White	3–8	F, P	M, Mo
June–Sept.	*Rudbeckia hirta*	Blackeyed Susan	1'–3'	Yellow	3–10	F, P	D, M, Mo
June–Sept.	*Spiraea alba*	Meadowsweet	2'–4'	White/pink	3–7	F, P	Mo, W
June–July	*Thalictrum dasycarpum*	Tall Meadow-Rue	3'–6'	White	3–9	F, P	Mo, W
June–July	*Tradescantia ohiensis*	Ohio Spiderwort	2'–4'	Blue	3–9	F, P	D, M, Mo
July–Oct.	*Conoclinium coelestinum*	Blue Mistflower	1'–3'	Blue/purple	4–10	F, P	M, Mo
July–Sept.	*Coreopsis tripteris*	Tall Coreopsis	3'–7'	Yellow	3–8	F, P	M, Mo, W
July–Aug.	*Desmodium canadense*	Canada Ticktrefoil	3'–5'	Purple	3–7	F	M, Mo
July–Sept.	*Hibiscus moscheutos*	Rosemallow	3'–7'	White/pink/red	5–9	F, P	M, Mo, W
July–Aug.	*Hypericum ascyron*	Great St. Johnswort	4'–6'	Yellow	3–5	F, P	Mo, W
July–Aug.	*Liatris pycnostachya*	Prairie Blazing Star	3'–5'	Purple/pink	3–9	F	M, Mo
July–Aug.	*Lilium michiganense*	Michigan Lily	4'–6'	Orange	3–7	F, P	Mo
July–Sept.	*Lobelia cardinalis*	Cardinal Flower	2'–5'	Red	3–9	F, P	Mo, W
July–Sept.	*Lobelia siphilitica*	Great Blue Lobelia	1'–4'	Blue	3–9	F, P	M, Mo
July–Sept.	*Monarda fistulosa*	Bergamot	2'–5'	Lavender	3–9	F, P	D, M, Mo
July–Sept.	*Pycnanthemum virginianum*	Virginia Mountainmint	1'–3'	White	3–7	F, P	Mo, W
July–Sept.	*Ratibida pinnata*	Yellow Coneflower	3'–6'	Yellow	3–9	F	D, M, Mo
July–Sept.	*Rudbeckia fulgida*	Orange Coneflower	2'–4'	Yellow	4–8	F, P	M, Mo
July–Oct.	*Rudbeckia triloba*	Browneyed Susan	2'–5'	Yellow	4–9	F, P	M, Mo
July–Aug.	*Senna hebecarpa*	Wild Senna	4'–6'	Yellow	4–7	F	M, Mo, W
July–Sept.	*Silphium integrifolium*	Rosinweed	2'–6'	Yellow	4–8	F	M, Mo

Soil texture	pH range	Plant spacing in prairie gardens	Plant spacing in flower beds	Root type	Plant life expectancy in years	Aggressiveness and method	Cut flowers	Dried arrangements	Deer palatability
S, L, C	?–8.0	6"	9"	FIB	3–5	L–S			L
S, L, C	?–8.0	2'–3'	3'	FIB	20+	L–S		SP	L
S, L, C	?	NR	NR	R	10–20	H–R			L
S, L, C	4.5–8.0	6"–1'	1'	FIB	20+	L–S		SP	L
S, L	?	6"–1'	1.5'	FIB	3–5	L–S	x		L
S, L, C	5.0–8.0	2'	5'	FIB	10–20	L–S	x	RH	L
S, L	?–7.5	1'	2'	FIB	5–10	L–S			L
S, L, C	?–8.0	1'	3'	FIB	5–10	L–S			L
S, L, C	6.5–8.0	6"–1'	1'	B	10–20	L–S	x	FL/SH	L
S, L, C	5.0–8.0	1'–2'	3'	FIB	5–10	L–S	x	SP	L
L, C	?	1'–2'	3'	R	10–20	L–R		SP	L
S, L, C	?–7.5	3'	3'	TAP	20+	L–S	x	SP	L
S, L, C	6.0–8.0	1'	1.5'	TAP	5–10	L–S		SP	M
S, L, C	?–8.0	2'	3'	R	20+	M–R	x		L
S, L, C	6.0–8.0	1'–2'	3'	FIB	5–10	H–S	x		L
S, L, C	4.8–7.3	1'	2'	R/CO	10–20	L–R	x	SP	L
S, L, C	?	2'	3'	FIB	2	M–S	x		L
S, L, C	?–8.0	1'–2'	3'	TAP	10–20	L–S	x		M
S, L, C	5.5–7.5	1'	2'	FIB	5–10	L–S	x		L
S, L, C	5.0–8.0	1'	2'	FIB	2	M–S	x	SH	L
S, L, C	4.3–7.2	2'	3'	FIB	10–20	L–S	x		L
S, L, C	?	1'–2'	2'	FIB	5–10	L–S			L
S, L, C	?–7.5	1'	2'	FIB	20+	M–S			H
S, L, C	5.0–8.0	1'–2'	2'	R	3–5	M–R			L
S, L, C	?	1'–2'	NR	FIB	10–20	L–S	x		L
S, L, C	?–7.5	NR	NR	TAP	10–20	M–S			H
S, L, C	4.0–8.0	2'	4'	R	10–20	L–R/S			L
S, L, C	5.7–7.1	1'–2'	3'	FIB	5–10	L–S	x	SP	L
S, L, C	6.0–8.0	6"–1'	1'	CO	10–20	L–S	x	FL/SH	L
S, L, C	?–6.8	2'	2'	B	20+	L–S	x	SP	H
S, L	5.8–7.8	6"–1'	2'	FIB	3–5	L–S	x		M
S, L, C	?–8.0	6"–1'	2'	FIB	5–10	M–S	x		L
S, L, C	6.0–8.0	1'–2'	3'	R	10–20	M–R	x	SH	L
S, L, C	?–8.0	1'–2'	3'	R	5–10	L–R		SH	L
S, L, C	5.5–8.0	1'–2'	2'	FIB	10–20	M–S	x	SH	L
L, C	?–8.0	1'–2'	3'	FIB	5–10	L–S	x	SH	L
S, L	?–8.0	1'–2'	3'	FIB	2	M–S	x	SH	L
S, L, C	?	1.5'–2'	3'–4'	FIB/TAP	10–20	L–S	x	FL/SH	M
S, L, C	?–8.0	2'	3'	TAP	20+	M–S	x		L

(continued)

TABLE 12.5. *(continued)*

Bloom time	Scientific name (4/1/2020)	Common name	Height	Flower color	USDA zones	Sun	Soil moisture
July–Sept.	*Silphium perfoliatum*	Cup Plant	3'–10'	Yellow	3–8	F, P	M, Mo, W
July–Sept.	*Silphium terebinthinaceum*	Prairie Dock	3'–10'	Yellow	4–7	F	M, Mo
July–Sept.	*Verbena hastata*	Blue Vervain	3'–6'	Blue	3–8	F, P	M, Mo, W
July–Sept.	*Vernonia fasciculata*	Ironweed	4'–6'	Red/pink	3–7	F	Mo
July–Aug.	*Veronicastrum virginicum*	Culver's Root	3'–6'	White	3–8	F, P, S	M, Mo
Aug.–Oct.	*Gentiana andrewsii*	Bottle Gentian	1'–2'	Blue	3–6	F, P	Mo, W
Aug.–Oct.	*Rudbeckia subtomentosa*	Sweet Blackeyed Susan	4'–6'	Yellow	4–8	F, P	M, Mo
Aug.–Oct.	*Symphyotrichum novae-angliae*	New England Aster	3'–6'	Pink/purple/blue	3–7	F, P	M, Mo
Aug.–Sept.	*Chelone glabra*	White Turtlehead	2'–4'	White	3–8	F, P	Mo, W
Aug.–Sept.	*Eupatorium perfoliatum*	Boneset	3'–4'	White	3–8	F	M, Mo, W
Aug.–Sept.	*Eutrochium fistulosum*	Tall Joe Pye Weed	5'–8'	Purple/pink	4–9	F, P	M, Mo, W
Aug.–Sept.	*Eutrochium maculatum*	Joe Pye Weed	4'–6'	Pink	3–6	F, P	Mo, W
Aug.–Sept.	*Helenium autumnale*	Sneezeweed	4'–5'	Yellow	3–9	F	Mo, W
Aug.–Sept.	*Liatris ligulistylis*	Meadow Blazing Star	3'–5'	Purple/pink	3–6	F	M, Mo
Aug.–Sept.	*Liatris spicata*	Dense Blazing Star	3'–6'	Purple/pink	4–10	F	M, Mo
Aug.–Sept.	*Oligoneuron ohioense*	Ohio Goldenrod	3'–4'	Yellow	4–5	F, P	M, Mo
Aug.–Sept.	*Oligoneuron rigidum*	Stiff Goldenrod	3'–5'	Yellow	3–9	F	D, M, Mo
Aug.–Sept.	*Physostegia virginiana*	Obedient Plant	1'–2'	Pink	3–9	F	M, Mo
Aug.–Sept.	*Rudbeckia laciniata*	Green-Headed Coneflower	5'–8'	Yellow	3–8	F, P	Mo, W
Aug.–Sept.	*Spiraea tomentosa*	Steeplebush	2'–4'	Pink	3–8	F, P	Mo, W
Aug.–Sept.	*Vernonia gigantea*	Tall Ironweed	5'–8'	Red/pink	4–9	F, P	Mo, W

Bloom time	Scientific name (4/1/2020)	Common name	Height	Fall leaf color	USDA zones	Sun	Soil moisture
Grasses and sedges							
May–June	*Carex brevior*	Shortbeak Sedge	1'–4'	Straw	3–7	F, P	D, M, Mo
May–June	*Hierochloe hirta*	Vanilla Sweetgrass	1'–2'	Straw	3–9	F, P	M, Mo, W
June–July	*Calamagrostis canadensis*	Bluejoint Grass	4'–6'	Straw	3–6	F	Mo, W
July–Aug.	*Elymus canadensis*	Canada Wildrye	4'–5'	Straw	2–9	F	D, M, Mo
July–Aug.	*Elymus virginicus*	Virginia Wildrye	4'–5'	Straw	2–9	F, P, S	M, Mo
July–Sept.	*Tripsacum dactyloides*	Eastern Gamagrass	4'–8'	Gold	4–10	F, P	M, Mo, W
Aug.–Oct.	*Andropogon gerardii*	Big Bluestem	5'–8'	Bronze/red	3–9	F	D, M, Mo
Aug.–Oct.	*Andropogon virginicus*	Broomsedge	2'–5'	Reddish brown	5–10	F	D, M, Mo
Aug.–Sept.	*Panicum virgatum*	Switchgrass	3'–6'	Gold	3–10	F	D, M, Mo
Aug.–Sept.	*Sorghastrum nutans*	Indiangrass	5'–7'	Gold	3–9	F	D, M, Mo
Aug.–Sept.	*Spartina pectinata*	Prairie Cordgrass	6'–9'	Gold	3–8	F	Mo, W

Notes: For soil moisture, D = dry, M = medium, Mo = moist, and W = wet. For soil texture, S = sand, L = loam, C = clay, and G = gravel. For sun, F = full sun, or at least a half day of full sun; P = partial sun, or a quarter to a half day of sun; and S = shade, or less than a quarter day of sun. For plant spacing, NR = not recommended. For root type, B = bulb, CO = corm, FIB = fibrous, R = rhizome, TAP =

Soil texture	pH range	Plant spacing in prairie gardens	Plant spacing in flower beds	Root type	Plant life expectancy in years	Aggressiveness and method	Cut flowers	Dried arrangements	Deer palatability
S, L, C	4.5–8.0	2'–3'	NR	FIB	20+	H–S	x		L
S, L, C	?–8.0	2'	4'	TAP	20+	L–S	x		L
S, L, C	?	1'	2'	FIB	2	M–S	x	SH	L
S, L, C	?	1'	3'	FIB	5–10	L–S	x		L
S, L, C	?–8.0	1'	3'	FIB	10–20	L–S	x	SH	L
S, L, C	5.8–7.2	1'	1.5'	FIB	5–10	L–S	x		L
S, L, C	?–8.0	1'–2'	2'	FIB	10–20	L–S	x	SH	L
S, L, C	5.0–8.0	1'	3'	FIB	5–10	L–S	x	Empty SH	H
S, L, C	?	1'	1'–2'	FIB	5–10	L–S			M
S, L, C	?–7.5	1'	NR	FIB	5–10	L–S			L
S, L, C	4.5–7.5	2'	3'–4'	FIB	5–10	L–S			H
S, L, C	5.5–7.5	1'–2'	3'–4'	FIB	5–10	L–S			L
S, L, C	4.0–7.5	1'–2'	2'–3'	FIB	5–10	L–S	x	SH	L
L	?	6"–1'	1'	CO	3–5	L–S	x	FL/SH	L
S, L, C	5.5–7.5	6"–1'	1'	CO	10–20	L–S	x	FL/SH	L
S, L, C	?–8.0	1'–2'	3'	FIB	5–10	L–S	x	SH	L
S, L, C	5.5–8.0	1'–1.5'	3'	FIB	5–10	M–S	x	SH	L
S, L, C	6.0–8.0	1'–2'	NR	R	10–20	H–R	x	SH	M
S, L, C	4.5–7.5	2'	2'	FIB	3–5	M–S	x	SH	L
S, L, C	4.5–7.0	2'	3'	R	10–20	L–R/S	x		L
S, L, C	?–8.0	1'–2'	3'	FIB	10–20	L–S	x		L

Soil texture	pH range	Plant spacing in prairie gardens	Plant spacing in flower beds	Root type	Plant life expectancy in years	Aggressiveness and method	Cut flower stalks	Dried seed arrangements	Deer palatability
S, L, C	4.5–7.5	1'	1'	FIB	5–10	L–S			L
S, L, C	5.7–8.0	1'	NR	R	10–20	H–R			L
S, L, C	4.5–8.0	1'	3'	FIB	10–20	M–R/S		x	L
S, L, C	5.0–8.0	1'	3'	FIB	3–5	M–S		x	L
S, L, C	5.0–7.5	1'	3'	FIB	3–5	M–S		x	L
S, L, C	4.5–7.5	2'	3'	FIB	20+	H–S		x	M
S, L, C	5.0–7.5	2'	NR	FIB	20+	M–S		x	L
S, L, C	4.0–7.0	1'–1.5'	NR	FIB	20+	H–S		x	L
S, L, C	4.5–8.0	1.5'–2'	NR	FIB	20+	M–S		x	L
S, L, C	5.0–8.0	1'–2'	3'	FIB	10–20	M–S	x	x	L
S, L, C	6.0–8.5	2'	NR	R	20+	H–R		x	L

taproot, and TAPCO = taprooted corm. For aggressiveness and method, L = low, M = medium, H = high, R = by rhizome, S = by seed, and R/S = rhizome and seed. For dried arrangements, FL = flowers, SH = seedheads, SP = seedpods, SD = seed, and RH = rose hips. For deer palatability, L = low, M = medium, and H = high.

TABLE 12.6.   PRAIRIE SPECIES FOR SEEDING ON SUBSOIL CLAY

Scientific name (4/1/2020)	Common name	Height	Flower color	Bloom time	USDA zones	Sun	Soil moisture
Flowers and shrubs							
Allium cernuum	Nodding Pink Onion	1′–2′	White/pink	June–Aug.	3–8	F, P	M, Mo
Asclepias syriaca	Common Milkweed	2′–4′	Lavender	June–Aug.	3–8	F, P	D, M
Baptisia alba	White False Indigo	3′–5′	White	June–July	4–9	F, P	M, Mo
Desmanthus illinoensis	Illinois Bundleflower	2′–6′	White	June–Aug.	5–8	F	D, M, Mo
Desmodium canadense	Canada Ticktrefoil	3′–5′	Purple	July–Aug.	3–7	F	M, Mo
Echinacea pallida	Pale Purple Coneflower	3′–5′	Purple	June–July	4–8	F	D, M
Echinacea paradoxa	Ozark Coneflower	3′–5′	Yellow	June–July	4–7	F, P	D, M
Echinacea purpurea	Purple Coneflower	3′–4′	Purple	July–Sept.	4–8	F, P	D, M
Eryngium yuccifolium	Rattlesnake Master	3′–5′	White	June–Aug.	4–9	F	D, M
Helianthus ×laetiflorus	Showy Sunflower	3′–6′	Yellow	Aug.–Sept.	3–8	F	D, M
Heliopsis helianthoides	Oxeye Sunflower	3′–6′	Yellow	June–Sept.	3–8	F	D, M, Mo
Monarda fistulosa	Bergamot	2′–5′	Lavender	July–Sept.	3–9	F, P	D, M, Mo
Oligoneuron rigidum	Stiff Goldenrod	3′–5′	Yellow	Aug.–Sept.	3–9	F	D, M, Mo
Parthenium integrifolium	Wild Quinine	3′–5′	White	June–Sept.	4–8	F	M, Mo
Penstemon digitalis	Foxglove Beardtongue	2′–3′	White	June–July	3–8	F, P	M, Mo
Ratibida pinnata	Yellow Coneflower	3′–6′	Yellow	July–Sept.	3–9	F	D, M, Mo
Rudbeckia hirta	Blackeyed Susan	1′–3′	Yellow	June–Sept.	3–10	F, P	D, M, Mo
Rudbeckia subtomentosa	Sweet Blackeyed Susan	4′–6′	Yellow	Aug.–Oct.	4–8	F, P	M, Mo
Senna hebecarpa	Wild Senna	4′–6′	Yellow	July–Aug.	4–7	F	M, Mo, W
Silphium integrifolium	Rosinweed	2′–6′	Yellow	July–Sept.	4–8	F	M, Mo
Silphium laciniatum	Compassplant	3′–10′	Yellow	June–Sept.	4–9	F	D, M
Silphium perfoliatum	Cup Plant	3′–10′	Yellow	July–Sept.	3–8	F, P	M, Mo, W
Silphium terebinthinaceum	Prairie Dock	3′–10′	Yellow	July–Sept.	4–7	F	M, Mo
Symphyotrichum novae-angliae	New England Aster	3′–6′	Pink/purple/blue	Aug.–Oct.	3–7	F, P	M, Mo
Verbena hastata	Blue Vervain	3′–6′	Blue	July–Sept.	3–8	F, P	M, Mo, W
Vernonia fasciculata	Ironweed	4′–6′	Red/pink	July–Sept.	3–7	F	Mo
Veronicastrum virginicum	Culver's Root	3′–6′	White	July–Aug.	3–8	F, P, S	M, Mo

Scientific name (4/1/2020)	Common name	Height	Fall leaf color	Bloom time	USDA zones	Sun	Soil moisture
Grasses and sedges							
Andropogon gerardii	Big Bluestem	5′–8′	Bronze/red	Aug.–Oct.	3–9	F	D, M, Mo
Andropogon virginicus	Broomsedge	2′–5′	Reddish brown	Aug.–Oct.	5–10	F	D, M, Mo
Elymus canadensis	Canada Wildrye	4′–5′	Straw	July–Aug.	2–9	F	D, M, Mo
Elymus virginicus	Virginia Wildrye	4′–5′	Straw	July–Aug.	2–9	F, P, S	M, Mo
Panicum virgatum	Switchgrass	3′–6′	Gold	Aug.–Sept.	3–10	F	D, M, Mo
Sorghastrum nutans	Indiangrass	5′–7′	Gold	Aug.–Sept.	3–9	F	D, M, Mo
Tridens flavus	Purpletop	3′–6′	Straw	July–Sept.	5–10	F	D, M
Tripsacum dactyloides	Eastern Gamagrass	4′–8′	Gold	July–Sept.	4–10	F, P	M, Mo, W

Notes: For soil moisture, D = dry, M = medium, Mo = moist, and W = wet. For soil texture, S = sand, L = loam, C = clay, and G = gravel. For sun, F = full sun, or at least a half day of full sun; P = partial sun, or a quarter to a half day of sun; and S = shade, or less than a quarter day of sun. For plant spacing, NR = not recommended. For root type, B = bulb, CO = corm, FIB = fibrous, R = rhizome, TAP =

Soil texture	pH range	Plant spacing in prairie gardens	Plant spacing in flower beds	Root type	Plant life expectancy in years	Aggressiveness and method	Cut flowers	Dried arrangements	Deer palatability
S, L, C	6.5–8.0	6"–1'	1'	B	10–20	L–S	x	FL/SH	L
S, L, C	5.0–8.0	NR	NR	R	10–20	H–R		SP	L
S, L, C	?–7.5	3'	3'	TAP	20+	L–S	x	SP	L
S, L, C	6.0–8.0	1'	1.5'	TAP	5–10	L–S		SP	M
S, L, C	?–7.5	NR	NR	TAP	10–20	M–S			H
S, L, C	6.0–7.5	1'	2'–3'	TAP	10–20	L–S	x	SH	M
S, L, C	?–7.5	1'	2'–3'	TAP	10–20	L–S	x	SH	M
S, L, C	5.5–8.0	1'–2'	3'	FIB	5–10	M–S	x	SH	M
S, L, C	6.0–8.0	1'	2'	FIB	5–10	M–S	x	SH	L
S, G, L, C	?	NR	NR	R	20+	H–R	x	SH	L
S, L, C	6.0–8.0	1'–2'	3'	FIB	5–10	H–S	x		L
S, L, C	6.0–8.0	1'–2'	3'	R	10–20	M–R	x	SH	L
S, L, C	5.5–8.0	1'–1.5'	3'	FIB	5–10	M–S	x	SH	L
S, L, C	?–8.0	1'–2'	3'	TAP	10–20	L–S	x		M
S, L, C	5.5–7.5	1'	2'	FIB	5–10	L–S	x		L
S, L, C	5.5–8.0	1'–2'	2'	FIB	10–20	M–S	x	SH	L
S, L, C	5.0–8.0	1'	2'	FIB	2	M–S	x	SH	L
S, L, C	?–8.0	1'–2'	2'	FIB	10–20	L–S	x	SH	L
S, L, C	?	1.5'–2'	3'–4'	FIB/TAP	10–20	L–S	x	FL/SH	M
S, L, C	?–8.0	2'	3'	TAP	20+	M–S	x		L
S, L, C	?–8.0	2'	4'	TAP	20+	L–S	x		L
S, L, C	4.5–8.0	2'–3'	NR	FIB	20+	H–S	x		L
S, L, C	?–8.0	2'	4'	TAP	20+	L–S	x		L
S, L, C	5.0–8.0	1'	3'	FIB	5–10	L–S	x	Empty SH	H
S, L, C	?	1'	2'	FIB	2	M–S	x	SH	L
S, L, C	?	1'	3'	FIB	5–10	L–S	x		L
S, L, C	?–8.0	1'	3'	FIB	10–20	L–S	x	SH	L

Soil texture	pH range	Plant spacing in prairie gardens	Plant spacing in flower beds	Root type	Plant life expectancy in years	Aggressiveness and method	Cut flower stalks	Dried seed arrangements	Deer palatability
S, L, C	5.0–7.5	2'	NR	FIB	20+	M–S		x	L
S, L, C	4.0–7.0	1'–1.5'	NR	FIB	20+	H–S		x	L
S, L, C	5.0–8.0	1'	3'	FIB	3–5	M–S		x	L
S, L, C	5.0–7.5	1'	3'	FIB	3–5	M–S		x	L
S, L, C	4.5–8.0	1.5'–2'	NR	FIB	20+	M–S		x	L
S, L, C	5.0–8.0	1'–2'	3'	FIB	10–20	M–S	x	x	L
S, L, C	5.0–7.5	1'	1.5'	FIB	5–10	M–S		x	L
S, L, C	4.5–7.5	2'	3'	FIB	20+	H–S		x	M

taproot, and TAPCO = taprooted corm. For aggressiveness and method, L = low, M = medium, H = high, R = by rhizome, S = by seed, and R/S = rhizome and seed. For dried arrangements, FL = flowers, SH = seedheads, SP = seedpods, SD = seed, and RH = rose hips. For deer palatability, L = low, M = medium, and H = high.

TABLE 12.7. PRAIRIE PLANTS FOR SEMISHADE

Scientific name (4/1/2020)	Common name	Height	Flower color	Bloom time	USDA zones	Sun	Soil moisture
Flowers and shrubs							
Agastache foeniculum	Lavender Hyssop	1'–3'	Purple	July–Sept.	2–6	F, P	D, M
Allium cernuum	Nodding Pink Onion	1'–2'	White/pink	June–Aug.	3–8	F, P	M, Mo
Allium stellatum	Prairie Onion	1'–2'	Lavender	July–Aug.	3–8	F, P	D, M
Amsonia tabernaemontana	Eastern Bluestar	2'–3'	Blue/white	May–June	4–8	F, P	M, Mo
Anemone canadensis	Canada Anemone	1'–2'	White	May–July	3–6	F, P	M, Mo
Arnoglossum atriplicifolium	Pale Indian Plantain	5'–10'	White	July–Sept.	4–8	F, P	D, M, Mo
Asclepias syriaca	Common Milkweed	2'–4'	Lavender	June–Aug.	3–8	F, P	D, M
Baptisia alba	White False Indigo	3'–5'	White	June–July	4–9	F, P	M, Mo
Baptisia australis	Blue False Indigo	3'–5'	Blue	June–July	3–10	F, P	M
Baptisia bracteata	Cream False Indigo	1'–2'	Cream	May–June	4–9	F, P	D, M
Blephilia ciliata	Downy Wood Mint	1'–2'	White/purple/violet	May–July	4–8	F, P	D, M
Callirhoe involucrata	Winecups	1'	Magenta	May–June	4–9	F, P	D, M
Ceanothus americanus	New Jersey Tea	2'–3'	White	July–Aug.	3–9	F, P	D, M
Chelone glabra	White Turtlehead	2'–4'	White	Aug.–Sept.	3–8	F, P	Mo, W
Conoclinium coelestinum	Blue Mistflower	1'–3'	Blue/purple	July–Oct.	4–10	F, P	M, Mo
Coreopsis tripteris	Tall Coreopsis	3'–7'	Yellow	July–Sept.	3–8	F, P	M, Mo, W
Dodecatheon meadia	Shooting Star	1'–2'	White/pink	May–June	4–8	F, P	M, Mo
Echinacea paradoxa	Ozark Coneflower	3'–5'	Yellow	June–July	4–7	F, P	D, M
Echinacea purpurea	Purple Coneflower	3'–4'	Purple	July–Sept.	4–8	F, P	D, M
Eutrochium fistulosum	Tall Joe Pye Weed	5'–8'	Purple/pink	Aug.–Sept.	4–9	F, P	M, Mo, W
Eutrochium maculatum	Joe Pye Weed	4'–6'	Pink	Aug.–Sept.	3–6	F, P	Mo, W
Gentiana alba	Cream Gentian	1'–2'	White	July–Aug.	3–7	F, P	M
Gentiana andrewsii	Bottle Gentian	1'–2'	Blue	Aug.–Oct.	3–6	F, P	Mo, W
Hibiscus moscheutos	Rosemallow	3'–7'	White/pink/red	July–Sept.	5–9	F, P	M, Mo, W
Hypericum ascyron	Great St. Johnswort	4'–6'	Yellow	July–Aug.	3–5	F, P	Mo, W
Iris virginica var. shrevei	Wild Iris	2'–3'	Blue	June–July	3–8	F, P	W
Lespedeza capitata	Roundhead Bush Clover	3'–5'	White	Aug.–Sept.	3–8	F, P	D, M
Lilium michiganense	Michigan Lily	4'–6'	Orange	July–Aug.	3–7	F, P	Mo
Lithospermum canescens	Hoary Puccoon	1'	Orange	May–June	3–7	F, P	D, M
Lobelia cardinalis	Cardinal Flower	2'–5'	Red	July–Sept.	3–9	F, P	Mo, W
Lobelia siphilitica	Great Blue Lobelia	1'–4'	Blue	July–Sept.	3–9	F, P	M, Mo
Lupinus perennis	Wild Lupine	1'–2'	Blue	May–June	3–8	F, P	D
Monarda fistulosa	Bergamot	2'–5'	Lavender	July–Sept.	3–9	F, P	D, M, Mo
Oenothera gaura	Biennial Beeblossom	3'–6'	White/pink	June–Sept.	4–8	F, P	D, M, Mo
Oligoneuron ohioense	Ohio Goldenrod	3'–4'	Yellow	Aug.–Sept.	4–5	F, P	M, Mo
Penstemon digitalis	Foxglove Beardtongue	2'–3'	White	June–July	3–8	F, P	M, Mo
Penstemon gracilis	Slender Beardtongue	1'–2'	Lavender	May–June	3–6	F, P	D
Penstemon hirsutus	Hairy Penstemon	1'–2'	Lavender	May–June	3–6	F, P	D
Polygonatum biflorum var. commutatum	Great Solomon's Seal	2'–4'	Cream	May–June	3–9	F, P	D, M

Soil texture	pH range	Plant spacing in prairie gardens	Plant spacing in flower beds	Root type	Plant life expectancy in years	Aggressiveness and method	Cut flowers	Dried arrangements	Deer palatability
S, L	?-8.0	1'	1'-2'	FIB	3-5	M-S	x	FL/SH	L
S, L, C	6.5-8.0	6"-1'	1'	B	10-20	L-S	x	FL/SH	L
S, L	5.5-8.0	6"-1'	1'	B	10-20	L-S	x	FL/SH	L
S, L, C	?-8.0	2'-3'	3'	FIB	20+	L-S		SP	L
S, L, C	?	NR	NR	R	10-20	H-R			L
S, L, C	?-8.0	2'	3'-4'	FIB	3-20	H-S (north); L-S (south)	x	FL	L
S, L, C	5.0-8.0	NR	NR	R	10-20	H-R		SP	L
S, L, C	?-7.5	3'	3'	TAP	20+	L-S	x	SP	L
S, L, C	?-8.0	3'	3'	TAP	20+	L-S	x	SP	L
S, L	?	2'	3'	TAP	20+	L-S	x	SP	L
L, C, G	?-8.00	1'	1'-2'	TAP/short Rs	3-5	L-S		FL/SH	L
S, L	5.5-7.5	1'-2'	2'	TAP	10-20	L-S			L
S, L	4.5-7.0	3'	3'	TAP	20+	L-S		SD	L
S, L, C	?	1'	1'-2'	FIB	5-10	L-S			M
S, L, C	5.0-8.0	1'-2'	2'	R	3-5	M-R			L
S, L, C	?	1'-2'	NR	FIB	10-20	L-S	x		L
S, L, C	4.5-8.0	6"-1'	1'	FIB	20+	L-S		SP	L
S, L, C	?-7.5	1'	2'-3'	TAP	10-20	L-S	x	SH	M
S, L, C	5.5-8.0	1'-2'	3'	FIB	5-10	M-S	x	SH	M
S, L, C	4.5-7.5	2'	3'-4'	FIB	5-10	L-S			H
S, L, C	5.5-7.5	1'-2'	3'-4'	FIB	5-10	L-S			L
S, L, C	?-7.5	1'	1.5'	FIB	10-20	L-S	x		L
S, L, C	5.8-7.2	1'	1.5'	FIB	5-10	L-S	x		L
S, L, C	4.0-8.0	2'	4'	R	10-20	L-R/S			L
S, L, C	5.7-7.1	1'-2'	3'	FIB	5-10	L-S	x	SP	L
S, L, C	4.8-7.3	1'	2'	R/CO	10-20	L-R	x	SP	L
S, L, G	5.5-8.0	1'	2'	TAP	10-20	L-S		SH	M
S, L, C	?-6.8	2'	2'	B	20+	L-S	x	SP	H
S, L	?	1'	1'	FIB	5-10	L-S			L
S, L	5.8-7.8	6"-1'	2'	FIB	3-5	L-S	x		M
S, L, C	?-8.0	6"-1'	2'	FIB	5-10	M-S	x		L
S	5.0-?	1'	1.5'	TAP	10-20	M-S			L
S, L, C	6.0-8.0	1'-2'	3'	R	10-20	M-R	x	SH	L
S, L, C	?	2'	3'	FIB	2	M-S	x		L
S, L, C	?-8.0	1'-2'	3'	FIB	5-10	L-S	x	SH	L
S, L, C	5.5-7.5	1'	2'	FIB	5-10	L-S	x		L
S, G	?-7.5	6"-1'	1'	FIB	5-10	L-S	x	SP	L
S, G	5.0-8.0	1'	1.5'	FIB	3-5	L-S	x	SP	L
S, L, C	?	1'-2'	3'	R	10-20	L-R/S			H

(continued)

TABLE 12.7. *(continued)*

Scientific name (4/1/2020)	Common name	Height	Flower color	Bloom time	USDA zones	Sun	Soil moisture
Pycnanthemum tenuifolium	Narrowleaf Mountainmint	1'–3'	White	July–Sept.	4–8	F, P	D, M
Pycnanthemum virginianum	Virginia Mountainmint	1'–3'	White	July–Sept.	3–7	F, P	Mo, W
Rosa setigera	Climbing Prairie Rose	6'–15'	Pink/white	May–July	4–9	F, P	M, Mo
Rudbeckia fulgida	Orange Coneflower	2'–4'	Yellow	July–Sept.	4–8	F, P	M, Mo
Rudbeckia hirta	Blackeyed Susan	1'–3'	Yellow	June–Sept.	3–10	F, P	D, M, Mo
Rudbeckia laciniata	Green-Headed Coneflower	5'–8'	Yellow	Aug.–Sept.	3–8	F, P	Mo, W
Rudbeckia subtomentosa	Sweet Blackeyed Susan	4'–6'	Yellow	Aug.–Oct.	4–8	F, P	M, Mo
Rudbeckia triloba	Browneyed Susan	2'–5'	Yellow	July–Oct.	4–9	F, P	M, Mo
Salvia azurea	Blue Sage	3'–5'	Blue	June–Aug.	4–9	F, P	D, M
Silene regia	Royal Catchfly	2'–4'	Red	July–Aug.	4–9	F, P	M
Silphium perfoliatum	Cup Plant	3'–10'	Yellow	July–Sept.	3–8	F, P	M, Mo, W
Sisyrinchium angustifolium	Narrow-Leaved Blue-Eyed Grass	6"–2'	Blue/purple	Apr.–June	3–10	F, P	M, Mo
Sisyrinchium campestre	Prairie Blue-Eyed Grass	6"–1'	Blue/white	Apr.–June	3–8	F, P	D, M
Spiraea alba	Meadowsweet	2'–4'	White/pink	June–Sept.	3–7	F, P	Mo, W
Spiraea tomentosa	Steeplebush	2'–4'	Pink	Aug.–Sept.	3–8	F, P	Mo, W
Symphyotrichum novae-angliae	New England Aster	3'–6'	Pink/purple/blue	Aug.–Oct.	3–7	F, P	M, Mo
Symphyotrichum oolentangiense	Skyblue Aster	2'–3'	Blue	Aug.–Oct.	3–8	F, P	D, M
Tephrosia virginiana	Goatsrue	1'–2'	Pink/yellow	June–Aug.	4–9	F, P	D
Thalictrum dasycarpum	Tall Meadow-Rue	3'–6'	White	June–July	3–9	F, P	Mo, W
Tradescantia bracteata	Prairie Spiderwort	1'–2'	Blue	June–July	3–7	F, P	D, M
Tradescantia ohiensis	Ohio Spiderwort	2'–4'	Blue	June–July	3–9	F, P	D, M, Mo
Verbena hastata	Blue Vervain	3'–6'	Blue	July–Sept.	3–8	F, P	M, Mo, W
Vernonia gigantea	Tall Ironweed	5'–8'	Red/pink	Aug.–Sept.	4–9	F, P	Mo, W
Veronicastrum virginicum	Culver's Root	3'–6'	White	July–Aug.	3–8	F, P, S	M, Mo
Viola pedata	Birdfoot Violet	6"	Blue/purple	Apr.–June	3–9	F, P	D
Zizia aurea	Golden Alexanders	1'–2'	Yellow	May–July	3–8	F, P, S	M, Mo, W

Scientific name (4/1/2020)	Common name	Height	Fall leaf color	Bloom time	USDA zones	Sun	Soil moisture
Grasses and sedges							
Carex brevior	Shortbeak Sedge	1'–4'	Straw	May–June	3–7	F, P	D, M, Mo
Elymus virginicus	Virginia Wildrye	4'–5'	Straw	July–Aug.	2–9	F, P, S	M, Mo
Eragrostis spectabilis	Purple Lovegrass	1'–2'	Purple/pink	July–Sept.	3–10	F, P	D
Hierochloe hirta	Vanilla Sweetgrass	1'–2'	Straw	May–June	3–9	F, P	M, Mo, W
Tripsacum dactyloides	Eastern Gamagrass	4'–8'	Gold	July–Sept.	4–10	F, P	M, Mo, W

Notes: For soil moisture, D = dry, M = medium, Mo = moist, and W = wet. For soil texture, S = sand, L = loam, C = clay, and G = gravel. For sun, F = full sun, or at least a half day of full sun; P = partial sun, or a quarter to a half day of sun; and S = shade, or less than a quarter day of sun. For plant spacing, NR = not recommended. For root type, B = bulb, CO = corm, FIB = fibrous, R = rhizome, TAP =

Soil texture	pH range	Plant spacing in prairie gardens	Plant spacing in flower beds	Root type	Plant life expectancy in years	Aggressiveness and method	Cut flowers	Dried arrangements	Deer palatability
S, L, C, G	?	1'	1.5'	TAP/R	5–10	L–S		SH	M
S, L, C	?–8.0	1'–2'	3'	R	5–10	L–R		SH	L
S, L, C	5.0–8.0	2'	5'	FIB	10–20	L–S	x	RH	L
L, C	?–8.0	1'–2'	3'	FIB	5–10	L–S	x	SH	L
S, L, C	5.0–8.0	1'	2'	FIB	2	M–S	x	SH	L
S, L, C	4.5–7.5	2'	2'	FIB	3–5	M–S	x	SH	L
S, L, C	?–8.0	1'–2'	2'	FIB	10–20	L–S	x	SH	L
S, L	?–8.0	1'–2'	3'	FIB	2	M–S	x	SH	L
S, L, G	?	1'	1.5'	FIB	5–10	L–S	x	SH	L
L	?	1'	1.5'	TAP	5–10	L–S	x	SH	L
S, L, C	4.5–8.0	2'–3'	NR	FIB	20+	H–S	x		L
S, L, C	?–8.0	6"	9"	FIB	3–5	L–S			L
S, L	?–7.5	6"	6"	FIB	3–5	L–S			L
S, L, C	4.3–7.2	2'	3'	FIB	10–20	L–S	x		L
S, L, C	4.5–7.0	2'	3'	R	10–20	L–R/S	x		L
S, L, C	5.0–8.0	1'	3'	FIB	5–10	L–S	x	Empty SH	H
S, L	5.0–8.0	1'	2'	FIB	5–10	L–S	x	Empty SH	M
S, G	4.5–?	2'	3'	TAP	10–20	L–S			L
S, L, C	?	1'–2'	2'	FIB	5–10	L–S			L
S, L, C	?	1'	2'	R	20+	M–R			M
S, L, C	?–7.5	1'	2'	FIB	20+	M–S			H
S, L, C	?	1'	2'	FIB	2	M–S	x	SH	L
S, L, C	?–8.0	1'–2'	3'	FIB	10–20	L–S	x		L
S, L, C	?–8.0	1'	3'	FIB	10–20	L–S	x	SH	L
S, G	?	6"	1'	R	5–10	L–S			L
S, L, C	?–8.0	1'	3'	FIB	5–10	L–S			L

Soil texture	pH range	Plant spacing in prairie gardens	Plant spacing in flower beds	Root type	Plant life expectancy in years	Aggressiveness and method	Cut flower stalks	Dried seed arrangements	Deer palatability
S, L, C	4.5–7.5	1'	1'	FIB	5–10	L–S			L
S, L, C	5.0–7.5	1'	3'	FIB	3–5	M–S		x	L
S, G	4.0–7.5	1'–1.5'	2'	FIB	5–10	M–S	x	x	L
S, L, C	5.7–8.0	1'	NR	R	10–20	H–R			L
S, L, C	4.5–7.5	2'	3'	FIB	20+	H–S		x	M

taproot, and TAPCO = taprooted corm. For aggressiveness and method, L = low, M = medium, H = high, R = by rhizome, S = by seed, and R/S = rhizome and seed. For dried arrangements, FL = flowers, SH = seedheads, SP = seedpods, SD = seed, and RH = rose hips. For deer palatability, L = low, M = medium, and H = high.

TABLE 12.8.   WHITE FLOWERS, BY SOIL MOISTURE

Soil moisture	Scientific name (4/1/2020)	Common name	Height	Flower color	Bloom time	USDA zones	Sun
D	*Antennaria neglecta*	Pussytoes	6″–1′	White	Apr.–May	3–7	F
D	*Pulsatilla patens*	Pasque Flower	6″–1′	White	Apr.–May	3–6	F
D	*Solidago ptarmicoides*	White Aster	1′–2′	White	Aug.–Sept.	3–7	F
D, M	*Anemone cylindrica*	Thimbleweed	1′–3′	White	May–June	3–6	F
D, M	*Blephilia ciliata*	Downy Wood Mint	1′–2′	White/purple/violet	May–July	4–8	F, P
D, M	*Ceanothus americanus*	New Jersey Tea	2′–3′	White	July–Aug.	3–9	F, P
D, M	*Dalea candida*	White Prairie Clover	1′–2′	White	July–Aug.	3–9	F
D, M	*Delphinium carolinianum* subsp. *virescens*	Prairie Larkspur	1′–4′	White	June–July	3–8	F
D, M	*Eryngium yuccifolium*	Rattlesnake Master	3′–5′	White	June–Aug.	4–9	F
D, M	*Euphorbia corollata*	Flowering Spurge	2′–4′	White	July–Aug.	3–9	F
D, M	*Lespedeza capitata*	Roundhead Bush Clover	3′–5′	White	Aug.–Sept.	3–8	F, P
D, M	*Onosmodium bejariense* var. *occidentale*	Marbleseed (Western Marbleseed)	1′–3′	White	May–June	3–8	F
D, M	*Pycnanthemum tenuifolium*	Narrowleaf Mountainmint	1′–3′	White	July–Sept.	4–8	F, P
D, M	*Sisyrinchium albidum*	White Blue-Eyed Grass	6″–1′	White	Apr.–May	4–9	F
D, M	*Symphyotrichum ericoides*	Heath Aster	1′–3′	White	Aug.–Oct.	3–9	F
D, M, Mo	*Arnoglossum atriplicifolium*	Pale Indian Plantain	5′–10′	White	July–Sept.	4–8	F, P
D, M, Mo	*Desmanthus illinoensis*	Illinois Bundleflower	2′–6′	White	June–Aug.	5–8	F
D, M, Mo	*Oenothera gaura*	Biennial Beeblossom	3′–6′	White/pink	June–Sept.	4–8	F, P
M, Mo	*Allium cernuum*	Nodding Pink Onion	1′–2′	White/pink	June–Aug.	3–8	F, P
M, Mo	*Anemone canadensis*	Canada Anemone	1′–2′	White	May–July	3–6	F, P
M, Mo	*Baptisia alba*	White False Indigo	3′–5′	White	June–July	4–9	F, P
M, Mo	*Dodecatheon meadia*	Shooting Star	1′–2′	White/pink	May–June	4–8	F, P
M	*Gentiana alba*	Cream Gentian	1′–2′	White	July–Aug.	3–7	F, P
M, Mo	*Parthenium integrifolium*	Wild Quinine	3′–5′	White	June–Sept.	4–8	F
M, Mo	*Penstemon digitalis*	Foxglove Beardtongue	2′–3′	White	June–July	3–8	F, P
M, Mo	*Veronicastrum virginicum*	Culver's Root	3′–6′	White	July–Aug.	3–8	F, P, S
M, Mo, W	*Eupatorium perfoliatum*	Boneset	3′–4′	White	Aug.–Sept.	3–8	F
M, Mo, W	*Hibiscus moscheutos*	Rosemallow	3′–7′	White/pink/red	July–Sept.	5–9	F, P
Mo, W	*Chelone glabra*	White Turtlehead	2′–4′	White	Aug.–Sept.	3–8	F, P
Mo, W	*Pycnanthemum virginianum*	Virginia Mountainmint	1′–3′	White	July–Sept.	3–7	F, P
Mo, W	*Spiraea alba*	Meadowsweet	2′–4′	White/pink	June–Sept.	3–7	F, P
Mo, W	*Thalictrum dasycarpum*	Tall Meadow-Rue	3′–6′	White	June–July	3–9	F, P

Notes: For soil moisture, D = dry, M = medium, Mo = moist, and W = wet. For soil texture, S = sand, L = loam, C = clay, and G = gravel. For sun, F = full sun, or at least a half day of full sun; P = partial sun, or a quarter to a half day of sun; and S = shade, or less than a quarter day of sun. For plant spacing, NR = not recommended. For root type, B = bulb, CO = corm, FIB = fibrous, R = rhizome, TAP =

Soil texture	pH range	Plant spacing in prairie gardens	Plant spacing in flower beds	Root type	Plant life expectancy in years	Aggressiveness and method	Cut flowers	Dried arrangements	Deer palatability
S, G	5.5–7.5	1'	1'	R	5–10	L–R	x	FL	L
S, G	?–8.0	6"–1'	1'	FIB	5–10	L–S		SD	H
S, G	6.5–8.0	1'	2'	FIB	5–10	L–S		Empty SH	L
S, L	?	1'	1'	FIB	10–20	L–S		SD	L
L, C, G	?–8.00	1'	1'–2'	TAP/short Rs	3–5	L–S		FL/SH	L
S, L	4.5–7.0	3'	3'	TAP	20+	L–S		SD	L
S, L	?	6"–1'	1'–2'	TAP	10–20	L–S			H
S, L	?	1'	1.5'	FIB	5–10	L–S	x	SP	L
S, L, C	6.0–8.0	1'	2'	FIB	5–10	M–S	x	SH	L
S, L	?	1'–2'	1'–2'	R	10–20	M–R			L
S, L, G	5.5–8.0	1'	2'	TAP	10–20	L–S		SH	M
S, L	?–7.5	1'–1.5'	2'	TAP	10–20	L–S			L
S, L, C, G	?	1'	1.5'	TAP/R	5–10	L–S		SH	M
S, L	?–8.0	6"	6"	FIB	3–5	L–S			L
S, G, L	?	1'–2'	2'	R	10–20	M–R		Empty SH	L
S, L, C	?–8.0	2'	3'–4'	FIB	3–20	H–S (north); L–S (south)	x	FL	L
S, L, C	6.0–8.0	1'	1.5'	TAP	5–10	L–S		SP	M
S, L, C	?	2'	3'	FIB	2	M–S	x		L
S, L, C	6.5–8.0	6"–1'	1'	B	10–20	L–S	x	FL/SH	L
S, L, C	?	NR	NR	R	10–20	H–R			L
S, L, C	?–7.5	3'	3'	TAP	20+	L–S	x	SP	L
S, L, C	4.5–8.0	6"–1'	1'	FIB	20+	L–S		SP	L
S, L, C	?–7.5	1'	1.5'	FIB	10–20	L–S	x		L
S, L, C	?–8.0	1'–2'	3'	TAP	10–20	L–S	x		M
S, L, C	5.5–7.5	1'	2'	FIB	5–10	L–S	x		L
S, L, C	?–8.0	1'	3'	FIB	10–20	L–S	x	SH	L
S, L, C	?–7.5	1'	NR	FIB	5–10	L–S			L
S, L, C	4.0–8.0	2'	4'	R	10–20	L–R/S			L
S, L, C	?	1'	1'–2'	FIB	5–10	L–S			M
S, L, C	?–8.0	1'–2'	3'	R	5–10	L–R		SH	L
S, L, C	4.3–7.2	2'	3'	FIB	10–20	L–S	x		L
S, L, C	?	1'–2'	2'	FIB	5–10	L–S			L

taproot, and TAPCO = taprooted corm. For aggressiveness and method, L = low, M = medium, H = high, R = by rhizome, S = by seed, and R/S = rhizome and seed. For dried arrangements, FL = flowers, SH = seedheads, SP = seedpods, SD = seed, and RH = rose hips. For deer palatability, L = low, M = medium, and H = high.

TABLE 12.9. YELLOW FLOWERS, BY SOIL MOISTURE

Soil moisture	Scientific name (4/1/2020)	Common name	Height	Flower color	Bloom time	USDA zones	Sun
D	Lithospermum caroliniense	Hairy Puccoon	1'–2'	Yellow	June–July	3–8	F
D	Lithospermum incisum	Fringed Puccoon	1'–2'	Yellow	June–July	3–9	F
D, M	Astragalus canadensis	Canada Milkvetch	2'–3'	Yellow	July–Aug.	3–8	F
D, M	Coreopsis lanceolata	Lanceleaf Coreopsis	1'–2'	Yellow	June–July	3–9	F
D, M	Coreopsis palmata	Stiff Coreopsis	2'–3'	Yellow	June–Aug.	3–8	F
D, M	Echinacea paradoxa	Ozark Coneflower	3'–5'	Yellow	June–July	4–7	F, P
D, M	Helianthus mollis	Downy Sunflower	4'–6'	Yellow	Aug.–Sept.	4–9	F
D, M	Helianthus occidentalis	Western Sunflower	2'–3'	Yellow	July–Aug.	3–9	F
D, M	Helianthus ×laetiflorus	Showy Sunflower	3'–6'	Yellow	Aug.–Sept.	3–8	F
D, M	Silphium laciniatum	Compassplant	3'–10'	Yellow	June–Sept.	4–9	F
D, M	Sisyrinchium albidum	White Blue-Eyed Grass	6"–1'	White	Apr.–May	4–9	F
D, M	Solidago speciosa	Showy Goldenrod	1'–3'	Yellow	Aug.–Sept.	3–8	F
D, M, Mo	Heliopsis helianthoides	Oxeye Sunflower	3'–6'	Yellow	June–Sept.	3–8	F
D, M, Mo	Oligoneuron rigidum	Stiff Goldenrod	3'–5'	Yellow	Aug.–Sept.	3–9	F
D, M, Mo	Ratibida pinnata	Yellow Coneflower	3'–6'	Yellow	July–Sept.	3–9	F
D, M, Mo	Rudbeckia hirta	Blackeyed Susan	1'–3'	Yellow	June–Sept.	3–10	F, P
M, Mo	Oligoneuron ohioense	Ohio Goldenrod	3'–4'	Yellow	Aug.–Sept.	4–5	F, P
M, Mo	Rudbeckia fulgida	Orange Coneflower	2'–4'	Yellow	July–Sept.	4–8	F, P
M, Mo	Rudbeckia subtomentosa	Sweet Blackeyed Susan	4'–6'	Yellow	Aug.–Oct.	4–8	F, P
M, Mo	Rudbeckia triloba	Browneyed Susan	2'–5'	Yellow	July–Oct.	4–9	F, P
M, Mo	Silphium integrifolium	Rosinweed	2'–6'	Yellow	July–Sept.	4–8	F
M, Mo	Silphium terebinthinaceum	Prairie Dock	3'–10'	Yellow	July–Sept.	4–7	F
M, Mo	Zizia aptera	Heartleaf Golden Alexanders	1'–2'	Yellow	May–June	3–8	F
M, Mo, W	Coreopsis tripteris	Tall Coreopsis	3'–7'	Yellow	July–Sept.	3–8	F, P
M, Mo, W	Senna hebecarpa	Wild Senna	4'–6'	Yellow	July–Aug.	4–7	F
M, Mo, W	Silphium perfoliatum	Cup Plant	3'–10'	Yellow	July–Sept.	3–8	F, P
M, Mo, W	Zizia aurea	Golden Alexanders	1'–2'	Yellow	May–July	3–8	F, P, S
Mo, W	Helenium autumnale	Sneezeweed	4'–5'	Yellow	Aug.–Sept.	3–9	F
Mo, W	Hypericum ascyron	Great St. Johnswort	4'–6'	Yellow	July–Aug.	3–5	F, P
Mo, W	Rudbeckia laciniata	Green-Headed Coneflower	5'–8'	Yellow	Aug.–Sept.	3–8	F, P

Notes: For soil moisture, D = dry, M = medium, Mo = moist, and W = wet. For soil texture, S = sand, L = loam, C = clay, and G = gravel. For sun, F = full sun, or at least a half day of full sun; P = partial sun, or a quarter to a half day of sun; and S = shade, or less than a quarter day of sun. For plant spacing, NR = not recommended. For root type, B = bulb, CO = corm, FIB = fibrous, R = rhizome, TAP =

Soil texture	pH range	Plant spacing in prairie gardens	Plant spacing in flower beds	Root type	Plant life expectancy in years	Aggressiveness and method	Cut flowers	Dried arrangements	Deer palatability
S	?	1'	1'	TAP	10-20	L-S			L
S, G	?	1'	1'	TAP	5-10	L-S			L
S, L	6.0-8.0	1'-2'	3'	TAP	10-20	M-S		SP	M
S, L	5.0-7.5	1'	1'-2'	FIB	3-5	M-S	x		M
S, L	?	1'-2'	2'	R	20+	M-R			L
S, L, C	?-7.5	1'	2'-3'	TAP	10-20	L-S	x	SH	M
S, L, C	6.0-8.0	2'	3'	R	20+	M-R	x	SH	L
S, L	?	2'	3'	R	20+	M-R	x		L
S, G, L, C	?	NR	NR	R	20+	H-R	x	SH	L
S, L, C	?-8.0	2'	4'	TAP	20+	L-S	x		L
S, L	?-8.0	6"	6"	FIB	?	L-S			?
S, L	?	1'	3'	FIB	5-10	L-S	x	SH	L
S, L, C	6.0-8.0	1'-2'	3'	FIB	5-10	H-S	x		L
S, L, C	5.5-8.0	1'-1.5'	3'	FIB	5-10	M-S	x	SH	L
S, L, C	5.5-8.0	1'-2'	2'	FIB	10-20	M-S	x	SH	L
S, L, C	5.0-8.0	1'	2'	FIB	2	M-S	x	SH	L
S, L, C	?-8.0	1'-2'	3'	FIB	5-10	L-S	x	SH	L
L, C	?-8.0	1'-2'	3'	FIB	5-10	L-S	x	SH	L
S, L, C	?-8.0	1'-2'	2'	FIB	10-20	L-S	x	SH	L
S, L	?-8.0	1'-2'	3'	FIB	2	M-S	x	SH	L
S, L, C	?-8.0	2'	3'	TAP	20+	M-S	x		L
S, L, C	?-8.0	2'	4'	TAP	20+	L-S	x		L
S, L	?-7.5	1'	2'	FIB	5-10	L-S			L
S, L, C	?	1'-2'	NR	FIB	10-20	L-S	x		L
S, L, C	?	1.5'-2'	3'-4'	FIB/TAP	10-20	L-S	x	FL/SH	M
S, L, C	4.5-8.0	2'-3'	NR	FIB	20+	H-S	x		L
S, L, C	?-8.0	1'	3'	FIB	5-10	L-S			L
S, L, C	4.0-7.5	1'-2'	2'-3'	FIB	5-10	L-S	x	SH	L
S, L, C	5.7-7.1	1'-2'	3'	FIB	5-10	L-S	x	SP	L
S, L, C	4.5-7.5	2'	2'	FIB	3-5	M-S	x	SH	L

taproot, and TAPCO = taprooted corm. For aggressiveness and method, L = low, M = medium, H = high, R = by rhizome, S = by seed, and R/S = rhizome and seed. For dried arrangements, FL = flowers, SH = seedheads, SP = seedpods, SD = seed, and RH = rose hips. For deer palatability, L = low, M = medium, and H = high.

TABLE 12.10. BLUE FLOWERS, BY SOIL MOISTURE

Soil moisture	Scientific name (4/1/2020)	Common name	Height	Flower color	Bloom time	USDA zones	Sun
D	*Lupinus perennis*	Wild Lupine	1'–2'	Blue	May–June	3–8	F, P
D	*Pediomelum esculentum*	Prairie Turnip	1'–2'	Blue	June–July	3–7	F
D	*Viola pedata*	Birdfoot Violet	6"	Blue/purple	Apr.–June	3–9	F, P
D, M	*Salvia azurea*	Blue Sage	3'–5'	Blue	June–Aug.	4–9	F, P
D, M	*Sisyrinchium campestre*	Prairie Blue-Eyed Grass	6"–1'	Blue/white	Apr.–June	3–8	F, P
D, M	*Symphyotrichum laeve*	Smooth Aster	2'–4'	Blue	Aug.–Oct.	4–8	F
D, M	*Symphyotrichum oolentangiense*	Skyblue Aster	2'–3'	Blue	Aug.–Oct.	3–8	F, P
D, M	*Tradescantia bracteata*	Prairie Spiderwort	1'–2'	Blue	June–July	3–7	F, P
D, M	*Verbena stricta*	Hoary Vervain	2'–4'	Blue	July–Sept.	3–8	F
D, M, Mo	*Tradescantia ohiensis*	Ohio Spiderwort	2'–4'	Blue	June–July	3–9	F, P
M	*Baptisia australis*	Blue False Indigo	3'–5'	Blue	June–July	3–10	F, P
M, Mo	*Amsonia tabernaemontana*	Eastern Bluestar	2'–3'	Blue/white	May–June	4–8	F, P
M, Mo	*Conoclinium coelestinum*	Blue Mistflower	1'–3'	Blue/purple	July–Oct.	4–10	F, P
M, Mo	*Lobelia siphilitica*	Great Blue Lobelia	1'–4'	Blue	July–Sept.	3–9	F, P
M, Mo	*Sisyrinchium angustifolium*	Narrow-Leaved Blue-Eyed Grass	6"–2'	Blue/purple	Apr.–June	3–10	F, P
M, Mo	*Symphyotrichum novae-angliae*	New England Aster	3'–6'	Pink/purple/blue	Aug.–Oct.	3–7	F, P
Mo, W	*Gentiana andrewsii*	Bottle Gentian	1'–2'	Blue	Aug.–Oct.	3–6	F, P
M, Mo, W	*Verbena hastata*	Blue Vervain	3'–6'	Blue	July–Sept.	3–8	F, P
W	*Iris virginica* var. *shrevei*	Wild Iris	2'–3'	Blue	June–July	3–8	F, P

Notes: For soil moisture, D = dry, M = medium, Mo = moist, and W = wet. For soil texture, S = sand, L = loam, C = clay, and G = gravel. For sun, F = full sun, or at least a half day of full sun; P = partial sun, or a quarter to a half day of sun; and S = shade, or less than a quarter day of sun. For plant spacing, NR = not recommended. For root type, B = bulb, CO = corm, FIB = fibrous, R = rhizome, TAP =

Soil texture	pH range	Plant spacing in prairie gardens	Plant spacing in flower beds	Root type	Plant life expectancy in years	Aggressiveness and method	Cut flowers	Dried arrangements	Deer palatability
S	5.0-?	1'	1.5'	TAP	10-20	M-S			L
S, G	?-8.5	1'-1.5'	2'	TAP	20+	L-S			L
S, G	?	6"	1'	R	5-10	L-S			L
S, G, L	?	1'	1.5'	FIB	5-10	L-S	x	SH	L
S, L	?-7.5	6"	6"	FIB	3-5	L-S			L
S, L	?	1'	3'	FIB	5-10	M-S	x	Empty SH	H
S, L	5.0-8.0	1'	2'	FIB	5-10	L-S	x	Empty SH	M
S, L, C	?	1'	2'	R	20+	M-R			M
S, L	?	1'	2'	FIB	3-5	M-S	x	SH	L
S, L, C	?-7.5	1'	2'	FIB	20+	M-S			H
S, L, C	?-8.0	3'	3'	TAP	20+	L-S	x	SP	L
S, L, C	?-8.0	2'-3'	3'	FIB	20+	L-S		SP	L
S, L, C	5.0-8.0	1'-2'	2'	R	3-5	M-R			L
S, L, C	?-8.0	6"-1'	2'	FIB	5-10	M-S	x		L
S, L, C	?-8.0	6"	9"	FIB	3-5	L-S			L
S, L, C	5.0-8.0	1'	3'	FIB	5-10	L-S	x	Empty SH	H
S, L, C	5.8-7.2	1'	1.5'	FIB	5-10	L-S	x		L
S, L, C	?	1'	2'	FIB	2	M-S	x	SH	L
S, L, C	4.8-7.3	1'	2'	R/CO	10-20	L-R	x	SP	L

taproot, and TAPCO = taprooted corm. For aggressiveness and method, L = low, M = medium, H = high, R = by rhizome, S = by seed, and R/S = rhizome and seed. For dried arrangements, FL = flowers, SH = seedheads, SP = seedpods, SD = seed, and RH = rose hips. For deer palatability, L = low, M = medium, and H = high.

TABLE 12.11.   PURPLE AND LAVENDER FLOWERS, BY SOIL MOISTURE

Soil moisture	Scientific name (4/1/2020)	Common name	Height	Flower color	Bloom time	USDA zones	Sun
D	Liatris punctata	Dotted Blazing Star	1'–2'	Purple/pink	Aug.–Oct.	3–9	F
D	Monarda punctata	Dotted Mint	1'–2'	Lavender	July–Sept.	3–10	F
D	Penstemon cobaea	Cobaea Beardtongue	2'–3'	Lavender	Apr.–June	5–8	F
D	Penstemon gracilis	Slender Beardtongue	1'–2'	Lavender	May–June	3–6	F, P
D	Penstemon grandiflorus	Large-Flowered Beardtongue	2'–4'	Lavender	May–June	3–7	F
D	Penstemon hirsutus	Hairy Penstemon	1'–2'	Lavender	May–June	3–6	F, P
D	Viola pedata	Birdfoot Violet	6"	Blue/purple	Apr.–June	3–9	F, P
D, M	Agastache foeniculum	Lavender Hyssop	1'–3'	Purple	July–Sept.	2–6	F, P
D, M	Allium stellatum	Prairie Onion	1'–2'	Lavender	July–Aug.	3–8	F, P
D, M	Amorpha canescens	Leadplant	2'–3'	Purple	June–July	3–8	F, P
D, M	Asclepias syriaca	Common Milkweed	2'–4'	Lavender	June–Aug.	3–8	F, P
D, M	Blephilia ciliata	Downy Wood Mint	1'–2'	White/purple/violet	May–July	4–8	F, P
D, M	Callirhoe involucrata	Winecups	1'	Magenta	May–June	4–9	F, P
D, M	Dalea purpurea	Purple Prairie Clover	1'–2'	Purple/yellow	July–Aug.	3–8	F
D, M	Echinacea pallida	Pale Purple Coneflower	3'–5'	Purple	June–July	4–8	F
D, M	Echinacea purpurea	Purple Coneflower	3'–4'	Purple	July–Sept.	4–8	F, P
D, M	Liatris aspera	Rough Blazing Star	2'–5'	Purple/pink	Aug.–Sept.	3–8	F
D, M	Liatris squarrosa	Scaly Blazing Star	1'–2'	Purple/pink	Aug.–Sept.	4–8	F
D, M	Ruellia humilis	Wild Petunia	1'–2'	Violet	June–Aug.	4–9	F
D, M	Symphyotrichum oblongifolium	Aromatic Aster	1'–2'	Purple/violet	Sept.–Oct.	3–8	F
D, M, Mo	Monarda fistulosa	Bergamot	2'–5'	Lavender	July–Sept.	3–9	F, P
M, Mo	Conoclinium coelestinum	Blue Mistflower	1'–3'	Blue/purple	July–Oct.	4–10	F, P
M, Mo	Desmodium canadense	Canada Ticktrefoil	3'–5'	Purple	July–Aug.	3–7	F
M, Mo	Liatris ligulistylis	Meadow Blazing Star	3'–5'	Purple/pink	Aug.–Sept.	3–6	F
M, Mo	Liatris pycnostachya	Prairie Blazing Star	3'–5'	Purple/pink	July–Aug.	3–9	F
M, Mo	Liatris spicata	Dense Blazingstar	3'–6'	Purple/pink	Aug.–Sept.	4–10	F
M, Mo	Sisyrinchium angustifolium	Narrow-Leaved Blue-Eyed Grass	6"–2'	Blue/purple	Apr.–June	3–10	F, P
M, Mo	Symphyotrichum novae-angliae	New England Aster	3'–6'	Pink/purple/blue	Aug.–Oct.	3–7	F, P
M, Mo, W	Eutrochium fistulosum	Tall Joe Pye Weed	5'–8'	Purple/pink	Aug.–Sept.	4–9	F, P

Notes: For soil moisture, D = dry, M = medium, Mo = moist, and W = wet. For soil texture, S = sand, L = loam, C = clay, and G = gravel. For sun, F = full sun, or at least a half day of full sun; P = partial sun, or a quarter to a half day of sun; and S = shade, or less than a quarter day of sun. For plant spacing, NR = not recommended. For root type, B = bulb, CO = corm, FIB = fibrous, R = rhizome, TAP =

Soil texture	pH range	Plant spacing in prairie gardens	Plant spacing in flower beds	Root type	Plant life expectancy in years	Aggressiveness and method	Cut flowers	Dried arrangements	Deer palatability
S, L	6.0–8.0	1'	1.5'–2'	TAPCO	10–20	L–S			L
S, G	5.0–?	1'	1'–2'	FIB	2	M–S	x	SH	L
S, G	?–7.5	1'	1'	FIB	3–5	L–S	x	SP	L
S, G	?–7.5	6"–1'	1'	FIB	5–10	L–S	x	SP	L
S, G	?	1'	2'–3'	FIB	3–5	L–S		SP	L
S, G	5.0–8.0	1'	1.5'	FIB	3–5	L–S	x	SP	L
S, G	?	6"	1'	R	5–10	L–S			L
S, L	?–8.0	1'	1'–2'	FIB	3–5	M–S	x	FL/SH	L
S, L	5.5–8.0	6"–1'	1'	B	10–20	L–S	x	FL/SH	L
S, L	5.5–8.0	2'	3'	TAP	20+	L–S		Leaves	L
S, L, C	5.0–8.0	NR	NR	R	10–20	H–R		SP	L
G, L, C	?–8.00	1'	1'–2'	TAP/short Rs	3–5	L–S		FL/SH	L
S, L	5.5–7.5	1'–2'	2'	TAP	10–20	L–S			L
S, L, C	?	6"–1'	1'–2'	TAP	10–20	L–S	x		H
S, L, C	6.0–7.5	1'	2'–3'	TAP	10–20	L–S	x	SH	M
S, L, C	5.5–8.0	1'–2'	3'	FIB	5–10	M–S	x	SH	M
S, L	?	6"–1'	1'	CO	10–20	L–S	x	FL/SH	L
S, L	?	1'	1.5'–2'	CO	10–20	L–S	x	FL/SH	L
S, L	4.5–7.5	1'	2'	FIB	5–10	L–S			L
S, L, C	?–8.0	1'	3'	R	5–10	M–R?		Empty SH	L
S, L, C	6.0–8.0	1'–2'	3'	R	10–20	M–R	x	SH	L
S, L, C	5.0–8.0	1'–2'	2'	R	3–5	M–R			L
S, L, C	?–7.5	NR	NR	TAP	10–20	M–S			H
L	?	6"–1'	1'	CO	3–5	L–S	x	FL/SH	L
S, L, C	6.0–8.0	6"–1'	1'	CO	10–20	L–S	x	FL/SH	L
S, L, C	5.5–7.5	6"–1'	1'	CO	10–20	L–S	x	FL/SH	L
S, L, C	?–8.0	6"	9"	FIB	3–5	L–S			L
S, L, C	5.0–8.0	1'	3'	FIB	5–10	L–S	x	Empty SH	H
S, L, C	4.5–7.5	2'	3'–4'	FIB	5–10	L–S			H

taproot, and TAPCO = taprooted corm. For aggressiveness and method, L = low, M = medium, H = high, R = by rhizome, S = by seed, and R/S = rhizome and seed. For dried arrangements, FL = flowers, SH = seedheads, SP = seedpods, SD = seed, and RH = rose hips. For deer palatability, L = low, M = medium, and H = high.

TABLE 12.12. RED, PINK, AND ORANGE FLOWERS, BY SOIL MOISTURE

Soil moisture	Scientific name (4/1/2020)	Common name	Height	Flower color	Bloom time	USDA zones	Sun
D	Callirhoe triangulata	Poppymallow	1'–2'	Magenta	July–Aug.	4–8	F
D	Liatris punctata	Dotted Blazing Star	1'–2'	Purple/pink	Aug.–Oct.	3–9	F
D	Tephrosia virginiana	Goatsrue	1'–2'	Pink/yellow	June–Aug.	4–9	F, P
D, M	Asclepias tuberosa	Butterflyweed	2'–3'	Orange	June–Aug.	3–10	F
D, M	Callirhoe involucrata	Winecups	1'	Magenta	May–June	4–9	F, P
D, M	Geum triflorum	Prairie Smoke	6"	Pink	May–June	3–6	F
D, M	Liatris aspera	Rough Blazing Star	2'–5'	Purple/pink	Aug.–Sept.	3–8	F
D, M	Liatris squarrosa	Scaly Blazing Star	1'–2'	Purple/pink	Aug.–Sept.	4–8	F
D, M	Lithospermum canescens	Hoary Puccoon	1'	Orange	May–June	3–7	F, P
D, M	Rosa carolina	Carolina Rose	1'–2'	Pink	June–July	3–8	F
D, M	Tradescantia occidentalis	Western Spiderwort	1'	Pink	June–July	3–9	F
D, M, Mo	Oenothera gaura	Biennial Beeblossom	3'–6'	White/pink	June–Sept.	4–8	F, P
D, M, Mo	Phlox pilosa	Downy Phlox	1'–2'	Pink	May–June	3–9	F
M	Silene regia	Royal Catchfly	2'–4'	Red	July–Aug.	4–9	F, P
M, Mo	Allium cernuum	Nodding Pink Onion	1'–2'	White/pink	June–Aug.	3–8	F, P
M, Mo	Asclepias sullivantii	Sullivant's Milkweed	3'–5'	Pink/yellow	June–Aug.	4–7	F
M, Mo	Dodecatheon meadia	Shooting Star	1'–2'	White/pink	May–June	4–8	F, P
M, Mo	Filipendula rubra	Queen of the Prairie	4'–5'	Pink	June–July	3–6	F
M, Mo	Liatris ligulistylis	Meadow Blazing Star	3'–5'	Purple/pink	Aug.–Sept.	3–6	F
M, Mo	Liatris pycnostachya	Prairie Blazing Star	3'–5'	Purple/pink	July–Aug.	3–9	F
M, Mo	Liatris spicata	Dense Blazing Star	3'–6'	Purple/pink	Aug.–Sept.	4–10	F
M, Mo	Physostegia virginiana	Obedient Plant	1'–2'	Pink	Aug.–Sept.	3–9	F
M, Mo	Rosa setigera	Climbing Prairie Rose	6'–15'	Pink/white	May–July	4–9	F, P
M, Mo	Symphyotrichum novae-angliae	New England Aster	3'–6'	Pink/purple/blue	Aug.–Oct.	3–7	F, P
M, Mo, W	Eutrochium fistulosum	Tall Joe Pye Weed	5'–8'	Purple/pink	Aug.–Sept.	4–9	F, P
M, Mo, W	Hibiscus moscheutos	Rosemallow	3'–7'	White/pink/red	July–Sept.	5–9	F, P
Mo	Lilium michiganense	Michigan Lily	4'–6'	Orange	July–Aug.	3–7	F, P
Mo	Vernonia fasciculata	Ironweed	4'–6'	Red/pink	July–Sept.	3–7	F
Mo, W	Asclepias incarnata	Marsh Milkweed	3'–5'	Red/pink	June–July	3–9	F
Mo, W	Eutrochium maculatum	Joe Pye Weed	4'–6'	Pink	Aug.–Sept.	3–6	F, P
Mo, W	Lobelia cardinalis	Cardinal Flower	2'–5'	Red	July–Sept.	3–9	F, P
Mo, W	Spiraea alba	Meadowsweet	2'–4'	White/pink	June–Sept.	3–7	F, P
Mo, W	Spiraea tomentosa	Steeplebush	2'–4'	Pink	Aug.–Sept.	3–8	F, P
Mo, W	Vernonia gigantea	Tall Ironweed	5'–8'	Red/pink	Aug.–Sept.	4–9	F, P

Notes: For soil moisture, D = dry, M = medium, Mo = moist, and W = wet. For soil texture, S = sand, L = loam, C = clay, and G = gravel. For sun, F = full sun, or at least a half day of full sun; P = partial sun, or a quarter to a half day of sun; and S = shade, or less than a quarter day of sun. For plant spacing, NR = not recommended. For root type, B = bulb, CO = corm, FIB = fibrous, R = rhizome, TAP =

Soil texture	pH range	Plant spacing in prairie gardens	Plant spacing in flower beds	Root type	Plant life expectancy in years	Aggressiveness and method	Cut flowers	Dried arrangements	Deer palatability
S	?	1'–2'	2'	TAP	10–20	L–S			L
S, L	6.0–8.0	1'	1.5'–2'	TAPCO	10–20	L–S			L
S, G	4.5–?	2'	3'	TAP	10–20	L–S			L
S, L	5.0–7.5	2'	3'	TAP	10–20	L–S	x	SP	L
S, L	5.5–7.5	1'–2'	2'	TAP	10–20	L–S			L
S, L	6.0–7.5	6"	1'	R	20+	L–R		SH	L
S, L	?	6"–1'	1'	CO	10–20	L–S	x	FL/SH	L
S, L	?	1'	1.5'–2'	CO	10–20	L–S	x	FL/SH	L
S, L	?	1'	1'	FIB	5–10	L–S			L
S, L	4.0–7.0	2'	3'	R	10–20	L–R/S	x	RH	L
S, L	?	1'	NR	R	20+	H–R			M
S, L, C	?	2'	3'	FIB	2	M–S	x		L
S, L	?	6"–1'	1.5'	FIB	3–5	L–S	x		L
L	?	1'	1.5'	TAP	5–10	L–S	x	SH	L
S, L, C	6.5–8.0	6"–1'	1'	B	10–20	L–S	x	FL/SH	L
L, C	?	1'–2'	3'	R	10–20	L–R		SP	L
S, L, C	4.5–8.0	6"–1'	1'	FIB	20+	L–S		SP	L
S, L, C	?–8.0	2'	3'	R	20+	M–R	x		L
L	?	6"–1'	1'	CO	3–5	L–S	x	FL/SH	L
S, L, C	6.0–8.0	6"–1'	1'	CO	10–20	L–S	x	FL/SH	L
S, L, C	5.5–7.5	6"–1'	1'	CO	10–20	L–S	x	FL/SH	L
S, L, C	6.0–8.0	1'–2'	NR	R	10–20	H–R	x	SH	M
S, L, C	5.0–8.0	2'	5'	FIB	10–20	L–S	x	RH	L
S, L, C	5.0–8.0	1'	3'	FIB	5–10	L–S	x	Empty SH	H
S, L, C	4.5–7.5	2'	3'–4'	FIB	5–10	L–S			H
S, L, C	4.0–8.0	2'	4'	R	10–20	L–R/S			L
S, L, C	?–6.8	2'	2'	B	20+	L–S	x	SP	H
S, L, C	?	1'	3'	FIB	5–10	L–S	x		L
S, L, C	5.0–8.0	1'–2'	3'	FIB	5–10	L–S	x	SP	L
S, L, C	5.5–7.5	1'–2'	3'–4'	FIB	5–10	L–S			L
S, L	5.8–7.8	6"–1'	2'	FIB	3–5	L–S	x		M
S, L, C	4.3–7.2	2'	3'	FIB	10–20	L–S	x		L
S, L, C	4.5–7.0	2'	3'	R	10–20	L–R/S	x		L
S, L, C	?–8.0	1'–2'	3'	FIB	10–20	L–S	x		L

taproot, and TAPCO = taprooted corm. For aggressiveness and method, L = low, M = medium, H = high, R = by rhizome, S = by seed, and R/S = rhizome and seed. For dried arrangements, FL = flowers, SH = seedheads, SP = seedpods, SD = seed, and RH = rose hips. For deer palatability, L = low, M = medium, and H = high.

TABLE 12.13. PLANTS LISTED BY HEIGHT

Height	Scientific name (4/1/2020)	Common name	Flower color	Bloom time	USDA zones	Sun	Soil moisture
Flowers and shrubs							
6″	*Geum triflorum*	Prairie Smoke	Pink	May–June	3–6	F	D, M
6″	*Viola pedata*	Birdfoot Violet	Blue/purple	Apr.–June	3–9	F, P	D
6″–1′	*Antennaria neglecta*	Pussytoes	White	Apr.–May	3–7	F	D
6″–1′	*Pulsatilla patens*	Pasque Flower	White	Apr.–May	3–6	F	D
6″–1′	*Sisyrinchium albidum*	White Blue-Eyed Grass	White	Apr.–May	4–9	F	D, M
6″–1′	*Sisyrinchium campestre*	Prairie Blue-Eyed Grass	Blue/white	Apr.–June	3–8	F, P	D, M
6″–2′	*Sisyrinchium angustifolium*	Narrow-Leaved Blue-Eyed Grass	Blue/purple	Apr.–June	3–10	F, P	M, Mo
1′	*Callirhoe involucrata*	Winecups	Magenta	May–June	4–9	F, P	D, M
1′	*Lithospermum canescens*	Hoary Puccoon	Orange	May–June	3–7	F, P	D, M
1′	*Tradescantia occidentalis*	Western Spiderwort	Pink	June–July	3–9	F	D, M
1′–2′	*Allium cernuum*	Nodding Pink Onion	White/pink	June–Aug.	3–8	F, P	M, Mo
1′–2′	*Allium stellatum*	Prairie Onion	Lavender	July–Aug.	3–8	F, P	D, M
1′–2′	*Anemone canadensis*	Canada Anemone	White	May–July	3–6	F, P	M, Mo
1′–2′	*Baptisia bracteata*	Cream False Indigo	Cream	May–June	4–9	F, P	D, M
1′–2′	*Blephilia ciliata*	Downy Wood Mint	White/purple/violet	May–July	4–8	F, P	D, M
1′–2′	*Callirhoe triangulata*	Poppymallow	Magenta	July–Aug.	4–8	F	D
1′–2′	*Coreopsis lanceolata*	Lanceleaf Coreopsis	Yellow	June–July	3–9	F	D, M
1′–2′	*Dalea candida*	White Prairie Clover	White	July–Aug.	3–9	F	D, M
1′–2′	*Dalea purpurea*	Purple Prairie Clover	Purple/yellow	July–Aug.	3–8	F	D, M
1′–2′	*Dodecatheon meadia*	Shooting Star	White/pink	May–June	4–8	F, P	M, Mo
1′–2′	*Gentiana alba*	Cream Gentian	White	July–Aug.	3–7	F, P	M
1′–2′	*Gentiana andrewsii*	Bottle Gentian	Blue	Aug.–Oct.	3–6	F, P	Mo, W
1′–2′	*Liatris punctata*	Dotted Blazing Star	Purple/pink	Aug.–Oct.	3–9	F	D
1′–2′	*Liatris squarrosa*	Scaly Blazing Star	Purple/pink	Aug.–Sept.	4–8	F	D, M
1′–2′	*Lithospermum caroliniense*	Hairy Puccoon	Yellow	June–July	3–8	F	D
1′–2′	*Lithospermum incisum*	Fringed Puccoon	Yellow	June–July	3–9	F	D
1′–2′	*Lupinus perennis*	Wild Lupine	Blue	May–June	3–8	F, P	D
1′–2′	*Monarda punctata*	Dotted Mint	Lavender	July–Sept.	3–10	F	D
1′–2′	*Pediomelum esculentum*	Prairie Turnip	Blue	June–July	3–7	F	D
1′–2′	*Penstemon gracilis*	Slender Beardtongue	Lavender	May–June	3–6	F, P	D
1′–2′	*Penstemon hirsutus*	Hairy Penstemon	Lavender	May–June	3–6	F, P	D
1′–2′	*Phlox pilosa*	Downy Phlox	Pink	May–June	3–9	F	D, M, Mo
1′–2′	*Physostegia virginiana*	Obedient Plant	Pink	Aug.–Sept.	3–9	F	M, Mo
1′–2′	*Rosa carolina*	Carolina Rose	Pink	June–July	3–8	F	D, M
1′–2′	*Ruellia humilis*	Wild Petunia	Violet	June–Aug.	4–9	F	D, M
1′–2′	*Solidago ptarmicoides*	White Aster	White	Aug.–Sept.	3–7	F	D
1′–2′	*Symphyotrichum oblongifolium*	Aromatic Aster	Purple/violet	Sept.–Oct.	3–8	F	D, M
1′–2′	*Tephrosia virginiana*	Goatsrue	Pink/yellow	June–Aug.	4–9	F, P	D
1′–2′	*Tradescantia bracteata*	Prairie Spiderwort	Blue	June–July	3–7	F, P	D, M

Soil texture	pH range	Plant spacing in prairie gardens	Plant spacing in flower beds	Root type	Plant life expectancy in years	Aggressiveness and method	Cut flowers	Dried arrangements	Deer palatability
S, L	6.0–7.5	6"	1'	R	20+	L–R		SH	L
S, G	?	6"	1'	R	5–10	L–S			L
S, G	5.5–7.5	1'	1'	R	5–10	L–R	x	FL	L
S, G	?–8.0	6"–1'	1'	FIB	5–10	L–S		SD	H
S, L	?–8.0	6"	6"	FIB	3–5	L–S			L
S, L	?–7.5	6"	6"	FIB	3–5	L–S			L
S, L, C	?–8.0	6"	9"	FIB	3–5	L–S			L
S, L	5.5–7.5	1'–2'	2'	TAP	10–20	L–S			L
S, L	?	1'	1'	FIB	5–10	L–S			L
S, L	?	1'	NR	R	20+	H–R			M
S, L, C	6.5–8.0	6"–1'	1'	B	10–20	L–S	x	FL/SH	L
S, L	5.5–8.0	6"–1'	1'	B	10–20	L–S	x	FL/SH	L
S, L, C	?	NR	NR	R	10–20	H–R			L
S, L	?	2'	3'	TAP	20+	L–S	x	SP	L
G, L, C	?–8.00	1'	1'–2'	TAP/short Rs	3–5	L–S		FL/SH	L
S	?	1'–2'	2'	TAP	10–20	L–S			L
S, L	5.0–7.5	1'	1'–2'	FIB	3–5	M–S	x		M
S, L	?	6"–1'	1'–2'	TAP	10–20	L–S			H
S, L, C	?	6"–1'	1'–2'	TAP	10–20	L–S	x		H
S, L, C	4.5–8.0	6"–1'	1'	FIB	20+	L–S		SP	L
S, L, C	?–7.5	1'	1.5'	FIB	10–20	L–S	x		L
S, L, C	5.8–7.2	1'	1.5'	FIB	5–10	L–S	x		L
S, L	6.0–8.0	1'	1.5'–2'	TAPCO	10–20	L–S			L
S, L	?	1'	1.5'–2'	CO	10–20	L–S	x	FL/SH	L
S	?	1'	1'	TAP	10–20	L–S			L
S, G	?	1'	1'	TAP	5–10	L–S			L
S	5.0–?	1'	1.5'	TAP	10–20	M–S			L
S, G	5.0–?	1'	1'–2'	FIB	2	M–S	x	SH	L
S, G	?–8.5	1'–1.5'	2'	TAP	20+	L–S			L
S, G	?–7.5	6"–1'	1'	FIB	5–10	L–S	x	SP	L
S, G	5.0–8.0	1'	1.5'	FIB	3–5	L–S	x	SP	L
S, L	?	6"–1'	1.5'	FIB	3–5	L–S	x		L
S, L, C	6.0–8.0	1'–2'	NR	R	10–20	H–R	x	SH	M
S, L	4.0–7.0	2'	3'	R	10–20	L–R/S	x	RH	L
S, L	4.5–7.5	1'	2'	FIB	5–10	L–S			L
S, G	6.5–8.0	1'	2'	FIB	5–10	L–S		Empty SH	L
S, L, C	?–8.0	1'	3'	R	5–10	M–R?		Empty SH	L
S, G	4.5–?	2'	3'	TAP	10–20	L–S			L
S, L, C	?	1'	2'	R	20+	M–R			M

(continued)

TABLE 12.13. (*continued*)

Height	Scientific name (4/1/2020)	Common name	Flower color	Bloom time	USDA zones	Sun	Soil moisture
1'–2'	*Zizia aptera*	Heartleaf Golden Alexanders	Yellow	May–June	3–8	F	M, Mo
1'–2'	*Zizia aurea*	Golden Alexanders	Yellow	May–July	3–8	F, P, S	M, Mo, W
1'–3'	*Agastache foeniculum*	Lavender Hyssop	Purple	July–Sept.	2–6	F, P	D, M
1'–3'	*Anemone cylindrica*	Thimbleweed	White	May–June	3–6	F	D, M
1'–3'	*Conoclinium coelestinum*	Blue Mistflower	Blue/purple	July–Oct.	4–10	F, P	M, Mo
1'–3'	*Onosmodium bejariense* var. *occidentale*	Marbleseed (Western Marbleseed)	White	May–June	3–8	F	D, M
1'–3'	*Pycnanthemum tenuifolium*	Narrowleaf Mountainmint	White	July–Sept.	4–8	F, P	D, M
1'–3'	*Pycnanthemum virginianum*	Virginia Mountainmint	White	July–Sept.	3–7	F, P	Mo, W
1'–3'	*Rudbeckia hirta*	Blackeyed Susan	Yellow	June–Sept.	3–10	F, P	D, M, Mo
1'–3'	*Solidago speciosa*	Showy Goldenrod	Yellow	Aug.–Sept.	3–8	F	D, M
1'–3'	*Symphyotrichum ericoides*	Heath Aster	White	Aug.–Oct.	3–9	F	D, M
1'–4'	*Delphinium carolinianum* subsp. *virescens*	Prairie Larkspur	White	June–July	3–8	F	D, M
1'–4'	*Lobelia siphilitica*	Great Blue Lobelia	Blue	July–Sept.	3–9	F, P	M, Mo
2'–3'	*Amorpha canescens*	Leadplant	Purple	June–July	3–8	F, P	D, M
2'–3'	*Amsonia tabernaemontana*	Eastern Bluestar	Blue/white	May–June	4–8	F, P	M, Mo
2'–3'	*Asclepias tuberosa*	Butterflyweed	Orange	June–Aug.	3–10	F	D, M
2'–3'	*Astragalus canadensis*	Canada Milkvetch	Yellow	July–Aug.	3–8	F	D, M
2'–3'	*Ceanothus americanus*	New Jersey Tea	White	July–Aug.	3–9	F, P	D, M
2'–3'	*Coreopsis palmata*	Stiff Coreopsis	Yellow	June–Aug.	3–8	F	D, M
2'–3'	*Helianthus occidentalis*	Western Sunflower	Yellow	July–Aug.	3–9	F	D, M
2'–3'	*Iris virginica* var. *shrevei*	Wild Iris	Blue	June–July	3–8	F, P	W
2'–3'	*Penstemon cobaea*	Cobaea Beardtongue	Lavender	Apr.–June	5–8	F	D
2'–3'	*Penstemon digitalis*	Foxglove Beardtongue	White	June–July	3–8	F, P	M, Mo
2'–3'	*Symphyotrichum oolentangiense*	Skyblue Aster	Blue	Aug.–Oct.	3–8	F, P	D, M
2'–4'	*Asclepias syriaca*	Common Milkweed	Lavender	June–Aug.	3–8	F, P	D, M
2'–4'	*Chelone glabra*	White Turtlehead	White	Aug.–Sept.	3–8	F, P	Mo, W
2'–4'	*Euphorbia corollata*	Flowering Spurge	White	July–Aug.	3–9	F	D, M
2'–4'	*Penstemon grandiflorus*	Large-Flowered Beardtongue	Lavender	May–June	3–7	F	D
2'–4'	*Polygonatum biflorum* var. *commutatum*	Great Solomon's Seal	Cream	May–June	3–9	F, P	D, M
2'–4'	*Rudbeckia fulgida*	Orange Coneflower	Yellow	July–Sept.	4–8	F, P	M, Mo
2'–4'	*Silene regia*	Royal Catchfly	Red	July–Aug.	4–9	F, P	M
2'–4'	*Spiraea alba*	Meadowsweet	White/pink	June–Sept.	3–7	F, P	Mo, W
2'–4'	*Spiraea tomentosa*	Steeplebush	Pink	Aug.–Sept.	3–8	F, P	Mo, W
2'–4'	*Symphyotrichum laeve*	Smooth Aster	Blue	Aug.–Oct.	4–8	F	D, M
2'–4'	*Tradescantia ohiensis*	Ohio Spiderwort	Blue	June–July	3–9	F, P	D, M, Mo
2'–4'	*Verbena stricta*	Hoary Vervain	Blue	July–Sept.	3–8	F	D, M

Soil texture	pH range	Plant spacing in prairie gardens	Plant spacing in flower beds	Root type	Plant life expectancy in years	Aggressiveness and method	Cut flowers	Dried arrangements	Deer palatability
S, L	?-7.5	1'	2'	FIB	5-10	L-S			L
S, L, C	?-8.0	1'	3'	FIB	5-10	L-S			L
S, L	?-8.0	1'	1'-2'	FIB	3-5	M-S	x	FL/SH	L
S, L	?	1'	1'	FIB	10-20	L-S		SD	L
S, L, C	5.0-8.0	1'-2'	2'	R	3-5	M-R			L
S, L	?-7.5	1'-1.5'	2'	TAP	10-20	L-S			L
S, G, L, C	?	1'	1.5'	TAP/R	5-10	L-S		SH	M
S, L, C	?-8.0	1'-2'	3'	R	5-10	L-R		SH	L
S, L, C	5.0-8.0	1'	2'	FIB	2	M-S	x	SH	L
S, L	?	1'	3'	FIB	5-10	L-S	x	SH	L
S, G, L	?	1'-2'	2'	R	10-20	M-R		Empty SH	L
S, L	?	1'	1.5'	FIB	5-10	L-S	x	SP	L
S, L, C	?-8.0	6"-1'	2'	FIB	5-10	M-S	x		L
S, L	5.5-8.0	2'	3'	TAP	20+	L-S		Leaves	L
S, L, C	?-8.0	2'-3'	3'	FIB	20+	L-S		SP	L
S, L	5.0-7.5	2'	3'	TAP	10-20	L-S	x	SP	L
S, L	6.0-8.0	1'-2'	3'	TAP	10-20	M-S		SP	M
S, L	4.5-7.0	3'	3'	TAP	20+	L-S		SD	L
S, L	?	1'-2'	2'	R	20+	M-R			L
S, L	?	2'	3'	R	20+	M-R	x		L
S, L, C	4.8-7.3	1'	2'	R/CO	10-20	L-R	x	SP	L
S, G	?-7.5	1'	1'	FIB	3-5	L-S	x	SP	L
S, L, C	5.5-7.5	1'	2'	FIB	5-10	L-S	x		L
S, L	5.0-8.0	1'	2'	FIB	5-10	L-S	x	Empty SH	M
S, L, C	5.0-8.0	NR	NR	R	10-20	H-R		SP	L
S, L, C	?	1'	1'-2'	FIB	5-10	L-S			M
S, L	?	1'-2'	1'-2'	R	10-20	M-R			L
S, G	?	1'	2'-3'	FIB	3-5	L-S		SP	L
S, L, C	?	1'-2'	3'	R	10-20	L-R/S			H
L, C	?-8.0	1'-2'	3'	FIB	5-10	L-S	x	SH	L
L	?	1'	1.5'	TAP	5-10	L-S	x	SH	L
S, L, C	4.3-7.2	2'	3'	FIB	10-20	L-S	x		L
S, L, C	4.5-7.0	2'	3'	R	10-20	L-R/S	x		L
S, L	?	1'	3'	FIB	5-10	M-S	x	Empty SH	H
S, L, C	?-7.5	1'	2'	FIB	20+	M-S			H
S, L	?	1'	2'	FIB	3-5	M-S	x	SH	L

(continued)

TABLE 12.13. *(continued)*

Height	Scientific name (4/1/2020)	Common name	Flower color	Bloom time	USDA zones	Sun	Soil moisture
2'-5'	*Liatris aspera*	Rough Blazing Star	Purple/pink	Aug.-Sept.	3-8	F	D, M
2'-5'	*Lobelia cardinalis*	Cardinal Flower	Red	July-Sept.	3-9	F, P	Mo, W
2'-5'	*Monarda fistulosa*	Bergamot	Lavender	July-Sept.	3-9	F, P	D, M, Mo
2'-5'	*Rudbeckia triloba*	Browneyed Susan	Yellow	July-Oct.	4-9	F, P	M, Mo
2'-6'	*Desmanthus illinoensis*	Illinois Bundleflower	White	June-Aug.	5-8	F	D, M, Mo
2'-6'	*Silphium integrifolium*	Rosinweed	Yellow	July-Sept.	4-8	F	M, Mo
3'-10'	*Silphium laciniatum*	Compassplant	Yellow	June-Sept.	4-9	F	D, M
3'-10'	*Silphium perfoliatum*	Cup Plant	Yellow	July-Sept.	3-8	F, P	M, Mo, W
3'-10'	*Silphium terebinthinaceum*	Prairie Dock	Yellow	July-Sept.	4-7	F	M, Mo
3'-4'	*Echinacea purpurea*	Purple Coneflower	Purple	July-Sept.	4-8	F, P	D, M
3'-4'	*Eupatorium perfoliatum*	Boneset	White	Aug.-Sept.	3-8	F	M, Mo, W
3'-4'	*Oligoneuron ohioense*	Ohio Goldenrod	Yellow	Aug.-Sept.	4-5	F, P	M, Mo
3'-5'	*Asclepias incarnata*	Marsh Milkweed	Red/pink	June-July	3-9	F	Mo, W
3'-5'	*Asclepias sullivantii*	Sullivant's Milkweed	Pink/yellow	June-Aug.	4-7	F	M, Mo
3'-5'	*Baptisia alba*	White False Indigo	White	June-July	4-9	F, P	M, Mo
3'-5'	*Baptisia australis*	Blue False Indigo	Blue	June-July	3-10	F, P	M
3'-5'	*Desmodium canadense*	Canada Ticktrefoil	Purple	July-Aug.	3-7	F	M, Mo
3'-5'	*Echinacea pallida*	Pale Purple Coneflower	Purple	June-July	4-8	F	D, M
3'-5'	*Echinacea paradoxa*	Ozark Coneflower	Yellow	June-July	4-7	F, P	D, M
3'-5'	*Eryngium yuccifolium*	Rattlesnake Master	White	June-Aug.	4-9	F	D, M
3'-5'	*Lespedeza capitata*	Roundhead Bush Clover	White	Aug.-Sept.	3-8	F, P	D, M
3'-5'	*Liatris ligulistylis*	Meadow Blazing Star	Purple/pink	Aug.-Sept.	3-6	F	M, Mo
3'-5'	*Liatris pycnostachya*	Prairie Blazing Star	Purple/pink	July-Aug.	3-9	F	M, Mo
3'-5'	*Oligoneuron rigidum*	Stiff Goldenrod	Yellow	Aug.-Sept.	3-9	F	D, M, Mo
3'-5'	*Parthenium integrifolium*	Wild Quinine	White	June-Sept.	4-8	F	M, Mo
3'-5'	*Salvia azurea*	Blue Sage	Blue	June-Aug.	4-9	F, P	D, M
3'-6'	*Helianthus ×laetiflorus*	Showy Sunflower	Yellow	Aug.-Sept.	3-8	F	D, M
3'-6'	*Heliopsis helianthoides*	Oxeye Sunflower	Yellow	June-Sept.	3-8	F	D, M, Mo
3'-6'	*Liatris spicata*	Dense Blazing Star	Purple/pink	Aug.-Sept.	4-10	F	M, Mo
3'-6'	*Oenothera gaura*	Biennial Beeblossom	White/pink	June-Sept.	4-8	F, P	D, M, Mo
3'-6'	*Ratibida pinnata*	Yellow Coneflower	Yellow	July-Sept.	3-9	F	D, M, Mo
3'-6'	*Symphyotrichum novae-angliae*	New England Aster	Pink/purple/blue	Aug.-Oct.	3-7	F, P	M, Mo
3'-6'	*Thalictrum dasycarpum*	Tall Meadow-Rue	White	June-July	3-9	F, P	Mo, W
3'-6'	*Verbena hastata*	Blue Vervain	Blue	July-Sept.	3-8	F, P	M, Mo, W
3'-6'	*Veronicastrum virginicum*	Culver's Root	White	July-Aug.	3-8	F, P, S	M, Mo
3'-7'	*Coreopsis tripteris*	Tall Coreopsis	Yellow	July-Sept.	3-8	F, P	M, Mo, W
3'-7'	*Hibiscus moscheutos*	Rosemallow	White/pink/red	July-Sept.	5-9	F, P	M, Mo, W
4'-5'	*Filipendula rubra*	Queen of the Prairie	Pink	June-July	3-6	F	M, Mo
4'-5'	*Helenium autumnale*	Sneezeweed	Yellow	Aug.-Sept.	3-9	F	Mo, W
4'-6'	*Eutrochium maculatum*	Joe Pye Weed	Pink	Aug.-Sept.	3-6	F, P	Mo, W
4'-6'	*Helianthus mollis*	Downy Sunflower	Yellow	Aug.-Sept.	4-9	F	D, M
4'-6'	*Hypericum ascyron*	Great St. Johnswort	Yellow	July-Aug.	3-5	F, P	Mo, W

Soil texture	pH range	Plant spacing in prairie gardens	Plant spacing in flower beds	Root type	Plant life expectancy in years	Aggressiveness and method	Cut flowers	Dried arrangements	Deer palatability
S, L	?	6"–1'	1'	CO	10–20	L–S	x	FL/SH	L
S, L	5.8–7.8	6"–1'	2'	FIB	3–5	L–S	x		M
S, L, C	6.0–8.0	1'–2'	3'	R	10–20	M–R	x	SH	L
S, L	?–8.0	1'–2'	3'	FIB	2	M–S	x	SH	L
S, L, C	6.0–8.0	1'	1.5'	TAP	5–10	L–S		SP	M
S, L, C	?–8.0	2'	3'	TAP	20+	M–S	x		L
S, L, C	?–8.0	2'	4'	TAP	20+	L–S	x		L
S, L, C	4.5–8.0	2'–3'	NR	FIB	20+	H–S	x		L
S, L, C	?–8.0	2'	4'	TAP	20+	L–S	x		L
S, L, C	5.5–8.0	1'–2'	3'	FIB	5–10	M–S	x	SH	M
S, L, C	?–7.5	1'	NR	FIB	5–10	L–S			L
S, L, C	?–8.0	1'–2'	3'	FIB	5–10	L–S	x	SH	L
S, L, C	5.0–8.0	1'–2'	3'	FIB	5–10	L–S	x	SP	L
L, C	?	1'–2'	3'	R	10–20	L–R		SP	L
S, L, C	?–7.5	3'	3'	TAP	20+	L–S	x	SP	L
S, L, C	?–8.0	3'	3'	TAP	20+	L–S	x	SP	L
S, L, C	?–7.5	NR	NR	TAP	10–20	M–S			H
S, L, C	6.0–7.5	1'	2'–3'	TAP	10–20	L–S	x	SH	M
S, L, C	?–7.5	1'	2'–3'	TAP	10–20	L–S	x	SH	M
S, L, C	6.0–8.0	1'	2'	FIB	5–10	M–S	x	SH	L
S, G, L	5.5–8.0	1'	2'	TAP	10–20	L–S		SH	M
L	?	6"–1'	1'	CO	3–5	L–S	x	FL/SH	L
S, L, C	6.0–8.0	6"–1'	1'	CO	10–20	L–S	x	FL/SH	L
S, L, C	5.5–8.0	1'–1.5'	3'	FIB	5–10	M–S	x	SH	L
S, L, C	?–8.0	1'–2'	3'	TAP	10–20	L–S	x		M
S, G, L	?	1'	1.5'	FIB	5–10	L–S	x	SH	L
S, G, L, C	?	NR	NR	R	20+	H–R	x	SH	L
S, L, C	6.0–8.0	1'–2'	3'	FIB	5–10	H–S	x		L
S, L, C	5.5–7.5	6"–1'	1'	CO	10–20	L–S	x	FL/SH	L
S, L, C	?	2'	3'	FIB	2	M–S	x		L
S, L, C	5.5–8.0	1'–2'	2'	FIB	10–20	M–S	x	SH	L
S, L, C	5.0–8.0	1'	3'	FIB	5–10	L–S	x	Empty SH	H
S, L, C	?	1'–2'	2'	FIB	5–10	L–S			L
S, L, C	?	1'	2'	FIB	2	M–S	x	SH	L
S, L, C	?–8.0	1'	3'	FIB	10–20	L–S	x	SH	L
S, L, C	?	1'–2'	NR	FIB	10–20	L–S	x		L
S, L, C	4.0–8.0	2'	4'	R	10–20	L–R/S			L
S, L, C	?–8.0	2'	3'	R	20+	M–R	x		L
S, L, C	4.0–7.5	1'–2'	2'–3'	FIB	5–10	L–S	x	SH	L
S, L, C	5.5–7.5	1'–2'	3'–4'	FIB	5–10	L–S			L
S, L, C	6.0–8.0	2'	3'	R	20+	M–R	x	SH	L
S, L, C	5.7–7.1	1'–2'	3'	FIB	5–10	L–S	x	SP	L

(continued)

TABLE 12.13. *(continued)*

Height	Scientific name (4/1/2020)	Common name	Flower color	Bloom time	USDA zones	Sun	Soil moisture
4'–6'	*Lilium michiganense*	Michigan Lily	Orange	July–Aug.	3–7	F, P	Mo
4'–6'	*Rudbeckia subtomentosa*	Sweet Blackeyed Susan	Yellow	Aug.–Oct.	4–8	F, P	M, Mo
4'–6'	*Senna hebecarpa*	Wild Senna	Yellow	July–Aug.	4–7	F	M, Mo, W
4'–6'	*Vernonia fasciculata*	Ironweed	Red/pink	July–Sept.	3–7	F	Mo
5'–10'	*Arnoglossum atriplicifolium*	Pale Indian Plantain	White	July–Sept.	4–8	F, P	D, M, Mo
5'–8'	*Eutrochium fistulosum*	Tall Joe Pye Weed	Purple/pink	Aug.–Sept.	4–9	F, P	M, Mo, W
5'–8'	*Rudbeckia laciniata*	Green-Headed Coneflower	Yellow	Aug.–Sept.	3–8	F, P	Mo, W
5'–8'	*Vernonia gigantea*	Tall Ironweed	Red/pink	Aug.–Sept.	4–9	F, P	Mo, W
6'–15'	*Rosa setigera*	Climbing Prairie Rose	Pink/white	May–July	4–9	F, P	M, Mo

Height	Scientific name (4/1/2020)	Common name	Fall leaf color	Bloom time	USDA zones	Sun	Soil moisture
Grasses and sedges							
1'–2'	*Eragrostis spectabilis*	Purple Lovegrass	Purple/pink	July–Sept.	3–10	F, P	D
1'–2'	*Hierochloe hirta*	Vanilla Sweetgrass	Straw	May–June	3–9	F, P	M, Mo, W
1'–4'	*Carex brevior*	Shortbeak Sedge	Straw	May–June	3–7	F, P	D, M, Mo
2'–3'	*Bouteloua curtipendula*	Sideoats Grama	Straw	Aug.–Sept.	3–9	F	D, M
2'–3'	*Koeleria macrantha*	Junegrass	Gold	May–June	3–8	F	D
2'–4'	*Bromus kalmii*	Prairie Bromegrass	Straw	June–July	3–6	F	D
2'–4'	*Hesperostipa spartea*	Needlegrass	Gold	Apr.–June	3–6	F	D, M
2'–4'	*Sporobolus heterolepis*	Prairie Dropseed	Gold	Aug.–Sept.	3–8	F	D, M
2'–5'	*Andropogon virginicus*	Broomsedge	Reddish brown	Aug.–Oct.	5–10	F	D, M, Mo
2'–5'	*Schizachyrium scoparium*	Little Bluestem	Crimson red	Aug.–Oct.	3–10	F	D, M
3'–6'	*Panicum virgatum*	Switchgrass	Gold	Aug.–Sept.	3–10	F	D, M, Mo
3'–6'	*Tridens flavus*	Purpletop	Straw	July–Sept.	5–10	F	D, M
4'–5'	*Elymus canadensis*	Canada Wildrye	Straw	July–Aug.	2–9	F	D, M, Mo
4'–5'	*Elymus virginicus*	Virginia Wildrye	Straw	July–Aug.	2–9	F, P, S	M, Mo
4'–6'	*Calamagrostis canadensis*	Bluejoint Grass	Straw	June–July	3–6	F	Mo, W
4'–8'	*Tripsacum dactyloides*	Eastern Gamagrass	Gold	July–Sept.	4–10	F, P	M, Mo, W
5'–7'	*Sorghastrum nutans*	Indiangrass	Gold	Aug.–Sept.	3–9	F	D, M, Mo
5'–8'	*Andropogon gerardii*	Big Bluestem	Bronze/red	Aug.–Oct.	3–9	F	D, M, Mo
6'–9'	*Spartina pectinata*	Prairie Cordgrass	Gold	Aug.–Sept.	3–8	F	Mo, W

Notes: For soil moisture, D = dry, M = medium, Mo = moist, and W = wet. For soil texture, S = sand, L = loam, C = clay, and G = gravel. For sun, F = full sun, or at least a half day of full sun; P = partial sun, or a quarter to a half day of sun; and S = shade, or less than a quarter day of sun. For plant spacing, NR = not recommended. For root type, B = bulb, CO = corm, FIB = fibrous, R = rhizome, TAP =

Soil texture	pH range	Plant spacing in prairie gardens	Plant spacing in flower beds	Root type	Plant life expectancy in years	Aggressiveness and method	Cut flowers	Dried arrangements	Deer palatability
S, L, C	?-6.8	2'	2'	B	20+	L-S	x	SP	H
S, L, C	?-8.0	1'-2'	2'	FIB	10-20	L-S	x	SH	L
S, L, C	?	1.5'-2'	3'-4'	FIB/TAP	10-20	L-S	x	FL/SH	M
S, L, C	?	1'	3'	FIB	5-10	L-S	x		L
S, L, C	?-8.0	2'	3'-4'	FIB	3-20	H-S (north); L-S (south)	x	FL	L
S, L, C	4.5-7.5	2'	3'-4'	FIB	5-10	L-S			H
S, L, C	4.5-7.5	2'	2'	FIB	3-5	M-S	x	SH	L
S, L, C	?-8.0	1'-2'	3'	FIB	10-20	L-S	x		L
S, L, C	5.0-8.0	2'	5'	FIB	10-20	L-S	x	RH	L

Soil texture	pH range	Plant spacing in prairie gardens	Plant spacing in flower beds	Root type	Plant life expectancy in years	Aggressiveness and method	Cut flower stalks	Dried seed arrangements	Deer palatability
S, G	4.0-7.5	1'-1.5'	2'	FIB	5-10	M-S	x	x	L
S, L, C	5.7-8.0	1'	NR	R	10-20	H-R			L
S, L, C	4.5-7.5	1'	1'	FIB	5-10	L-S			L
S, L	5.5-8.5	1'	2'	FIB	10-20	L-S		x	L
S, G	6.0-8.0	6"-1'	1'	FIB	5-10	L-S		x	L
S, G	5.7-7.0	1'	2'	FIB	3-5	L-S		x	L
S, G, L	5.0-7.5	1'	NR	FIB	10-20	L-S		x	L
S, L	6.0-8.0	2'	2'	FIB	20+	L-S	x	x	L
S, L, C	4.0-7.0	1'-1.5'	NR	FIB	20+	H-S		x	L
S, L	5.0-8.5	1'-1.5'	3'	FIB	20+	L-S		x	L
S, L, C	4.5-8.0	1.5'-2'	NR	FIB	20+	M-S		x	L
S, L, C	5.0-7.5	1'	1.5'	FIB	5-10	M-S		x	L
S, L, C	5.0-8.0	1'	3'	FIB	3-5	M-S		x	L
S, L, C	5.0-7.5	1'	3'	FIB	3-5	M-S		x	L
S, L, C	4.5-8.0	1'	3'	FIB	10-20	M-R/S		x	L
S, L, C	4.5-7.5	2'	3'	FIB	20+	H-S		x	M
S, L, C	5.0-8.0	1'-2'	3'	FIB	10-20	M-S	x	x	L
S, L, C	5.0-7.5	2'	NR	FIB	20+	M-S		x	L
S, L, C	6.0-8.5	2'	NR	R	20+	H-R		x	L

taproot, and TAPCO = taprooted corm. For aggressiveness and method, L = low, M = medium, H = high, R = by rhizome, S = by seed, and R/S = rhizome and seed. For dried arrangements, FL = flowers, SH = seedheads, SP = seedpods, SD = seed, and RH = rose hips. For deer palatability, L = low, M = medium, and H = high.

TABLE 12.14. PLANTS LISTED BY ROOT TYPE

Root type	Scientific name (4/1/2020)	Common name	Height	Flower color	Bloom time	USDA zones	Sun
Flowers and shrubs							
B	*Allium cernuum*	Nodding Pink Onion	1'–2'	White/pink	June–Aug.	3–8	F, P
B	*Allium stellatum*	Prairie Onion	1'–2'	Lavender	July–Aug.	3–8	F, P
B	*Lilium michiganense*	Michigan Lily	4'–6'	Orange	July–Aug.	3–7	F, P
CO	*Liatris aspera*	Rough Blazing Star	2'–5'	Purple/pink	Aug.–Sept.	3–8	F
CO	*Liatris ligulistylis*	Meadow Blazing Star	3'–5'	Purple/pink	Aug.–Sept.	3–6	F
CO	*Liatris pycnostachya*	Prairie Blazing Star	3'–5'	Purple/pink	July–Aug.	3–9	F
CO	*Liatris spicata*	Dense Blazing Star	3'–6'	Purple/pink	Aug.–Sept.	4–10	F
CO	*Liatris squarrosa*	Scaly Blazing Star	1'–2'	Purple/pink	Aug.–Sept.	4–8	F
FIB	*Agastache foeniculum*	Lavender Hyssop	1'–3'	Purple	July–Sept.	2–6	F, P
FIB	*Amsonia tabernaemontana*	Eastern Bluestar	2'–3'	Blue/white	May–June	4–8	F, P
FIB	*Anemone cylindrica*	Thimbleweed	1–3'	White	May–June	3–6	F
FIB	*Arnoglossum atriplicifolium*	Pale Indian Plantain	5'–10'	White	July–Sept.	4–8	F, P
FIB	*Asclepias incarnata*	Marsh Milkweed	3'–5'	Red/pink	June–July	3–9	F
FIB	*Chelone glabra*	White Turtlehead	2'–4'	White	Aug.–Sept.	3–8	F, P
FIB	*Coreopsis lanceolata*	Lanceleaf Coreopsis	1'–2'	Yellow	June–July	3–9	F
FIB	*Coreopsis tripteris*	Tall Coreopsis	3'–7'	Yellow	July–Sept.	3–8	F, P
FIB	*Delphinium carolinianum* subsp. *virescens*	Prairie Larkspur	1'–4'	White	June–July	3–8	F
FIB	*Dodecatheon meadia*	Shooting Star	1'–2'	White/pink	May–June	4–8	F, P
FIB	*Echinacea purpurea*	Purple Coneflower	3'–4'	Purple	July–Sept.	4–8	F, P
FIB	*Eryngium yuccifolium*	Rattlesnake Master	3'–5'	White	June–Aug.	4–9	F
FIB	*Eupatorium perfoliatum*	Boneset	3'–4'	White	Aug.–Sept.	3–8	F
FIB	*Eutrochium fistulosum*	Tall Joe Pye Weed	5'–8'	Purple/pink	Aug.–Sept.	4–9	F, P
FIB	*Eutrochium maculatum*	Joe Pye Weed	4'–6'	Pink	Aug.–Sept.	3–6	F, P
FIB	*Gentiana alba*	Cream Gentian	1'–2'	White	July–Aug.	3–7	F, P
FIB	*Gentiana andrewsii*	Bottle Gentian	1'–2'	Blue	Aug.–Oct.	3–6	F, P
FIB	*Helenium autumnale*	Sneezeweed	4'–5'	Yellow	Aug.–Sept.	3–9	F
FIB	*Heliopsis helianthoides*	Oxeye Sunflower	3'–6'	Yellow	June–Sept.	3–8	F
FIB	*Hypericum ascyron*	Great St. Johnswort	4'–6'	Yellow	July–Aug.	3–5	F, P
FIB	*Lithospermum canescens*	Hoary Puccoon	1'	Orange	May–June	3–7	F, P
FIB	*Lobelia cardinalis*	Cardinal Flower	2'–5'	Red	July–Sept.	3–9	F, P
FIB	*Lobelia siphilitica*	Great Blue Lobelia	1'–4'	Blue	July–Sept.	3–9	F, P
FIB	*Monarda punctata*	Dotted Mint	1'–2'	Lavender	July–Sept.	3–10	F
FIB	*Oenothera gaura*	Biennial Beeblossom	3'–6'	White/pink	June–Sept.	4–8	F, P
FIB	*Oligoneuron ohioense*	Ohio Goldenrod	3'–4'	Yellow	Aug.–Sept.	4–5	F, P
FIB	*Oligoneuron rigidum*	Stiff Goldenrod	3'–5'	Yellow	Aug.–Sept.	3–9	F
FIB	*Penstemon cobaea*	Cobaea Beardtongue	2'–3'	Lavender	Apr.–June	5–8	F
FIB	*Penstemon digitalis*	Foxglove Beardtongue	2'–3'	White	June–July	3–8	F, P
FIB	*Penstemon gracilis*	Slender Beardtongue	1'–2'	Lavender	May–June	3–6	F, P

Soil moisture	Soil texture	pH range	Plant spacing in prairie gardens	Plant spacing in flower beds	Plant life expectancy in years	Aggressiveness and method	Cut flowers	Dried arrangements	Deer palatability
M, Mo	S, L, C	6.5–8.0	6"–1'	1'	10–20	L–S	x	FL/SH	L
D, M	S, L	5.5–8.0	6"–1'	1'	10–20	L–S	x	FL/SH	L
Mo	S, L, C	?–6.8	2'	2'	20+	L–S	x	SP	H
D, M	S, L	?	6"–1'	1'	10–20	L–S	x	FL/SH	L
M, Mo	L	?	6"–1'	1'	3–5	L–S	x	FL/SH	L
M, Mo	S, L, C	6.0–8.0	6"–1'	1'	10–20	L–S	x	FL/SH	L
M, Mo	S, L, C	5.5–7.5	6"–1'	1'	10–20	L–S	x	FL/SH	L
D, M	S, L	?	1'	1.5'–2'	10–20	L–S	x	FL/SH	L
D, M	S, L	?–8.0	1'	1'–2'	3–5	M–S	x	FL/SH	L
M, Mo	S, L, C	?–8.0	2'–3'	3'	20+	L–S		SP	L
D, M	S, L	?	1'	1'	10–20	L–S		SD	L
D, M, Mo	S, L, C	?–8.0	2'	3'–4'	3–20	H–S (north); L–S (south)	x	FL	L
Mo, W	S, L, C	5.0–8.0	1'–2'	3'	5–10	L–S	x	SP	L
Mo, W	S, L, C	?	1'	1'–2'	5–10	L–S			M
D, M	S, L	5.0–7.5	1'	1'–2'	3–5	M–S	x		M
M, Mo, W	S, L, C	?	1'–2'	NR	10–20	L–S	x		L
D, M	S, L	?	1'	1.5'	5–10	L–S	x	SP	L
M, Mo	S, L, C	4.5–8.0	6"–1'	1'	20+	L–S		SP	L
D, M	S, L, C	5.5–8.0	1'–2'	3'	5–10	M–S	x	SH	M
D, M	S, L, C	6.0–8.0	1'	2'	5–10	M–S	x	SH	L
M, Mo, W	S, L, C	?–7.5	1'	NR	5–10	L–S			L
M, Mo, W	S, L, C	4.5–7.5	2'	3'–4'	5–10	L–S			H
Mo, W	S, L, C	5.5–7.5	1'–2'	3'–4'	5–10	L–S			L
M	S, L, C	?–7.5	1'	1.5'	10–20	L–S	x		L
Mo, W	S, L, C	5.8–7.2	1'	1.5'	5–10	L–S	x		L
Mo, W	S, L, C	4.0–7.5	1'–2'	2'–3'	5–10	L–S	x	SH	L
D, M, Mo	S, L, C	6.0–8.0	1'–2'	3'	5–10	H–S	x		L
Mo, W	S, L, C	5.7–7.1	1'–2'	3'	5–10	L–S	x	SP	L
D, M	S, L	?	1'	1'	5–10	L–S			L
Mo, W	S, L	5.8–7.8	6"–1'	2'	3–5	L–S	x		M
M, Mo	S, L, C	?–8.0	6"–1'	2'	5–10	M–S	x		L
D	S, G	5.0–?	1'	1'–2'	2	M–S	x	SH	L
D, M, Mo	S, L, C	?	2'	3'	2	M–S	x		L
M, Mo	S, L, C	?–8.0	1'–2'	3'	5–10	L–S	x	SH	L
D, M, Mo	S, L, C	5.5–8.0	1'–1.5'	3'	5–10	M–S	x	SH	L
D	S, G	?–7.5	1'	1'	3–5	L–S	x	SP	L
M, Mo	S, L, C	5.5–7.5	1'	2'	5–10	L–S	x		L
D	S, G	?–7.5	6"–1'	1'	5–10	L–S	x	SP	L

(continued)

TABLE 12.14. *(continued)*

Root type	Scientific name (4/1/2020)	Common name	Height	Flower color	Bloom time	USDA zones	Sun
FIB	*Penstemon grandiflorus*	Large-Flowered Beardtongue	2'–4'	Lavender	May–June	3–7	F
FIB	*Penstemon hirsutus*	Hairy Penstemon	1'–2'	Lavender	May–June	3–6	F, P
FIB	*Phlox pilosa*	Downy Phlox	1'–2'	Pink	May–June	3–9	F
FIB	*Pulsatilla patens*	Pasque Flower	6"–1'	White	Apr.–May	3–6	F
FIB	*Ratibida pinnata*	Yellow Coneflower	3'–6'	Yellow	July–Sept.	3–9	F
FIB	*Rosa setigera*	Climbing Prairie Rose	6'–15'	Pink/white	May–July	4–9	F, P
FIB	*Rudbeckia fulgida*	Orange Coneflower	2'–4'	Yellow	July–Sept.	4–8	F, P
FIB	*Rudbeckia hirta*	Blackeyed Susan	1'–3'	Yellow	June–Sept.	3–10	F, P
FIB	*Rudbeckia laciniata*	Green-Headed Coneflower	5'–8'	Yellow	Aug.–Sept.	3–8	F, P
FIB	*Rudbeckia subtomentosa*	Sweet Blackeyed Susan	4'–6'	Yellow	Aug.–Oct.	4–8	F, P
FIB	*Rudbeckia triloba*	Browneyed Susan	2'–5'	Yellow	July–Oct.	4–9	F, P
FIB	*Ruellia humilis*	Wild Petunia	1'–2'	Violet	June–Aug.	4–9	F
FIB	*Salvia azurea*	Blue Sage	3'–5'	Blue	June–Aug.	4–9	F, P
FIB	*Silphium perfoliatum*	Cup Plant	3'–10'	Yellow	July–Sept.	3–8	F, P
FIB	*Sisyrinchium albidum*	White Blue-Eyed Grass	6"–1'	White	Apr.–May	4–9	F
FIB	*Sisyrinchium angustifolium*	Narrow-Leaved Blue-Eyed Grass	6"–2'	Blue/purple	Apr.–June	3–10	F, P
FIB	*Sisyrinchium campestre*	Prairie Blue-Eyed Grass	6"–1'	Blue/white	Apr.–June	3–8	F, P
FIB	*Solidago ptarmicoides*	White Aster	1'–2'	White	Aug.–Sept.	3–7	F
FIB	*Solidago speciosa*	Showy Goldenrod	1'–3'	Yellow	Aug.–Sept.	3–8	F
FIB	*Spiraea alba*	Meadowsweet	2'–4'	White/pink	June–Sept.	3–7	F, P
FIB	*Symphyotrichum laeve*	Smooth Aster	2'–4'	Blue	Aug.–Oct.	4–8	F
FIB	*Symphyotrichum novae-angliae*	New England Aster	3'–6'	Pink/purple/blue	Aug.–Oct.	3–7	F, P
FIB	*Symphyotrichum oolentangiense*	Skyblue Aster	2'–3'	Blue	Aug.–Oct.	3–8	F, P
FIB	*Thalictrum dasycarpum*	Tall Meadow-Rue	3'–6'	White	June–July	3–9	F, P
FIB	*Tradescantia ohiensis*	Ohio Spiderwort	2'–4'	Blue	June–July	3–9	F, P
FIB	*Verbena hastata*	Blue Vervain	3'–6'	Blue	July–Sept.	3–8	F, P
FIB	*Verbena stricta*	Hoary Vervain	2'–4'	Blue	July–Sept.	3–8	F
FIB	*Vernonia fasciculata*	Ironweed	4'–6'	Red/pink	July–Sept.	3–7	F
FIB	*Vernonia gigantea*	Tall Ironweed	5'–8'	Red/pink	Aug.–Sept.	4–9	F, P
FIB	*Veronicastrum virginicum*	Culver's Root	3'–6'	White	July–Aug.	3–8	F, P, S
FIB	*Zizia aptera*	Heartleaf Golden Alexanders	1'–2'	Yellow	May–June	3–8	F
FIB	*Zizia aurea*	Golden Alexanders	1'–2'	Yellow	May–July	3–8	F, P, S
FIB/TAP	*Senna hebecarpa*	Wild Senna	4'–6'	Yellow	July–Aug.	4–7	F
TAP	*Amorpha canescens*	Leadplant	2'–3'	Purple	June–July	3–8	F, P
TAP	*Asclepias tuberosa*	Butterflyweed	2'–3'	Orange	June–Aug.	3–10	F
TAP	*Astragalus canadensis*	Canada Milkvetch	2'–3'	Yellow	July–Aug.	3–8	F
TAP	*Baptisia alba*	White False Indigo	3'–5'	White	June–July	4–9	F, P
TAP	*Baptisia australis*	Blue False Indigo	3'–5'	Blue	June–July	3–10	F, P

Soil moisture	Soil texture	pH range	Plant spacing in prairie gardens	Plant spacing in flower beds	Plant life expectancy in years	Aggressiveness and method	Cut flowers	Dried arrangements	Deer palatability
D	S, G	?	1'	2'-3'	3-5	L-S		SP	L
D	S, G	5.0-8.0	1'	1.5'	3-5	L-S	x	SP	L
D, M, Mo	S, L	?	6"-1'	1.5'	3-5	L-S	x		L
D	S, G	?-8.0	6"-1'	1'	5-10	L-S		SD	H
D, M, Mo	S, L, C	5.5-8.0	1'-2'	2'	10-20	M-S	x	SH	L
M, Mo	S, L, C	5.0-8.0	2'	5'	10-20	L-S	x	RH	L
M, Mo	L, C	?-8.0	1'-2'	3'	5-10	L-S	x	SH	L
D, M, Mo	S, L, C	5.0-8.0	1'	2'	2	M-S	x	SH	L
Mo, W	S, L, C	4.5-7.5	2'	2'	3-5	M-S	x	SH	L
M, Mo	S, L, C	?-8.0	1'-2'	2'	10-20	L-S	x	SH	L
M, Mo	S, L	?-8.0	1'-2'	3'	2	M-S	x	SH	L
D, M	S, L	4.5-7.5	1'	2'	5-10	L-S			L
D, M	S, G, L	?	1'	1.5'	5-10	L-S	x	SH	L
M, Mo, W	S, L, C	4.5-8.0	2'-3'	NR	20+	H-S	x		L
D, M	S, L	?-8.0	6"	6"	3-5	L-S			L
M, Mo	S, L, C	?-8.0	6"	9"	3-5	L-S			L
D, M	S, L	?-7.5	6"	6"	3-5	L-S			L
D	S, G	6.5-8.0	1'	2'	5-10	L-S		Empty SH	L
D, M	S, L	?	1'	3'	5-10	L-S	x	SH	L
Mo, W	S, L, C	4.3-7.2	2'	3'	10-20	L-S	x		L
D, M	S, L	?	1'	3'	5-10	M-S	x	Empty SH	H
M, Mo	S, L, C	5.0-8.0	1'	3'	5-10	L-S	x	Empty SH	H
D, M	S, L	5.0-8.0	1'	2'	5-10	L-S	x	Empty SH	M
Mo, W	S, L, C	?	1'-2'	2'	5-10	L-S			L
D, M, Mo	S, L, C	?-7.5	1'	2'	20+	M-S			H
M, Mo, W	S, L, C	?	1'	2'	2	M-S	x	SH	L
D, M	S, L	?	1'	2'	3-5	M-S	x	SH	L
Mo	S, L, C	?	1'	3'	5-10	L-S	x		L
Mo, W	S, L, C	?-8.0	1'-2'	3'	10-20	L-S	x		L
M, Mo	S, L, C	?-8.0	1'	3'	10-20	L-S	x	SH	L
M, Mo	S, L	?-7.5	1'	2'	5-10	L-S			L
M, Mo, W	S, L, C	?-8.0	1'	3'	5-10	L-S			L
M, Mo, W	S, L, C	?	1.5'-2'	3'-4'	10-20	L-S	x	FL/SH	M
D, M	S, L	5.5-8.0	2'	3'	20+	L-S		Leaves	L
D, M	S, L	5.0-7.5	2'	3'	10-20	L-S	x	SP	L
D, M	S, L	6.0-8.0	1'-2'	3'	10-20	M-S		SP	M
M, Mo	S, L, C	?-7.5	3'	3'	20+	L-S	x	SP	L
M	S, L, C	?-8.0	3'	3'	20+	L-S	x	SP	L

(continued)

TABLE 12.14. *(continued)*

Root type	Scientific name (4/1/2020)	Common name	Height	Flower color	Bloom time	USDA zones	Sun
TAP	*Baptisia bracteata*	Cream False Indigo	1'–2'	Cream	May–June	4–9	F, P
TAP	*Callirhoe involucrata*	Winecups	1'	Magenta	May–June	4–9	F, P
TAP	*Callirhoe triangulata*	Poppymallow	1'–2'	Magenta	July–Aug.	4–8	F
TAP	*Ceanothus americanus*	New Jersey Tea	2'–3'	White	July–Aug.	3–9	F, P
TAP	*Dalea candida*	White Prairie Clover	1'–2'	White	July–Aug.	3–9	F
TAP	*Dalea purpurea*	Purple Prairie Clover	1'–2'	Purple/yellow	July–Aug.	3–8	F
TAP	*Desmanthus illinoensis*	Illinois Bundleflower	2'–6'	White	June–Aug.	5–8	F
TAP	*Desmodium canadense*	Canada Ticktrefoil	3'–5'	Purple	July–Aug.	3–7	F
TAP	*Echinacea pallida*	Pale Purple Coneflower	3'–5'	Purple	June–July	4–8	F
TAP	*Echinacea paradoxa*	Ozark Coneflower	3'–5'	Yellow	June–July	4–7	F, P
TAP	*Lespedeza capitata*	Roundhead Bush Clover	3'–5'	White	Aug.–Sept.	3–8	F, P
TAP	*Lithospermum caroliniense*	Hairy Puccoon	1'–2'	Yellow	June–July	3–8	F
TAP	*Lithospermum incisum*	Fringed Puccoon	1'–2'	Yellow	June–July	3–9	F
TAP	*Lupinus perennis*	Wild Lupine	1'–2'	Blue	May–June	3–8	F, P
TAP	*Onosmodium bejariense* var. *occidentale*	Marbleseed (Western Marbleseed)	1'–3'	White	May–June	3–8	F
TAP	*Parthenium integrifolium*	Wild Quinine	3'–5'	White	June–Sept.	4–8	F
TAP	*Pediomelum esculentum*	Prairie Turnip	1'–2'	Blue	June–July	3–7	F
TAP	*Silene regia*	Royal Catchfly	2'–4'	Red	July–Aug.	4–9	F, P
TAP	*Silphium integrifolium*	Rosinweed	2'–6'	Yellow	July–Sept.	4–8	F
TAP	*Silphium laciniatum*	Compassplant	3'–10'	Yellow	June–Sept.	4–9	F
TAP	*Silphium terebinthinaceum*	Prairie Dock	3'–10'	Yellow	July–Sept.	4–7	F
TAP	*Tephrosia virginiana*	Goatsrue	1'–2'	Pink/yellow	June–Aug.	4–9	F, P
TAP/R	*Pycnanthemum tenuifolium*	Narrowleaf Mountainmint	1'–3'	White	July–Sept.	4–8	F, P
TAP/short Rs	*Blephilia ciliata*	Downy Wood Mint	1'–2'	White/purple/violet	May–July	4–8	F, P
TAPCO	*Liatris punctata*	Dotted Blazing Star	1'–2'	Purple/pink	Aug.–Oct.	3–9	F
R	*Anemone canadensis*	Canada Anemone	1'–2'	White	May–July	3–6	F, P
R	*Antennaria neglecta*	Pussytoes	6"–1'	White	Apr.–May	3–7	F
R	*Asclepias sullivantii*	Sullivant's Milkweed	3'–5'	Pink/yellow	June–Aug.	4–7	F
R	*Asclepias syriaca*	Common Milkweed	2'–4'	Lavender	June–Aug.	3–8	F, P
R	*Conoclinium coelestinum*	Blue Mistflower	1'–3'	Blue/purple	July–Oct.	4–10	F, P
R	*Coreopsis palmata*	Stiff Coreopsis	2'–3'	Yellow	June–Aug.	3–8	F
R	*Euphorbia corollata*	Flowering Spurge	2'–4'	White	July–Aug.	3–9	F
R	*Filipendula rubra*	Queen of the Prairie	4'–5'	Pink	June–July	3–6	F
R	*Geum triflorum*	Prairie Smoke	6"	Pink	May–June	3–6	F
R	*Helianthus mollis*	Downy Sunflower	4'–6'	Yellow	Aug.–Sept.	4–9	F
R	*Helianthus occidentalis*	Western Sunflower	2'–3'	Yellow	July–Aug.	3–9	F
R	*Helianthus* ×*laetiflorus*	Showy Sunflower	3'–6'	Yellow	Aug.–Sept.	3–8	F
R	*Hibiscus moscheutos*	Rosemallow	3'–7'	White/pink/red	July–Sept.	5–9	F, P
R	*Monarda fistulosa*	Bergamot	2'–5'	Lavender	July–Sept.	3–9	F, P
R	*Physostegia virginiana*	Obedient Plant	1'–2'	Pink	Aug.–Sept.	3–9	F

Soil moisture	Soil texture	pH range	Plant spacing in prairie gardens	Plant spacing in flower beds	Plant life expectancy in years	Aggressiveness and method	Cut flowers	Dried arrangements	Deer palatability
D, M	S, L	?	2'	3'	20+	L–S	x	SP	L
D, M	S, L	5.5–7.5	1'–2'	2'	10–20	L–S			L
D	S	?	1'–2'	2'	10–20	L–S			L
D, M	S, L	4.5–7.0	3'	3'	20+	L–S		SD	L
D, M	S, L	?	6"–1'	1'–2'	10–20	L–S			H
D, M	S, L, C	?	6"–1'	1'–2'	10–20	L–S	x		H
D, M, Mo	S, L, C	6.0–8.0	1'	1.5'	5–10	L–S		SP	M
M, Mo	S, L, C	?–7.5	NR	NR	10–20	M–S			H
D, M	S, L, C	6.0–7.5	1'	2'–3'	10–20	L–S	x	SH	M
D, M	S, L, C	?–7.5	1'	2'–3'	10–20	L–S	x	SH	M
D, M	S, G, L	5.5–8.0	1'	2'	10–20	L–S		SH	M
D	S	?	1'	1'	10–20	L–S			L
D	S, G	?	1'	1'	5–10	L–S			L
D	S	5.0–?	1'	1.5'	10–20	M–S			L
D, M	S, L	?–7.5	1'–1.5'	2'	10–20	L–S			L
M, Mo	S, L, C	?–8.0	1'–2'	3'	10–20	L–S	x		M
D	S, G	?–8.5	1'–1.5'	2'	20+	L–S			L
M	L	?	1'	1.5'	5–10	L–S	x	SH	L
M, Mo	S, L, C	?–8.0	2'	3'	20+	M–S	x		L
D, M	S, L, C	?–8.0	2'	4'	20+	L–S	x		L
M, Mo	S, L, C	?–8.0	2'	4'	20+	L–S	x		L
D	S, G	4.5–?	2'	3'	10–20	L–S			L
D, M	S, G, L, C	?	1'	1.5'	5–10	L–S		SH	M
D, M	G, L, C	?–8.00	1'	1'–2'	3–5	L–S		FL/SH	L
D	S, L	6.0–8.0	1'	1.5'–2'	10–20	L–S			L
M, Mo	S, L, C	?	NR	NR	10–20	H–R			L
D	S, G	5.5–7.5	1'	1'	5–10	L–R	x	FL	L
M, Mo	L, C	?	1'–2'	3'	10–20	L–R		SP	L
D, M	S, L, C	5.0–8.0	NR	NR	10–20	H–R		SP	L
M, Mo	S, L, C	5.0–8.0	1'–2'	2'	3–5	M–R			L
D, M	S, L	?	1'–2'	2'	20+	M–R			L
D, M	S, L	?	1'–2'	1'–2'	10–20	M–R			L
M, Mo	S, L, C	?–8.0	2'	3'	20+	M–R	x		L
D, M	S, L	6.0–7.5	6"	1'	20+	L–R		SH	L
D, M	S, L, C	6.0–8.0	2'	3'	20+	M–R	x	SH	L
D, M	S, L	?	2'	3'	20+	M–R	x		L
D, M	S, G, L, C	?	NR	NR	20+	H–R	x	SH	L
M, Mo, W	S, L, C	4.0–8.0	2'	4'	10–20	L–R/S			L
D, M, Mo	S, L, C	6.0–8.0	1'–2'	3'	10–20	M–R	x	SH	L
M, Mo	S, L, C	6.0–8.0	1'–2'	NR	10–20	H–R	x	SH	M

(continued)

TABLE 12.14. *(continued)*

Root type	Scientific name (4/1/2020)	Common name	Height	Flower color	Bloom time	USDA zones	Sun
R	*Polygonatum biflorum* var. *commutatum*	Great Solomon's Seal	2'–4'	Cream	May–June	3–9	F, P
R	*Pycnanthemum virginianum*	Virginia Mountainmint	1'–3'	White	July–Sept.	3–7	F, P
R	*Rosa carolina*	Carolina Rose	1'–2'	Pink	June–July	3–8	F
R	*Spiraea tomentosa*	Steeplebush	2'–4'	Pink	Aug.–Sept.	3–8	F, P
R	*Symphyotrichum ericoides*	Heath Aster	1'–3'	White	Aug.–Oct.	3–9	F
R	*Symphyotrichum oblongifolium*	Aromatic Aster	1'–2'	Purple/violet	Sept.–Oct.	3–8	F
R	*Tradescantia bracteata*	Prairie Spiderwort	1'–2'	Blue	June–July	3–7	F, P
R	*Tradescantia occidentalis*	Western Spiderwort	1'	Pink	June–July	3–9	F
R	*Viola pedata*	Birdfoot Violet	6"	Blue/purple	Apr.–June	3–9	F, P
R/CO	*Iris virginica* var. *shrevei*	Wild Iris	2'–3'	Blue	June–July	3–8	F, P

Root type	Scientific name (4/1/2020)	Common name	Height	Fall leaf color	Bloom time	USDA zones	Sun
Grasses and sedges							
FIB	*Andropogon gerardii*	Big Bluestem	5'–8'	Bronze/red	Aug.–Oct.	3–9	F
FIB	*Andropogon virginicus*	Broomsedge	2'–5'	Reddish brown	Aug.–Oct.	5–10	F
FIB	*Bouteloua curtipendula*	Sideoats Grama	2'–3'	Straw	Aug.–Sept.	3–9	F
FIB	*Bromus kalmii*	Prairie Bromegrass	2'–4'	Straw	June–July	3–6	F
FIB	*Calamagrostis canadensis*	Bluejoint Grass	4'–6'	Straw	June–July	3–6	F
FIB	*Carex brevior*	Shortbeak Sedge	1'–4'	Straw	May–June	3–7	F, P
FIB	*Elymus canadensis*	Canada Wildrye	4'–5'	Straw	July–Aug.	2–9	F
FIB	*Elymus virginicus*	Virginia Wildrye	4'–5'	Straw	July–Aug.	2–9	F, P, S
FIB	*Eragrostis spectabilis*	Purple Lovegrass	1'–2'	Purple/pink	July–Sept.	3–10	F, P
FIB	*Hesperostipa spartea*	Needlegrass	2'–4'	Gold	Apr.–June	3–6	F
FIB	*Koeleria macrantha*	Junegrass	2'–3'	Gold	May–June	3–8	F
FIB	*Panicum virgatum*	Switchgrass	3'–6'	Gold	Aug.–Sept.	3–10	F
FIB	*Schizachyrium scoparium*	Little Bluestem	2'–5'	Crimson red	Aug.–Oct.	3–10	F
FIB	*Sorghastrum nutans*	Indiangrass	5'–7'	Gold	Aug.–Sept.	3–9	F
FIB	*Sporobolus heterolepis*	Prairie Dropseed	2'–4'	Gold	Aug.–Sept.	3–8	F
FIB	*Tridens flavus*	Purpletop	3'–6'	Straw	July–Sept.	5–10	F
FIB	*Tripsacum dactyloides*	Eastern Gamagrass	4'–8'	Gold	July–Sept.	4–10	F, P
R	*Hierochloe hirta*	Vanilla Sweetgrass	1'–2'	Straw	May–June	3–9	F, P
R	*Spartina pectinata*	Prairie Cordgrass	6'–9'	Gold	Aug.–Sept.	3–8	F

Notes: For soil moisture, D = dry, M = medium, Mo = moist, and W = wet. For soil texture, S = sand, L = loam, C = clay, and G = gravel. For sun, F = full sun, or at least a half day of full sun; P = partial sun, or a quarter to a half day of sun; and S = shade, or less than a quarter day of sun. For plant spacing, NR = not recommended. For root type, B = bulb, CO = corm, FIB = fibrous, R = rhizome, TAP =

Soil moisture	Soil texture	pH range	Plant spacing in prairie gardens	Plant spacing in flower beds	Plant life expectancy in years	Aggressiveness and method	Cut flowers	Dried arrangements	Deer palatability
D, M	S, L, C	?	1'-2'	3'	10-20	L-R/S			H
Mo, W	S, L, C	?-8.0	1'-2'	3'	5-10	L-R		SH	L
D, M	S, L	4.0-7.0	2'	3'	10-20	L-R/S	x	RH	L
Mo, W	S, L, C	4.5-7.0	2'	3'	10-20	L-R/S	x		L
D, M	S, G, L	?	1'-2'	2'	10-20	M-R		Empty SH	L
D, M	S, L, C	?-8.0	1'	3'	5-10	M-R?		Empty SH	L
D, M	S, L, C	?	1'	2'	20+	M-R			M
D, M	S, L	?	1'	NR	20+	H-R			M
D	S, G	?	6"	1'	5-10	L-S			L
W	S, L, C	4.8-7.3	1'	2'	10-20	L-R	x	SP	L

Soil moisture	Soil texture	pH range	Plant spacing in prairie gardens	Plant spacing in flower beds	Plant life expectancy in years	Aggressiveness and method	Cut flower stalks	Dried seed arrangements	Deer palatability
D, M, Mo	S, L, C	5.0-7.5	2'	NR	20+	M-S		x	L
D, M, Mo	S, L, C	4.0-7.0	1'-1.5'	NR	20+	H-S		x	L
D, M	S, L	5.5-8.5	1'	2'	10-20	L-S		x	L
D	S, G	5.7-7.0	1'	2'	3-5	L-S		x	L
Mo, W	S, L, C	4.5-8.0	1'	3'	10-20	M-R/S		x	L
D, M, Mo	S, L, C	4.5-7.5	1'	1'	5-10	L-S			L
D, M, Mo	S, L, C	5.0-8.0	1'	3'	3-5	M-S		x	L
M, Mo	S, L, C	5.0-7.5	1'	3'	3-5	M-S		x	L
D	S, G	4.0-7.5	1'-1.5'	2'	5-10	M-S	x	x	L
D, M	S, G, L	5.0-7.5	1'	NR	10-20	L-S		x	L
D	S, G	6.0-8.0	6"-1'	1'	5-10	L-S		x	L
D, M, Mo	S, L, C	4.5-8.0	1.5'-2'	NR	20+	M-S		x	L
D, M	S, L	5.0-8.5	1'-1.5'	3'	20+	L-S		x	L
D, M, Mo	S, L, C	5.0-8.0	1'-2'	3'	10-20	M-S	x	x	L
D, M	S, L	6.0-8.0	2'	2'	20+	L-S	x	x	L
D, M	S, L, C	5.0-7.5	1'	1.5'	5-10	M-S		x	L
M, Mo, W	S, L, C	4.5-7.5	2'	3'	20+	H-S		x	M
M, Mo, W	S, L, C	5.7-8.0	1'	NR	10-20	H-R			L
Mo, W	S, L, C	6.0-8.5	2'	NR	20+	H-R		x	L

taproot, and TAPCO = taprooted corm. For aggressiveness and method, L = low, M = medium, H = high, R = by rhizome, S = by seed, and R/S = rhizome and seed. For dried arrangements, FL = flowers, SH = seedheads, SP = seedpods, SD = seed, and RH = rose hips. For deer palatability, L = low, M = medium, and H = high.

TABLE 12.15. DEER-RESISTANT PRAIRIE PLANTS

Deer palatability	Scientific name (4/1/2020)	Common name	Height	Flower color	Bloom time	USDA zones	Sun
Flowers and shrubs							
L	*Agastache foeniculum*	Lavender Hyssop	1'–3'	Purple	July–Sept.	2–6	F, P
L	*Allium cernuum*	Nodding Pink Onion	1'–2'	White/pink	June–Aug.	3–8	F, P
L	*Allium stellatum*	Prairie Onion	1'–2'	Lavender	July–Aug.	3–8	F, P
L	*Amorpha canescens*	Leadplant	2'–3'	Purple	June–July	3–8	F, P
L	*Amsonia tabernaemontana*	Eastern Bluestar	2'–3'	Blue/white	May–June	4–8	F, P
L	*Anemone canadensis*	Canada Anemone	1'–2'	White	May–July	3–6	F, P
L	*Anemone cylindrica*	Thimbleweed	1'–3'	White	May–June	3–6	F
L	*Antennaria neglecta*	Pussytoes	6"–1'	White	Apr.–May	3–7	F
L	*Arnoglossum atriplicifolium*	Pale Indian Plantain	5'–10'	White	July–Sept.	4–8	F, P
L	*Asclepias incarnata*	Marsh Milkweed	3'–5'	Red/pink	June–July	3–9	F
L	*Asclepias sullivantii*	Sullivant's Milkweed	3'–5'	Pink/yellow	June–Aug.	4–7	F
L	*Asclepias syriaca*	Common Milkweed	2'–4'	Lavender	June–Aug.	3–8	F, P
L	*Asclepias tuberosa*	Butterflyweed	2'–3'	Orange	June–Aug.	3–10	F
L	*Baptisia alba*	White False Indigo	3'–5'	White	June–July	4–9	F, P
L	*Baptisia australis*	Blue False Indigo	3'–5'	Blue	June–July	3–10	F, P
L	*Baptisia bracteata*	Cream False Indigo	1'–2'	Cream	May–June	4–9	F, P
L	*Blephilia ciliata*	Downy Wood Mint	1'–2'	White/purple/violet	May–July	4–8	F, P
L	*Callirhoe involucrata*	Winecups	1'	Magenta	May–June	4–9	F, P
L	*Callirhoe triangulata*	Poppymallow	1'–2'	Magenta	July–Aug.	4–8	F
L	*Ceanothus americanus*	New Jersey Tea	2'–3'	White	July–Aug.	3–9	F, P
L	*Conoclinium coelestinum*	Blue Mistflower	1'–3'	Blue/purple	July–Oct.	4–10	F, P
L	*Coreopsis palmata*	Stiff Coreopsis	2'–3'	Yellow	June–Aug.	3–8	F
L	*Coreopsis tripteris*	Tall Coreopsis	3'–7'	Yellow	July–Sept.	3–8	F, P
L	*Delphinium carolinianum* subsp. *virescens*	Prairie Larkspur	1'–4'	White	June–July	3–8	F
L	*Dodecatheon meadia*	Shooting Star	1'–2'	White/pink	May–June	4–8	F, P
L	*Eryngium yuccifolium*	Rattlesnake Master	3'–5'	White	June–Aug.	4–9	F
L	*Eupatorium perfoliatum*	Boneset	3'–4'	White	Aug.–Sept.	3–8	F
L	*Euphorbia corollata*	Flowering Spurge	2'–4'	White	July–Aug.	3–9	F
L	*Eutrochium maculatum*	Joe Pye Weed	4'–6'	Pink	Aug.–Sept.	3–6	F, P
L	*Filipendula rubra*	Queen of the Prairie	4'–5'	Pink	June–July	3–6	F
L	*Gentiana alba*	Cream Gentian	1'–2'	White	July–Aug.	3–7	F, P
L	*Gentiana andrewsii*	Bottle Gentian	1'–2'	Blue	Aug.–Oct.	3–6	F, P
L	*Geum triflorum*	Prairie Smoke	6"	Pink	May–June	3–6	F
L	*Helenium autumnale*	Sneezeweed	4'–5'	Yellow	Aug.–Sept.	3–9	F
L	*Helianthus mollis*	Downy Sunflower	4'–6'	Yellow	Aug.–Sept.	4–9	F
L	*Helianthus occidentalis*	Western Sunflower	2'–3'	Yellow	July–Aug.	3–9	F
L	*Helianthus ×laetiflorus*	Showy Sunflower	3'–6'	Yellow	Aug.–Sept.	3–8	F
L	*Heliopsis helianthoides*	Oxeye Sunflower	3'–6'	Yellow	June–Sept.	3–8	F

Soil moisture	Soil texture	pH range	Plant spacing in prairie gardens	Plant spacing in flower beds	Root type	Plant life expectancy in years	Aggressiveness and method	Cut flowers	Dried arrangements
D, M	S, L	?-8.0	1'	1'-2'	FIB	3-5	M-S	x	FL/SH
M, Mo	S, L, C	6.5-8.0	6"-1'	1'	B	10-20	L-S	x	FL/SH
D, M	S, L	5.5-8.0	6"-1'	1'	B	10-20	L-S	x	FL/SH
D, M	S, L	5.5-8.0	2'	3'	TAP	20+	L-S		Leaves
M, Mo	S, L, C	?-8.0	2'-3'	3'	FIB	20+	L-S		SP
M, Mo	S, L, C	?	NR	NR	R	10-20	H-R		
D, M	S, L	?	1'	1'	FIB	10-20	L-S		SD
D	S, G	5.5-7.5	1'	1'	R	5-10	L-R	x	FL
D, M, Mo	S, L, C	?-8.0	2'	3'-4'	FIB	3-20	H-S (north); L-S (south)	x	FL
Mo, W	S, L, C	5.0-8.0	1'-2'	3'	FIB	5-10	L-S	x	SP
M, Mo	L, C	?	1'-2'	3'	R	10-20	L-R		SP
D, M	S, L, C	5.0-8.0	NR	NR	R	10-20	H-R		SP
D, M	S, L	5.0-7.5	2'	3'	TAP	10-20	L-S	x	SP
M, Mo	S, L, C	?-7.5	3'	3'	TAP	20+	L-S	x	SP
M	S, L, C	?-8.0	3'	3'	TAP	20+	L-S	x	SP
D, M	S, L	?	2'	3'	TAP	20+	L-S	x	SP
D, M	G, L, C	?-8.00	1'	1'-2'	TAP/short Rs	3-5	L-S		FL/SH
D, M	S, L	5.5-7.5	1'-2'	2'	TAP	10-20	L-S		
D	S	?	1'-2'	2'	TAP	10-20	L-S		
D, M	S, L	4.5-7.0	3'	3'	TAP	20+	L-S		SD
M, Mo	S, L, C	5.0-8.0	1'-2'	2'	R	3-5	M-R		
D, M	S, L	?	1'-2'	2'	R	20+	M-R		
M, Mo, W	S, L, C	?	1'-2'	NR	FIB	10-20	L-S	x	
D, M	S, L	?	1'	1.5'	FIB	5-10	L-S	x	SP
M, Mo	S, L, C	4.5-8.0	6"-1'	1'	FIB	20+	L-S		SP
D, M	S, L, C	6.0-8.0	1'	2'	FIB	5-10	M-S	x	SH
M, Mo, W	S, L, C	?-7.5	1'	NR	FIB	5-10	L-S		
D, M	S, L	?	1'-2'	1'-2'	R	10-20	M-R		
Mo, W	S, L, C	5.5-7.5	1'-2'	3'-4'	FIB	5-10	L-S		
M, Mo	S, L, C	?-8.0	2'	3'	R	20+	M-R	x	
M	S, L, C	?-7.5	1'	1.5'	FIB	10-20	L-S	x	
Mo, W	S, L, C	5.8-7.2	1'	1.5'	FIB	5-10	L-S	x	
D, M	S, L	6.0-7.5	6"	1'	R	20+	L-R		SH
Mo, W	S, L, C	4.0-7.5	1'-2'	2'-3'	FIB	5-10	L-S	x	SH
D, M	S, L, C	6.0-8.0	2'	3'	R	20+	M-R	x	SH
D, M	S, L	?	2'	3'	R	20+	M-R	x	
D, M	S, G, L, C	?	NR	NR	R	20+	H-R	x	SH
D, M, Mo	S, L, C	6.0-8.0	1'-2'	3'	FIB	5-10	H-S	x	

(continued)

TABLE 12.15. *(continued)*

Deer palatability	Scientific name (4/1/2020)	Common name	Height	Flower color	Bloom time	USDA zones	Sun
L	*Hibiscus moscheutos*	Rosemallow	3'–7'	White/pink/red	July–Sept.	5–9	F, P
L	*Hypericum ascyron*	Great St. Johnswort	4'–6'	Yellow	July–Aug.	3–5	F, P
L	*Iris virginica var. shrevei*	Wild Iris	2'–3'	Blue	June–July	3–8	F, P
L	*Liatris aspera*	Rough Blazing Star	2'–5'	Purple/pink	Aug.–Sept.	3–8	F
L	*Liatris ligulistylis*	Meadow Blazing Star	3'–5'	Purple/pink	Aug.–Sept.	3–6	F
L	*Liatris punctata*	Dotted Blazing Star	1'–2'	Purple/pink	Aug.–Oct.	3–9	F
L	*Liatris pycnostachya*	Prairie Blazing Star	3'–5'	Purple/pink	July–Aug.	3–9	F
L	*Liatris spicata*	Dense Blazing Star	3'–6'	Purple/pink	Aug.–Sept.	4–10	F
L	*Liatris squarrosa*	Scaly Blazing Star	1'–2'	Purple/pink	Aug.–Sept.	4–8	F
L	*Lithospermum canescens*	Hoary Puccoon	1'	Orange	May–June	3–7	F, P
L	*Lithospermum caroliniense*	Hairy Puccoon	1'–2'	Yellow	June–July	3–8	F
L	*Lithospermum incisum*	Fringed Puccoon	1'–2'	Yellow	June–July	3–9	F
L	*Lobelia siphilitica*	Great Blue Lobelia	1'–4'	Blue	July–Sept.	3–9	F, P
L	*Lupinus perennis*	Wild Lupine	1'–2'	Blue	May–June	3–8	F, P
L	*Monarda fistulosa*	Bergamot	2'–5'	Lavender	July–Sept.	3–9	F, P
L	*Monarda punctata*	Dotted Mint	1'–2'	Lavender	July–Sept.	3–10	F
L	*Oenothera gaura*	Biennial Beeblossom	3'–6'	White/pink	June–Sept.	4–8	F, P
L	*Oligoneuron ohioense*	Ohio Goldenrod	3'–4'	Yellow	Aug.–Sept.	4–5	F, P
L	*Oligoneuron rigidum*	Stiff Goldenrod	3'–5'	Yellow	Aug.–Sept.	3–9	F
L	*Onosmodium bejariense var. occidentale*	Marbleseed (Western Marbleseed)	1'–3'	White	May–June	3–8	F
L	*Pediomelum esculentum*	Prairie Turnip	1'–2'	Blue	June–July	3–7	F
L	*Penstemon cobaea*	Cobaea Beardtongue	2'–3'	Lavender	Apr.–June	5–8	F
L	*Penstemon digitalis*	Foxglove Beardtongue	2'–3'	White	June–July	3–8	F, P
L	*Penstemon gracilis*	Slender Beardtongue	1'–2'	Lavender	May–June	3–6	F, P
L	*Penstemon grandiflorus*	Large-Flowered Beardtongue	2'–4'	Lavender	May–June	3–7	F
L	*Penstemon hirsutus*	Hairy Penstemon	1'–2'	Lavender	May–June	3–6	F, P
L	*Phlox pilosa*	Downy Phlox	1'–2'	Pink	May–June	3–9	F
L	*Pycnanthemum virginianum*	Virginia Mountainmint	1'–3'	White	July–Sept.	3–7	F, P
L	*Ratibida pinnata*	Yellow Coneflower	3'–6'	Yellow	July–Sept.	3–9	F
L	*Rosa carolina*	Carolina Rose	1'–2'	Pink	June–July	3–8	F
L	*Rosa setigera*	Climbing Prairie Rose	6'–15'	Pink/white	May–July	4–9	F, P
L	*Rudbeckia fulgida*	Orange Coneflower	2'–4'	Yellow	July–Sept.	4–8	F, P
L	*Rudbeckia hirta*	Blackeyed Susan	1'–3'	Yellow	June–Sept.	3–10	F, P
L	*Rudbeckia laciniata*	Green-Headed Coneflower	5'–8'	Yellow	Aug.–Sept.	3–8	F, P
L	*Rudbeckia subtomentosa*	Sweet Blackeyed Susan	4'–6'	Yellow	Aug.–Oct.	4–8	F, P
L	*Rudbeckia triloba*	Browneyed Susan	2'–5'	Yellow	July–Oct.	4–9	F, P

Soil moisture	Soil texture	pH range	Plant spacing in prairie gardens	Plant spacing in flower beds	Root type	Plant life expectancy in years	Aggressiveness and method	Cut flowers	Dried arrangements
M, Mo, W	S, L, C	4.0–8.0	2'	4'	R	10–20	L–R/S		
Mo, W	S, L, C	5.7–7.1	1'–2'	3'	FIB	5–10	L–S	x	SP
W	S, L, C	4.8–7.3	1'	2'	R/CO	10–20	L–R	x	SP
D, M	S, L	?	6"–1'	1'	CO	10–20	L–S	x	FL/SH
M, Mo	L	?	6"–1'	1'	CO	3–5	L–S	x	FL/SH
D	S, L	6.0–8.0	1'	1.5'–2'	TAPCO	10–20	L–S		
M, Mo	S, L, C	6.0–8.0	6"–1'	1'	CO	10–20	L–S	x	FL/SH
M, Mo	S, L, C	5.5–7.5	6"–1'	1'	CO	10–20	L–S	x	FL/SH
D, M	S, L	?	1'	1.5'–2'	CO	10–20	L–S	x	FL/SH
D, M	S, L	?	1'	1'	FIB	5–10	L–S		
D	S	?	1'	1'	TAP	10–20	L–S		
D	S, G	?	1'	1'	TAP	5–10	L–S		
M, Mo	S, L, C	?–8.0	6"–1'	2'	FIB	5–10	M–S	x	
D	S	5.0–?	1'	1.5'	TAP	10–20	M–S		
D, M, Mo	S, L, C	6.0–8.0	1'–2'	3'	R	10–20	M–R	x	SH
D	S, G	5.0–?	1'	1'–2'	FIB	2	M–S	x	SH
D, M, Mo	S, L, C	?	2'	3'	FIB	2	M–S	x	
M, Mo	S, L, C	?–8.0	1'–2'	3'	FIB	5–10	L–S	x	SH
D, M, Mo	S, L, C	5.5–8.0	1'–1.5'	3'	FIB	5–10	M–S	x	SH
D, M	S, L	?–7.5	1'–1.5'	2'	TAP	10–20	L–S		
D	S, G	?–8.5	1'–1.5'	2'	TAP	20+	L–S		
D	S, G	?–7.5	1'	1'	FIB	3–5	L–S	x	SP
M, Mo	S, L, C	5.5–7.5	1'	2'	FIB	5–10	L–S	x	
D	S, G	?–7.5	6"–1'	1'	FIB	5–10	L–S	x	SP
D	S, G	?	1'	2'–3'	FIB	3–5	L–S		SP
D	S, G	5.0–8.0	1'	1.5'	FIB	3–5	L–S	x	SP
D, M, Mo	S, L	?	6"–1'	1.5'	FIB	3–5	L–S	x	
Mo, W	S, L, C	?–8.0	1'–2'	3'	R	5–10	L–R		SH
D, M, Mo	S, L, C	5.5–8.0	1'–2'	2'	FIB	10–20	M–S	x	SH
D, M	S, L	4.0–7.0	2'	3'	R	10–20	L–R/S	x	RH
M, Mo	S, L, C	5.0–8.0	2'	5'	FIB	10–20	L–S	x	RH
M, Mo	L, C	?–8.0	1'–2'	3'	FIB	5–10	L–S	x	SH
D, M, Mo	S, L, C	5.0–8.0	1'	2'	FIB	2	M–S	x	SH
Mo, W	S, L, C	4.5–7.5	2'	2'	FIB	3–5	M–S	x	SH
M, Mo	S, L, C	?–8.0	1'–2'	2'	FIB	10–20	L–S	x	SH
M, Mo	S, L	?–8.0	1'–2'	3'	FIB	2	M–S	x	SH

(continued)

TABLE 12.15. (*continued*)

Deer palatability	Scientific name (4/1/2020)	Common name	Height	Flower color	Bloom time	USDA zones	Sun
L	*Ruellia humilis*	Wild Petunia	1'–2'	Violet	June–Aug.	4–9	F
L	*Salvia azurea*	Blue Sage	3'–5'	Blue	June–Aug.	4–9	F, P
L	*Silene regia*	Royal Catchfly	2'–4'	Red	July–Aug.	4–9	F, P
L	*Silphium integrifolium*	Rosinweed	2'–6'	Yellow	July–Sept.	4–8	F
L	*Silphium laciniatum*	Compassplant	3'–10'	Yellow	June–Sept.	4–9	F
L	*Silphium perfoliatum*	Cup Plant	3'–10'	Yellow	July–Sept.	3–8	F, P
L	*Silphium terebinthinaceum*	Prairie Dock	3'–10'	Yellow	July–Sept.	4–7	F
L	*Sisyrinchium albidum*	White Blue-Eyed Grass	6"–1'	White	Apr.–May	4–9	F
L	*Sisyrinchium angustifolium*	Narrow-Leaved Blue-Eyed Grass	6"–2'	Blue/purple	Apr.–June	3–10	F, P
L	*Sisyrinchium campestre*	Prairie Blue-Eyed Grass	6"–1'	Blue/white	Apr.–June	3–8	F, P
L	*Solidago ptarmicoides*	White Aster	1'–2'	White	Aug.–Sept.	3–7	F
L	*Solidago speciosa*	Showy Goldenrod	1'–3'	Yellow	Aug.–Sept.	3–8	F
L	*Spiraea alba*	Meadowsweet	2'–4'	White/pink	June–Sept.	3–7	F, P
L	*Spiraea tomentosa*	Steeplebush	2'–4'	Pink	Aug.–Sept.	3–8	F, P
L	*Symphyotrichum ericoides*	Heath Aster	1'–3'	White	Aug.–Oct.	3–9	F
L	*Symphyotrichum oblongifolium*	Aromatic Aster	1'–2'	Purple/violet	Sept.–Oct.	3–8	F
L	*Tephrosia virginiana*	Goatsrue	1'–2'	Pink/yellow	June–Aug.	4–9	F, P
L	*Thalictrum dasycarpum*	Tall Meadow-Rue	3'–6'	White	June–July	3–9	F, P
L	*Verbena hastata*	Blue Vervain	3'–6'	Blue	July–Sept.	3–8	F, P
L	*Verbena stricta*	Hoary Vervain	2'–4'	Blue	July–Sept.	3–8	F
L	*Vernonia fasciculata*	Ironweed	4'–6'	Red/pink	July–Sept.	3–7	F
L	*Vernonia gigantea*	Tall Ironweed	5'–8'	Red/pink	Aug.–Sept.	4–9	F, P
L	*Veronicastrum virginicum*	Culver's Root	3'–6'	White	July–Aug.	3–8	F, P, S
L	*Viola pedata*	Birdfoot Violet	6"	Blue/purple	Apr.–June	3–9	F, P
L	*Zizia aptera*	Heartleaf Golden Alexanders	1'–2'	Yellow	May–June	3–8	F
L	*Zizia aurea*	Golden Alexanders	1'–2'	Yellow	May–July	3–8	F, P, S
M	*Astragalus canadensis*	Canada Milkvetch	2'–3'	Yellow	July–Aug.	3–8	F
M	*Chelone glabra*	White Turtlehead	2'–4'	White	Aug.–Sept.	3–8	F, P
M	*Coreopsis lanceolata*	Lanceleaf Coreopsis	1'–2'	Yellow	June–July	3–9	F
M	*Desmanthus illinoensis*	Illinois Bundleflower	2'–6'	White	June–Aug.	5–8	F
M	*Echinacea pallida*	Pale Purple Coneflower	3'–5'	Purple	June–July	4–8	F
M	*Echinacea paradoxa*	Ozark Coneflower	3'–5'	Yellow	June–July	4–7	F, P
M	*Echinacea purpurea*	Purple Coneflower	3'–4'	Purple	July–Sept.	4–8	F, P
M	*Lespedeza capitata*	Roundhead Bush Clover	3'–5'	White	Aug.–Sept.	3–8	F, P
M	*Lobelia cardinalis*	Cardinal Flower	2'–5'	Red	July–Sept.	3–9	F, P
M	*Parthenium integrifolium*	Wild Quinine	3'–5'	White	June–Sept.	4–8	F
M	*Physostegia virginiana*	Obedient Plant	1'–2'	Pink	Aug.–Sept.	3–9	F
M	*Pycnanthemum tenuifolium*	Narrowleaf Mountainmint	1'–3'	White	July–Sept.	4–8	F, P
M	*Senna hebecarpa*	Wild Senna	4'–6'	Yellow	July–Aug.	4–7	F

Soil moisture	Soil texture	pH range	Plant spacing in prairie gardens	Plant spacing in flower beds	Root type	Plant life expectancy in years	Aggressiveness and method	Cut flowers	Dried arrangements
D, M	S, L	4.5-7.5	1'	2'	FIB	5-10	L-S		
D, M	S, G, L	?	1'	1.5'	FIB	5-10	L-S	x	SH
M	L	?	1'	1.5'	TAP	5-10	L-S	x	SH
M, Mo	S, L, C	?-8.0	2'	3'	TAP	20+	M-S	x	
D, M	S, L, C	?-8.0	2'	4'	TAP	20+	L-S	x	
M, Mo, W	S, L, C	4.5-8.0	2'-3'	NR	FIB	20+	H-S	x	
M, Mo	S, L, C	?-8.0	2'	4'	TAP	20+	L-S	x	
D, M	S, L	?-8.0	6"	6"	FIB	3-5	L-S		
M, Mo	S, L, C	?-8.0	6"	9"	FIB	3-5	L-S		
D, M	S, L	?-7.5	6"	6"	FIB	3-5	L-S		
D	S, G	6.5-8.0	1'	2'	FIB	5-10	L-S		Empty SH
D, M	S, L	?	1'	3'	FIB	5-10	L-S	x	SH
Mo, W	S, L, C	4.3-7.2	2'	3'	FIB	10-20	L-S	x	
Mo, W	S, L, C	4.5-7.0	2'	3'	R	10-20	L-R/S	x	
D, M	S, G, L	?	1'-2'	2'	R	10-20	M-R		Empty SH
D, M	S, L, C	?-8.0	1'	3'	R	5-10	M-R?		Empty SH
D	S, G	4.5-?	2'	3'	TAP	10-20	L-S		
Mo, W	S, L, C	?	1'-2'	2'	FIB	5-10	L-S		
M, Mo, W	S, L, C	?	1'	2'	FIB	2	M-S	x	SH
D, M	S, L	?	1'	2'	FIB	3-5	M-S	x	SH
Mo	S, L, C	?	1'	3'	FIB	5-10	L-S	x	
Mo, W	S, L, C	?-8.0	1'-2'	3'	FIB	10-20	L-S	x	
M, Mo	S, L, C	?-8.0	1'	3'	FIB	10-20	L-S	x	SH
D	S, G	?	6"	1'	R	5-10	L-S		
M, Mo	S, L	?-7.5	1'	2'	FIB	5-10	L-S		
M, Mo, W	S, L, C	?-8.0	1'	3'	FIB	5-10	L-S		
D, M	S, L	6.0-8.0	1'-2'	3'	TAP	10-20	M-S		SP
Mo, W	S, L, C	?	1'	1'-2'	FIB	5-10	L-S		
D, M	S, L	5.0-7.5	1'	1'-2'	FIB	3-5	M-S	x	
D, M, Mo	S, L, C	6.0-8.0	1'	1.5'	TAP	5-10	L-S		SP
D, M	S, L, C	6.0-7.5	1'	2'-3'	TAP	10-20	L-S	x	SH
D, M	S, L, C	?-7.5	1'	2'-3'	TAP	10-20	L-S	x	SH
D, M	S, L, C	5.5-8.0	1'-2'	3'	FIB	5-10	M-S	x	SH
D, M	S, G, L	5.5-8.0	1'	2'	TAP	10-20	L-S		SH
Mo, W	S, L	5.8-7.8	6"-1'	2'	FIB	3-5	L-S	x	
M, Mo	S, L, C	?-8.0	1'-2'	3'	TAP	10-20	L-S	x	
M, Mo	S, L, C	6.0-8.0	1'-2'	NR	R	10-20	H-R	x	SH
D, M	S, G, L, C	?	1'	1.5'	TAP/R	5-10	L-S		SH
M, Mo, W	S, L, C	?	1.5'-2'	3'-4'	FIB/TAP	10-20	L-S	x	FL/SH

(continued)

TABLE 12.15. (continued)

Deer palatability	Scientific name (4/1/2020)	Common name	Height	Flower color	Bloom time	USDA zones	Sun
M	Symphyotrichum oolentangiense	Skyblue Aster	2'–3'	Blue	Aug.–Oct.	3–8	F, P
M	Tradescantia bracteata	Prairie Spiderwort	1'–2'	Blue	June–July	3–7	F, P
M	Tradescantia occidentalis	Western Spiderwort	1'	Pink	June–July	3–9	F

Deer palatability	Scientific name (4/1/2020)	Common name	Height	Fall leaf color	Bloom time	USDA zones	Sun
Grasses and sedges							
L	Andropogon gerardii	Big Bluestem	5'–8'	Bronze/red	Aug.–Oct.	3–9	F
L	Andropogon virginicus	Broomsedge	2'–5'	Reddish brown	Aug.–Oct.	5–10	F
L	Bouteloua curtipendula	Sideoats Grama	2'–3'	Straw	Aug.–Sept.	3–9	F
L	Bromus kalmii	Prairie Bromegrass	2'–4'	Straw	June–July	3–6	F
L	Calamagrostis canadensis	Bluejoint Grass	4'–6'	Straw	June–July	3–6	F
L	Carex brevior	Shortbeak Sedge	1'–4'	Straw	May–June	3–7	F, P
L	Elymus canadensis	Canada Wildrye	4'–5'	Straw	July–Aug.	2–9	F
L	Elymus virginicus	Virginia Wildrye	4'–5'	Straw	July–Aug.	2–9	F, P, S
L	Eragrostis spectabilis	Purple Lovegrass	1'–2'	Purple/pink	July–Sept.	3–10	F, P
L	Hesperostipa spartea	Needlegrass	2'–4'	Gold	Apr.–June	3–6	F
L	Hierochloe hirta	Vanilla Sweetgrass	1'–2'	Straw	May–June	3–9	F, P
L	Koeleria macrantha	Junegrass	2'–3'	Gold	May–June	3–8	F
L	Panicum virgatum	Switchgrass	3'–6'	Gold	Aug.–Sept.	3–10	F
L	Schizachyrium scoparium	Little Bluestem	2'–5'	Crimson red	Aug.–Oct.	3–10	F
L	Sorghastrum nutans	Indiangrass	5'–7'	Gold	Aug.–Sept.	3–9	F
L	Spartina pectinata	Prairie Cordgrass	6'–9'	Gold	Aug.–Sept.	3–8	F
L	Sporobolus heterolepis	Prairie Dropseed	2'–4'	Gold	Aug.–Sept.	3–8	F
L	Tridens flavus	Purpletop	3'–6'	Straw	July–Sept.	5–10	F
M	Tripsacum dactyloides	Eastern Gamagrass	4'–8'	Gold	July–Sept.	4–10	F, P

Notes: For soil moisture, D = dry, M = medium, Mo = moist, and W = wet. For soil texture, S = sand, L = loam, C = clay, and G = gravel. For sun, F = full sun, or at least a half day of full sun; P = partial sun, or a quarter to a half day of sun; and S = shade, or less than a quarter day of sun. For plant spacing, NR = not recommended. For root type, B = bulb, CO = corm, FIB = fibrous, R = rhizome, TAP =

Soil moisture	Soil texture	pH range	Plant spacing in prairie gardens	Plant spacing in flower beds	Root type	Plant life expectancy in years	Aggressiveness and method	Cut flowers	Dried arrangements
D, M	S, L	5.0–8.0	1'	2'	FIB	5–10	L–S	x	Empty SH
D, M	S, L, C	?	1'	2'	R	20+	M–R		
D, M	S, L	?	1'	NR	R	20+	H–R		

Soil moisture	Soil texture	pH range	Plant spacing in prairie gardens	Plant spacing in flower beds	Root type	Plant life expectancy in years	Aggressiveness and method	Cut flower stalks	Dried seed arrangements
D, M, Mo	S, L, C	5.0–7.5	2'	NR	FIB	20+	M–S		x
D, M, Mo	S, L, C	4.0–7.0	1'–1.5'	NR	FIB	20+	H–S		x
D, M	S, L	5.5–8.5	1'	2'	FIB	10–20	L–S		x
D	S, G	5.7–7.0	1'	2'	FIB	3–5	L–S		x
Mo, W	S, L, C	4.5–8.0	1'	3'	FIB	10–20	M–R/S		x
D, M, Mo	S, L, C	4.5–7.5	1'	1'	FIB	5–10	L–S		
D, M, Mo	S, L, C	5.0–8.0	1'	3'	FIB	3–5	M–S		x
M, Mo	S, L, C	5.0–7.5	1'	3'	FIB	3–5	M–S		x
D	S, G	4.0–7.5	1'–1.5'	2'	FIB	5–10	M–S	x	x
D, M	S, G, L	5.0–7.5	1'	NR	FIB	10–20	L–S		x
M, Mo, W	S, L, C	5.7–8.0	1'	NR	R	10–20	H–R		
D	S, G	6.0–8.0	6"–1'	1'	FIB	5–10	L–S		x
D, M, Mo	S, L, C	4.5–8.0	1.5'–2'	NR	FIB	20+	M–S		x
D, M	S, L	5.0–8.5	1'–1.5'	3'	FIB	20+	L–S		x
D, M, Mo	S, L, C	5.0–8.0	1'–2'	3'	FIB	10–20	M–S	x	x
Mo, W	S, L, C	6.0–8.5	2'	NR	R	20+	H–R		x
D, M	S, L	6.0–8.0	2'	2'	FIB	20+	L–S	x	x
D, M	S, L, C	5.0–7.5	1'	1.5'	FIB	5–10	M–S		x
M, Mo, W	S, L, C	4.5–7.5	2'	3'	FIB	20+	H–S		x

taproot, and TAPCO = taprooted corm. For aggressiveness and method, L = low, M = medium, H = high, R = by rhizome, S = by seed, and R/S = rhizome and seed. For dried arrangements, FL = flowers, SH = seedheads, SP = seedpods, SD = seed, and RH = rose hips. For deer palatability, L = low, M = medium, and H = high.

TABLE 12.16.   PLANTS LISTED BY AGGRESSIVENESS AND METHOD

Aggressiveness and method	Scientific name (4/1/2020)	Common name	Height	Flower color	Bloom time	USDA zones	Sun
Flowers and shrubs							
H–R	*Anemone canadensis*	Canada Anemone	1'–2'	White	May–July	3–6	F, P
H–R	*Asclepias syriaca*	Common Milkweed	2'–4'	Lavender	June–Aug.	3–8	F, P
H–R	*Helianthus ×laetiflorus*	Showy Sunflower	3'–6'	Yellow	Aug.–Sept.	3–8	F
H–R	*Physostegia virginiana*	Obedient Plant	1'–2'	Pink	Aug.–Sept.	3–9	F
H–R	*Tradescantia occidentalis*	Western Spiderwort	1'	Pink	June–July	3–9	F
M–R	*Conoclinium coelestinum*	Blue Mistflower	1'–3'	Blue/purple	July–Oct.	4–10	F, P
M–R	*Coreopsis palmata*	Stiff Coreopsis	2'–3'	Yellow	June–Aug.	3–8	F
M–R	*Euphorbia corollata*	Flowering Spurge	2'–4'	White	July–Aug.	3–9	F
M–R	*Filipendula rubra*	Queen of the Prairie	4'–5'	Pink	June–July	3–6	F
M–R	*Helianthus mollis*	Downy Sunflower	4'–6'	Yellow	Aug.–Sept.	4–9	F
M–R	*Helianthus occidentalis*	Western Sunflower	2'–3'	Yellow	July–Aug.	3–9	F
M–R	*Monarda fistulosa*	Bergamot	2'–5'	Lavender	July–Sept.	3–9	F, P
M–R	*Symphyotrichum ericoides*	Heath Aster	1'–3'	White	Aug.–Oct.	3–9	F
M–R	*Symphyotrichum oblongifolium*	Aromatic Aster	1'–2'	Purple/violet	Sept.–Oct.	3–8	F
M–R	*Tradescantia bracteata*	Prairie Spiderwort	1'–2'	Blue	June–July	3–7	F, P
L–R	*Antennaria neglecta*	Pussytoes	6"–1'	White	Apr.–May	3–7	F
L–R	*Asclepias sullivantii*	Sullivant's Milkweed	3'–5'	Pink/yellow	June–Aug.	4–7	F
L–R	*Geum triflorum*	Prairie Smoke	6"	Pink	May–June	3–6	F
L–R	*Iris virginica var. shrevei*	Wild Iris	2'–3'	Blue	June–July	3–8	F, P
L–R	*Pycnanthemum virginianum*	Virginia Mountainmint	1'–3'	White	July–Sept.	3–7	F, P
L–R/S	*Hibiscus moscheutos*	Rosemallow	3'–7'	White/pink/red	July–Sept.	5–9	F, P
L–R/S	*Polygonatum biflorum var. commutatum*	Great Solomon's Seal	2'–4'	Cream	May–June	3–9	F, P
L–R/S	*Rosa carolina*	Carolina Rose	1'–2'	Pink	June–July	3–8	F
L–R/S	*Spiraea tomentosa*	Steeplebush	2'–4'	Pink	Aug.–Sept.	3–8	F, P
H–S	*Heliopsis helianthoides*	Oxeye Sunflower	3'–6'	Yellow	June–Sept.	3–8	F
H–S	*Silphium perfoliatum*	Cup Plant	3'–10'	Yellow	July–Sept.	3–8	F, P
M–S	*Agastache foeniculum*	Lavender Hyssop	1'–3'	Purple	July–Sept.	2–6	F, P
M–S	*Astragalus canadensis*	Canada Milkvetch	2'–3'	Yellow	July–Aug.	3–8	F
M–S	*Coreopsis lanceolata*	Lanceleaf Coreopsis	1'–2'	Yellow	June–July	3–9	F
M–S	*Desmodium canadense*	Canada Ticktrefoil	3'–5'	Purple	July–Aug.	3–7	F
M–S	*Echinacea purpurea*	Purple Coneflower	3'–4'	Purple	July–Sept.	4–8	F, P
M–S	*Eryngium yuccifolium*	Rattlesnake Master	3'–5'	White	June–Aug.	4–9	F
M–S	*Lobelia siphilitica*	Great Blue Lobelia	1'–4'	Blue	July–Sept.	3–9	F, P
M–S	*Lupinus perennis*	Wild Lupine	1'–2'	Blue	May–June	3–8	F, P
M–S	*Monarda punctata*	Dotted Mint	1'–2'	Lavender	July–Sept.	3–10	F
M–S	*Oenothera gaura*	Biennial Beeblossom	3'–6'	White/pink	June–Sept.	4–8	F, P
M–S	*Oligoneuron rigidum*	Stiff Goldenrod	3'–5'	Yellow	Aug.–Sept.	3–9	F
M–S	*Ratibida pinnata*	Yellow Coneflower	3'–6'	Yellow	July–Sept.	3–9	F
M–S	*Rudbeckia hirta*	Blackeyed Susan	1'–3'	Yellow	June–Sept.	3–10	F, P

Soil moisture	Soil texture	pH range	Plant spacing in prairie gardens	Plant spacing in flower beds	Root type	Plant life expectancy in years	Cut flowers	Dried arrangements	Deer palatability
M, Mo	S, L, C	?	NR	NR	R	10–20			L
D, M	S, L, C	5.0–8.0	NR	NR	R	10–20		SP	L
D, M	S, G, L, C	?	NR	NR	R	20+	x	SH	L
M, Mo	S, L, C	6.0–8.0	1′–2′	NR	R	10–20	x	SH	M
D, M	S, L	?	1′	NR	R	20+			M
M, Mo	S, L, C	5.0–8.0	1′–2′	2′	R	3–5			L
D, M	S, L	?	1′–2′	2′	R	20+			L
D, M	S, L	?	1′–2′	1′–2′	R	10–20			L
M, Mo	S, L, C	?–8.0	2′	3′	R	20+	x		L
D, M	S, L, C	6.0–8.0	2′	3′	R	20+	x	SH	L
D, M	S, L	?	2′	3′	R	20+	x		L
D, M, Mo	S, L, C	6.0–8.0	1′–2′	3′	R	10–20	x	SH	L
D, M	S, G, L	?	1′–2′	2′	R	10–20		Empty SH	L
D, M	S, L, C	?–8.0	1′	3′	R	5–10		Empty SH	L
D, M	S, L, C	?	1′	2′	R	20+			M
D	S, G	5.5–7.5	1′	1′	R	5–10	x	FL	L
M, Mo	L, C	?	1′–2′	3′	R	10–20		SP	L
D, M	S, L	6.0–7.5	6″	1′	R	20+		SH	L
W	S, L, C	4.8–7.3	1′	2′	R/CO	10–20	x	SP	L
Mo, W	S, L, C	?–8.0	1′–2′	3′	R	5–10		SH	L
M, Mo, W	S, L, C	4.0–8.0	2′	4′	R	10–20			L
D, M	S, L, C	?	1′–2′	3′	R	10–20			H
D, M	S, L	4.0–7.0	2′	3′	R	10–20	x	RH	L
Mo, W	S, L, C	4.5–7.0	2′	3′	R	10–20	x		L
D, M, Mo	S, L, C	6.0–8.0	1′–2′	3′	FIB	5–10	x		L
M, Mo, W	S, L, C	4.5–8.0	2′–3′	NR	FIB	20+	x		L
D, M	S, L	?–8.0	1′	1′–2′	FIB	3–5	x	FL/SH	L
D, M	S, L	6.0–8.0	1′–2′	3′	TAP	10–20		SP	M
D, M	S, L	5.0–7.5	1′	1′–2′	FIB	3–5	x		M
M, Mo	S, L, C	?–7.5	NR	NR	TAP	10–20			H
D, M	S, L, C	5.5–8.0	1′–2′	3′	FIB	5–10	x	SH	M
D, M	S, L, C	6.0–8.0	1′	2′	FIB	5–10	x	SH	L
M, Mo	S, L, C	?–8.0	6″–1′	2′	FIB	5–10	x		L
D	S	5.0–?	1′	1.5′	TAP	10–20			L
D	S, G	5.0–?	1′	1′–2′	FIB	2	x	SH	L
D, M, Mo	S, L, C	?	2′	3′	FIB	2	x		L
D, M, Mo	S, L, C	5.5–8.0	1′–1.5′	3′	FIB	5–10	x	SH	L
D, M, Mo	S, L, C	5.5–8.0	1′–2′	2′	FIB	10–20	x	SH	L
D, M, Mo	S, L, C	5.0–8.0	1′	2′	FIB	2	x	SH	L

(continued)

TABLE 12.16. *(continued)*

Aggressiveness and method	Scientific name (4/1/2020)	Common name	Height	Flower color	Bloom time	USDA zones	Sun
M–S	*Rudbeckia laciniata*	Green-Headed Coneflower	5'–8'	Yellow	Aug.–Sept.	3–8	F, P
M–S	*Rudbeckia triloba*	Browneyed Susan	2'–5'	Yellow	July–Oct.	4–9	F, P
M–S	*Silphium integrifolium*	Rosinweed	2'–6'	Yellow	July–Sept.	4–8	F
M–S	*Symphyotrichum laeve*	Smooth Aster	2'–4'	Blue	Aug.–Oct.	4–8	F
M–S	*Tradescantia ohiensis*	Ohio Spiderwort	2'–4'	Blue	June–July	3–9	F, P
M–S	*Verbena hastata*	Blue Vervain	3'–6'	Blue	July–Sept.	3–8	F, P
M–S	*Verbena stricta*	Hoary Vervain	2'–4'	Blue	July–Sept.	3–8	F
H–S (north); L–S (south)	*Arnoglossum atriplicifolium*	Pale Indian Plantain	5'–10'	White	July–Sept.	4–8	F, P
L–S	*Allium cernuum*	Nodding Pink Onion	1'–2'	White/pink	June–Aug.	3–8	F, P
L–S	*Allium stellatum*	Prairie Onion	1'–2'	Lavender	July–Aug.	3–8	F, P
L–S	*Amorpha canescens*	Leadplant	2'–3'	Purple	June–July	3–8	F, P
L–S	*Amsonia tabernaemontana*	Eastern Bluestar	2'–3'	Blue/white	May–June	4–8	F, P
L–S	*Anemone cylindrica*	Thimbleweed	1'–3'	White	May–June	3–6	F
L–S	*Asclepias incarnata*	Marsh Milkweed	3'–5'	Red/pink	June–July	3–9	F
L–S	*Asclepias tuberosa*	Butterflyweed	2'–3'	Orange	June–Aug.	3–10	F
L–S	*Baptisia alba*	White False Indigo	3'–5'	White	June–July	4–9	F, P
L–S	*Baptisia australis*	Blue False Indigo	3'–5'	Blue	June–July	3–10	F, P
L–S	*Baptisia bracteata*	Cream False Indigo	1'–2'	Cream	May–June	4–9	F, P
L–S	*Blephilia ciliata*	Downy Wood Mint	1'–2'	White/purple/ violet	May–July	4–8	F, P
L–S	*Callirhoe involucrata*	Winecups	1'	Magenta	May–June	4–9	F, P
L–S	*Callirhoe triangulata*	Poppymallow	1'–2'	Magenta	July–Aug.	4–8	F
L–S	*Ceanothus americanus*	New Jersey Tea	2'–3'	White	July–Aug.	3–9	F, P
L–S	*Chelone glabra*	White Turtlehead	2'–4'	White	Aug.–Sept.	3–8	F, P
L–S	*Coreopsis tripteris*	Tall Coreopsis	3'–7'	Yellow	July–Sept.	3–8	F, P
L–S	*Dalea candida*	White Prairie Clover	1'–2'	White	July–Aug.	3–9	F
L–S	*Dalea purpurea*	Purple Prairie Clover	1'–2'	Purple/yellow	July–Aug.	3–8	F
L–S	*Delphinium carolinianum* subsp. *virescens*	Prairie Larkspur	1'–4'	White	June–July	3–8	F
L–S	*Desmanthus illinoensis*	Illinois Bundleflower	2'–6'	White	June–Aug.	5–8	F
L–S	*Dodecatheon meadia*	Shooting Star	1'–2'	White/pink	May–June	4–8	F, P
L–S	*Echinacea pallida*	Pale Purple Coneflower	3'–5'	Purple	June–July	4–8	F
L–S	*Echinacea paradoxa*	Ozark Coneflower	3'–5'	Yellow	June–July	4–7	F, P
L–S	*Eupatorium perfoliatum*	Boneset	3'–4'	White	Aug.–Sept.	3–8	F
L–S	*Eutrochium fistulosum*	Tall Joe Pye Weed	5'–8'	Purple/pink	Aug.–Sept.	4–9	F, P
L–S	*Eutrochium maculatum*	Joe Pye Weed	4'–6'	Pink	Aug.–Sept.	3–6	F, P
L–S	*Gentiana alba*	Cream Gentian	1'–2'	White	July–Aug.	3–7	F, P
L–S	*Gentiana andrewsii*	Bottle Gentian	1'–2'	Blue	Aug.–Oct.	3–6	F, P
L–S	*Helenium autumnale*	Sneezeweed	4'–5'	Yellow	Aug.–Sept.	3–9	F
L–S	*Hypericum ascyron*	Great St. Johnswort	4'–6'	Yellow	July–Aug.	3–5	F, P
L–S	*Lespedeza capitata*	Roundhead Bush Clover	3'–5'	White	Aug.–Sept.	3–8	F, P

Soil moisture	Soil texture	pH range	Plant spacing in prairie gardens	Plant spacing in flower beds	Root type	Plant life expectancy in years	Cut flowers	Dried arrangements	Deer palatability
Mo, W	S, L, C	4.5–7.5	2'	2'	FIB	3–5	x	SH	L
M, Mo	S, L	?–8.0	1'–2'	3'	FIB	2	x	SH	L
M, Mo	S, L, C	?–8.0	2'	3'	TAP	20+	x		L
D, M	S, L	?	1'	3'	FIB	5–10	x	Empty SH	H
D, M, Mo	S, L, C	?–7.5	1'	2'	FIB	20+			H
M, Mo, W	S, L, C	?	1'	2'	FIB	2	x	SH	L
D, M	S, L	?	1'	2'	FIB	3–5	x	SH	L
D, M, Mo	S, L, C	?–8.0	2'	3'–4'	FIB	3–20	x	FL	L
M, Mo	S, L, C	6.5–8.0	6"–1'	1'	B	10–20	x	FL/SH	L
D, M	S, L	5.5–8.0	6"–1'	1'	B	10–20	x	FL/SH	L
D, M	S, L	5.5–8.0	2'	3'	TAP	20+		Leaves	L
M, Mo	S, L, C	?–8.0	2'–3'	3'	FIB	20+		SP	L
D, M	S, L	?	1'	1'	FIB	10–20		SD	L
Mo, W	S, L, C	5.0–8.0	1'–2'	3'	FIB	5–10	x	SP	L
D, M	S, L	5.0–7.5	2'	3'	TAP	10–20	x	SP	L
M, Mo	S, L, C	?–7.5	3'	3'	TAP	20+	x	SP	L
M	S, L, C	?–8.0	3'	3'	TAP	20+	x	SP	L
D, M	S, L	?	2'	3'	TAP	20+	x	SP	L
D, M	G, L, C	?–8.00	1'	1'–2'	TAP/short Rs	3–5		FL/SH	L
D, M	S, L	5.5–7.5	1'–2'	2'	TAP	10–20			L
D	S	?	1'–2'	2'	TAP	10–20			L
D, M	S, L	4.5–7.0	3'	3'	TAP	20+		SD	L
Mo, W	S, L, C	?	1'	1'–2'	FIB	5–10			M
M, Mo, W	S, L, C	?	1'–2'	NR	FIB	10–20	x		L
D, M	S, L	?	6"–1'	1'–2'	TAP	10–20			H
D, M	S, L, C	?	6"–1'	1'–2'	TAP	10–20	x		H
D, M	S, L	?	1'	1.5'	FIB	5–10	x	SP	L
D, M, Mo	S, L, C	6.0–8.0	1'	1.5'	TAP	5–10		SP	M
M, Mo	S, L, C	4.5–8.0	6"–1'	1'	FIB	20+		SP	L
D, M	S, L, C	6.0–7.5	1'	2'–3'	TAP	10–20	x	SH	M
D, M	S, L, C	?–7.5	1'	2'–3'	TAP	10–20	x	SH	M
M, Mo, W	S, L, C	?–7.5	1'	NR	FIB	5–10			L
M, Mo, W	S, L, C	4.5–7.5	2'	3'–4'	FIB	5–10			H
Mo, W	S, L, C	5.5–7.5	1'–2'	3'–4'	FIB	5–10			L
M	S, L, C	?–7.5	1'	1.5'	FIB	10–20	x		L
Mo, W	S, L, C	5.8–7.2	1'	1.5'	FIB	5–10	x		L
Mo, W	S, L, C	4.0–7.5	1'–2'	2'–3'	FIB	5–10	x	SH	L
Mo, W	S, L, C	5.7–7.1	1'–2'	3'	FIB	5–10	x	SP	L
D, M	S, G, L	5.5–8.0	1'	2'	TAP	10–20		SH	M

(continued)

TABLE 12.16. *(continued)*

Aggressiveness and method	Scientific name (4/1/2020)	Common name	Height	Flower color	Bloom time	USDA zones	Sun
L–S	*Liatris aspera*	Rough Blazing Star	2'–5'	Purple/pink	Aug.–Sept.	3–8	F
L–S	*Liatris ligulistylis*	Meadow Blazing Star	3'–5'	Purple/pink	Aug.–Sept.	3–6	F
L–S	*Liatris punctata*	Dotted Blazing Star	1'–2'	Purple/pink	Aug.–Oct.	3–9	F
L–S	*Liatris pycnostachya*	Prairie Blazing Star	3'–5'	Purple/pink	July–Aug.	3–9	F
L–S	*Liatris spicata*	Dense Blazing Star	3'–6'	Purple/pink	Aug.–Sept.	4–10	F
L–S	*Liatris squarrosa*	Scaly Blazing Star	1'–2'	Purple/pink	Aug.–Sept.	4–8	F
L–S	*Lilium michiganense*	Michigan Lily	4'–6'	Orange	July–Aug.	3–7	F, P
L–S	*Lithospermum canescens*	Hoary Puccoon	1'	Orange	May–June	3–7	F, P
L–S	*Lithospermum caroliniense*	Hairy Puccoon	1'–2'	Yellow	June–July	3–8	F
L–S	*Lithospermum incisum*	Fringed Puccoon	1'–2'	Yellow	June–July	3–9	F
L–S	*Lobelia cardinalis*	Cardinal Flower	2'–5'	Red	July–Sept.	3–9	F, P
L–S	*Oligoneuron ohioense*	Ohio Goldenrod	3'–4'	Yellow	Aug.–Sept.	4–5	F, P
L–S	*Onosmodium bejariense* var. *occidentale*	Marbleseed (Western Marbleseed)	1'–3'	White	May–June	3–8	F
L–S	*Parthenium integrifolium*	Wild Quinine	3'–5'	White	June–Sept.	4–8	F
L–S	*Pediomelum esculentum*	Prairie Turnip	1'–2'	Blue	June–July	3–7	F
L–S	*Penstemon cobaea*	Cobaea Beardtongue	2'–3'	Lavender	Apr.–June	5–8	F
L–S	*Penstemon digitalis*	Foxglove Beardtongue	2'–3'	White	June–July	3–8	F, P
L–S	*Penstemon gracilis*	Slender Beardtongue	1'–2'	Lavender	May–June	3–6	F, P
L–S	*Penstemon grandiflorus*	Large-Flowered Beardtongue	2'–4'	Lavender	May–June	3–7	F
L–S	*Penstemon hirsutus*	Hairy Penstemon	1'–2'	Lavender	May–June	3–6	F, P
L–S	*Phlox pilosa*	Downy Phlox	1'–2'	Pink	May–June	3–9	F
L–S	*Pulsatilla patens*	Pasque Flower	6"–1'	White	Apr.–May	3–6	F
L–S	*Pycnanthemum tenuifolium*	Narrowleaf Mountainmint	1'–3'	White	July–Sept.	4–8	F, P
L–S	*Rosa setigera*	Climbing Prairie Rose	6'–15'	Pink/white	May–July	4–9	F, P
L–S	*Rudbeckia fulgida*	Orange Coneflower	2'–4'	Yellow	July–Sept.	4–8	F, P
L–S	*Rudbeckia subtomentosa*	Sweet Blackeyed Susan	4'–6'	Yellow	Aug.–Oct.	4–8	F, P
L–S	*Ruellia humilis*	Wild Petunia	1'–2'	Violet	June–Aug.	4–9	F
L–S	*Salvia azurea*	Blue Sage	3'–5'	Blue	June–Aug.	4–9	F, P
L–S	*Senna hebecarpa*	Wild Senna	4'–6'	Yellow	July–Aug.	4–7	F
L–S	*Silene regia*	Royal Catchfly	2'–4'	Red	July–Aug.	4–9	F, P
L–S	*Silphium laciniatum*	Compassplant	3'–10'	Yellow	June–Sept.	4–9	F
L–S	*Silphium terebinthinaceum*	Prairie Dock	3'–10'	Yellow	July–Sept.	4–7	F
L–S	*Sisyrinchium albidum*	White Blue-Eyed Grass	6"–1'	White	Apr.–May	4–9	F
L–S	*Sisyrinchium angustifolium*	Narrow-Leaved Blue-Eyed Grass	6"–2'	Blue/purple	Apr.–June	3–10	F, P
L–S	*Sisyrinchium campestre*	Prairie Blue-Eyed Grass	6"–1'	Blue/white	Apr.–June	3–8	F, P
L–S	*Solidago ptarmicoides*	White Aster	1'–2'	White	Aug.–Sept.	3–7	F
L–S	*Solidago speciosa*	Showy Goldenrod	1'–3'	Yellow	Aug.–Sept.	3–8	F
L–S	*Spiraea alba*	Meadowsweet	2'–4'	White/pink	June–Sept.	3–7	F, P

Soil moisture	Soil texture	pH range	Plant spacing in prairie gardens	Plant spacing in flower beds	Root type	Plant life expectancy in years	Cut flowers	Dried arrangements	Deer palatability
D, M	S, L	?	6"–1'	1'	CO	10–20	x	FL/SH	L
M, Mo	L	?	6"–1'	1'	CO	3–5	x	FL/SH	L
D	S, L	6.0–8.0	1'	1.5'–2'	TAPCO	10–20			L
M, Mo	S, L, C	6.0–8.0	6"–1'	1'	CO	10–20	x	FL/SH	L
M, Mo	S, L, C	5.5–7.5	6"–1'	1'	CO	10–20	x	FL/SH	L
D, M	S, L	?	1'	1.5'–2'	CO	10–20	x	FL/SH	L
Mo	S, L, C	?–6.8	2'	2'	B	20+	x	SP	H
D, M	S, L	?	1'	1'	FIB	5–10			L
D	S	?	1'	1'	TAP	10–20			L
D	S, G	?	1'	1'	TAP	5–10			L
Mo, W	S, L	5.8–7.8	6"–1'	2'	FIB	3–5	x		M
M, Mo	S, L, C	?–8.0	1'–2'	3'	FIB	5–10	x	SH	L
D, M	S, L	?–7.5	1'–1.5'	2'	TAP	10–20			L
M, Mo	S, L, C	?–8.0	1'–2'	3'	TAP	10–20	x		M
D	S, G	?–8.5	1'–1.5'	2'	TAP	20+			L
D	S, G	?–7.5	1'	1'	FIB	3–5	x	SP	L
M, Mo	S, L, C	5.5–7.5	1'	2'	FIB	5–10	x		L
D	S, G	?–7.5	6"–1'	1'	FIB	5–10	x	SP	L
D	S, G	?	1'	2'–3'	FIB	3–5		SP	L
D	S, G	5.0–8.0	1'	1.5'	FIB	3–5	x	SP	L
D, M, Mo	S, L	?	6"–1'	1.5'	FIB	3–5	x		L
D	S, G	?–8.0	6"–1'	1'	FIB	5–10		SD	H
D, M	S, G, L, C	?	1'	1.5'	TAP/R	5–10		SH	M
M, Mo	S, L, C	5.0–8.0	2'	5'	FIB	10–20	x	RH	L
M, Mo	L, C	?–8.0	1'–2'	3'	FIB	5–10	x	SH	L
M, Mo	S, L, C	?–8.0	1'–2'	2'	FIB	10–20	x	SH	L
D, M	S, L	4.5–7.5	1'	2'	FIB	5–10			L
D, M	S, G, L	?	1'	1.5'	FIB	5–10	x	SH	L
M, Mo, W	S, L, C	?	1.5'–2'	3'–4'	FIB/TAP	10–20	x	FL/SH	M
M	L	?	1'	1.5'	TAP	5–10	x	SH	L
D, M	S, L, C	?–8.0	2'	4'	TAP	20+	x		L
M, Mo	S, L, C	?–8.0	2'	4'	TAP	20+	x		L
D, M	S, L	?–8.0	6"	6"	FIB	3–5			L
M, Mo	S, L, C	?–8.0	6"	9"	FIB	3–5			L
D, M	S, L	?–7.5	6"	6"	FIB	3–5			L
D	S, G	6.5–8.0	1'	2'	FIB	5–10		Empty SH	L
D, M	S, L	?	1'	3'	FIB	5–10	x	SH	L
Mo, W	S, L, C	4.3–7.2	2'	3'	FIB	10–20	x		L

(continued)

TABLE 12.16. (*continued*)

Aggressiveness and method	Scientific name (4/1/2020)	Common name	Height	Flower color	Bloom time	USDA zones	Sun
L–S	*Symphyotrichum novae-angliae*	New England Aster	3'–6'	Pink/purple/ blue	Aug.–Oct.	3–7	F, P
L–S	*Symphyotrichum oolentangiense*	Skyblue Aster	2'–3'	Blue	Aug.–Oct.	3–8	F, P
L–S	*Tephrosia virginiana*	Goatsrue	1'–2'	Pink/yellow	June–Aug.	4–9	F, P
L–S	*Thalictrum dasycarpum*	Tall Meadow-Rue	3'–6'	White	June–July	3–9	F, P
L–S	*Vernonia fasciculata*	Ironweed	4'–6'	Red/pink	July–Sept.	3–7	F
L–S	*Vernonia gigantea*	Tall Ironweed	5'–8'	Red/pink	Aug.–Sept.	4–9	F, P
L–S	*Veronicastrum virginicum*	Culver's Root	3'–6'	White	July–Aug.	3–8	F, P, S
L–S	*Viola pedata*	Birdfoot Violet	6"	Blue/purple	Apr.–June	3–9	F, P
L–S	*Zizia aptera*	Heartleaf Golden Alexanders	1'–2'	Yellow	May–June	3–8	F
L–S	*Zizia aurea*	Golden Alexanders	1'–2'	Yellow	May–July	3–8	F, P, S

Aggressiveness and method	Scientific name (4/1/2020)	Common name	Height	Fall leaf color	Bloom time	USDA zones	Sun
Grasses and sedges							
H–R	*Hierochloe hirta*	Vanilla Sweetgrass	1'–2'	Straw	May–June	3–9	F, P
H–R	*Spartina pectinata*	Prairie Cordgrass	6'–9'	Gold	Aug.–Sept.	3–8	F
M–R/S	*Calamagrostis canadensis*	Bluejoint Grass	4'–6'	Straw	June–July	3–6	F
H–S	*Andropogon virginicus*	Broomsedge	2'–5'	Reddish brown	Aug.–Oct.	5–10	F
H–S	*Tripsacum dactyloides*	Eastern Gamagrass	4'–8'	Gold	July–Sept.	4–10	F, P
M–S	*Andropogon gerardii*	Big Bluestem	5'–8'	Bronze/red	Aug.–Oct.	3–9	F
M–S	*Elymus canadensis*	Canada Wildrye	4'–5'	Straw	July–Aug.	2–9	F
M–S	*Elymus virginicus*	Virginia Wildrye	4'–5'	Straw	July–Aug.	2–9	F, P, S
M–S	*Eragrostis spectabilis*	Purple Lovegrass	1'–2'	Purple/pink	July–Sept.	3–10	F, P
M–S	*Panicum virgatum*	Switchgrass	3'–6'	Gold	Aug.–Sept.	3–10	F
M–S	*Sorghastrum nutans*	Indiangrass	5'–7'	Gold	Aug.–Sept.	3–9	F
M–S	*Tridens flavus*	Purpletop	3'–6'	Straw	July–Sept.	5–10	F
L–S	*Bouteloua curtipendula*	Sideoats Grama	2'–3'	Straw	Aug.–Sept.	3–9	F
L–S	*Bromus kalmii*	Prairie Bromegrass	2'–4'	Straw	June–July	3–6	F
L–S	*Carex brevior*	Shortbeak Sedge	1'–4'	Straw	May–June	3–7	F, P
L–S	*Hesperostipa spartea*	Needlegrass	2'–4'	Gold	Apr.–June	3–6	F
L–S	*Koeleria macrantha*	Junegrass	2'–3'	Gold	May–June	3–8	F
L–S	*Schizachyrium scoparium*	Little Bluestem	2'–5'	Crimson red	Aug.–Oct.	3–10	F
L–S	*Sporobolus heterolepis*	Prairie Dropseed	2'–4'	Gold	Aug.–Sept.	3–8	F

Notes: For soil moisture, D = dry, M = medium, Mo = moist, and W = wet. For soil texture, S = sand, L = loam, C = clay, and G = gravel. For sun, F = full sun, or at least a half day of full sun; P = partial sun, or a quarter to a half day of sun; and S = shade, or less than a quarter day of sun. For plant spacing, NR = not recommended. For root type, B = bulb, CO = corm, FIB = fibrous, R = rhizome, TAP =

Soil moisture	Soil texture	pH range	Plant spacing in prairie gardens	Plant spacing in flower beds	Root type	Plant life expectancy in years	Cut flowers	Dried arrangements	Deer palatability
M, Mo	S, L, C	5.0–8.0	1'	3'	FIB	5–10	x	Empty SH	H
D, M	S, L	5.0–8.0	1'	2'	FIB	5–10	x	Empty SH	M
D	S, G	4.5–?	2'	3'	TAP	10–20			L
Mo, W	S, L, C	?	1'–2'	2'	FIB	5–10			L
Mo	S, L, C	?	1'	3'	FIB	5–10	x		L
Mo, W	S, L, C	?–8.0	1'–2'	3'	FIB	10–20	x		L
M, Mo	S, L, C	?–8.0	1'	3'	FIB	10–20	x	SH	L
D	S, G	?	6"	1'	R	5–10			L
M, Mo	S, L	?–7.5	1'	2'	FIB	5–10			L
M, Mo, W	S, L, C	?–8.0	1'	3'	FIB	5–10			L

Soil moisture	Soil texture	pH range	Plant spacing in prairie gardens	Plant spacing in flower beds	Root type	Plant life expectancy in years	Cut flower stalks	Dried seed arrangements	Deer palatability
M, Mo, W	S, L, C	5.7–8.0	1'	NR	R	10–20			L
Mo, W	S, L, C	6.0–8.5	2'	NR	R	20+		x	L
Mo, W	S, L, C	4.5–8.0	1'	3'	FIB	10–20		x	L
D, M, Mo	S, L, C	4.0–7.0	1'–1.5'	NR	FIB	20+		x	L
M, Mo, W	S, L, C	4.5–7.5	2'	3'	FIB	20+		x	M
D, M, Mo	S, L, C	5.0–7.5	2'	NR	FIB	20+		x	L
D, M, Mo	S, L, C	5.0–8.0	1'	3'	FIB	3–5		x	L
M, Mo	S, L, C	5.0–7.5	1'	3'	FIB	3–5		x	L
D	S, G	4.0–7.5	1'–1.5'	2'	FIB	5–10	x	x	L
D, M, Mo	S, L, C	4.5–8.0	1.5'–2'	NR	FIB	20+		x	L
D, M, Mo	S, L, C	5.0–8.0	1'–2'	3'	FIB	10–20	x	x	L
D, M	S, L, C	5.0–7.5	1'	1.5'	FIB	5–10		x	L
D, M	S, L	5.5–8.5	1'	2'	FIB	10–20		x	L
D	S, G	5.7–7.0	1'	2'	FIB	3–5		x	L
D, M, Mo	S, L, C	4.5–7.5	1'	1'	FIB	5–10			L
D, M	S, G, L	5.0–7.5	1'	NR	FIB	10–20		x	L
D	S, G	6.0–8.0	6"–1'	1'	FIB	5–10		x	L
D, M	S, L	5.0–8.5	1'–1.5'	3'	FIB	20+		x	L
D, M	S, L	6.0–8.0	2'	2'	FIB	20+	x	x	L

taproot, and TAPCO = taprooted corm. For aggressiveness and method, L = low, M = medium, H = high, R = by rhizome, S = by seed, and R/S = rhizome and seed. For dried arrangements, FL = flowers, SH = seedheads, SP = seedpods, SD = seed, and RH = rose hips. For deer palatability, L = low, M = medium, and H = high.

TABLE 12.17. BEST PRAIRIE GARDEN PLANTS

Scientific name (4/1/2020)	Common name	Height	Flower color	Bloom time	USDA zones	Sun	Soil moisture
Flowers and shrubs							
Agastache foeniculum	Lavender Hyssop	1'–3'	Purple	July–Sept.	2–6	F, P	D, M
Allium cernuum	Nodding Pink Onion	1'–2'	White/pink	June–Aug.	3–8	F, P	M, Mo
Allium stellatum	Prairie Onion	1'–2'	Lavender	July–Aug.	3–8	F, P	D, M
Amorpha canescens	Leadplant	2'–3'	Purple	June–July	3–8	F, P	D, M
Amsonia tabernaemontana	Eastern Bluestar	2'–3'	Blue/white	May–June	4–8	F, P	M, Mo
Anemone cylindrica	Thimbleweed	1'–3'	White	May–June	3–6	F	D, M
Antennaria neglecta	Pussytoes	6"–1'	White	Apr.–May	3–7	F	D
Arnoglossum atriplicifolium	Pale Indian Plantain	5'–10'	White	July–Sept.	4–8	F, P	D, M, Mo
Asclepias incarnata	Marsh Milkweed	3'–5'	Red/pink	June–July	3–9	F	Mo, W
Asclepias sullivantii	Sullivant's Milkweed	3'–5'	Pink/yellow	June–Aug.	4–7	F	M, Mo
Asclepias tuberosa	Butterflyweed	2'–3'	Orange	June–Aug.	3–10	F	D, M
Astragalus canadensis	Canada Milkvetch	2'–3'	Yellow	July–Aug.	3–8	F	D, M
Baptisia alba	White False Indigo	3'–5'	White	June–July	4–9	F, P	M, Mo
Baptisia australis	Blue False Indigo	3'–5'	Blue	June–July	3–10	F, P	M
Baptisia bracteata	Cream False Indigo	1'–2'	Cream	May–June	4–9	F, P	D, M
Blephilia ciliata	Downy Wood Mint	1'–2'	White/purple/violet	May–July	4–8	F, P	D, M
Callirhoe involucrata	Winecups	1'	Magenta	May–June	4–9	F, P	D, M
Callirhoe triangulata	Poppymallow	1'–2'	Magenta	July–Aug.	4–8	F	D
Ceanothus americanus	New Jersey Tea	2'–3'	White	July–Aug.	3–9	F, P	D, M
Chelone glabra	White Turtlehead	2'–4'	White	Aug.–Sept.	3–8	F, P	Mo, W
Conoclinium coelestinum	Blue Mistflower	1'–3'	Blue/purple	July–Oct.	4–10	F, P	M, Mo
Coreopsis lanceolata	Lanceleaf Coreopsis	1'–2'	Yellow	June–July	3–9	F	D, M
Coreopsis palmata	Stiff Coreopsis	2'–3'	Yellow	June–Aug.	3–8	F	D, M
Coreopsis tripteris	Tall Coreopsis	3'–7'	Yellow	July–Sept.	3–8	F, P	M, Mo, W
Dalea candida	White Prairie Clover	1'–2'	White	July–Aug.	3–9	F	D, M
Dalea purpurea	Purple Prairie Clover	1'–2'	Purple/yellow	July–Aug.	3–8	F	D, M
Delphinium carolinianum subsp. *virescens*	Prairie Larkspur	1'–4'	White	June–July	3–8	F	D, M
Desmanthus illinoensis	Illinois Bundleflower	2'–6'	White	June–Aug.	5–8	F	D, M, Mo
Dodecatheon meadia	Shooting Star	1'–2'	White/pink	May–June	4–8	F, P	M, Mo
Echinacea pallida	Pale Purple Coneflower	3'–5'	Purple	June–July	4–8	F	D, M
Echinacea paradoxa	Ozark Coneflower	3'–5'	Yellow	June–July	4–7	F, P	D, M
Echinacea purpurea	Purple Coneflower	3'–4'	Purple	July–Sept.	4–8	F, P	D, M
Eryngium yuccifolium	Rattlesnake Master	3'–5'	White	June–Aug.	4–9	F	D, M
Eupatorium perfoliatum	Boneset	3'–4'	White	Aug.–Sept.	3–8	F	M, Mo, W
Euphorbia corollata	Flowering Spurge	2'–4'	White	July–Aug.	3–9	F	D, M
Eutrochium fistulosum	Tall Joe Pye Weed	5'–8'	Purple/pink	Aug.–Sept.	4–9	F, P	M, Mo, W
Eutrochium maculatum	Joe Pye Weed	4'–6'	Pink	Aug.–Sept.	3–6	F, P	Mo, W
Filipendula rubra	Queen of the Prairie	4'–5'	Pink	June–July	3–6	F	M, Mo

Soil texture	pH range	Plant spacing in prairie gardens	Plant spacing in flower beds	Root type	Plant life expectancy in years	Aggressiveness and method	Cut flowers	Dried arrangements	Deer palatability
S, L	?-8.0	1'	1'-2'	FIB	3-5	M-S	x	FL/SH	L
S, L, C	6.5-8.0	6"-1'	1'	B	10-20	L-S	x	FL/SH	L
S, L	5.5-8.0	6"-1'	1'	B	10-20	L-S	x	FL/SH	L
S, L	5.5-8.0	2'	3'	TAP	20+	L-S		Leaves	L
S, L, C	?-8.0	2'-3'	3'	FIB	20+	L-S		SP	L
S, L	?	1'	1'	FIB	10-20	L-S		SD	L
S, G	5.5-7.5	1'	1'	R	5-10	L-R	x	FL	L
S, L, C	?-8.0	2'	3'-4'	FIB	3-20	H-S (north); L-S (south)	x	FL	L
S, L, C	5.0-8.0	1'-2'	3'	FIB	5-10	L-S	x	SP	L
L, C	?	1'-2'	3'	R	10-20	L-R		SP	L
S, L	5.0-7.5	2'	3'	TAP	10-20	L-S	x	SP	L
S, L	6.0-8.0	1'-2'	3'	TAP	10-20	M-S		SP	M
S, L, C	?-7.5	3'	3'	TAP	20+	L-S	x	SP	L
S, L, C	?-8.0	3'	3'	TAP	20+	L-S	x	SP	L
S, L	?	2'	3'	TAP	20+	L-S	x	SP	L
G, L, C	?-8.00	1'	1'-2'	TAP/short Rs	3-5	L-S		FL/SH	L
S, L	5.5-7.5	1'-2'	2'	TAP	10-20	L-S			L
S	?	1'-2'	2'	TAP	10-20	L-S			L
S, L	4.5-7.0	3'	3'	TAP	20+	L-S		SD	L
S, L, C	?	1'	1'-2'	FIB	5-10	L-S			M
S, L, C	5.0-8.0	1'-2'	2'	R	3-5	M-R			L
S, L	5.0-7.5	1'	1'-2'	FIB	3-5	M-S	x		M
S, L	?	1'-2'	2'	R	20+	M-R			L
S, L, C	?	1'-2'	NR	FIB	10-20	L-S	x		L
S, L	?	6"-1'	1'-2'	TAP	10-20	L-S			H
S, L, C	?	6"-1'	1'-2'	TAP	10-20	L-S	x		H
S, L	?	1'	1.5'	FIB	5-10	L-S	x	SP	L
S, L, C	6.0-8.0	1'	1.5'	TAP	5-10	L-S		SP	M
S, L, C	4.5-8.0	6"-1'	1'	FIB	20+	L-S		SP	L
S, L, C	6.0-7.5	1'	2'-3'	TAP	10-20	L-S	x	SH	M
S, L, C	?-7.5	1'	2'-3'	TAP	10-20	L-S	x	SH	M
S, L, C	5.5-8.0	1'-2'	3'	FIB	5-10	M-S	x	SH	M
S, L, C	6.0-8.0	1'	2'	FIB	5-10	M-S	x	SH	L
S, L, C	?-7.5	1'	NR	FIB	5-10	L-S			L
S, L	?	1'-2'	1'-2'	R	10-20	M-R			L
S, L, C	4.5-7.5	2'	3'-4'	FIB	5-10	L-S			H
S, L, C	5.5-7.5	1'-2'	3'-4'	FIB	5-10	L-S			L
S, L, C	?-8.0	2'	3'	R	20+	M-R	x		L

(continued)

TABLE 12.17. *(continued)*

Scientific name (4/1/2020)	Common name	Height	Flower color	Bloom time	USDA zones	Sun	Soil moisture
Gentiana alba	Cream Gentian	1'–2'	White	July–Aug.	3–7	F, P	M
Gentiana andrewsii	Bottle Gentian	1'–2'	Blue	Aug.–Oct.	3–6	F, P	Mo, W
Geum triflorum	Prairie Smoke	6"	Pink	May–June	3–6	F	D, M
Helenium autumnale	Sneezeweed	4'–5'	Yellow	Aug.–Sept.	3–9	F	Mo, W
Helianthus mollis	Downy Sunflower	4'–6'	Yellow	Aug.–Sept.	4–9	F	D, M
Helianthus occidentalis	Western Sunflower	2'–3'	Yellow	July–Aug.	3–9	F	D, M
Heliopsis helianthoides	Oxeye Sunflower	3'–6'	Yellow	June–Sept.	3–8	F	D, M, Mo
Hibiscus moscheutos	Rosemallow	3'–7'	White/pink/red	July–Sept.	5–9	F, P	M, Mo, W
Hypericum ascyron	Great St. Johnswort	4'–6'	Yellow	July–Aug.	3–5	F, P	Mo, W
Iris virginica var. *shrevei*	Wild Iris	2'–3'	Blue	June–July	3–8	F, P	W
Lespedeza capitata	Roundhead Bush Clover	3'–5'	White	Aug.–Sept.	3–8	F, P	D, M
Liatris aspera	Rough Blazing Star	2'–5'	Purple/pink	Aug.–Sept.	3–8	F	D, M
Liatris ligulistylis	Meadow Blazing Star	3'–5'	Purple/pink	Aug.–Sept.	3–6	F	M, Mo
Liatris punctata	Dotted Blazing Star	1'–2'	Purple/pink	Aug.–Oct.	3–9	F	D
Liatris pycnostachya	Prairie Blazing Star	3'–5'	Purple/pink	July–Aug.	3–9	F	M, Mo
Liatris spicata	Dense Blazing Star	3'–6'	Purple/pink	Aug.–Sept.	4–10	F	M, Mo
Liatris squarrosa	Scaly Blazing Star	1'–2'	Purple/pink	Aug.–Sept.	4–8	F	D, M
Lilium michiganense	Michigan Lily	4'–6'	Orange	July–Aug.	3–7	F, P	Mo
Lithospermum canescens	Hoary Puccoon	1'	Orange	May–June	3–7	F, P	D, M
Lithospermum caroliniense	Hairy Puccoon	1'–2'	Yellow	June–July	3–8	F	D
Lithospermum incisum	Fringed Puccoon	1'–2'	Yellow	June–July	3–9	F	D
Lobelia cardinalis	Cardinal Flower	2'–5'	Red	July–Sept.	3–9	F, P	Mo, W
Lobelia siphilitica	Great Blue Lobelia	1'–4'	Blue	July–Sept.	3–9	F, P	M, Mo
Lupinus perennis	Wild Lupine	1'–2'	Blue	May–June	3–8	F, P	D
Monarda fistulosa	Bergamot	2'–5'	Lavender	July–Sept.	3–9	F, P	D, M, Mo
Monarda punctata	Dotted Mint	1'–2'	Lavender	July–Sept.	3–10	F	D
Oenothera gaura	Biennial Beeblossom	3'–6'	White/pink	June–Sept.	4–8	F, P	D, M, Mo
Oligoneuron ohioense	Ohio Goldenrod	3'–4'	Yellow	Aug.–Sept.	4–5	F, P	M, Mo
Oligoneuron rigidum	Stiff Goldenrod	3'–5'	Yellow	Aug.–Sept.	3–9	F	D, M, Mo
Onosmodium bejariense var. occidentale	Marbleseed (Western Marbleseed)	1'–3'	White	May–June	3–8	F	D, M
Parthenium integrifolium	Wild Quinine	3'–5'	White	June–Sept.	4–8	F	M, Mo
Pediomelum esculentum	Prairie Turnip	1'–2'	Blue	June–July	3–7	F	D
Penstemon cobaea	Cobaea Beardtongue	2'–3'	Lavender	Apr.–June	5–8	F	D
Penstemon digitalis	Foxglove Beardtongue	2'–3'	White	June–July	3–8	F, P	M, Mo
Penstemon gracilis	Slender Beardtongue	1'–2'	Lavender	May–June	3–6	F, P	D
Penstemon grandiflorus	Large-Flowered Beardtongue	2'–4'	Lavender	May–June	3–7	F	D
Penstemon hirsutus	Hairy Penstemon	1'–2'	Lavender	May–June	3–6	F, P	D
Phlox pilosa	Downy Phlox	1'–2'	Pink	May–June	3–9	F	D, M, Mo
Polygonatum biflorum var. commutatum	Great Solomon's Seal	2'–4'	Cream	May–June	3–9	F, P	D, M

Soil texture	pH range	Plant spacing in prairie gardens	Plant spacing in flower beds	Root type	Plant life expectancy in years	Aggressiveness and method	Cut flowers	Dried arrangements	Deer palatability
S, L, C	?–7.5	1'	1.5'	FIB	10–20	L–S	x		L
S, L, C	5.8–7.2	1'	1.5'	FIB	5–10	L–S	x		L
S, L	6.0–7.5	6"	1'	R	20+	L–R		SH	L
S, L, C	4.0–7.5	1'–2'	2'–3'	FIB	5–10	L–S	x	SH	L
S, L, C	6.0–8.0	2'	3'	R	20+	M–R	x	SH	L
S, L	?	2'	3'	R	20+	M–R	x		L
S, L, C	6.0–8.0	1'–2'	3'	FIB	5–10	H–S	x		L
S, L, C	4.0–8.0	2'	4'	R	10–20	L–R/S			L
S, L, C	5.7–7.1	1'–2'	3'	FIB	5–10	L–S	x	SP	L
S, L, C	4.8–7.3	1'	2'	R/CO	10–20	L–R	x	SP	L
S, G, L	5.5–8.0	1'	2'	TAP	10–20	L–S		SH	M
S, L	?	6"–1'	1'	CO	10–20	L–S	x	FL/SH	L
L	?	6"–1'	1'	CO	3–5	L–S	x	FL/SH	L
S, L	6.0–8.0	1'	1.5'–2'	TAPCO	10–20	L–S			L
S, L, C	6.0–8.0	6"–1'	1'	CO	10–20	L–S	x	FL/SH	L
S, L, C	5.5–7.5	6"–1'	1'	CO	10–20	L–S	x	FL/SH	L
S, L	?	1'	1.5'–2'	CO	10–20	L–S	x	FL/SH	L
S, L, C	?–6.8	2'	2'	B	20+	L–S	x	SP	H
S, L	?	1'	1'	FIB	5–10	L–S			L
S	?	1'	1'	TAP	10–20	L–S			L
S, G	?	1'	1'	TAP	5–10	L–S			L
S, L	5.8–7.8	6"–1'	2'	FIB	3–5	L–S	x		M
S, L, C	?–8.0	6"–1'	2'	FIB	5–10	M–S	x		L
S	5.0–?	1'	1.5'	TAP	10–20	M–S			L
S, L, C	6.0–8.0	1'–2'	3'	R	10–20	M–R	x	SH	L
S, G	5.0–?	1'	1'–2'	FIB	2	M–S	x	SH	L
S, L, C	?	2'	3'	FIB	2	M–S	x		L
S, L, C	?–8.0	1'–2'	3'	FIB	5–10	L–S	x	SH	L
S, L, C	5.5–8.0	1'–1.5'	3'	FIB	5–10	M–S	x	SH	L
S, L	?–7.5	1'–1.5'	2'	TAP	10–20	L–S			L
S, L, C	?–8.0	1'–2'	3'	TAP	10–20	L–S	x		M
S, G	?–8.5	1'–1.5'	2'	TAP	20+	L–S			L
S, G	?–7.5	1'	1'	FIB	3–5	L–S	x	SP	L
S, L, C	5.5–7.5	1'	2'	FIB	5–10	L–S	x		L
S, G	?–7.5	6"–1'	1'	FIB	5–10	L–S	x	SP	L
S, G	?	1'	2'–3'	FIB	3–5	L–S		SP	L
S, G	5.0–8.0	1'	1.5'	FIB	3–5	L–S	x	SP	L
S, L	?	6"–1'	1.5'	FIB	3–5	L–S	x		L
S, L, C	?	1'–2'	3'	R	10–20	L–R/S			H

(continued)

TABLE 12.17. *(continued)*

Scientific name (4/1/2020)	Common name	Height	Flower color	Bloom time	USDA zones	Sun	Soil moisture
Pulsatilla patens	Pasque Flower	6"–1'	White	Apr.–May	3–6	F	D
Pycnanthemum tenuifolium	Narrowleaf Mountainmint	1'–3'	White	July–Sept.	4–8	F, P	D, M
Pycnanthemum virginianum	Virginia Mountainmint	1'–3'	White	July–Sept.	3–7	F, P	Mo, W
Ratibida pinnata	Yellow Coneflower	3'–6'	Yellow	July–Sept.	3–9	F	D, M, Mo
Rosa carolina	Carolina Rose	1'–2'	Pink	June–July	3–8	F	D, M
Rosa setigera	Climbing Prairie Rose	6'–15'	Pink/white	May–July	4–9	F, P	M, Mo
Rudbeckia fulgida	Orange Coneflower	2'–4'	Yellow	July–Sept.	4–8	F, P	M, Mo
Rudbeckia hirta	Blackeyed Susan	1'–3'	Yellow	June–Sept.	3–10	F, P	D, M, Mo
Rudbeckia laciniata	Green-Headed Coneflower	5'–8'	Yellow	Aug.–Sept.	3–8	F, P	Mo, W
Rudbeckia subtomentosa	Sweet Blackeyed Susan	4'–6'	Yellow	Aug.–Oct.	4–8	F, P	M, Mo
Rudbeckia triloba	Browneyed Susan	2'–5'	Yellow	July–Oct.	4–9	F, P	M, Mo
Ruellia humilis	Wild Petunia	1'–2'	Violet	June–Aug.	4–9	F	D, M
Salvia azurea	Blue Sage	3'–5'	Blue	June–Aug.	4–9	F, P	D, M
Senna hebecarpa	Wild Senna	4'–6'	Yellow	July–Aug.	4–7	F	M, Mo, W
Silene regia	Royal Catchfly	2'–4'	Red	July–Aug.	4–9	F, P	M
Silphium integrifolium	Rosinweed	2'–6'	Yellow	July–Sept.	4–8	F	M, Mo
Silphium laciniatum	Compassplant	3'–10'	Yellow	June–Sept.	4–9	F	D, M
Silphium terebinthinaceum	Prairie Dock	3'–10'	Yellow	July–Sept.	4–7	F	M, Mo
Sisyrinchium albidum	White Blue-Eyed Grass	6"–1'	White	Apr.–May	4–9	F	D, M
Sisyrinchium angustifolium	Narrow-Leaved Blue-Eyed Grass	6"–2'	Blue/purple	Apr.–June	3–10	F, P	M, Mo
Sisyrinchium campestre	Prairie Blue-Eyed Grass	6"–1'	Blue/white	Apr.–June	3–8	F, P	D, M
Solidago ptarmicoides	White Aster	1'–2'	White	Aug.–Sept.	3–7	F	D
Solidago speciosa	Showy Goldenrod	1'–3'	Yellow	Aug.–Sept.	3–8	F	D, M
Spiraea alba	Meadowsweet	2'–4'	White/pink	June–Sept.	3–7	F, P	Mo, W
Spiraea tomentosa	Steeplebush	2'–4'	Pink	Aug.–Sept.	3–8	F, P	Mo, W
Symphyotrichum ericoides	Heath Aster	1'–3'	White	Aug.–Oct.	3–9	F	D, M
Symphyotrichum laeve	Smooth Aster	2'–4'	Blue	Aug.–Oct.	4–8	F	D, M
Symphyotrichum novae-angliae	New England Aster	3'–6'	Pink/purple/blue	Aug.–Oct.	3–7	F, P	M, Mo
Symphyotrichum oblongifolium	Aromatic Aster	1'–2'	Purple/violet	Sept.–Oct.	3–8	F	D, M
Symphyotrichum oolentangiense	Skyblue Aster	2'–3'	Blue	Aug.–Oct.	3–8	F, P	D, M
Tephrosia virginiana	Goatsrue	1'–2'	Pink/yellow	June–Aug.	4–9	F, P	D
Thalictrum dasycarpum	Tall Meadow-Rue	3'–6'	White	June–July	3–9	F, P	Mo, W
Tradescantia bracteata	Prairie Spiderwort	1'–2'	Blue	June–July	3–7	F, P	D, M
Tradescantia ohiensis	Ohio Spiderwort	2'–4'	Blue	June–July	3–9	F, P	D, M, Mo
Verbena hastata	Blue Vervain	3'–6'	Blue	July–Sept.	3–8	F, P	M, Mo, W
Verbena stricta	Hoary Vervain	2'–4'	Blue	July–Sept.	3–8	F	D, M
Vernonia fasciculata	Ironweed	4'–6'	Red/pink	July–Sept.	3–7	F	Mo
Vernonia gigantea	Tall Ironweed	5'–8'	Red/pink	Aug.–Sept.	4–9	F, P	Mo, W

Soil texture	pH range	Plant spacing in prairie gardens	Plant spacing in flower beds	Root type	Plant life expectancy in years	Aggressiveness and method	Cut flowers	Dried arrangements	Deer palatability
S, G	?–8.0	6"–1'	1'	FIB	5–10	L–S		SD	H
S, G, L, C	?	1'	1.5'	TAP/R	5–10	L–S		SH	M
S, L, C	?–8.0	1'–2'	3'	R	5–10	L–R		SH	L
S, L, C	5.5–8.0	1'–2'	2'	FIB	10–20	M–S	x	SH	L
S, L	4.0–7.0	2'	3'	R	10–20	L–R/S	x	RH	L
S, L, C	5.0–8.0	2'	5'	FIB	10–20	L–S	x	RH	L
L, C	?–8.0	1'–2'	3'	FIB	5–10	L–S	x	SH	L
S, L, C	5.0–8.0	1'	2'	FIB	2	M–S	x	SH	L
S, L, C	4.5–7.5	2'	2'	FIB	3–5	M–S	x	SH	L
S, L, C	?–8.0	1'–2'	2'	FIB	10–20	L–S	x	SH	L
S, L	?–8.0	1'–2'	3'	FIB	2	M–S	x	SH	L
S, L	4.5–7.5	1'	2'	FIB	5–10	L–S			L
S, G, L	?	1'	1.5'	FIB	5–10	L–S	x	SH	L
S, L, C	?	1.5'–2'	3'–4'	FIB/TAP	10–20	L–S	x	FL/SH	M
L	?	1'	1.5'	TAP	5–10	L–S	x	SH	L
S, L, C	?–8.0	2'	3'	TAP	20+	M–S	x		L
S, L, C	?–8.0	2'	4'	TAP	20+	L–S	x		L
S, L, C	?–8.0	2'	4'	TAP	20+	L–S	x		L
S, L	?–8.0	6"	6"	FIB	3–5	L–S			L
S, L, C	?–8.0	6"	9"	FIB	3–5	L–S			L
S, L	?–7.5	6"	6"	FIB	3–5	L–S			L
S, G	6.5–8.0	1'	2'	FIB	5–10	L–S		Empty SH	L
S, L	?	1'	3'	FIB	5–10	L–S	x	SH	L
S, L, C	4.3–7.2	2'	3'	FIB	10–20	L–S	x		L
S, L, C	4.5–7.0	2'	3'	R	10–20	L–R/S	x		L
S, G, L	?	1'–2'	2'	R	10–20	M–R		Empty SH	L
S, L	?	1'	3'	FIB	5–10	M–S	x	Empty SH	H
S, L, C	5.0–8.0	1'	3'	FIB	5–10	L–S	x	Empty SH	H
S, L, C	?–8.0	1'	3'	R	5–10	M–R?		Empty SH	L
S, L	5.0–8.0	1'	2'	FIB	5–10	L–S	x	Empty SH	M
S, G	4.5–?	2'	3'	TAP	10–20	L–S			L
S, L, C	?	1'–2'	2'	FIB	5–10	L–S			L
S, L, C	?	1'	2'	R	20+	M–R			M
S, L, C	?–7.5	1'	2'	FIB	20+	M–S			H
S, L, C	?	1'	2'	FIB	2	M–S	x	SH	L
S, L	?	1'	2'	FIB	3–5	M–S	x	SH	L
S, L, C	?	1'	3'	FIB	5–10	L–S	x		L
S, L, C	?–8.0	1'–2'	3'	FIB	10–20	L–S	x		L

(continued)

TABLE 12.17. *(continued)*

Scientific name (4/1/2020)	Common name	Height	Flower color	Bloom time	USDA zones	Sun	Soil moisture
Veronicastrum virginicum	Culver's Root	3'–6'	White	July–Aug.	3–8	F, P, S	M, Mo
Viola pedata	Birdfoot Violet	6"	Blue/purple	Apr.–June	3–9	F, P	D
Zizia aptera	Heartleaf Golden Alexanders	1'–2'	Yellow	May–June	3–8	F	M, Mo
Zizia aurea	Golden Alexanders	1'–2'	Yellow	May–July	3–8	F, P, S	M, Mo, W

Scientific name (4/1/2020)	Common name	Height	Fall leaf color	Bloom time	USDA zones	Sun	Soil moisture
Grasses and sedges							
Bouteloua curtipendula	Sideoats Grama	2'–3'	Straw	Aug.–Sept.	3–9	F	D, M
Bromus kalmii	Prairie Bromegrass	2'–4'	Straw	June–July	3–6	F	D
Calamagrostis canadensis	Bluejoint Grass	4'–6'	Straw	June–July	3–6	F	Mo, W
Carex brevior	Shortbeak Sedge	1'–4'	Straw	May–June	3–7	F, P	D, M, Mo
Elymus canadensis	Canada Wildrye	4'–5'	Straw	July–Aug.	2–9	F	D, M, Mo
Elymus virginicus	Virginia Wildrye	4'–5'	Straw	July–Aug.	2–9	F, P, S	M, Mo
Eragrostis spectabilis	Purple Lovegrass	1'–2'	Purple/pink	July–Sept.	3–10	F, P	D
Koeleria macrantha	Junegrass	2'–3'	Gold	May–June	3–8	F	D
Panicum virgatum	Switchgrass	3'–6'	Gold	Aug.–Sept.	3–10	F	D, M, Mo
Schizachyrium scoparium	Little Bluestem	2'–5'	Crimson red	Aug.–Oct.	3–10	F	D, M
Sorghastrum nutans	Indiangrass	5'–7'	Gold	Aug.–Sept.	3–9	F	D, M, Mo
Sporobolus heterolepis	Prairie Dropseed	2'–4'	Gold	Aug.–Sept.	3–8	F	D, M
Tridens flavus	Purpletop	3'–6'	Straw	July–Sept.	5–10	F	D, M
Tripsacum dactyloides	Eastern Gamagrass	4'–8'	Gold	July–Sept.	4–10	F, P	M, Mo, W

Notes: For soil moisture, D = dry, M = medium, Mo = moist, and W = wet. For soil texture, S = sand, L = loam, C = clay, and G = gravel. For sun, F = full sun, or at least a half day of full sun; P = partial sun, or a quarter to a half day of sun; and S = shade, or less than a quarter day of sun. For plant spacing, NR = not recommended. For root type, B = bulb, CO = corm, FIB = fibrous, R = rhizome, TAP =

Soil texture	pH range	Plant spacing in prairie gardens	Plant spacing in flower beds	Root type	Plant life expectancy in years	Aggressiveness and method	Cut flowers	Dried arrangements	Deer palatability
S, L, C	?–8.0	1'	3'	FIB	10–20	L–S	x	SH	L
S, G	?	6"	1'	R	5–10	L–S			L
S, L	?–7.5	1'	2'	FIB	5–10	L–S			L
S, L, C	?–8.0	1'	3'	FIB	5–10	L–S			L

Soil texture	pH range	Plant spacing in prairie gardens	Plant spacing in flower beds	Root type	Plant life expectancy in years	Aggressiveness and method	Cut flower stalks	Dried seed arrangements	Deer palatability
S, L	5.5–8.5	1'	2'	FIB	10–20	L–S		x	L
S, G	5.7–7.0	1'	2'	FIB	3–5	L–S		x	L
S, L, C	4.5–8.0	1'	3'	FIB	10–20	M–R/S		x	L
S, L, C	4.5–7.5	1'	1'	FIB	5–10	L–S			L
S, L, C	5.0–8.0	1'	3'	FIB	3–5	M–S		x	L
S, L, C	5.0–7.5	1'	3'	FIB	3–5	M–S		x	L
S, G	4.0–7.5	1'–1.5'	2'	FIB	5–10	M–S	x	x	L
S, G	6.0–8.0	6"–1'	1'	FIB	5–10	L–S		x	L
S, L, C	4.5–8.0	1.5'–2'	NR	FIB	20+	M–S		x	L
S, L	5.0–8.5	1'–1.5'	3'	FIB	20+	L–S		x	L
S, L, C	5.0–8.0	1'–2'	3'	FIB	10–20	M–S	x	x	L
S, L	6.0–8.0	2'	2'	FIB	20+	L–S	x	x	L
S, L, C	5.0–7.5	1'	1.5'	FIB	5–10	M–S		x	L
S, L, C	4.5–7.5	2'	3'	FIB	20+	H–S		x	M

taproot, and TAPCO = taprooted corm. For aggressiveness and method, L = low, M = medium, H = high, R = by rhizome, S = by seed, and R/S = rhizome and seed. For dried arrangements, FL = flowers, SH = seedheads, SP = seedpods, SD = seed, and RH = rose hips. For deer palatability, L = low, M = medium, and H = high.

TABLE 12.18. PRAIRIE SPECIMEN PLANTS

Scientific name (4/1/2020)	Common name	Height	Flower color	Bloom time	USDA zones	Sun	Soil moisture
Flowers and shrubs							
Amorpha canescens	Leadplant	2'–3'	Purple	June–July	3–8	F, P	D, M
Amsonia tabernaemontana	Eastern Bluestar	2'–3'	Blue/white	May–June	4–8	F, P	M, Mo
Arnoglossum atriplicifolium	Pale Indian Plantain	5'–10'	White	July–Sept.	4–8	F, P	D, M, Mo
Asclepias incarnata	Marsh Milkweed	3'–5'	Red/pink	June–July	3–9	F	Mo, W
Asclepias tuberosa	Butterflyweed	2'–3'	Orange	June–Aug.	3–10	F	D, M
Baptisia alba	White False Indigo	3'–5'	White	June–July	4–9	F, P	M, Mo
Baptisia australis	Blue False Indigo	3'–5'	Blue	June–July	3–10	F, P	M
Baptisia bracteata	Cream False Indigo	1'–2'	Cream	May–June	4–9	F, P	D, M
Ceanothus americanus	New Jersey Tea	2'–3'	White	July–Aug.	3–9	F, P	D, M
Coreopsis tripteris	Tall Coreopsis	3'–7'	Yellow	July–Sept.	3–8	F, P	M, Mo, W
Echinacea pallida	Pale Purple Coneflower	3'–5'	Purple	June–July	4–8	F	D, M
Echinacea paradoxa	Ozark Coneflower	3'–5'	Yellow	June–July	4–7	F, P	D, M
Echinacea purpurea	Purple Coneflower	3'–4'	Purple	July–Sept.	4–8	F, P	D, M
Eryngium yuccifolium	Rattlesnake Master	3'–5'	White	June–Aug.	4–9	F	D, M
Eutrochium fistulosum	Tall Joe Pye Weed	5'–8'	Purple/pink	Aug.–Sept.	4–9	F, P	M, Mo, W
Eutrochium maculatum	Joe Pye Weed	4'–6'	Pink	Aug.–Sept.	3–6	F, P	Mo, W
Filipendula rubra	Queen of the Prairie	4'–5'	Pink	June–July	3–6	F	M, Mo
Hibiscus moscheutos	Rosemallow	3'–7'	White/pink/red	July–Sept.	5–9	F, P	M, Mo, W
Hypericum ascyron	Great St. Johnswort	4'–6'	Yellow	July–Aug.	3–5	F, P	Mo, W
Liatris pycnostachya	Prairie Blazing Star	3'–5'	Purple/pink	July–Aug.	3–9	F	M, Mo
Liatris spicata	Dense Blazing Star	3'–6'	Purple/pink	Aug.–Sept.	4–10	F	M, Mo
Lilium michiganense	Michigan Lily	4'–6'	Orange	July–Aug.	3–7	F, P	Mo
Lobelia cardinalis	Cardinal Flower	2'–5'	Red	July–Sept.	3–9	F, P	Mo, W
Oligoneuron ohioense	Ohio Goldenrod	3'–4'	Yellow	Aug.–Sept.	4–5	F, P	M, Mo
Penstemon cobaea	Cobaea Beardtongue	2'–3'	Lavender	Apr.–June	5–8	F	D
Penstemon grandiflorus	Large-Flowered Beardtongue	2'–4'	Lavender	May–June	3–7	F	D
Rudbeckia fulgida	Orange Coneflower	2'–4'	Yellow	July–Sept.	4–8	F, P	M, Mo
Rudbeckia laciniata	Green-Headed Coneflower	5'–8'	Yellow	Aug.–Sept.	3–8	F, P	Mo, W
Rudbeckia subtomentosa	Sweet Blackeyed Susan	4'–6'	Yellow	Aug.–Oct.	4–8	F, P	M, Mo
Rudbeckia triloba	Browneyed Susan	2'–5'	Yellow	July–Oct.	4–9	F, P	M, Mo
Senna hebecarpa	Wild Senna	4'–6'	Yellow	July–Aug.	4–7	F	M, Mo, W
Silene regia	Royal Catchfly	2'–4'	Red	July–Aug.	4–9	F, P	M
Silphium laciniatum	Compassplant	3'–10'	Yellow	June–Sept.	4–9	F	D, M
Silphium perfoliatum	Cup Plant	3'–10'	Yellow	July–Sept.	3–8	F, P	M, Mo, W
Silphium terebinthinaceum	Prairie Dock	3'–10'	Yellow	July–Sept.	4–7	F	M, Mo
Spiraea alba	Meadowsweet	2'–4'	White/pink	June–Sept.	3–7	F, P	Mo, W
Spiraea tomentosa	Steeplebush	2'–4'	Pink	Aug.–Sept.	3–8	F, P	Mo, W
Symphyotrichum novae-angliae	New England Aster	3'–6'	Pink/purple/blue	Aug.–Oct.	3–7	F, P	M, Mo

Soil texture	pH range	Plant spacing in prairie gardens	Plant spacing in flower beds	Root type	Plant life expectancy in years	Aggressiveness and method	Cut flowers	Dried arrangements	Deer palatability
S, L	5.5–8.0	2'	3'	TAP	20+	L–S		Leaves	L
S, L, C	?–8.0	2'–3'	3'	FIB	20+	L–S		SP	L
S, L, C	?–8.0	2'	3'–4'	FIB	3–20	H–S (north); L–S (south)	x	FL	L
S, L, C	5.0–8.0	1'–2'	3'	FIB	5–10	L–S	x	SP	L
S, L	5.0–7.5	2'	3'	TAP	10–20	L–S	x	SP	L
S, L, C	?–7.5	3'	3'	TAP	20+	L–S	x	SP	L
S, L, C	?–8.0	3'	3'	TAP	20+	L–S	x	SP	L
S, L	?	2'	3'	TAP	20+	L–S	x	SP	L
S, L	4.5–7.0	3'	3'	TAP	20+	L–S		SD	L
S, L, C	?	1'–2'	NR	FIB	10–20	L–S	x		L
S, L, C	6.0–7.5	1'	2'–3'	TAP	10–20	L–S	x	SH	M
S, L, C	?–7.5	1'	2'–3'	TAP	10–20	L–S	x	SH	M
S, L, C	5.5–8.0	1'–2'	3'	FIB	5–10	M–S	x	SH	M
S, L, C	6.0–8.0	1'	2'	FIB	5–10	M–S	x	SH	L
S, L, C	4.5–7.5	2'	3'–4'	FIB	5–10	L–S			H
S, L, C	5.5–7.5	1'–2'	3'–4'	FIB	5–10	L–S			L
S, L, C	?–8.0	2'	3'	R	20+	M–R	x		L
S, L, C	4.0–8.0	2'	4'	R	10–20	L–R/S			L
S, L, C	5.7–7.1	1'–2'	3'	FIB	5–10	L–S	x	SP	L
S, L, C	6.0–8.0	6"–1'	1'	CO	10–20	L–S	x	FL/SH	L
S, L, C	5.5–7.5	6"–1'	1'	CO	10–20	L–S	x	FL/SH	L
S, L, C	?–6.8	2'	2'	B	20+	L–S	x	SP	H
S, L	5.8–7.8	6"–1'	2'	FIB	3–5	L–S	x		M
S, L, C	?–8.0	1'–2'	3'	FIB	5–10	L–S	x	SH	L
S, G	?–7.5	1'	1'	FIB	3–5	L–S	x	SP	L
S, G	?	1'	2'–3'	FIB	3–5	L–S		SP	L
L, C	?–8.0	1'–2'	3'	FIB	5–10	L–S	x	SH	L
S, L, C	4.5–7.5	2'	2'	FIB	3–5	M–S	x	SH	L
S, L, C	?–8.0	1'–2'	2'	FIB	10–20	L–S	x	SH	L
S, L	?–8.0	1'–2'	3'	FIB	2	M–S	x	SH	L
S, L, C	?	1.5'–2'	3'–4'	FIB/TAP	10–20	L–S	x	FL/SH	M
L	?	1'	1.5'	TAP	5–10	L–S	x	SH	L
S, L, C	?–8.0	2'	4'	TAP	20+	L–S	x		L
S, L, C	4.5–8.0	2'–3'	NR	FIB	20+	H–S	x		L
S, L, C	?–8.0	2'	4'	TAP	20+	L–S	x		L
S, L, C	4.3–7.2	2'	3'	FIB	10–20	L–S	x		L
S, L, C	4.5–7.0	2'	3'	R	10–20	L–R/S	x		L
S, L, C	5.0–8.0	1'	3'	FIB	5–10	L–S	x	Empty SH	H

(continued)

TABLE 12.18. *(continued)*

Scientific name (4/1/2020)	Common name	Height	Flower color	Bloom time	USDA zones	Sun	Soil moisture
Symphyotrichum oblongifolium	Aromatic Aster	1'–2'	Purple/violet	Sept.–Oct.	3–8	F	D, M
Tephrosia virginiana	Goatsrue	1'–2'	Pink/yellow	June–Aug.	4–9	F, P	D
Thalictrum dasycarpum	Tall Meadow-Rue	3'–6'	White	June–July	3–9	F, P	Mo, W
Vernonia fasciculata	Ironweed	4'–6'	Red/pink	July–Sept.	3–7	F	Mo
Vernonia gigantea	Tall Ironweed	5'–8'	Red/pink	Aug.–Sept.	4–9	F, P	Mo, W
Veronicastrum virginicum	Culver's Root	3'–6'	White	July–Aug.	3–8	F, P, S	M, Mo

Scientific name (4/1/2020)	Common name	Height	Fall leaf color	Bloom time	USDA zones	Sun	Soil moisture
Grasses and sedges							
Andropogon gerardii	Big Bluestem	5'–8'	Bronze/red	Aug.–Oct.	3–9	F	D, M, Mo
Andropogon virginicus	Broomsedge	2'–5'	Reddish brown	Aug.–Oct.	5–10	F	D, M, Mo
Eragrostis spectabilis	Purple Lovegrass	1'–2'	Purple/pink	July–Sept.	3–10	F, P	D
Hesperostipa spartea	Needlegrass	2'–4'	Gold	Apr.–June	3–6	F	D, M
Panicum virgatum	Switchgrass	3'–6'	Gold	Aug.–Sept.	3–10	F	D, M, Mo
Schizachyrium scoparium	Little Bluestem	2'–5'	Crimson red	Aug.–Oct.	3–10	F	D, M
Sorghastrum nutans	Indiangrass	5'–7'	Gold	Aug.–Sept.	3–9	F	D, M, Mo
Spartina pectinata	Prairie Cordgrass	6'–9'	Gold	Aug.–Sept.	3–8	F	Mo, W
Sporobolus heterolepis	Prairie Dropseed	2'–4'	Gold	Aug.–Sept.	3–8	F	D, M
Tridens flavus	Purpletop	3'–6'	Straw	July–Sept.	5–10	F	D, M
Tripsacum dactyloides	Eastern Gamagrass	4'–8'	Gold	July–Sept.	4–10	F, P	M, Mo, W

Notes: For soil moisture, D = dry, M = medium, Mo = moist, and W = wet. For soil texture, S = sand, L = loam, C = clay, and G = gravel. For sun, F = full sun, or at least a half day of full sun; P = partial sun, or a quarter to a half day of sun; and S = shade, or less than a quarter day of sun. For plant spacing, NR = not recommended. For root type, B = bulb, CO = corm, FIB = fibrous, R = rhizome, TAP =

Soil texture	pH range	Plant spacing in prairie gardens	Plant spacing in flower beds	Root type	Plant life expectancy in years	Aggressiveness and method	Cut flowers	Dried arrangements	Deer palatability
S, L, C	?-8.0	1'	3'	R	5-10	M-R?		Empty SH	L
S, G	4.5-?	2'	3'	TAP	10-20	L-S			L
S, L, C	?	1'-2'	2'	FIB	5-10	L-S			L
S, L, C	?	1'	3'	FIB	5-10	L-S	x		L
S, L, C	?-8.0	1'-2'	3'	FIB	10-20	L-S	x		L
S, L, C	?-8.0	1'	3'	FIB	10-20	L-S	x	SH	L

Soil texture	pH range	Plant spacing in prairie gardens	Plant spacing in flower beds	Root type	Plant life expectancy in years	Aggressiveness and method	Cut flower stalks	Dried seed arrangements	Deer palatability
S, L, C	5.0-7.5	2'	NR	FIB	20+	M-S		x	L
S, L, C	4.0-7.0	1'-1.5'	NR	FIB	20+	H-S		x	L
S, G	4.0-7.5	1'-1.5'	2'	FIB	5-10	M-S	x	x	L
S, G, L	5.0-7.5	1'	NR	FIB	10-20	L-S		x	L
S, L, C	4.5-8.0	1.5'-2'	NR	FIB	20+	M-S		x	L
S, L	5.0-8.5	1'-1.5'	3'	FIB	20+	L-S		x	L
S, L, C	5.0-8.0	1'-2'	3'	FIB	10-20	M-S	x	x	L
S, L, C	6.0-8.5	2'	NR	R	20+	H-R		x	L
S, L	6.0-8.0	2'	2'	FIB	20+	L-S	x	x	L
S, L, C	5.0-7.5	1'	1.5'	FIB	5-10	M-S		x	L
S, L, C	4.5-7.5	2'	3'	FIB	20+	H-S		x	M

taproot, and TAPCO = taprooted corm. For aggressiveness and method, L = low, M = medium, H = high, R = by rhizome, S = by seed, and R/S = rhizome and seed. For dried arrangements, FL = flowers, SH = seedheads, SP = seedpods, SD = seed, and RH = rose hips. For deer palatability, L = low, M = medium, and H = high.

TABLE 12.19.   PRAIRIE PLANT GROUNDCOVERS

Scientific name (4/1/2020)	Common name	Height	Flower color	Bloom time	USDA zones	Sun	Soil moisture
Flowers and shrubs							
Short groundcovers, under 2 feet tall							
Anemone canadensis	Canada Anemone	1'–2'	White	May–July	3–6	F, P	M, Mo
Antennaria neglecta	Pussytoes	6"–1'	White	Apr.–May	3–7	F	D
Callirhoe involucrata	Winecups	1'	Magenta	May–June	4–9	F, P	D, M
Callirhoe triangulata	Poppymallow	1'–2'	Magenta	July–Aug.	4–8	F	D
Conoclinium coelestinum	Blue Mistflower	1'–3'	Blue/purple	July–Oct.	4–10	F, P	M, Mo
Geum triflorum	Prairie Smoke	6"	Pink	May–June	3–6	F	D, M
Physostegia virginiana	Obedient Plant	1'–2'	Pink	Aug.–Sept.	3–9	F	M, Mo
Symphyotrichum ericoides	Heath Aster	1'–3'	White	Aug.–Oct.	3–9	F	D, M
Tradescantia occidentalis	Western Spiderwort	1'	Pink	June–July	3–9	F	D, M
Taller groundcovers, 2–6 feet tall							
Coreopsis palmata	Stiff Coreopsis	2'–3'	Yellow	June–Aug.	3–8	F	D, M
Filipendula rubra	Queen of the Prairie	4'–5'	Pink	June–July	3–6	F	M, Mo
Helianthus mollis	Downy Sunflower	4'–6'	Yellow	Aug.–Sept.	4–9	F	D, M
Helianthus occidentalis	Western Sunflower	2'–3'	Yellow	July–Aug.	3–9	F	D, M
Helianthus ×laetiflorus	Showy Sunflower	3'–6'	Yellow	Aug.–Sept.	3–8	F	D, M

Scientific name (4/1/2020)	Common name	Height	Fall leaf color	Bloom time	USDA zones	Sun	Soil moisture
Grasses and sedges							
Eragrostis spectabilis	Purple Lovegrass	1'–2'	Purple/pink	July–Sept.	3–10	F, P	D
Sporobolus heterolepis	Prairie Dropseed	2'–4'	Gold	Aug.–Sept.	3–8	F	D, M

Notes: For soil moisture, D = dry, M = medium, Mo = moist, and W = wet. For soil texture, S = sand, L = loam, C = clay, and G = gravel. For sun, F = full sun, or at least a half day of full sun; P = partial sun, or a quarter to a half day of sun; and S = shade, or less than a quarter day of sun. For plant spacing, NR = not recommended. For root type, B = bulb, CO = corm, FIB = fibrous, R = rhizome, TAP =

Soil texture	pH range	Plant spacing in prairie gardens	Plant spacing in flower beds	Root type	Plant life expectancy in years	Aggressiveness and method	Cut flowers	Dried arrangements	Deer palatability
S, L, C	?	NR	NR	R	10–20	H–R			L
S, G	5.5–7.5	1'	1'	R	5–10	L–R	x	FL	L
S, L	5.5–7.5	1'–2'	2'	TAP	10–20	L–S			L
S	?	1'–2'	2'	TAP	10–20	L–S			L
S, L, C	5.0–8.0	1'–2'	2'	R	3–5	M–R			L
S, L	6.0–7.5	6"	1'	R	20+	L–R		SH	L
S, L, C	6.0–8.0	1'–2'	NR	R	10–20	H–R	x	SH	M
S, G, L	?	1'–2'	2'	R	10–20	M–R		Empty SH	L
S, L	?	1'	NR	R	20+	H–R			M
S, L	?	1'–2'	2'	R	20+	M–R			L
S, L, C	?–8.0	2'	3'	R	20+	M–R	x		L
S, L, C	6.0–8.0	2'	3'	R	20+	M–R	x	SH	L
S, L	?	2'	3'	R	20+	M–R	x		L
S, G, L, C	?	NR	NR	R	20+	H–R	x	SH	L

Soil texture	pH range	Plant spacing in prairie gardens	Plant spacing in flower beds	Root type	Plant life expectancy in years	Aggressiveness and method	Cut flower stalks	Dried seed arrangements	Deer palatability
S, G	4.0–7.5	1'–1.5'	2'	FIB	5–10	M–S	x	x	L
S, L	6.0–8.0	2'	2'	FIB	20+	L–S	x	x	L

taproot, and TAPCO = taprooted corm. For aggressiveness and method, L = low, M = medium, H = high, R = by rhizome, S = by seed, and R/S = rhizome and seed. For dried arrangements, FL = flowers, SH = seedheads, SP = seedpods, SD = seed, and RH = rose hips. For deer palatability, L = low, M = medium, and H = high.

TABLE 12.20. SPRING-BLOOMING PRAIRIE FLOWERS

Bloom time	Scientific name (4/1/2020)	Common name	Height	Flower color	USDA zones	Sun	Soil moisture
Apr.–May	*Antennaria neglecta*	Pussytoes	6"–1'	White	3–7	F	D
Apr.–May	*Pulsatilla patens*	Pasque Flower	6"–1'	White	3–6	F	D
Apr.–May	*Sisyrinchium albidum*	White Blue-Eyed Grass	6"–1'	White	4–9	F	D, M
Apr.–June	*Penstemon cobaea*	Cobaea Beardtongue	2'–3'	Lavender	5–8	F	D
Apr.–June	*Sisyrinchium angustifolium*	Narrow-Leaved Blue-Eyed Grass	6"–2'	Blue/purple	3–10	F, P	M, Mo
Apr.–June	*Sisyrinchium campestre*	Prairie Blue-Eyed Grass	6"–1'	Blue/white	3–8	F, P	D, M
Apr.–June	*Viola pedata*	Birdfoot Violet	6"	Blue/purple	3–9	F, P	D
May–June	*Amsonia tabernaemontana*	Eastern Bluestar	2'–3'	Blue/white	4–8	F, P	M, Mo
May–June	*Anemone cylindrica*	Thimbleweed	1'–3'	White	3–6	F	D, M
May–June	*Baptisia bracteata*	Cream False Indigo	1'–2'	Cream	4–9	F, P	D, M
May–June	*Callirhoe involucrata*	Winecups	1'	Magenta	4–9	F, P	D, M
May–June	*Dodecatheon meadia*	Shooting Star	1'–2'	White/pink	4–8	F, P	M, Mo
May–June	*Geum triflorum*	Prairie Smoke	6"	Pink	3–6	F	D, M
May–June	*Lithospermum canescens*	Hoary Puccoon	1'	Orange	3–7	F, P	D, M
May–June	*Lupinus perennis*	Wild Lupine	1'–2'	Blue	3–8	F, P	D
May–June	*Onosmodium bejariense* var. *occidentale*	Marbleseed (Western Marbleseed)	1'–3'	White	3–8	F	D, M
May–June	*Penstemon gracilis*	Slender Beardtongue	1'–2'	Lavender	3–6	F, P	D
May–June	*Penstemon grandiflorus*	Large-Flowered Beardtongue	2'–4'	Lavender	3–7	F	D
May–June	*Penstemon hirsutus*	Hairy Penstemon	1'–2'	Lavender	3–6	F, P	D
May–June	*Phlox pilosa*	Downy Phlox	1'–2'	Pink	3–9	F	D, M, Mo
May–June	*Polygonatum biflorum* var. *commutatum*	Great Solomon's Seal	2'–4'	Cream	3–9	F, P	D, M
May–June	*Zizia aptera*	Heartleaf Golden Alexanders	1'–2'	Yellow	3–8	F	M, Mo
May–July	*Anemone canadensis*	Canada Anemone	1'–2'	White	3–6	F, P	M, Mo
May–July	*Blephilia ciliata*	Downy Wood Mint	1'–2'	White/purple/violet	4–8	F, P	D, M
May–July	*Rosa setigera*	Climbing Prairie Rose	6'–15'	Pink/white	4–9	F, P	M, Mo
May–July	*Zizia aurea*	Golden Alexanders	1'–2'	Yellow	3–8	F, P, S	M, Mo, W
June–July	*Baptisia alba*	White False Indigo	3'–5'	White	4–9	F, P	M, Mo
June–July	*Baptisia australis*	Blue False Indigo	3'–5'	Blue	3–10	F, P	M
June–July	*Coreopsis lanceolata*	Lanceleaf Coreopsis	1'–2'	Yellow	3–9	F	D, M
June–July	*Delphinium carolinianum* subsp. *virescens*	Prairie Larkspur	1'–4'	White	3–8	F	D, M
June–July	*Echinacea pallida*	Pale Purple Coneflower	3'–5'	Purple	4–8	F	D, M
June–July	*Echinacea paradoxa*	Ozark Coneflower	3'–5'	Yellow	4–7	F, P	D, M
June–July	*Iris virginica* var. *shrevei*	Wild Iris	2'–3'	Blue	3–8	F, P	W
June–July	*Lithospermum caroliniense*	Hairy Puccoon	1'–2'	Yellow	3–8	F	D
June–July	*Lithospermum incisum*	Fringed Puccoon	1'–2'	Yellow	3–9	F	D

Soil texture	pH range	Plant spacing in prairie gardens	Plant spacing in flower beds	Root type	Plant life expectancy in years	Aggressiveness and method	Cut flowers	Dried arrangements	Deer palatability
S, G	5.5–7.5	1'	1'	R	5–10	L–R	x	FL	L
S, G	?–8.0	6"–1'	1'	FIB	5–10	L–S		SD	H
S, L	?–8.0	6"	6"	FIB	3–5	L–S			L
S, G	?–7.5	1'	1'	FIB	3–5	L–S	x	SP	L
S, L, C	?–8.0	6"	9"	FIB	3–5	L–S			L
S, L	?–7.5	6"	6"	FIB	3–5	L–S			L
S, G	?	6"	1'	R	5–10	L–S			L
S, L, C	?–8.0	2'–3'	3'	FIB	20+	L–S		SP	L
S, L	?	1'	1'	FIB	10–20	L–S		SD	L
S, L	?	2'	3'	TAP	20+	L–S	x	SP	L
S, L	5.5–7.5	1'–2'	2'	TAP	10–20	L–S			L
S, L, C	4.5–8.0	6"–1'	1'	FIB	20+	L–S		SP	L
S, L	6.0–7.5	6"	1'	R	20+	L–R		SH	L
S, L	?	1'	1'	FIB	5–10	L–S			L
S	5.0–?	1'	1.5'	TAP	10–20	M–S			L
S, L	?–7.5	1'–1.5'	2'	TAP	10–20	L–S			L
S, G	?–7.5	6"–1'	1'	FIB	5–10	L–S	x	SP	L
S, G	?	1'	2'–3'	FIB	3–5	L–S		SP	L
S, G	5.0–8.0	1'	1.5'	FIB	3–5	L–S	x	SP	L
S, L	?	6"–1'	1.5'	FIB	3–5	L–S	x		L
S, L, C	?	1'–2'	3'	R	10–20	L–R/S			H
S, L	?–7.5	1'	2'	FIB	5–10	L–S			L
S, L, C	?	NR	NR	R	10–20	H–R			L
G, L, C	?–8.00	1'	1'–2'	TAP/short Rs	3–5	L–S		FL/SH	L
S, L, C	5.0–8.0	2'	5'	FIB	10–20	L–S	x	RH	L
S, L, C	?–8.0	1'	3'	FIB	5–10	L–S			L
S, L, C	?–7.5	3'	3'	TAP	20+	L–S	x	SP	L
S, L, C	?–8.0	3'	3'	TAP	20+	L–S	x	SP	L
S, L	5.0–7.5	1'	1'–2'	FIB	3–5	M–S	x		M
S, L	?	1'	1.5'	FIB	5–10	L–S	x	SP	L
S, L, C	6.0–7.5	1'	2'–3'	TAP	10–20	L–S	x	SH	M
S, L, C	?–7.5	1'	2'–3'	TAP	10–20	L–S	x	SH	M
S, L, C	4.8–7.3	1'	2'	R/CO	10–20	L–R	x	SP	L
S	?	1'	1'	TAP	10–20	L–S			L
S, G	?	1'	1'	TAP	5–10	L–S			L

(continued)

TABLE 12.20. (*continued*)

Bloom time	Scientific name (4/1/2020)	Common name	Height	Flower color	USDA zones	Sun	Soil moisture
June–July	*Penstemon digitalis*	Foxglove Beardtongue	2'–3'	White	3–8	F, P	M, Mo
June–July	*Thalictrum dasycarpum*	Tall Meadow-Rue	3'–6'	White	3–9	F, P	Mo, W
June–July	*Tradescantia bracteata*	Prairie Spiderwort	1'–2'	Blue	3–7	F, P	D, M
June–July	*Tradescantia occidentalis*	Western Spiderwort	1'	Pink	3–9	F	D, M
June–July	*Tradescantia ohiensis*	Ohio Spiderwort	2'–4'	Blue	3–9	F, P	D, M, Mo

Notes: For soil moisture, D = dry, M = medium, Mo = moist, and W = wet. For soil texture, S = sand, L = loam, C = clay, and G = gravel. For sun, F = full sun, or at least a half day of full sun; P = partial sun, or a quarter to a half day of sun; and S = shade, or less than a quarter day of sun. For plant spacing, NR = not recommended. For root type, B = bulb, CO = corm, FIB = fibrous, R = rhizome, TAP =

Soil texture	pH range	Plant spacing in prairie gardens	Plant spacing in flower beds	Root type	Plant life expectancy in years	Aggressiveness and method	Cut flowers	Dried arrangements	Deer palatability
S, L, C	5.5–7.5	1'	2'	FIB	5–10	L–S	x		L
S, L, C	?	1'–2'	2'	FIB	5–10	L–S			L
S, L, C	?	1'	2'	R	20+	M–R			M
S, L	?	1'	NR	R	20+	H–R			M
S, L, C	?–7.5	1'	2'	FIB	20+	M–S			H

taproot, and TAPCO = taprooted corm. For aggressiveness and method, L = low, M = medium, H = high, R = by rhizome, S = by seed, and R/S = rhizome and seed. For dried arrangements, FL = flowers, SH = seedheads, SP = seedpods, SD = seed, and RH = rose hips. For deer palatability, L = low, M = medium, and H = high.

TABLE 12.21. TAPROOTED PRAIRIE FLOWERS FOR MIXING WITH GRASSES

Scientific name (4/1/2020)	Common name	Height	Flower color	Bloom time	USDA zones	Sun	Soil moisture
Amorpha canescens	Leadplant	2'–3'	Purple	June–July	3–8	F, P	D, M
Asclepias tuberosa	Butterflyweed	2'–3'	Orange	June–Aug.	3–10	F	D, M
Astragalus canadensis	Canada Milkvetch	2'–3'	Yellow	July–Aug.	3–8	F	D, M
Baptisia alba	White False Indigo	3'–5'	White	June–July	4–9	F, P	M, Mo
Baptisia australis	Blue False Indigo	3'–5'	Blue	June–July	3–10	F, P	M
Baptisia bracteata	Cream False Indigo	1'–2'	Cream	May–June	4–9	F, P	D, M
Callirhoe involucrata	Winecups	1'	Magenta	May–June	4–9	F, P	D, M
Callirhoe triangulata	Poppymallow	1'–2'	Magenta	July–Aug.	4–8	F	D
Ceanothus americanus	New Jersey Tea	2'–3'	White	July–Aug.	3–9	F, P	D, M
Dalea candida	White Prairie Clover	1'–2'	White	July–Aug.	3–9	F	D, M
Dalea purpurea	Purple Prairie Clover	1'–2'	Purple/yellow	July–Aug.	3–8	F	D, M
Desmanthus illinoensis	Illinois Bundleflower	2'–6'	White	June–Aug.	5–8	F	D, M, Mo
Desmodium canadense	Canada Ticktrefoil	3'–5'	Purple	July–Aug.	3–7	F	M, Mo
Echinacea pallida	Pale Purple Coneflower	3'–5'	Purple	June–July	4–8	F	D, M
Echinacea paradoxa	Ozark Coneflower	3'–5'	Yellow	June–July	4–7	F, P	D, M
Lespedeza capitata	Roundhead Bush Clover	3'–5'	White	Aug.–Sept.	3–8	F, P	D, M
Liatris punctata	Dotted Blazing Star	1'–2'	Purple/pink	Aug.–Oct.	3–9	F	D
Lithospermum caroliniense	Hairy Puccoon	1'–2'	Yellow	June–July	3–8	F	D
Lithospermum incisum	Fringed Puccoon	1'–2'	Yellow	June–July	3–9	F	D
Lupinus perennis	Wild Lupine	1'–2'	Blue	May–June	3–8	F, P	D
Onosmodium bejariense var. occidentale	Marbleseed (Western Marbleseed)	1'–3'	White	May–June	3–8	F	D, M
Parthenium integrifolium	Wild Quinine	3'–5'	White	June–Sept.	4–8	F	M, Mo
Pediomelum esculentum	Prairie Turnip	1'–2'	Blue	June–July	3–7	F	D
Senna hebecarpa	Wild Senna	4'–6'	Yellow	July–Aug.	4–7	F	M, Mo, W
Silene regia	Royal Catchfly	2'–4'	Red	July–Aug.	4–9	F, P	M
Silphium integrifolium	Rosinweed	2'–6'	Yellow	July–Sept.	4–8	F	M, Mo
Silphium laciniatum	Compassplant	3'–10'	Yellow	June–Sept.	4–9	F	D, M
Silphium terebinthinaceum	Prairie Dock	3'–10'	Yellow	July–Sept.	4–7	F	M, Mo
Tephrosia virginiana	Goatsrue	1'–2'	Pink/yellow	June–Aug.	4–9	F, P	D

Notes: For soil moisture, D = dry, M = medium, Mo = moist, and W = wet. For soil texture, S = sand, L = loam, C = clay, and G = gravel. For sun, F = full sun, or at least a half day of full sun; P = partial sun, or a quarter to a half day of sun; and S = shade, or less than a quarter day of sun. For plant spacing, NR = not recommended. For root type, B = bulb, CO = corm, FIB = fibrous, R = rhizome, TAP =

Soil texture	pH range	Plant spacing in prairie gardens	Plant spacing in flower beds	Root type	Plant life expectancy in years	Aggressiveness and method	Cut flowers	Dried arrangements	Deer palatability
S, L	5.5–8.0	2'	3'	TAP	20+	L–S		Leaves	L
S, L	5.0–7.5	2'	3'	TAP	10–20	L–S	x	SP	L
S, L	6.0–8.0	1'–2'	3'	TAP	10–20	M–S		SP	M
S, L, C	?–7.5	3'	3'	TAP	20+	L–S	x	SP	L
S, L, C	?–8.0	3'	3'	TAP	20+	L–S	x	SP	L
S, L	?	2'	3'	TAP	20+	L–S	x	SP	L
S, L	5.5–7.5	1'–2'	2'	TAP	10–20	L–S			L
S	?	1'–2'	2'	TAP	10–20	L–S			L
S, L	4.5–7.0	3'	3'	TAP	20+	L–S		SD	L
S, L	?	6"–1'	1'–2'	TAP	10–20	L–S			H
S, L, C	?	6"–1'	1'–2'	TAP	10–20	L–S	x		H
S, L, C	6.0–8.0	1'	1.5'	TAP	5–10	L–S		SP	M
S, L, C	?–7.5	NR	NR	TAP	10–20	M–S			H
S, L, C	6.0–7.5	1'	2'–3'	TAP	10–20	L–S	x	SH	M
S, L, C	?–7.5	1'	2'–3'	TAP	10–20	L–S	x	SH	M
S, G, L	5.5–8.0	1'	2'	TAP	10–20	L–S		SH	M
S, L	6.0–8.0	1'	1.5'–2'	TAPCO	10–20	L–S			L
S	?	1'	1'	TAP	10–20	L–S			L
S, G	?	1'	1'	TAP	5–10	L–S			L
S	5.0–?	1'	1.5'	TAP	10–20	M–S			L
S, L	?–7.5	1'–1.5'	2'	TAP	10–20	L–S			L
S, L, C	?–8.0	1'–2'	3'	TAP	10–20	L–S	x		M
S, G	?–8.5	1'–1.5'	2'	TAP	20+	L–S			L
S, L, C	?	1.5'–2'	3'–4'	FIB/TAP	10–20	L–S	x	FL/SH	M
L	?	1'	1.5'	TAP	5–10	L–S	x	SH	L
S, L, C	?–8.0	2'	3'	TAP	20+	M–S	x		L
S, L, C	?–8.0	2'	4'	TAP	20+	L–S	x		L
S, L, C	?–8.0	2'	4'	TAP	20+	L–S	x		L
S, G	4.5–?	2'	3'	TAP	10–20	L–S			L

taproot, and TAPCO = taprooted corm. For aggressiveness and method, L = low, M = medium, H = high, R = by rhizome, S = by seed, and R/S = rhizome and seed. For dried arrangements, FL = flowers, SH = seedheads, SP = seedpods, SD = seed, and RH = rose hips. For deer palatability, L = low, M = medium, and H = high.

TABLE 12.22. EASY-TO-DIVIDE PRAIRIE PLANTS

Scientific name (4/1/2020)	Common name	Root type	Fall	Spring	Summer
Flowers and shrubs					
Agastache foeniculum	Lavender Hyssop	FIB	x	x	
Allium cernuum	Nodding Pink Onion	B	x	x	
Allium stellatum	Prairie Onion	B	x	x	
Amsonia tabernaemontana	Eastern Bluestar	FIB	x	x	
Anemone canadensis	Canada Anemone	R	x	x	
Antennaria neglecta	Pussytoes	R	x	x	June
Asclepias incarnata	Marsh Milkweed	FIB	x	x	
Asclepias sullivantii	Sullivant's Milkweed	R	x	x	
Asclepias syriaca	Common Milkweed	R	x	x	
Conoclinium coelestinum	Blue Mistflower	R	x	x	
Coreopsis palmata	Stiff Coreopsis	R	x	x	
Dodecatheon meadia	Shooting Star	FIB	x		August
Echinacea purpurea	Purple Coneflower	FIB	x	x	
Eryngium yuccifolium	Rattlesnake Master	FIB	x	x	
Eupatorium perfoliatum	Boneset	FIB	x	x	
Euphorbia corollata	Flowering Spurge	R	x	x	
Eutrochium fistulosum	Tall Joe Pye Weed	FIB	x	x	
Eutrochium maculatum	Joe Pye Weed	FIB	x	x	
Filipendula rubra	Queen of the Prairie	R	x	x	
Gentiana alba	Cream Gentian	FIB	x	x	
Gentiana andrewsii	Bottle Gentian	FIB	x	x	
Geum triflorum	Prairie Smoke	R	x	x	
Helenium autumnale	Sneezeweed	FIB	x	x	
Helianthus mollis	Downy Sunflower	R	x	x	
Helianthus occidentalis	Western Sunflower	R	x	x	
Helianthus ×laetiflorus	Showy Sunflower	R	x	x	
Hibiscus moscheutos	Rosemallow	R	x	x	
Iris virginica var. *shrevei*	Wild Iris	R/CO	x	x	x
Lilium michiganense	Michigan Lily	B	x		
Lobelia cardinalis	Cardinal Flower	FIB	x	x	
Lobelia siphilitica	Great Blue Lobelia	FIB	x	x	
Monarda fistulosa	Bergamot	R	x	x	
Oligoneuron ohioense	Ohio Goldenrod	FIB	x	x	
Oligoneuron rigidum	Stiff Goldenrod	FIB	x	x	
Penstemon digitalis	Foxglove Beardtongue	FIB	x	x	
Phlox pilosa	Downy Phlox	FIB	x	x	
Physostegia virginiana	Obedient Plant	R	x	x	
Polygonatum biflorum var. *commutatum*	Great Solomon's Seal	R	x	x	
Pycnanthemum tenuifolium	Narrowleaf Mountainmint	TAP/R	x	x	
Pycnanthemum virginianum	Virginia Mountainmint	R	x	x	
Rudbeckia fulgida	Orange Coneflower	FIB	x	x	
Rudbeckia laciniata	Green-Headed Coneflower	FIB	x	x	
Rudbeckia subtomentosa	Sweet Blackeyed Susan	FIB	x	x	

Division notes

Pull individual bulbs from clump
Pull individual bulbs from clump

Divide fine rhizomes; do not allow to dry out — plant immediately
Cut or break off 3"–4" pieces of rhizome
Tease apart individual root clumps, each with a bud
Cut rhizomes into 8"–12" segments with terminal bud
Cut rhizomes into 8"–12" segments with terminal bud

Cut rhizomes into 6" segments with terminal bud
Carefully tease apart individual root clumps, each with a bud; never divide when green
Cut root mass into a few sections, each with a bud; do not overdivide
Tease apart individual root clumps, each with a bud
Cut root mass into a few sections, each with a bud; do not overdivide
Cut rhizomes into 6" segments with at least one bud
Cut root mass into a few sections, each with a bud; do not overdivide
Cut root mass into a few sections, each with a bud; do not overdivide
Cut rhizomes into 8"–12" segments with terminal bud
Tease apart individual root clumps, each with a bud
Tease apart individual root clumps, each with a bud
Break root mass into individual 4"–6" rhizomes, each with a bud
Cut or break root mass apart into 1"–2" clumps, each with a bud
Cut rhizomes into 6" segments with terminal bud
Cut rhizomes into 6" segments with terminal bud
Cut rhizomes into 6" segments with terminal bud

Break root mass into individual 6" rhizomes, each with a bud — cut back leaves 50% in summer
Pull scales off bulb and plant 3" deep in rich, moist soil in fall after plant goes dormant
Carefully tease apart individual root clumps, each with a bud
Carefully tease apart individual root clumps, each with a bud
Divide root mass into 2"×2" clumps
Cut or break root mass apart into 1"–2" clumps, each with a bud
Cut or break root mass apart into 1"–2" clumps, each with a bud
Pull or cut root mass apart into 3-4 clumps, each with a bud
Carefully tease apart individual small root clumps, each with a bud
Cut rhizomes into 6" segments with terminal bud
Cut rhizomes into 8"–12" segments with terminal bud

Cut root mass into a few sections, each with a bud
Cut or break root mass apart into 1"–2" clumps, each with a bud
Cut or break root mass apart into 1"–2" clumps, each with a bud
Cut or break root mass apart into 1"–2" clumps, each with a bud

(continued)

TABLE 12.22. *(continued)*

Scientific name (4/1/2020)	Common name	Root type	Fall	Spring	Summer
Sisyrinchium albidum	White Blue-Eyed Grass	FIB	x	x	July
Sisyrinchium angustifolium	Narrow-Leaved Blue-Eyed Grass	FIB	x	x	July
Sisyrinchium campestre	Prairie Blue-Eyed Grass	FIB	x	x	July
Solidago ptarmicoides	White Aster	FIB	x	x	
Solidago speciosa	Showy Goldenrod	FIB	x	x	
Symphyotrichum ericoides	Heath Aster	R	x	x	
Symphyotrichum laeve	Smooth Aster	FIB	x	x	
Symphyotrichum novae-angliae	New England Aster	FIB	x	x	
Symphyotrichum oblongifolium	Aromatic Aster	R	x	x	
Tradescantia bracteata	Prairie Spiderwort	R	x	x	
Tradescantia occidentalis	Western Spiderwort	R	x	x	
Tradescantia ohiensis	Ohio Spiderwort	FIB	x	x	
Vernonia fasciculata	Ironweed	FIB	x	x	
Vernonia gigantea	Tall Ironweed	FIB	x	x	
Veronicastrum virginicum	Culver's Root	FIB	x	x	
Viola pedata	Birdfoot Violet	R	x	x	July

Grasses and sedges

Andropogon gerardii	Big Bluestem	FIB	x	x	
Andropogon virginicus	Broomsedge	FIB	x	x	
Bouteloua curtipendula	Sideoats Grama	FIB	x	x	
Calamagrostis canadensis	Bluejoint Grass	FIB	x	x	
Carex brevior	Shortbeak Sedge	FIB	x	x	
Eragrostis spectabilis	Purple Lovegrass	FIB	x	x	
Hesperostipa spartea	Needlegrass	FIB	x	x	
Hierochloe hirta	Vanilla Sweetgrass	R	x	x	
Koeleria macrantha	Junegrass	FIB	x	x	
Panicum virgatum	Switchgrass	FIB	x	x	
Schizachyrium scoparium	Little Bluestem	FIB	x	x	
Sorghastrum nutans	Indiangrass	FIB	x	x	
Spartina pectinata	Prairie Cordgrass	R	x	x	
Sporobolus heterolepis	Prairie Dropseed	FIB	x	x	
Tridens flavus	Purpletop	FIB	x	x	
Tripsacum dactyloides	Eastern Gamagrass	FIB	x	x	

Notes: For root type, B = bulb, CO = corm, FIB = fibrous, R = rhizome, TAP = taproot, and TAPCO = taprooted corm.

Division notes

Tear apart individual "fans" with buds; cut back leaves 50% in summer

Tear apart individual "fans" with buds; cut back leaves 50% in summer

Tear apart individual "fans" with buds; cut back leaves 50% in summer

Cut or break root mass apart into 1"–2" clumps, each with a bud

Cut or break root mass apart into 1"–2" clumps, each with a bud

Cut or break root mass apart into 1"–2" clumps, each with a bud

Cut or break root mass apart into 1"–2" clumps, each with a bud

Cut or break root mass apart into 1"–2" clumps, each with a bud

Cut or break root mass apart into 1"–2" clumps, each with a bud

Pull individual clumps from root mass, each with a bud

Break apart clumps and rhizomes, each with a bud

Pull individual clumps from root mass, each with a bud

Cut or break root mass apart into 1"–2" clumps, each with a bud

Cut or break root mass apart into 1"–2" clumps, each with a bud

Break apart clumps into 6" long pieces, each with a bud

Pull individual plants from root mass; divide in summer after seed has ripened

Cut with sharp spade or hatchet; can be divided into 2"×2" pieces

Cut with sharp spade or hatchet; can be divided into 2"×2" pieces

Tease apart individual loosely connected individual pieces

Cool-season grass — divide in fall or early spring; cut clumps with sharp spade into 2"×2" pieces

Cool-season sedge — divide in fall or early spring; carefully tease apart roots into pieces with buds

Cut with sharp spade or hatchet; can be divided into 2"×2" pieces

Cool-season grass — divide in fall or early spring; carefully tease apart roots into pieces with buds

Cut rhizomes into 6" segments with terminal bud

Cool-season grass — divide in fall or early spring; carefully tease apart roots into pieces with buds

Cut with sharp spade or hatchet; can be divided into 2"×2" pieces

Cut with sharp spade or hatchet; can be divided into 2"×2" pieces

Cut with sharp spade or hatchet; can be divided into 2"×2" pieces

Cut rhizomes into 8"–12" segments with terminal bud

Cut with sharp spade or hatchet into 2"–3" pieces; roots tear easily; do not divide too finely

Cut with sharp spade or hatchet; can be divided into 2"×2" pieces

Cut with sharp spade or hatchet; can be divided into 2"×2" pieces

TABLE 12.23. PLANT STEM-CUTTING PROPAGATION GUIDELINES

Seasonal timing of stem cuttings	Scientific name (4/1/2020)	Common name	Cutting success rate	Plant height	Flower color	Bloom time
Flowers and shrubs						
Basal stems– midspring	*Callirhoe involucrata*	Winecups	H	1'	Magenta	May–June
Early–mid	*Agastache foeniculum*	Lavender Hyssop	H	1'–3'	Purple	July–Sept.
Early–mid (softwood)	*Ceanothus americanus*	New Jersey Tea	H	2'–3'	White	July–Aug.
Early–mid	*Chelone glabra*	White Turtlehead	H	2'–4'	White	Aug.–Sept.
Early–mid	*Conoclinium coelestinum*	Blue Mistflower	H	1'–3'	Blue/purple	July–Oct.
Early–mid	*Hibiscus moscheutos*	Rosemallow	H	3'–7'	White/pink/red	July–Sept.
Early–mid	*Hypericum ascyron*	Great St. Johnswort	H	4'–6'	Yellow	July–Aug.
Early–mid	*Lobelia cardinalis*	Cardinal Flower	H	2'–5'	Red	July–Sept.
Early–mid	*Lobelia siphilitica*	Great Blue Lobelia	H	1'–4'	Blue	July–Sept.
Early–mid	*Oenothera gaura*	Biennial Beeblossom	H	3'–6'	White/pink	June–Sept.
Early–mid	*Oligoneuron ohioense*	Ohio Goldenrod	H	3'–4'	Yellow	Aug.–Sept.
Early–mid	*Oligoneuron rigidum*	Stiff Goldenrod	H	3'–5'	Yellow	Aug.–Sept.
Early–mid	*Phlox pilosa*	Downy Phlox	H	1'–2'	Pink	May–June
Early–mid (new wood)	*Rosa carolina*	Carolina Rose	H	1'–2'	Pink	June–July
Early–mid (new wood)	*Rosa setigera*	Climbing Prairie Rose	H	6'–15'	Pink/white	May–July
Early–mid	*Solidago ptarmicoides*	White Aster	H	1'–2'	White	Aug.–Sept.
Early–mid	*Solidago speciosa*	Showy Goldenrod	H	1'–3'	Yellow	Aug.–Sept.
Early–mid (new wood)	*Spiraea alba*	Meadowsweet	H	2'–4'	White/pink	June–Sept.
Early–mid (new wood)	*Spiraea tomentosa*	Steeplebush	H	2'–4'	Pink	Aug.–Sept.
Early–mid	*Symphyotrichum ericoides*	Heath Aster	H	1'–3'	White	Aug.–Oct.
Early–mid	*Symphyotrichum laeve*	Smooth Aster	H	2'–4'	Blue	Aug.–Oct.
Early–mid	*Symphyotrichum novae-angliae*	New England Aster	H	3'–6'	Pink/purple/ blue	Aug.–Oct.
Early–mid	*Symphyotrichum oblongifolium*	Aromatic Aster	H	1'–2'	Purple/violet	Sept.–Oct.
Early–mid	*Symphyotrichum oolentangiense*	Skyblue Aster	H	2'–3'	Blue	Aug.–Oct.
Early–mid	*Ruellia humilis*	Wild Petunia	H	1'–2'	Violet	June–Aug.
Early–mid	*Verbena hastata*	Blue Vervain	H	3'–6'	Blue	July–Sept.
Early–mid	*Verbena stricta*	Hoary Vervain	H	2'–4'	Blue	July–Sept.
Early summer	*Pycnanthemum tenuifolium*	Narrowleaf Mountainmint	H	1'–3'	White	July–Sept.
Early summer	*Pycnanthemum virginianum*	Virginia Mountainmint	H	1'–3'	White	July–Sept.
Summer	*Amorpha canescens*	Leadplant	H	2'–3'	Purple	June–July
Summer–small leaves	*Echinacea purpurea*	Purple Coneflower	M	3'–4'	Purple	July–Sept.
Summer	*Monarda fistulosa*	Bergamot	H	2'–5'	Lavender	July–Sept.
Summer	*Rudbeckia fulgida*	Orange Coneflower	M	2'–4'	Yellow	July–Sept.
New growth	*Amsonia tabernaemontana*	Eastern Bluestar	L	2'–3'	Blue/white	May–June
New growth	*Baptisia alba*	White False Indigo	L	3'–5'	White	June–July
New growth	*Baptisia australis*	Blue False Indigo	L	3'–5'	Blue	June–July

USDA zones	Sun	Soil moisture	Soil texture	pH range	Plant spacing in prairie gardens	Plant spacing in flower beds	Root type	Plant life expectancy in years	Aggressiveness and method	Deer palatability
4–9	F, P	D, M	S, L	5.5–7.5	1'–2'	2'	TAP	10–20	L–S	L
2–6	F, P	D, M	S, L	?–8.0	1'	1'–2'	FIB	3–5	M–S	L
3–9	F, P	D, M	S, L	4.5–7.0	3'	3'	TAP	20+	L–S	L
3–8	F, P	Mo, W	S, L, C	?	1'	1'–2'	FIB	5–10	L–S	M
4–10	F, P	M, Mo	S, L, C	5.0–8.0	1'–2'	2'	R	3–5	M–R	L
5–9	F, P	M, Mo, W	S, L, C	4.0–8.0	2'	4'	R	10–20	L–R/S	L
3–5	F, P	Mo, W	S, L, C	5.7–7.1	1'–2'	3'	FIB	5–10	L–S	L
3–9	F, P	Mo, W	S, L	5.8–7.8	6"–1'	2'	FIB	3–5	L–S	M
3–9	F, P	M, Mo	S, L, C	?–8.0	6"–1'	2'	FIB	5–10	M–S	L
4–8	F, P	D, M, Mo	S, L, C	?	2'	3'	FIB	2	M–S	L
4–5	F, P	M, Mo	S, L, C	?–8.0	1'–2'	3'	FIB	5–10	L–S	L
3–9	F	D, M, Mo	S, L, C	5.5–8.0	1'–1.5'	3'	FIB	5–10	M–S	L
3–9	F	D, M, Mo	S, L	?	6"–1'	1.5'	FIB	3–5	L–S	L
3–8	F	D, M	S, L	4.0–7.0	2'	3'	R	10–20	L–R/S	L
4–9	F, P	M, Mo	S, L, C	5.0–8.0	2'	5'	FIB	10–20	L–S	L
3–7	F	D	S, G	6.5–8.0	1'	2'	FIB	5–10	L–S	L
3–8	F	D, M	S, L	?	1'	3'	FIB	5–10	L–S	L
3–7	F, P	Mo, W	S, L, C	4.3–7.2	2'	3'	FIB	10–20	L–S	L
3–8	F, P	Mo, W	S, L, C	4.5–7.0	2'	3'	R	10–20	L–R/S	L
3–9	F	D, M	S, G, L	?	1'–2'	2'	R	10–20	M–R	L
4–8	F	D, M	S, L	?	1'	3'	FIB	5–10	M–S	H
3–7	F, P	M, Mo	S, L, C	5.0–8.0	1'	3'	FIB	5–10	L–S	H
3–8	F	D, M	S, L, C	?–8.0	1'	3'	R	5–10	M–R?	L
3–8	F, P	D, M	S, L	5.0–8.0	1'	2'	FIB	5–10	L–S	M
4–9	F	D, M	S, L	4.5–7.5	1'	2'	FIB	5–10	L–S	L
3–8	F, P	M, Mo, W	S, L, C	?	1'	2'	FIB	2	M–S	L
3–8	F	D, M	S, L	?	1'	2'	FIB	3–5	M–S	L
4–8	F, P	D, M	S, G, L, C	?	1'	1.5'	TAP/R	5–10	L–S	M
3–7	F, P	Mo, W	S, L, C	?–8.0	1'–2'	3'	R	5–10	L–R	L
3–8	F, P	D, M	S, L	5.5–8.0	2'	3'	TAP	20+	L–S	L
4–8	F, P	D, M	S, L, C	5.5–8.0	1'–2'	3'	FIB	5–10	M–S	M
3–9	F, P	D, M, Mo	S, L, C	6.0–8.0	1'–2'	3'	R	10–20	M–R	L
4–8	F, P	M, Mo	L, C	?–8.0	1'–2'	3'	FIB	5–10	L–S	L
4–8	F, P	M, Mo	S, L, C	?–8.0	2'–3'	3'	FIB	20+	L–S	L
4–9	F, P	M, Mo	S, L, C	?–7.5	3'	3'	TAP	20+	L–S	L
3–10	F, P	M	S, L, C	?–8.0	3'	3'	TAP	20+	L–S	L

(continued)

TABLE 12.23. *(continued)*

Seasonal timing of stem cuttings	Scientific name (4/1/2020)	Common name	Cutting success rate	Plant height	Flower color	Bloom time
New growth	*Baptisia bracteata*	Cream False Indigo	L	1'–2'	Cream	May–June
New growth	*Coreopsis lanceolata*	Lanceleaf Coreopsis	L	1'–2'	Yellow	June–July
New growth	*Coreopsis palmata*	Stiff Coreopsis	L	2'–3'	Yellow	June–Aug.
New growth	*Coreopsis tripteris*	Tall Coreopsis	L	3'–7'	Yellow	July–Sept.
New growth	*Eupatorium perfoliatum*	Boneset	L	3'–4'	White	Aug.–Sept.
New growth	*Euphorbia corollata*	Flowering Spurge	H	2'–4'	White	July–Aug.
New growth	*Eutrochium fistulosum*	Tall Joe Pye Weed	L	5'–8'	Purple/pink	Aug.–Sept.
New growth	*Eutrochium maculatum*	Joe Pye Weed	L	4'–6'	Pink	Aug.–Sept.
New growth	*Helenium autumnale*	Sneezeweed	L	4'–5'	Yellow	Aug.–Sept.
New growth	*Helianthus mollis*	Downy Sunflower	L	4'–6'	Yellow	Aug.–Sept.
New growth	*Helianthus occidentalis*	Western Sunflower	L	2'–3'	Yellow	July–Aug.
New growth	*Helianthus ×laetiflorus*	Showy Sunflower	L	3'–6'	Yellow	Aug.–Sept.
New growth	*Heliopsis helianthoides*	Oxeye Sunflower	L	3'–6'	Yellow	June–Sept.
New growth	*Liatris aspera*	Rough Blazing Star	L	2'–5'	Purple/pink	Aug.–Sept.
New growth	*Liatris ligulistylis*	Meadow Blazing Star	L	3'–5'	Purple/pink	Aug.–Sept.
New growth	*Liatris punctata*	Dotted Blazing Star	L	1'–2'	Purple/pink	Aug.–Oct.
New growth	*Liatris pycnostachya*	Prairie Blazing Star	L	3'–5'	Purple/pink	July–Aug.
New growth	*Liatris spicata*	Dense Blazing Star	L	3'–6'	Purple/pink	Aug.–Sept.
New growth	*Liatris squarrosa*	Scaly Blazing Star	L	1'–2'	Purple/pink	Aug.–Sept.
New growth	*Physostegia virginiana*	Obedient Plant	L	1'–2'	Pink	Aug.–Sept.
New growth	*Salvia azurea*	Blue Sage	H	3'–5'	Blue	June–Aug.
New growth	*Senna hebecarpa*	Wild Senna	L	4'–6'	Yellow	July–Aug.
New growth	*Vernonia fasciculata*	Ironweed	L	4'–6'	Red/pink	July–Sept.
New growth	*Vernonia gigantea*	Tall Ironweed	L	5'–8'	Red/pink	Aug.–Sept.
New growth	*Veronicastrum virginicum*	Culver's Root	L	3'–6'	White	July–Aug.
?	*Blephilia ciliata*	Downy Wood Mint		1'–2'	White/purple/violet	May–July
?	*Callirhoe triangulata*	Poppymallow		1'–2'	Magenta	July–Aug.
?	*Delphinium carolinianum subsp. virescens*	Prairie Larkspur		1'–4'	White	June–July
?	*Monarda punctata*	Dotted Mint		1'–2'	Lavender	July–Sept.

Notes: For soil moisture, D = dry, M = medium, Mo = moist, and W = wet. For soil texture, S = sand, L = loam, C = clay, and G = gravel. For sun, F = full sun, or at least a half day of full sun; P = partial sun, or a quarter to a half day of sun; and S = shade, or less than a quarter day of sun. For plant spacing, NR = not recommended. For root type, B = bulb, CO = corm, FIB = fibrous, R = rhizome, TAP =

USDA zones	Sun	Soil moisture	Soil texture	pH range	Plant spacing in prairie gardens	Plant spacing in flower beds	Root type	Plant life expectancy in years	Aggressiveness and method	Deer palatability
4–9	F, P	D, M	S, L	?	2'	3'	TAP	20+	L–S	L
3–9	F	D, M	S, L	5.0–7.5	1'	1'–2'	FIB	3–5	M–S	M
3–8	F	D, M	S, L	?	1'–2'	2'	R	20+	M–R	L
3–8	F, P	M, Mo, W	S, L, C	?	1'–2'	NR	FIB	10–20	L–S	L
3–8	F	M, Mo, W	S, L, C	?–7.5	1'	NR	FIB	5–10	L–S	L
3–9	F	D, M	S, L	?	1'–2'	1'–2'	R	10–20	M–R	L
4–9	F, P	M, Mo, W	S, L, C	4.5–7.5	2'	3'–4'	FIB	5–10	L–S	H
3–6	F, P	Mo, W	S, L, C	5.5–7.5	1'–2'	3'–4'	FIB	5–10	L–S	L
3–9	F	Mo, W	S, L, C	4.0–7.5	1'–2'	2'–3'	FIB	5–10	L–S	L
4–9	F	D, M	S, L, C	6.0–8.0	2'	3'	R	20+	M–R	L
3–9	F	D, M	S, L	?	2'	3'	R	20+	M–R	L
3–8	F	D, M	S, G, L, C	?	NR	NR	R	20+	H–R	L
3–8	F	D, M, Mo	S, L, C	6.0–8.0	1'–2'	3'	FIB	5–10	H–S	L
3–8	F	D, M	S, L	?	6"–1'	1'	CO	10–20	L–S	L
3–6	F	M, Mo	L	?	6"–1'	1'	CO	3–5	L–S	L
3–9	F	D	S, L	6.0–8.0	1'	1.5'–2'	TAPCO	10–20	L–S	L
3–9	F	M, Mo	S, L, C	6.0–8.0	6"–1'	1'	CO	10–20	L–S	L
4–10	F	M, Mo	S, L, C	5.5–7.5	6"–1'	1'	CO	10–20	L–S	L
4–8	F	D, M	S, L	?	1'	1.5'–2'	CO	10–20	L–S	L
3–9	F	M, Mo	S, L, C	6.0–8.0	1'–2'	NR	R	10–20	H–R	M
4–9	F, P	D, M	S, G, L	?	1'	1.5'	FIB	5–10	L–S	L
4–7	F	M, Mo, W	S, L, C	?	1.5'–2'	3'–4'	FIB/TAP	10–20	L–S	M
3–7	F	Mo	S, L, C	?	1'	3'	FIB	5–10	L–S	L
4–9	F, P	Mo, W	S, L, C	?–8.0	1'–2'	3'	FIB	10–20	L–S	L
3–8	F, P, S	M, Mo	S, L, C	?–8.0	1'	3'	FIB	10–20	L–S	L
4–8	F, P	D, M	G, L, C	?–8.00	1'	1'–2'	TAP/short Rs	3–5	L–S	L
4–8	F	D	S	?	1'–2'	2'	TAP	10–20	L–S	L
3–8	F	D, M	S, L	?	1'	1.5'	FIB	5–10	L–S	L
3–10	F	D	S, G	5.0–?	1'	1'–2'	FIB	2	M–S	L

taproot, and TAPCO = taprooted corm. For aggressiveness and method, L = low, M = medium, H = high, R = by rhizome, S = by seed, and R/S = rhizome and seed. For dried arrangements, FL = flowers, SH = seedheads, SP = seedpods, SD = seed, and RH = rose hips. For deer palatability, L = low, M = medium, and H = high. For cutting success rate, L = low, M = medium, and H = high.

TABLE 12.24. PRAIRIE PLANT COMBINATIONS FOR DRY SOILS

Scientific name (4/1/2020)	Common name	Soil moisture	Height	Flower color	Bloom time	USDA zones	Sun
Flowers and shrubs							
Spring							
Coreopsis lanceolata	Lanceleaf Coreopsis	D, M	1'–2'	Yellow	June–July	3–9	F
Lupinus perennis	Wild Lupine	D	1'–2'	Blue	May–June	3–8	F, P
Coreopsis lanceolata	Lanceleaf Coreopsis	D, M	1'–2'	Yellow	June–July	3–9	F
Penstemon grandiflorus	Large-Flowered Beardtongue	D	2'–4'	Lavender	May–June	3–7	F
Geum triflorum	Prairie Smoke	D, M	6"	Pink	May–June	3–6	F
Penstemon gracilis	Slender Beardtongue	D	1'–2'	Lavender	May–June	3–6	F, P
Geum triflorum	Prairie Smoke	D, M	6"	Pink	May–June	3–6	F
Penstemon hirsutus	Hairy Penstemon	D	1'–2'	Lavender	May–June	3–6	F, P
Lithospermum canescens	Hoary Puccoon	D, M	1'	Orange	May–June	3–7	F, P
Phlox pilosa	Downy Phlox	D, M, Mo	1'–2'	Pink	May–June	3–9	F
Lithospermum caroliniense	Hairy Puccoon	D	1'–2'	Yellow	June–July	3–8	F
Phlox pilosa	Downy Phlox	D, M, Mo	1'–2'	Pink	May–June	3–9	F
Summer							
Agastache foeniculum	Lavender Hyssop	D, M	1'–3'	Purple	July–Sept.	2–6	F, P
Callirhoe triangulata	Poppymallow	D	1'–2'	Magenta	July–Aug.	4–8	F
Agastache foeniculum	Lavender Hyssop	D, M	1'–3'	Purple	July–Sept.	2–6	F, P
Solidago speciosa	Showy Goldenrod	D, M	1'–3'	Yellow	Aug.–Sept.	3–8	F
Asclepias tuberosa	Butterflyweed	D, M	2'–3'	Orange	June–Aug.	3–10	F
Dalea purpurea	Purple Prairie Clover	D, M	1'–2'	Purple/yellow	July–Aug.	3–8	F
Asclepias tuberosa	Butterflyweed	D, M	2'–3'	Orange	June–Aug.	3–10	F
Echinacea pallida	Pale Purple Coneflower	D, M	3'–5'	Purple	June–July	4–8	F
Asclepias tuberosa	Butterflyweed	D, M	2'–3'	Orange	June–Aug.	3–10	F
Verbena stricta	Hoary Vervain	D, M	2'–4'	Blue	July–Sept.	3–8	F
Dalea candida	White Prairie Clover	D, M	1'–2'	White	July–Aug.	3–9	F
Dalea purpurea	Purple Prairie Clover	D, M	1'–2'	Purple/yellow	July–Aug.	3–8	F
Echinacea pallida	Pale Purple Coneflower	D, M	3'–5'	Purple	June–July	4–8	F
Echinacea paradoxa	Ozark Coneflower	D, M	3'–5'	Yellow	June–July	4–7	F, P

Soil texture	pH range	Plant spacing in prairie gardens	Plant spacing in flower beds	Root type	Plant life expectancy in years	Aggressiveness and method	Cut flowers	Dried arrangements	Deer palatability
S, L	5.0–7.5	1'	1'–2'	FIB	3–5	M–S	x		M
S	5.0–?	1'	1.5'	TAP	10–20	M–S			L
S, L	5.0–7.5	1'	1'–2'	FIB	3–5	M–S	x		M
S, G	?	1'	2'–3'	FIB	3–5	L–S		SP	L
S, L	6.0–7.5	6"	1'	R	20+	L–R		SH	L
S, G	?–7.5	6"–1'	1'	FIB	5–10	L–S	x	SP	L
S, L	6.0–7.5	6"	1'	R	20+	L–R		SH	L
S, G	5.0–8.0	1'	1.5'	FIB	3–5	L–S	x	SP	L
S, L	?	1'	1'	FIB	5–10	L–S			L
S, L	?	6"–1'	1.5'	FIB	3–5	L–S	x		L
S	?	1'	1'	TAP	10–20	L–S			L
S, L	?	6"–1'	1.5'	FIB	3–5	L–S	x		L
S, L	?–8.0	1'	1'–2'	FIB	3–5	M–S	x	FL/SH	L
S	?	1'–2'	2'	TAP	10–20	L–S			L
S, L	?–8.0	1'	1'–2'	FIB	3–5	M–S	x	FL/SH	L
S, L	?	1'	3'	FIB	5–10	L–S	x	SH	L
S, L	5.0–7.5	2'	3'	TAP	10–20	L–S	x	SP	L
S, L, C	?	6"–1'	1'–2'	TAP	10–20	L–S	x		H
S, L	5.0–7.5	2'	3'	TAP	10–20	L–S	x	SP	L
S, L, C	6.0–7.5	1'	2'–3'	TAP	10–20	L–S	x	SH	M
S, L	5.0–7.5	2'	3'	TAP	10–20	L–S	x	SP	L
S, L	?	1'	2'	FIB	3–5	M–S	x	SH	L
S, L	?	6"–1'	1'–2'	TAP	10–20	L–S			H
S, L, C	?	6"–1'	1'–2'	TAP	10–20	L–S	x		H
S, L, C	6.0–7.5	1'	2'–3'	TAP	10–20	L–S	x	SH	M
S, L, C	?–7.5	1'	2'–3'	TAP	10–20	L–S	x	SH	M

(continued)

TABLE 12.24. *(continued)*

Scientific name (4/1/2020)	Common name	Soil moisture	Height	Flower color	Bloom time	USDA zones	Sun
Autumn							
Liatris aspera	Rough Blazing Star	D, M	2'–5'	Purple/pink	Aug.–Sept.	3–8	F
Oligoneuron rigidum	Stiff Goldenrod	D, M, Mo	3'–5'	Yellow	Aug.–Sept.	3–9	F
Liatris aspera	Rough Blazing Star	D, M	2'–5'	Purple/pink	Aug.–Sept.	3–8	F
Solidago speciosa	Showy Goldenrod	D, M	1'–3'	Yellow	Aug.–Sept.	3–8	F
Liatris punctata	Dotted Blazing Star	D	1'–2'	Purple/pink	Aug.–Oct.	3–9	F
Solidago speciosa	Showy Goldenrod	D, M	1'–3'	Yellow	Aug.–Sept.	3–8	F
Oligoneuron rigidum	Stiff Goldenrod	D, M, Mo	3'–5'	Yellow	Aug.–Sept.	3–9	F
Symphyotrichum laeve	Smooth Aster	D, M	2'–4'	Blue	Aug.–Oct.	4–8	F
Solidago speciosa	Showy Goldenrod	D, M	1'–3'	Yellow	Aug.–Sept.	3–8	F
Symphyotrichum oolentangiense	Skyblue Aster	D, M	2'–3'	Blue	Aug.–Oct.	3–8	F, P
Symphyotrichum ericoides	Heath Aster	D, M	1'–3'	White	Aug.–Oct.	3–9	F
Symphyotrichum oolentangiense	Skyblue Aster	D, M	2'–3'	Blue	Aug.–Oct.	3–8	F, P
Grasses							
Schizachyrium scoparium	Little Bluestem	D, M	2'–5'	Crimson red	Aug.–Oct.	3–10	F
Sorghastrum nutans	Indiangrass	D, M, Mo	5'–7'	Gold	Aug.–Sept.	3–9	F

Notes: For soil moisture, D = dry, M = medium, Mo = moist, and W = wet. For soil texture, S = sand, L = loam, C = clay, and G = gravel. For sun, F = full sun, or at least a half day of full sun; P = partial sun, or a quarter to a half day of sun; and S = shade, or less than a quarter day of sun. For plant spacing, NR = not recommended. For root type, B = bulb, CO = corm, FIB = fibrous, R = rhizome, TAP =

Soil texture	pH range	Plant spacing in prairie gardens	Plant spacing in flower beds	Root type	Plant life expectancy in years	Aggressiveness and method	Cut flowers	Dried arrangements	Deer palatability
S, L	?	6"–1'	1'	CO	10–20	L–S	x	FL/SH	L
S, L, C	5.5–8.0	1'–1.5'	3'	FIB	5–10	M–S	x	SH	L
S, L	?	6"–1'	1'	CO	10–20	L–S	x	FL/SH	L
S, L	?	1'	3'	FIB	5–10	L–S	x	SH	L
S, L	6.0–8.0	1'	1.5'–2'	TAPCO	10–20	L–S			L
S, L	?	1'	3'	FIB	5–10	L–S	x	SH	L
S, L, C	5.5–8.0	1'–1.5'	3'	FIB	5–10	M–S	x	SH	L
S, L	?	1'	3'	FIB	5–10	M–S	x	Empty SH	H
S, L	?	1'	3'	FIB	5–10	L–S	x	SH	L
S, L	5.0–8.0	1'	2'	FIB	5–10	L–S	x	Empty SH	M
S, G, L	?	1'–2'	2'	R	10–20	M–R		Empty SH	L
S, L	5.0–8.0	1'	2'	FIB	5–10	L–S	x	Empty SH	M
S, L	5.0–8.5	1'–1.5'	3'	FIB	20+	L–S		x	L
S, L, C	5.0–8.0	1'–2'	3'	FIB	10–20	M–S	x	x	L

taproot, and TAPCO = taprooted corm. For aggressiveness and method, L = low, M = medium, H = high, R = by rhizome, S = by seed, and R/S = rhizome and seed. For dried arrangements, FL = flowers, SH = seedheads, SP = seedpods, SD = seed, and RH = rose hips. For deer palatability, L = low, M = medium, and H = high.

TABLE 12.25. PRAIRIE PLANT COMBINATIONS FOR MEDIUM SOILS

Scientific name (4/1/2020)	Common name	Soil moisture	Height	Flower color	Bloom time	USDA zones	Sun
Flowers and shrubs							
Spring							
Geum triflorum	Prairie Smoke	D, M	6"	Pink	May–June	3–6	F
Dodecatheon meadia	Shooting Star	M, Mo	1'–2'	White/pink	May–June	4–8	F, P
Penstemon digitalis	Foxglove Beardtongue	M, Mo	2'–3'	White	June–July	3–8	F, P
Tradescantia bracteata	Prairie Spiderwort	D, M	1'–2'	Blue	June–July	3–7	F, P
Penstemon digitalis	Foxglove Beardtongue	M, Mo	2'–3'	White	June–July	3–8	F, P
Tradescantia ohiensis	Ohio Spiderwort	D, M, Mo	2'–4'	Blue	June–July	3–9	F, P
Summer							
Agastache foeniculum	Lavender Hyssop	D, M	1'–3'	Purple	July–Sept.	2–6	F, P
Allium cernuum	Nodding Pink Onion	M, Mo	1'–2'	White/pink	June–Aug.	3–8	F, P
Agastache foeniculum	Lavender Hyssop	D, M	1'–3'	Purple	July–Sept.	2–6	F, P
Asclepias tuberosa	Butterflyweed	D, M	2'–3'	Orange	June–Aug.	3–10	F
Asclepias sullivantii	Sullivant's Milkweed	M, Mo	3'–5'	Pink/yellow	June–Aug.	4–7	F
Monarda fistulosa	Bergamot	D, M, Mo	2'–5'	Lavender	July–Sept.	3–9	F, P
Asclepias tuberosa	Butterflyweed	D, M	2'–3'	Orange	June–Aug.	3–10	F
Dalea purpurea	Purple Prairie Clover	D, M	1'–2'	Purple/yellow	July–Aug.	3–8	F
Asclepias tuberosa	Butterflyweed	D, M	2'–3'	Orange	June–Aug.	3–10	F
Echinacea pallida	Pale Purple Coneflower	D, M	3'–5'	Purple	June–July	4–8	F
Asclepias tuberosa	Butterflyweed	D, M	2'–3'	Orange	June–Aug.	3–10	F
Echinacea purpurea	Purple Coneflower	D, M	3'–4'	Purple	July–Sept.	4–8	F, P
Asclepias tuberosa	Butterflyweed	D, M	2'–3'	Orange	June–Aug.	3–10	F
Verbena stricta	Hoary Vervain	D, M	2'–4'	Blue	July–Sept.	3–8	F
Conoclinium coelestinum	Blue Mistflower	M, Mo	1'–3'	Blue/purple	July–Oct.	4–10	F, P
Liatris pycnostachya	Prairie Blazing Star	M, Mo	3'–5'	Purple/pink	July–Aug.	3–9	F
Conoclinium coelestinum	Blue Mistflower	M, Mo	1'–3'	Blue/purple	July–Oct.	4–10	F, P
Liatris spicata	Dense Blazing Star	M, Mo	3'–6'	Purple/pink	Aug.–Sept.	4–10	F
Conoclinium coelestinum	Blue Mistflower	M, Mo	1'–3'	Blue/purple	July–Oct.	4–10	F, P
Oligoneuron ohioense	Ohio Goldenrod	M, Mo	3'–4'	Yellow	Aug.–Sept.	4–5	F, P
Conoclinium coelestinum	Blue Mistflower	M, Mo	1'–3'	Blue/purple	July–Oct.	4–10	F, P
Silene regia	Royal Catchfly	M	2'–4'	Red	July–Aug.	4–9	F, P

Soil texture	pH range	Plant spacing in prairie gardens	Plant spacing in flower beds	Root type	Plant life expectancy in years	Aggressiveness and method	Cut flowers	Dried arrangements	Deer palatability
S, L	6.0–7.5	6"	1'	R	20+	L–R		SH	L
S, L, C	4.5–8.0	6"–1'	1'	FIB	20+	L–S		SP	L
S, L, C	5.5–7.5	1'	2'	FIB	5–10	L–S	x		L
S, L, C	?	1'	2'	R	20+	M–R			M
S, L, C	5.5–7.5	1'	2'	FIB	5–10	L–S	x		L
S, L, C	?–7.5	1'	2'	FIB	20+	M–S			H
S, L	?–8.0	1'	1'–2'	FIB	3–5	M–S	x	FL/SH	L
S, L, C	6.5–8.0	6"–1'	1'	B	10–20	L–S	x	FL/SH	L
S, L	?–8.0	1'	1'–2'	FIB	3–5	M–S	x	FL/SH	L
S, L	5.0–7.5	2'	3'	TAP	10–20	L–S	x	SP	L
L, C	?	1'–2'	3'	R	10–20	L–R		SP	L
S, L, C	6.0–8.0	1'–2'	3'	R	10–20	M–R	x	SH	L
S, L	5.0–7.5	2'	3'	TAP	10–20	L–S	x	SP	L
S, L, C	?	6"–1'	1'–2'	TAP	10–20	L–S	x		H
S, L	5.0–7.5	2'	3'	TAP	10–20	L–S	x	SP	L
S, L, C	6.0–7.5	1'	2'–3'	TAP	10–20	L–S	x	SH	M
S, L	5.0–7.5	2'	3'	TAP	10–20	L–S	x	SP	L
S, L, C	5.5–8.0	1'–2'	3'	FIB	5–10	M–S	x	SH	M
S, L	5.0–7.5	2'	3'	TAP	10–20	L–S	x	SP	L
S, L	?	1'	2'	FIB	3–5	M–S	x	SH	L
S, L, C	5.0–8.0	1'–2'	2'	R	3–5	M–R			L
S, L, C	6.0–8.0	6"–1'	1'	CO	10–20	L–S	x	FL/SH	L
S, L, C	5.0–8.0	1'–2'	2'	R	3–5	M–R			L
S, L, C	5.5–7.5	6"–1'	1'	CO	10–20	L–S	x	FL/SH	L
S, L, C	5.0–8.0	1'–2'	2'	R	3–5	M–R			L
S, L, C	?–8.0	1'–2'	3'	FIB	5–10	L–S	x	SH	L
S, L, C	5.0–8.0	1'–2'	2'	R	3–5	M–R			L
L	?	1'	1.5'	TAP	5–10	L–S	x	SH	L

(continued)

TABLE 12.25. *(continued)*

Scientific name (4/1/2020)	Common name	Soil moisture	Height	Flower color	Bloom time	USDA zones	Sun
Dalea candida	White Prairie Clover	D, M	1'–2'	White	July–Aug.	3–9	F
Dalea purpurea	Purple Prairie Clover	D, M	1'–2'	Purple/yellow	July–Aug.	3–8	F
Echinacea pallida	Pale Purple Coneflower	D, M	3'–5'	Purple	June–July	4–8	F
Echinacea paradoxa	Ozark Coneflower	D, M	3'–5'	Yellow	June–July	4–7	F, P
Echinacea purpurea	Purple Coneflower	D, M	3'–4'	Purple	July–Sept.	4–8	F, P
Agastache foeniculum	Lavender Hyssop	D, M	1'–3'	Purple	July–Sept.	2–6	F, P
Echinacea purpurea	Purple Coneflower	D, M	3'–4'	Purple	July–Sept.	4–8	F, P
Callirhoe involucrata	Winecups	D, M	1'	Magenta	May–June	4–9	F, P
Echinacea purpurea	Purple Coneflower	D, M	3'–4'	Purple	July–Sept.	4–8	F, P
Eryngium yuccifolium	Rattlesnake Master	D, M	3'–5'	White	June–Aug.	4–9	F
Echinacea purpurea	Purple Coneflower	D, M	3'–4'	Purple	July–Sept.	4–8	F, P
Monarda fistulosa	Bergamot	D, M, Mo	2'–5'	Lavender	July–Sept.	3–9	F, P
Echinacea purpurea	Purple Coneflower	D, M	3'–4'	Purple	July–Sept.	4–8	F, P
Parthenium integrifolium	Wild Quinine	M, Mo	3'–5'	White	June–Sept.	4–8	F
Echinacea purpurea	Purple Coneflower	D, M	3'–4'	Purple	July–Sept.	4–8	F, P
Rudbeckia fulgida	Orange Coneflower	M, Mo	2'–4'	Yellow	July–Sept.	4–8	F, P
Eryngium yuccifolium	Rattlesnake Master	D, M	3'–5'	White	June–Aug.	4–9	F
Liatris pycnostachya	Prairie Blazing Star	M, Mo	3'–5'	Purple/pink	July–Aug.	3–9	F
Filipendula rubra	Queen of the Prairie	M, Mo	4'–5'	Pink	June–July	3–6	F
Liatris pycnostachya	Prairie Blazing Star	M, Mo	3'–5'	Purple/pink	July–Aug.	3–9	F
Liatris spicata	Dense Blazing Star	M, Mo	3'–6'	Purple/pink	Aug.–Sept.	4–10	F
Oligoneuron ohioense	Ohio Goldenrod	M, Mo	3'–4'	Yellow	Aug.–Sept.	4–5	F, P
Monarda fistulosa	Bergamot	D, M, Mo	2'–5'	Lavender	July–Sept.	3–9	F, P
Liatris pycnostachya	Prairie Blazing Star	M, Mo	3'–5'	Purple/pink	July–Aug.	3–9	F
Parthenium integrifolium	Wild Quinine	M, Mo	3'–5'	White	June–Sept.	4–8	F
Liatris pycnostachya	Prairie Blazing Star	M, Mo	3'–5'	Purple/pink	July–Aug.	3–9	F
Vernonia fasciculata	Ironweed	Mo	4'–6'	Red/pink	July–Sept.	3–7	F
Veronicastrum virginicum	Culver's Root	M, Mo	3'–6'	White	July–Aug.	3–8	F, P, S
Autumn							
Liatris aspera	Rough Blazing Star	D, M	2'–5'	Purple/pink	Aug.–Sept.	3–8	F
Oligoneuron rigidum	Stiff Goldenrod	D, M, Mo	3'–5'	Yellow	Aug.–Sept.	3–9	F

Soil texture	pH range	Plant spacing in prairie gardens	Plant spacing in flower beds	Root type	Plant life expectancy in years	Aggressiveness and method	Cut flowers	Dried arrangements	Deer palatability
S, L	?	6"–1'	1'–2'	TAP	10–20	L–S			H
S, L, C	?	6"–1'	1'–2'	TAP	10–20	L–S	x		H
S, L, C	6.0–7.5	1'	2'–3'	TAP	10–20	L–S	x	SH	M
S, L, C	?–7.5	1'	2'–3'	TAP	10–20	L–S	x	SH	M
S, L, C	5.5–8.0	1'–2'	3'	FIB	5–10	M–S	x	SH	M
S, L	?–8.0	1'	1'–2'	FIB	3–5	M–S	x	FL/SH	L
S, L, C	5.5–8.0	1'–2'	3'	FIB	5–10	M–S	x	SH	M
S, L	5.5–7.5	1'–2'	2'	TAP	10–20	L–S			L
S, L, C	5.5–8.0	1'–2'	3'	FIB	5–10	M–S	x	SH	M
S, L, C	6.0–8.0	1'	2'	FIB	5–10	M–S	x	SH	L
S, L, C	5.5–8.0	1'–2'	3'	FIB	5–10	M–S	x	SH	M
S, L, C	6.0–8.0	1'–2'	3'	R	10–20	M–R	x	SH	L
S, L, C	5.5–8.0	1'–2'	3'	FIB	5–10	M–S	x	SH	M
S, L, C	?–8.0	1'–2'	3'	TAP	10–20	L–S	x		M
S, L, C	5.5–8.0	1'–2'	3'	FIB	5–10	M–S	x	SH	M
L, C	?–8.0	1'–2'	3'	FIB	5–10	L–S	x	SH	L
S, L, C	6.0–8.0	1'	2'	FIB	5–10	M–S	x	SH	L
S, L, C	6.0–8.0	6"–1'	1'	CO	10–20	L–S	x	FL/SH	L
S, L, C	?–8.0	2'	3'	R	20+	M–R	x		L
S, L, C	6.0–8.0	6"–1'	1'	CO	10–20	L–S	x	FL/SH	L
S, L, C	5.5–7.5	6"–1'	1'	CO	10–20	L–S	x	FL/SH	L
S, L, C	?–8.0	1'–2'	3'	FIB	5–10	L–S	x	SH	L
S, L, C	6.0–8.0	1'–2'	3'	R	10–20	M–R	x	SH	L
S, L, C	6.0–8.0	6"–1'	1'	CO	10–20	L–S	x	FL/SH	L
S, L, C	?–8.0	1'–2'	3'	TAP	10–20	L–S	x		M
S, L, C	6.0–8.0	6"–1'	1'	CO	10–20	L–S	x	FL/SH	L
S, L, C	?	1'	3'	FIB	5–10	L–S	x		L
S, L, C	?–8.0	1'	3'	FIB	10–20	L–S	x	SH	L
S, L	?	6"–1'	1'	CO	10–20	L–S	x	FL/SH	L
S, L, C	5.5–8.0	1'–1.5'	3'	FIB	5–10	M–S	x	SH	L

(continued)

TABLE 12.25. *(continued)*

Scientific name (4/1/2020)	Common name	Soil moisture	Height	Flower color	Bloom time	USDA zones	Sun
Liatris aspera	Rough Blazing Star	D, M	2'–5'	Purple/pink	Aug.–Sept.	3–8	F
Solidago speciosa	Showy Goldenrod	D, M	1'–3'	Yellow	Aug.–Sept.	3–8	F
Oligoneuron rigidum	Stiff Goldenrod	D, M, Mo	3'–5'	Yellow	Aug.–Sept.	3–9	F
Symphyotrichum laeve	Smooth Aster	D, M	2'–4'	Blue	Aug.–Oct.	4–8	F
Solidago speciosa	Showy Goldenrod	D, M	1'–3'	Yellow	Aug.–Sept.	3–8	F
Symphyotrichum oolentangiense	Skyblue Aster	D, M	2'–3'	Blue	Aug.–Oct.	3–8	F, P
Grasses							
Schizachyrium scoparium	Little Bluestem	D, M	2'–5'	Crimson red	Aug.–Oct.	3–10	F
Sorghastrum nutans	Indiangrass	D, M, Mo	5'–7'	Gold	Aug.–Sept.	3–9	F

Notes: For soil moisture, D = dry, M = medium, Mo = moist, and W = wet. For soil texture, S = sand, L = loam, C = clay, and G = gravel. For sun, F = full sun, or at least a half day of full sun; P = partial sun, or a quarter to a half day of sun; and S = shade, or less than a quarter day of sun. For plant spacing, NR = not recommended. For root type, B = bulb, CO = corm, FIB = fibrous, R = rhizome, TAP =

Soil texture	pH range	Plant spacing in prairie gardens	Plant spacing in flower beds	Root type	Plant life expectancy in years	Aggressiveness and method	Cut flowers	Dried arrangements	Deer palatability
S, L	?	6"–1'	1'	CO	10–20	L–S	x	FL/SH	L
S, L	?	1'	3'	FIB	5–10	L–S	x	SH	L
S, L, C	5.5–8.0	1'–1.5'	3'	FIB	5–10	M–S	x	SH	L
S, L	?	1'	3'	FIB	5–10	M–S	x	Empty SH	H
S, L	?	1'	3'	FIB	5–10	L–S	x	SH	L
S, L	5.0–8.0	1'	2'	FIB	5–10	L–S	x	Empty SH	M
S, L	5.0–8.5	1'–1.5'	3'	FIB	20+	L–S		x	L
S, L, C	5.0–8.0	1'–2'	3'	FIB	10–20	M–S	x	x	L

taproot, and TAPCO = taprooted corm. For aggressiveness and method, L = low, M = medium, H = high, R = by rhizome, S = by seed, and R/S = rhizome and seed. For dried arrangements, FL = flowers, SH = seedheads, SP = seedpods, SD = seed, and RH = rose hips. For deer palatability, L = low, M = medium, and H = high.

TABLE 12.26. PRAIRIE PLANT COMBINATIONS FOR MOIST SOILS

Scientific name (4/1/2020)	Common name	Soil moisture	Height	Flower color	Bloom time	USDA zones	Sun
Flowers and shrubs							
Spring							
Penstemon digitalis	Foxglove Beardtongue	M, Mo	2'–3'	White	June–July	3–8	F, P
Tradescantia bracteata	Prairie Spiderwort	D, M	1'–2'	Blue	June–July	3–7	F, P
Penstemon digitalis	Foxglove Beardtongue	M, Mo	2'–3'	White	June–July	3–8	F, P
Tradescantia ohiensis	Ohio Spiderwort	D, M, Mo	2'–4'	Blue	June–July	3–9	F, P
Summer							
Agastache foeniculum	Lavender Hyssop	D, M	1'–3'	Purple	July–Sept.	2–6	F, P
Allium cernuum	Nodding Pink Onion	M, Mo	1'–2'	White/pink	June–Aug.	3–8	F, P
Asclepias sullivantii	Sullivant's Milkweed	M, Mo	3'–5'	Pink/yellow	June–Aug.	4–7	F
Monarda fistulosa	Bergamot	D, M, Mo	2'–5'	Lavender	July–Sept.	3–9	F, P
Conoclinium coelestinum	Blue Mistflower	M, Mo	1'–3'	Blue/purple	July–Oct.	4–10	F, P
Liatris pycnostachya	Prairie Blazing Star	M, Mo	3'–5'	Purple/pink	July–Aug.	3–9	F
Conoclinium coelestinum	Blue Mistflower	M, Mo	1'–3'	Blue/purple	July–Oct.	4–10	F, P
Liatris spicata	Dense Blazing Star	M, Mo	3'–6'	Purple/pink	Aug.–Sept.	4–10	F
Conoclinium coelestinum	Blue Mistflower	M, Mo	1'–3'	Blue/purple	July–Oct.	4–10	F, P
Oligoneuron ohioense	Ohio Goldenrod	M, Mo	3'–4'	Yellow	Aug.–Sept.	4–5	F, P
Echinacea purpurea	Purple Coneflower	D, M	3'–4'	Purple	July–Sept.	4–8	F, P
Eryngium yuccifolium	Rattlesnake Master	D, M	3'–5'	White	June–Aug.	4–9	F
Echinacea purpurea	Purple Coneflower	D, M	3'–4'	Purple	July–Sept.	4–8	F, P
Monarda fistulosa	Bergamot	D, M, Mo	2'–5'	Lavender	July–Sept.	3–9	F, P
Echinacea purpurea	Purple Coneflower	D, M	3'–4'	Purple	July–Sept.	4–8	F, P
Parthenium integrifolium	Wild Quinine	M, Mo	3'–5'	White	June–Sept.	4–8	F
Echinacea purpurea	Purple Coneflower	D, M	3'–4'	Purple	July–Sept.	4–8	F, P
Rudbeckia fulgida	Orange Coneflower	M, Mo	2'–4'	Yellow	July–Sept.	4–8	F, P
Eryngium yuccifolium	Rattlesnake Master	D, M	3'–5'	White	June–Aug.	4–9	F
Liatris pycnostachya	Prairie Blazing Star	M, Mo	3'–5'	Purple/pink	July–Aug.	3–9	F
Filipendula rubra	Queen of the Prairie	M, Mo	4'–5'	Pink	June–July	3–6	F
Liatris pycnostachya	Prairie Blazing Star	M, Mo	3'–5'	Purple/pink	July–Aug.	3–9	F
Helenium autumnale	Sneezeweed	Mo, W	4'–5'	Yellow	Aug.–Sept.	3–9	F
Monarda fistulosa	Bergamot	D, M, Mo	2'–5'	Lavender	July–Sept.	3–9	F, P

Soil texture	pH range	Plant spacing in prairie gardens	Plant spacing in flower beds	Root type	Plant life expectancy in years	Aggressiveness and method	Cut flowers	Dried arrangements	Deer palatability
S, L, C	5.5–7.5	1'	2'	FIB	5–10	L–S	x		L
S, L, C	?	1'	2'	R	20+	M–R			M
S, L, C	5.5–7.5	1'	2'	FIB	5–10	L–S	x		L
S, L, C	?–7.5	1'	2'	FIB	20+	M–S			H
S, L	?–8.0	1'	1'–2'	FIB	3–5	M–S	x	FL/SH	L
S, L, C	6.5–8.0	6"–1'	1'	B	10–20	L–S	x	FL/SH	L
L, C	?	1'–2'	3'	R	10–20	L–R		SP	L
S, L, C	6.0–8.0	1'–2'	3'	R	10–20	M–R	x	SH	L
S, L, C	5.0–8.0	1'–2'	2'	R	3–5	M–R			L
S, L, C	6.0–8.0	6"–1'	1'	CO	10–20	L–S	x	FL/SH	L
S, L, C	5.0–8.0	1'–2'	2'	R	3–5	M–R			L
S, L, C	5.5–7.5	6"–1'	1'	CO	10–20	L–S	x	FL/SH	L
S, L, C	5.0–8.0	1'–2'	2'	R	3–5	M–R			L
S, L, C	?–8.0	1'–2'	3'	FIB	5–10	L–S	x	SH	L
S, L, C	5.5–8.0	1'–2'	3'	FIB	5–10	M–S	x	SH	M
S, L, C	6.0–8.0	1'	2'	FIB	5–10	M–S	x	SH	L
S, L, C	5.5–8.0	1'–2'	3'	FIB	5–10	M–S	x	SH	M
S, L, C	6.0–8.0	1'–2'	3'	R	10–20	M–R	x	SH	L
S, L, C	5.5–8.0	1'–2'	3'	FIB	5–10	M–S	x	SH	M
S, L, C	?–8.0	1'–2'	3'	TAP	10–20	L–S	x		M
S, L, C	5.5–8.0	1'–2'	3'	FIB	5–10	M–S	x	SH	M
L, C	?–8.0	1'–2'	3'	FIB	5–10	L–S	x	SH	L
S, L, C	6.0–8.0	1'	2'	FIB	5–10	M–S	x	SH	L
S, L, C	6.0–8.0	6"–1'	1'	CO	10–20	L–S	x	FL/SH	L
S, L, C	?–8.0	2'	3'	R	20+	M–R	x		L
S, L, C	6.0–8.0	6"–1'	1'	CO	10–20	L–S	x	FL/SH	L
S, L, C	4.0–7.5	1'–2'	2'–3'	FIB	5–10	L–S	x	SH	L
S, L, C	6.0–8.0	1'–2'	3'	R	10–20	M–R	x	SH	L

(continued)

TABLE 12.26. *(continued)*

Scientific name (4/1/2020)	Common name	Soil moisture	Height	Flower color	Bloom time	USDA zones	Sun
Liatris spicata	Dense Blazing Star	M, Mo	3'–6'	Purple/pink	Aug.–Sept.	4–10	F
Oligoneuron ohioense	Ohio Goldenrod	M, Mo	3'–4'	Yellow	Aug.–Sept.	4–5	F, P
Lobelia siphilitica	Great Blue Lobelia	M, Mo	1'–4'	Blue	July–Sept.	3–9	F, P
Physostegia virginiana	Obedient Plant	M, Mo	1'–2'	Pink	Aug.–Sept.	3–9	F
Monarda fistulosa	Bergamot	D, M, Mo	2'–5'	Lavender	July–Sept.	3–9	F, P
Liatris pycnostachya	Prairie Blazing Star	M, Mo	3'–5'	Purple/pink	July–Aug.	3–9	F
Parthenium integrifolium	Wild Quinine	M, Mo	3'–5'	White	June–Sept.	4–8	F
Liatris pycnostachya	Prairie Blazing Star	M, Mo	3'–5'	Purple/pink	July–Aug.	3–9	F
Vernonia fasciculata	Ironweed	Mo	4'–6'	Red/pink	July–Sept.	3–7	F
Veronicastrum virginicum	Culver's Root	M, Mo	3'–6'	White	July–Aug.	3–8	F, P, S
Autumn							
Coreopsis tripteris	Tall Coreopsis	M, Mo, W	3'–7'	Yellow	July–Sept.	3–8	F, P
Eutrochium maculatum	Joe Pye Weed	Mo, W	4'–6'	Pink	Aug.–Sept.	3–6	F, P
Eutrochium fistulosum	Tall Joe Pye Weed	M, Mo, W	5'–8'	Purple/pink	Aug.–Sept.	4–9	F, P
Vernonia gigantea	Tall Ironweed	Mo, W	5'–8'	Red/pink	Aug.–Sept.	4–9	F, P
Eutrochium maculatum	Joe Pye Weed	Mo, W	4'–6'	Pink	Aug.–Sept.	3–6	F, P
Vernonia fasciculata	Ironweed	Mo	4'–6'	Red/pink	July–Sept.	3–7	F
Oligoneuron ohioense	Ohio Goldenrod	M, Mo	3'–4'	Yellow	Aug.–Sept.	4–5	F, P
Symphyotrichum laeve	Smooth Aster	D, M	2'–4'	Blue	Aug.–Oct.	4–8	F
Oligoneuron rigidum	Stiff Goldenrod	D, M, Mo	3'–5'	Yellow	Aug.–Sept.	3–9	F
Symphyotrichum laeve	Smooth Aster	D, M	2'–4'	Blue	Aug.–Oct.	4–8	F
Grasses							
Schizachyrium scoparium	Little Bluestem	D, M	2'–5'	Crimson red	Aug.–Oct.	3–10	F
Sorghastrum nutans	Indiangrass	D, M, Mo	5'–7'	Gold	Aug.–Sept.	3–9	F

Notes: For soil moisture, D = dry, M = medium, Mo = moist, and W = wet. For soil texture, S = sand, L = loam, C = clay, and G = gravel. For sun, F = full sun, or at least a half day of full sun; P = partial sun, or a quarter to a half day of sun; and S = shade, or less than a quarter day of sun. For plant spacing, NR = not recommended. For root type, B = bulb, CO = corm, FIB = fibrous, R = rhizome, TAP =

Soil texture	pH range	Plant spacing in prairie gardens	Plant spacing in flower beds	Root type	Plant life expectancy in years	Aggressiveness and method	Cut flowers	Dried arrangements	Deer palatability
S, L, C	5.5–7.5	6"–1'	1'	CO	10–20	L–S	x	FL/SH	L
S, L, C	?–8.0	1'–2'	3'	FIB	5–10	L–S	x	SH	L
S, L, C	?–8.0	6"–1'	2'	FIB	5–10	M–S	x		L
S, L, C	6.0–8.0	1'–2'	NR	R	10–20	H–R	x	SH	M
S, L, C	6.0–8.0	1'–2'	3'	R	10–20	M–R	x	SH	L
S, L, C	6.0–8.0	6"–1'	1'	CO	10–20	L–S	x	FL/SH	L
S, L, C	?–8.0	1'–2'	3'	TAP	10–20	L–S	x		M
S, L, C	6.0–8.0	6"–1'	1'	CO	10–20	L–S	x	FL/SH	L
S, L, C	?	1'	3'	FIB	5–10	L–S	x		L
S, L, C	?–8.0	1'	3'	FIB	10–20	L–S	x	SH	L
S, L, C	?	1'–2'	NR	FIB	10–20	L–S	x		L
S, L, C	5.5–7.5	1'–2'	3'–4'	FIB	5–10	L–S			L
S, L, C	4.5–7.5	2'	3'–4'	FIB	5–10	L–S			H
S, L, C	?–8.0	1'–2'	3'	FIB	10–20	L–S	x		L
S, L, C	5.5–7.5	1'–2'	3'–4'	FIB	5–10	L–S			L
S, L, C	?	1'	3'	FIB	5–10	L–S	x		L
S, L, C	?–8.0	1'–2'	3'	FIB	5–10	L–S	x	SH	L
S, L	?	1'	3'	FIB	5–10	M–S	x	Empty SH	H
S, L, C	5.5–8.0	1'–1.5'	3'	FIB	5–10	M–S	x	SH	L
S, L	?	1'	3'	FIB	5–10	M–S	x	Empty SH	H
S, L	5.0–8.5	1'–1.5'	3'	FIB	20+	L–S		x	L
S, L, C	5.0–8.0	1'–2'	3'	FIB	10–20	M–S	x	x	L

taproot, and TAPCO = taprooted corm. For aggressiveness and method, L = low, M = medium, H = high, R = by rhizome, S = by seed, and R/S = rhizome and seed. For dried arrangements, FL = flowers, SH = seedheads, SP = seedpods, SD = seed, and RH = rose hips. For deer palatability, L = low, M = medium, and H = high.

TABLE 12.27.   SEED PROPAGATION GUIDELINES

Scientific name (4/1/2020)	Common name	Greenhouse propagation methods	
		Dry stratify	Moist stratify
Flowers and shrubs			
Agastache foeniculum	Lavender Hyssop	x	
Allium cernuum	Nodding Pink Onion	Low germ.	x
Allium stellatum	Prairie Onion	Low germ.	x
Amorpha canescens	Leadplant	Fair germ.	x
Amsonia tabernaemontana	Eastern Bluestar		
Anemone canadensis	Canada Anemone	No germ.	x
Anemone cylindrica	Thimbleweed	Low germ.	x
Antennaria neglecta	Pussytoes		
Arnoglossum atriplicifolium	Pale Indian Plantain	Low germ.	x
Asclepias incarnata	Marsh Milkweed	Fair germ.	x
Asclepias sullivantii	Sullivant's Milkweed	Low germ.	x
Asclepias syriaca	Common Milkweed	Fair germ.	x
Asclepias tuberosa	Butterflyweed	Fair germ.	x
Astragalus canadensis	Canada Milkvetch	Fair germ.	x
Baptisia alba	White False Indigo	Low germ.	x
Baptisia australis	Blue False Indigo	Fair germ.	x
Baptisia bracteata	Cream False Indigo	Low germ.	x
Blephilia ciliata	Downy Wood Mint	Fair germ.	
Callirhoe involucrata	Winecups	Low germ.	x
Callirhoe triangulata	Poppymallow	Low germ.	x
Ceanothus americanus	New Jersey Tea	Low germ.	x
Chelone glabra	White Turtlehead	Low germ.	x
Conoclinium coelestinum	Blue Mistflower	Low germ.	x
Coreopsis lanceolata	Lanceleaf Coreopsis	x	
Coreopsis palmata	Stiff Coreopsis	Low germ.	x
Coreopsis tripteris	Tall Coreopsis	Fair germ.	x
Dalea candida	White Prairie Clover	x	
Dalea purpurea	Purple Prairie Clover	x	
Delphinium carolinianum subsp. virescens	Prairie Larkspur	No germ.	x
Desmanthus illinoensis	Illinois Bundleflower	Fair germ.	x
Desmodium canadense	Canada Ticktrefoil	Fair germ.	x
Dodecatheon meadia	Shooting Star	No germ.	x
Echinacea pallida	Pale Purple Coneflower	Fair germ.	x
Echinacea paradoxa	Ozark Coneflower	Fair germ.	x
Echinacea purpurea	Purple Coneflower	x	
Eryngium yuccifolium	Rattlesnake Master	Low germ.	x
Eupatorium perfoliatum	Boneset	Fair germ.	x
Euphorbia corollata	Flowering Spurge	No germ.	x
Eutrochium fistulosum	Tall Joe Pye Weed	Fair germ.	x
Eutrochium maculatum	Joe Pye Weed	Fair germ.	x
Filipendula rubra	Queen of the Prairie	Low germ.	x
Gentiana alba	Cream Gentian	Low germ.	x
Gentiana andrewsii	Bottle Gentian	Low germ.	x

Greenhouse propagation methods		Direct sowing in field
Days to MS	Seed scarification helpful	Optimal seeding times
		Spring, early summer, fall
30		Fall
30		Fall
30	x	Fall
		Spring, early summer, fall
30		Fall
30		Fall
30		Fall
10		Spring, early summer, fall
10		Fall
10		Spring, early summer, fall
10		Spring, early summer, fall
30		Spring, early summer, fall
90	x	Fall
30	x	Fall
90	x	Fall
		Spring, early summer, fall
30		Fall
30		Fall
Hot H$_2$O Treatment; 30 days MS	x	Fall
30		Fall
30		Fall
		Spring, early summer, fall
30		Fall
30		Fall
		Spring, early summer, fall
		Spring, early summer, fall
30 days or plant fresh seed in June		Fall, fresh in summer
10		Spring, early summer, fall
10		Spring, early summer, fall
30		Fall
30		Fall
30		Fall
		Spring, early summer, fall
30		Fall
30		Fall
30		Fall
30		Fall
30		Fall
30		Fall
30		Fall
30		Fall

(*continued*)

TABLE 12.27. *(continued)*

Scientific name (4/1/2020)	Common name	Greenhouse propagation methods	
		Dry stratify	Moist stratify
Geum triflorum	Prairie Smoke	Fair germ.	x
Helenium autumnale	Sneezeweed	x	
Helianthus mollis	Downy Sunflower	Fair germ.	x
Helianthus occidentalis	Western Sunflower	Fair germ.	x
Helianthus ×laetiflorus	Showy Sunflower	Fair germ.	x
Heliopsis helianthoides	Oxeye Sunflower	x	
Hibiscus moscheutos	Rosemallow		
Hypericum ascyron	Great St. Johnswort	Fair germ.	x
Iris virginica var. *shrevei*	Wild Iris	No germ.	x
Lespedeza capitata	Roundhead Bush Clover	Fair germ.	x
Liatris aspera	Rough Blazing Star	Low germ.	x
Liatris ligulistylis	Meadow Blazing Star	Fair germ.	x
Liatris punctata	Dotted Blazing Star	Low germ.	x
Liatris pycnostachya	Prairie Blazing Star	Fair germ.	x
Liatris spicata	Dense Blazing Star	Fair germ.	x
Liatris squarrosa	Scaly Blazing Star	Low germ.	x
Lilium michiganense	Michigan Lily	Fair germ.	x
Lithospermum canescens	Hoary Puccoon (Hemi-parasitic)	No germ.	x
Lithospermum caroliniense	Hairy Puccoon	No germ.	x
Lithospermum incisum	Fringed Puccoon	No germ.	x
Lobelia cardinalis	Cardinal Flower	Fair germ.	
Lobelia siphilitica	Great Blue Lobelia	Fair germ.	
Lupinus perennis	Wild Lupine	Low germ.	x
Monarda fistulosa	Bergamot	x	
Monarda punctata	Dotted Mint	x	
Oenothera gaura	Biennial Beeblossom	x	
Oligoneuron ohioense	Ohio Goldenrod	Fair germ.	x
Oligoneuron rigidum	Stiff Goldenrod	x	
Onosmodium bejariense var. *occidentale*	Marbleseed (Western Marbleseed)	No germ.	x
Parthenium integrifolium	Wild Quinine	Fair germ.	x
Pediomelum esculentum	Prairie Turnip	Low germ.	x
Penstemon cobaea	Cobaea Beardtongue	Low germ.	x
Penstemon digitalis	Foxglove Beardtongue	Fair germ.	x
Penstemon gracilis	Slender Beardtongue	Low germ.	x
Penstemon grandiflorus	Large-Flowered Beardtongue	Low germ.	x
Penstemon hirsutus	Hairy Penstemon	Low germ.	x
Phlox pilosa	Downy Phlox	No germ.	x
Physostegia virginiana	Obedient Plant	Fair germ.	x
Polygonatum biflorum var. *commutatum*	Great Solomon's Seal	No germ.	x
Pulsatilla patens	Pasque Flower	x	
Pycnanthemum tenuifolium	Narrowleaf Mountainmint	?	x
Pycnanthemum virginianum	Virginia Mountainmint	?	x
Ratibida pinnata	Yellow Coneflower	x	

Greenhouse propagation methods		Direct sowing in field
Days to MS	Seed scarification helpful	Optimal seeding times
Plant fresh seed in June		Spring, fall, fresh in summer
		Spring, early summer, fall
30		Spring, early summer, fall
30		Spring, early summer, fall
30		Spring, early summer, fall
		Spring, early summer, fall
30		Fall
120	x	Fall
10	x	Spring, early summer, fall
30		Fall
30		Spring, early summer, fall
30		Fall
30		Spring, early summer, fall
30		Spring, early summer, fall
30		Fall
30		Fall
Plant into loosened prairie sod in fall	x	Fall
90	x	Fall
90	x	Fall
30		Fall
30		Fall
7 (sprouts!) or fresh in July		Fall, fresh in summer
		Spring, early summer, fall
		Spring, early summer, fall
		Spring, early summer, fall
30		Fall
		Spring, early summer, fall
30		Fall
30		Fall
30	x	Fall
30		Fall
30		Spring, early summer, fall
30		Fall
30		Fall
30		Fall
30		Fall
30		Fall
Double-dormant (4)		Fall
Plant fresh seed in June		Early spring, fresh or fall
30		Fall
30		Fall
		Spring, early summer, fall

(continued)

TABLE 12.27. *(continued)*

Scientific name (4/1/2020)	Common name	Greenhouse propagation methods	
		Dry stratify	Moist stratify
Rosa carolina	Carolina Rose	No germ.	
Rosa setigera	Climbing Prairie Rose	No germ.	
Rudbeckia fulgida	Orange Coneflower	Fair germ.	x
Rudbeckia hirta	Blackeyed Susan	x	
Rudbeckia laciniata	Green-Headed Coneflower	x	
Rudbeckia subtomentosa	Sweet Blackeyed Susan	x	
Rudbeckia triloba	Browneyed Susan	x	
Ruellia humilis	Wild Petunia	Low germ.	x
Salvia azurea	Blue Sage	Low germ.	x
Senna hebecarpa	Wild Senna	Fair germ.	x
Silene regia	Royal Catchfly	Low germ.	x
Silphium integrifolium	Rosinweed	Fair germ.	x
Silphium laciniatum	Compassplant	Fair germ.	x
Silphium perfoliatum	Cup Plant	Fair germ.	x
Silphium terebinthinaceum	Prairie Dock	Fair germ.	x
Sisyrinchium albidum	White Blue-Eyed Grass	No germ.	x
Sisyrinchium angustifolium	Narrow-Leaved Blue-Eyed Grass	No germ.	x
Sisyrinchium campestre	Prairie Blue-Eyed Grass	No germ.	x
Solidago ptarmicoides	White Aster	Fair germ.	x
Solidago speciosa	Showy Goldenrod	Fair germ.	x
Spiraea alba	Meadowsweet	Low germ.	x
Spiraea tomentosa	Steeplebush	Low germ.	x
Symphyotrichum ericoides	Heath Aster	Low germ.	x
Symphyotrichum laeve	Smooth Aster	Fair germ.	x
Symphyotrichum novae-angliae	New England Aster	Fair germ.	x
Symphyotrichum oblongifolium	Aromatic Aster	Fair germ.	x
Symphyotrichum oolentangiense	Skyblue Aster	Fair germ.	x
Tephrosia virginiana	Goatsrue	Low germ.	x
Thalictrum dasycarpum	Tall Meadow-Rue	Low germ.	x
Tradescantia bracteata	Prairie Spiderwort	Low germ.	x
Tradescantia occidentalis	Western Spiderwort	Low germ.	x
Tradescantia ohiensis	Ohio Spiderwort	Low germ.	x
Verbena hastata	Blue Vervain	x	
Verbena stricta	Hoary Vervain	Fair germ.	x
Vernonia fasciculata	Ironweed	Fair germ.	x
Vernonia gigantea	Tall Ironweed	Fair germ.	x
Veronicastrum virginicum	Culver's Root	Fair germ.	x
Viola pedata	Birdfoot Violet	x	
Zizia aptera	Heartleaf Golden Alexanders	Low germ.	x
Zizia aurea	Golden Alexanders	Low germ.	x

Greenhouse propagation methods		Direct sowing in field
Days to MS	Seed scarification helpful	Optimal seeding times
Double-dormant (4)	x	Fall
Double-dormant (4)	x	Fall
30		Fall
		Spring, early summer, fall
		Spring, early summer, fall
		Spring, early summer, fall
		Spring, early summer, fall
30		Fall
30		Fall
10	x	Spring, early summer, fall
30		Fall
30		Spring, early summer, fall
30		Spring, early summer, fall
30		Spring, early summer, fall
30		Spring, early summer, fall
30		Fall
30		Fall
30		Fall
30		Fall
30		Early spring, fall
30		Fall
30		Fall
30		Fall
30		Early spring, fall
30		Early spring, fall
30		Early spring, fall
30		Early spring, fall
15	x	Fall
30		Fall
Plant fresh seed in July		Fall, fresh in summer
Plant fresh seed in July		Fall, fresh in summer
Plant fresh seed in July		Fall, fresh in summer
		Spring, early summer, fall
30		Early spring, fall
30		Fall
30		Fall
30		Fall
Plant fresh seed in June		Fresh or fall
30		Fall
30		Fall

(continued)

TABLE 12.27.   *(continued)*

Scientific name (4/1/2020)	Common name	Greenhouse propagation methods	
		Dry stratify	Moist stratify
Grasses and sedges			
Andropogon gerardii	Big Bluestem	x	
Andropogon virginicus	Broomsedge	x	
Bouteloua curtipendula	Sideoats Grama	x	
Bromus kalmii	Prairie Bromegrass	x	
Calamagrostis canadensis	Bluejoint Grass		x
Carex brevior	Shortbeak Sedge		x
Elymus canadensis	Canada Wildrye	x	
Elymus virginicus	Virginia Wildrye	x	
Eragrostis spectabilis	Purple Lovegrass	x	
Hesperostipa spartea	Needlegrass		x
Hierochloe hirta	Vanilla Sweetgrass		x
Koeleria macrantha	Junegrass	x	
Panicum virgatum	Switchgrass	x	
Schizachyrium scoparium	Little Bluestem	x	
Sorghastrum nutans	Indiangrass	x	
Spartina pectinata	Prairie Cordgrass		x
Sporobolus heterolepis	Prairie Dropseed	x	
Tridens flavus	Purpletop	x	
Tripsacum dactyloides	Eastern Gamagrass		x

Notes: **Dry stratification (DS)** — Seed is subjected to cold temperatures around freezing or below for a period of one month or longer (often over winter in an unheated building). Seed should be stored in rodent-proof containers. Most prairie grasses require only 30 days of dry stratification. Some flowers will germinate well with 30 days or more of dry stratification, but most show improved germination when moist stratified for the recommended number of days. Some flower seeds will not germinate at all without this moist, cold pretreatment. **Moist Stratification (MS)** — Seed is mixed with a slightly damp inert material, such as fresh sawdust, peat moss, clean sand, etc., and placed in a plastic bag in the refrigerator at 33–36°F for the recommended number of days. Label the bag with the species and date. Most prairie flowers exhibit significantly higher germination with moist stratification. Flats can be sown first and then placed in a refrigerator or walk-in cooler. Cover each sown flat with waxed paper, and cover the flat with plastic wrap over the waxed paper to retain moisture. **Scarification** — Some flowers and shrubs with hard seed coats require scratching to allow water to penetrate the seed and initiate the germination process. Place seed in a wooden box (1′×1′ works well for small quantities). Gently rub the seed with 120- to 150-grain sandpaper on a sanding block to lightly scratch the seed surface. Seed that is to be moist stratified should be scarified prior to stratification. **Double-Dormant** — Many members of the Lily family and Rose family have complex dormancies in which the seed must experience 2 consecutive winters in the soil to break dormancy. When seeded directly in the soil in fall, they will germinate 18 months later in the second spring. Double-dormant seeds can be induced to germinate quicker by subjecting them to 75 days of moist stratification immediately after harvest (September).

Greenhouse propagation methods		Direct sowing in field
Days to MS	Seed scarification helpful	Optimal seeding times
		Late spring, early summer
		Late spring, early summer
		Late spring, early summer
		Early spring
30		Fall
30		Fall
		Anytime
		Anytime
		Spring, fall
30		Fall
30		Fall
		Early spring, fall
		Late spring, early summer
		Late spring, early summer
		Late spring, early summer
120		Fall
		Fall, early spring
		Spring, fall
30		Fall

Remove the seeds from the refrigerator in late December and leave at room temperature for 60 days. Place the seed back in the refrigerator in February for another 75 days. Remove the seeds from the refrigerator in late April or early May to germinate. **Hot-Water Treatment** — Some seeds require exposure to heat from a prairie fire or forest fire to break seed dormancy, including members of the genus *Ceanothus* (New Jersey Tea). Scarify the seed with sandpaper first, then place it in a bowl. Pour boiling water over the seed and allow to cool. This mimics the effects of fire. Pour off the water, and moist stratify the seed as per directions for moist stratification. **Hemi-parasitic** — Hoary Puccoon requires a grass host plant to grow. It should be seeded into existing grass cover in late summer or fall. The seed will germinate in bare soil without grass, but seedlings will die within a few days if no grass host is present. To sow in the field, disturb existing grass cover by digging down into the sod and then pushing the seed 1/2 inch deep into the loosened soil and roots in late summer or fall, using freshly harvested seed. Greenhouse propagation has yet to be successful (as of 2020). **Rhizobium inoculum for legumes** — There is some controversy regarding the effectiveness of treating members of the Pea family (*Fabaceae*) with symbiotic Rhizobium inoculum to provide immediate Nitrogen-fixing ability in the seedlings. Some practitioners have observed little or no difference in growth rates between inoculated and un-inoculated plants. Availability of genus-specific native inocula has historically been sporadic. Most native seed companies no longer offer inoculum for sale to their customers.

TABLE 12.28. COMMON WEEDS IN PRAIRIE SEEDINGS

Scientific name (4/1/2020)	Common name	Notes
Annual grasses		
Avena fatua	Wild Oat	Adapted to a wide variety of soil types; not typically persistent
Bromus arvensis (*B. japonicus*)	Field Brome/Japanese Brome	Fall-germinating annual; can be persistent in seedings until the prairie grasses fill in (3 years)
Bromus secalinus	Rye Brome/Common Chess	Fall-germinating annual; can be persistent in seedings until the prairie grasses fill in (3 years)
Bromus tectorum	Downy Brome	Fall-germinating annual; can be persistent in seedings until the prairie grasses fill in (3 years)
Cenchrus longispinus	Sandbur	Restricted to dry, sandy soils; burs are very painful!
Digitaria ischaemum	Smooth Crabgrass	Ubiquitous warm-season annual; cannot be controlled with mowing; fall seeding reduces density
Digitaria sanguinalis	Crabgrass	Ubiquitous warm-season annual; cannot be controlled with mowing; fall seeding reduces density
Echinocloa crus-galli	Barnyard Grass	Prefers moist soil; sometimes used as an annual nurse crop in wetland seed mixes
Eleusine indica	Goosegrass	Widely adapted to varying soils and moisture regimes; not typically a long-term problem
Eragrostis cilianensis	Stinkgrass	Most common in sandy soils; not a serious problem
Hordeum pusillim	Little Barley	Native; more common southward, absent in Upper Midwest
Microstegium vimineum	Japanese Stiltgrass	More common eastward; extremely aggressive by seed; can persist for many years
Panicum capillare	Witchgrass	Most common in sandy soils; not a serious long-term problem
Panicum dichotomiflorum	Fall Panicum	Prefers rich mesic to moist soils; not persistent once the prairie develops
Panicum milliaceum	Wild Proso Millet	Escaped grain crop that can be a weed in new prairie seedings
Setaria faberi	Giant Foxtail	All foxtails can persist for 3 years or longer; control with mowing
Setaria pumila	Yellow Foxtail	All foxtails can persist for 3 years or longer; control with mowing
Setaria viridis	Green Foxtail	All foxtails can persist for 3 years or longer; control with mowing
Perennial grasses		
Invasive rhizomatous perennial grasses and nutsedges		
Bromus inermis	Smooth Brome	Cool season; highly invasive and hard to control with fire
Cynodon dactylon	Bermuda Grass	Warm season; more common in southern Midwest and southeast US
Cyperus esculentus	Yellow Nutsedge	Prefers rich, moist soils; control with halosulfuron-methyl herbicide (Sedgehammer)
Elymus repens	Quackgrass	Cool season; highly invasive by rhizomes; easy to control with midspring burning
Festuca arundinacea	Tall Fescue	Cool season; highly invasive, esp. in central Midwest
Phalaris arundinacea	Reed Canary Grass	Cool season; prefers rich, moist soil; grows on upland clay and silt loam
Phragmites australis	Common Reed	Usually restricted to wetlands; can be a problem in wet prairies
Sorghum halepense	Johnson Grass	Warm season; cannot be controlled with spring burning

TABLE 12.28. (*continued*)

Scientific name (4/1/2020)	Common name	Notes
Nonrhizomatous perennial grasses		
Agrostis gigantea	Redtop	Cool season; easy to control with midspring burning; prefers moist, clay soil
Dactylis glomerata	Orchardgrass	Cool season; easy to control with midspring burning
Festuca elatior	Meadow Fescue	Cool season; easy to control with midspring burning
Festuca rubra	Fine/Red Fescue	Cool season; easy to control with midspring burning
Hordeum jubatum	Foxtail Barley	Ornamental "weedy" native grass with pink seedheads; prefers damp, alkaline soils
Lolium perenne	Perennial Rye	Cool season; short-lived and easy to control with midspring burning
Paspalum cetaceum	Thin Paspalum	Native warm-season grass; typically not a problem
Phleum pratense	Timothy	Cool season; easy to control with midspring burning
Annual broadleaf weeds		
Abutilon theophrasti	Velvetleaf	Usually appears in 1st year only and disappears in prairie seedings
Amaranthus albus	Tumble Pigweed	Typically not a long-term problem; more common westward
Amaranthus blitoides	Prostrate Pigweed	Typically not a long-term problem
Amaranthus hybridus	Smooth Pigweed	Usually appears in 1st year only and disappears in prairie seedings; eastern Midwest
Amaranthus retroflexus	Pigweed	Usually appears in 1st year only and disappears in prairie seedings
Ambrosia artemisiifolia	Common Ragweed	Often persists in reduced form for years
Ambrosia trifida	Giant Ragweed	Grows in rich, often damp soils; must be controlled or it can take over everything!
Anthemis cotula	Mayweed	Germinates in fall and blooms the following spring and early summer
Barbarea vulgaris	Yellow Rocket	Germinates in fall and blooms in spring; control by cutting young seedstalks just after flowering
Brassica kaber	Wild Mustard	Germinates in fall and blooms in spring; control by cutting young seedstalks just after flowering
Brassica nigra	Black Mustard	Germinates in fall and blooms in spring; not a long-term problem
Capsella bursa-pastoris	Shepherd's Purse	Germinates in fall and blooms in spring; not a long-term problem
Chenopodium album	Lambsquarters	Usually appears in 1st year only and disappears in prairie seedings
Chenopodium hybridum	Maple Leaf Goosefoot	Usually appears in 1st year only and disappears in prairie seedings
Datura stramonium	Jimsonweed	Toxic and hallucinogenic; prefers rich soils
Echinocystis lobata	Wild Cucumber	Twining vine; easily controlled with mowing in 1st year; seldom a long-term problem
Erigeron canadensis	Horseweed/Marestail	Germinates in fall and blooms the following late summer
Erigeron strigosus	Daisy Fleabane	Early successional native annual or biennial; seldom a problem
Euphorbia maculata	Prostrate Spurge	Usually appears in 1st year only and disappears in prairie seedings
Euphorbia preslii	Spotted Spurge	Usually appears in 1st year only and disappears in prairie seedings
Galium aparine	Catchweed Bedstraw	Occurs in moist soils and woodland edges; seldom a long-term problem

(*continued*)

TABLE 12.28. *(continued)*

Scientific name (4/1/2020)	Common name	Notes
Hibiscus trionum	Flower of an Hour	Attractive whitish-yellow flowers with red centers; seldom a long-term problem
Kochia scoparia	Kochia	More common westward
Lactuca serriola	Prickly Wild Lettuce	Typically not a long-term problem
Lamium amplexicaule	Henbit	Prefers rich soil; seldom a long-term problem
Lepidium campestre	Field Pepperweed	Germinates in fall and blooms in spring; not a long-term problem
Lepidium densiflorum	Greenflower Pepperweed	Germinates in fall and blooms in spring; not a long-term problem
Lepidium virginicum	Virginia Pepperweed	Germinates in fall and blooms in spring; not a long-term problem
Malva neglecta	Common Mallow/Cheeses	Annual to biennial; typically not a long-term problem
Matricaria matricarioides	Pineappleweed	Typically not a long-term problem
Medicago lupulina	Black Medic	Can be biennial or short-lived perennial
Mollugo verticillata	Carpetweed	Sandy, dry soils; beneficial — low-growing mat helps retain soil moisture
Polygonum aviculare	Prostrate Knotweed	Beneficial — low-growing, spreading mat helps retain soil moisture
Polygonum convolvulus	Wild Buckwheat	Twining vine; resembles the noxious perennial Field Bindweed
Polygonum pensylvanicum	Pennsylvania Smartweed	Grows in moist soils
Portulaca oleracea	Purslane	Beneficial — low-growing, spreading mat helps retain soil moisture
Rapahnus raphanistrum	Wild Radish	Germinates in fall and blooms in spring; control by cutting young seedstalks just after flowering
Salsola kali	Russian Thistle	More common westward
Sida spinosa	Prickly Sida	Typically not a long-term problem
Silene antirrhina	Sleepy Catchfly	Northern Midwest; seldom a long-term problem
Silene noctiflora	Night Flowering Catchfly	Typically not a long-term problem
Sisymbrium altissimum	Tumble Mustard	Germinates in fall and blooms in spring; control by cutting young seedstalks just after flowering
Sisymbrium officinale	Hedge Mustard	Germinates in fall and blooms in spring; control by cutting young seedstalks just after flowering
Solanum ptycanthum	Black Nightshade	Typically not a long-term problem
Sonchus oleraceus	Annual Sowthistle	Typically not a long-term problem
Stelaria meadia	Common Chickweed	Typically not a long-term problem
Thlaspi arvense	Field Pennycress	Germinates in fall and blooms in spring; attractive seedheads; seldom a long-term problem
Veronica arvensis	Corn Speedwell	Small, unobtrusive plant that presents little or no competitive threat to prairie seedlings
Veronica peregrina	Purslane Speedwell	Small, unobtrusive plant that presents little or no competitive threat to prairie seedlings
Vicia angustifolia	Common Vetch	Often a fall-germinating annual; not a long-term problem
Xanthium strumarium	Cocklebur	Typically not a long-term problem

Biennial broadleaf weeds

Alliaria petiolata	Garlic Mustard	Common in shade, but can creep into sunny prairies at woodland edges
Arctium lappa	Greater Burdock	Wide-spreading basal leaves can shade out prairie seedlings; seldom a long-term problem

TABLE 12.28. (*continued*)

Scientific name (4/1/2020)	Common name	Notes
Arctium minus	Lesser Burdock	Wide-spreading basal leaves can shade out prairie seedlings; seldom a long-term problem
Carduus acanthoides	Plumeless Thistle	Early successional weed; seldom a long-term problem; more common westward
Carduus nutans	Musk or Nodding Thistle	Early successional weed; seldom a long-term problem
Centaurea maculosa	Spotted Knapweed	Highly invasive biennial to short-lived perennial; dry soils
Cirsium vulgare	Bull Thistle	Early successional weed; seldom a long-term problem
Conium maculatum	Poison Hemlock	Acutely toxic if ingested, often causing death; requires moist soils
Cynoglossum officinale	Houndtongue	Early successional weed; seldom a long-term problem
Daucus carota	Queen Anne's Lace	Seeds remain viable in the soil for many years, making it hard to eradicate
Dipsacus fullonum	Common Teasel	Biennial to short-lived perennial
Grindelia squarrosa	Gumweed	Native primarily to western half of US, adventive eastward; usually not a problem
Hesperis matronlis	Dame's Rocket	Resembles Phlox, but has 4 petals instead of the 5 petals of Phlox
Lactuca canadensis	Tall Lettuce	Early successional weed; seldom a long-term problem
Melilotus alba	White Sweet Clover	Aggressive by seed which remain viable in soil for decades; fire stimulates germination
Melilotus officinale	Yellow Sweet Clover	Aggressive by seed which remain viable in soil for decades; fire stimulates germination
Pastinaca sativa	Wild Parsnip	Sap is phototoxic; can cause serious burns
Tragopogon dubius	Meadow Goatsbeard	Attractive large dandelion-like seedheads; seldom a long-term problem
Verbascum blattaria	Moth Mullein	Small leaves and less problematic than Common Mullein; not a long-term problem
Verbascum thapsus	Common Mullein	Wide-spreading basal leaves can shade out prairie seedlings; not a long-term problem

Invasive rhizomatous perennial broadleaf weeds

Achillea millefolium	Common Yarrow	Native to North America and Old World; allelopathic — forms clones that exclude of other species
Apocynum androsaemifolium	Spreading Dogbane	Native; outstanding nectar source for butterflies; forms monocultural clones, excluding others
Apocynum cannabinum	Indian Hemp	Native; outstanding nectar source for butterflies; forms monocultural clones, excluding others
Artemisia vulgaris	Mugwort	Highly invasive — more common eastward
Campanula rapunculoides	Creeping Bellflower	Upper Midwest; prefers semishaded woodland edges
Cirsium arvense	Canada Thistle	Highly invasive; resistant to glyphosate herbicide
Convolvulus arvensis	Field Bindweed	Highly invasive; resistant to glyphosate herbicide
Euphorbia cyparissias	Cypress Spurge	Low-growing colonies can suppress prairie plants; outcompeted by tall prairie species

(*continued*)

TABLE 12.28. *(continued)*

Scientific name (4/1/2020)	Common name	Notes
Euphorbia esula	Leafy Spurge	Highly invasive, low-growing creeping weed; western states only
Euthamia graminifolia	Grass-Leaved Goldenrod	Highly invasive native; unaffected by fall or spring burning
Glechoma hederacea	Creeping Charlie	More problematic in savanna and semishady areas than in open prairies
Helianthus grosseserratus	Sawtooth Sunflower	Highly invasive native; sometimes included in wet prairie-seed mixes — bad idea!
Helianthus tuberosus	Jerusalem Artichoke	Highly invasive native; tubers are difficult to eradicate by digging
Hieracium aurantiacum	Orange Hawkweed	Spreads on dry, sandy soils to form clones; MI, WI, MN
Linaria vulgaris	Butter and Eggs	Attractive yellow and orange flowers; low-growing and seldom a long-term problem
Polygonum cuspidatum	Japanese Knotweed	Forms large, tall colonies, smothering everything
Rumex acetosella	Sheep Sorrel	Occurs mostly on dry, acidic sandy and rocky soils; seldom a serious long-term threat
Securigera varia	Crown Vetch	Highly invasive; resistant to glyphosate herbicide
Solanum caroliniense	Carolina Horsenettle	Deep-rooted and rhizomatous; resistant to glyphosate herbicide
Solidago canadensis	Canada Goldenrod	Highly invasive native; unaffected by fall or spring burning.
Sonchus arvensis	Perennial Sowthistle	Invasive by both seeds and rhizomes
Tanacetum vulgare	Common Tansy	Common above 45 degrees north
Trifolium repens	White Clover	Can form dense carpet that outcompetes native seedlings; control with nitrogen fertilizer application
Urtica dioica	Stinging Nettle	Native; larval host plant for Red Admiral Butterfly (*Vanessa atalanta*)

Nonrhizomatous perennial broadleaf weeds

Berteroa incana	Hoary Alyssum	Early successional annual or perennial; often persists for 3–5 years, fading as prairie fills in
Cerastium vulgatum	Mouse-Eared Chickweed	Low-growing, early successional perennial; seldom a long-term problem
Cichorium intybus	Chicory	Attractive pastel blue flowers; common in compacted clay soils and roadsides
Dipsacus sylvestris	Teasel	Can form large colonies and suppress prairie species
Hypericum perforatum	St. Johnswort	Early successional plant; seldom a problem; restricted to dry, sandy soils
Leonurus cardiaca	Motherwort	Early successional plant; seldom a long-term problem
Lespedeza cuneata	Sericea Lespedeza	Can be a serious long-term problem; seed remain viable in soil for decades
Leucanthemum vulgare	Oxeye Daisy	Early successional plant; seldom a long-term problem
Lotus corniculatus	Birdsfoot Trefoil	Nonnative legume; can be very aggressive by seed
Lychnis alba	White Campion	Early successional plant; seldom a long-term problem
Mentha arvensis	Field Mint	Prefers rich moist soils; usually not a long-term threat in prairie seedings
Nepeta cataria	Catnip	Early successional plant; seldom a long-term problem
Oxalis stricta	Yellow Wood Sorrel	Early successional plant; seldom a long-term problem

TABLE 12.28. (*continued*)

Scientific name (4/1/2020)	Common name	Notes
Phytolacca americana	Pokeweed	Native to southeastern US, moving northward into Upper Midwest with climate change
Plantago lanceolata	Buckhorn Plantain	Narrow, upright leaves do not smother prairie seedlings; seldom a long-term problem
Plantago major	Broadleaf Plantain	Low-spreading clump growth habit can smother seedlings in first 2 years after seeding
Plantago rugelii	Blackseed Plantain	Low-spreading clump growth habit can smother seedlings in first 2 years after seeding
Potentilla argentea	Cinquefoil	Early successional plant; seldom a long-term problem
Potentilla norvegica	Rough Cinquefoil	Biennial or short-lived perennial; early successional plant; seldom a long-term problem
Potentilla recta	Sulphur Cinquefoil	Early successional plant; seldom a long-term problem
Prunella vulgaris	Heal-All	Native to North America and Old World; early successional perennial, seldom a long-term problem
Rumex crispus	Curly Dock	Prefers rich moist soils; usually not a long-term threat in prairie seedings
Saponaria officinalis	Soapwort	Can form large colonies and suppress prairie species over the long-term
Silene cucurbatus	Bladder Campion	Attractive white, balloon-like flowers; not usually invasive; MN, WI, upper MI
Solanum dulcamara	Bitter Nightshade	Bright red berries; early successional perennial, seldom a long-term problem
Taraxacum officinale	Dandelion	Early successional plant; seldom a long-term problem
Trifolium pratense	Red Clover	Short-lived; can form dense stands that shade out seedlings in 1st 3 years of seeding

Trees

Acer negundo	Boxelder	Native; highly invasive by seed
Acer platanoides	Norway Maple	Nonnative; invasive by seed, esp. eastward
Ailanthus altissima	Tree of Heaven	Nonnative; can be aggressive be seed, esp. east and south
Juglans nigra	Black Walnut	Native; seeds commonly planted by squirrels
Morus alba	White Mulberry	Nonnative; invasive by seed
Populus deltoides	Cottonwood	Native; highly invasive by seed
Populus tremuloides	Trembling (Quaking) Aspen	Native; aggressive by rhizomes
Robinia pseudoacacia	Black Locust	Native; aggressive by rhizomes
Ulmus pumila	Siberian Elm	Nonnative; more common westward; invasive by seed

Shrubs

Berberis thunbergii	Japanese Barberry	Invasive by seed
Elaeagnus angustifolia	Russian Olive	Invasive by seed
Elaeagnus umbellata	Autumn Olive	Invasive by seed
Euonymus alatus	Burning Bush	Invasive by seed
Lonicera mackii	Amur Honeysuckle	Invasive by seed
Lonicera tatarica	Tatarian Honeysuckle	Invasive by seed

(*continued*)

TABLE 12.28. *(continued)*

Scientific name (4/1/2020)	Common name	Notes
Rhamnus catharticus	Common Buckthorn	Invasive by seed
Rhus glabra	Smooth Sumac	Native; invasive by rhizomes
Rhus typhina	Staghorn Sumac	Native; invasive by rhizomes
Rosa multiflora	Multiflora Rose	Invasive by seed
Rubus spp.	Blackberry, Raspberry, Dewberry	Native; invasive by rhizomes and seeds
Zanthoxylum americanum	Prickly Ash	Native; invasive by rhizomes and seeds
Vines		
Ampelopsis brevipedunculata	Porcelain Berry/Amur Peppervine	Invasive by seed; more common east and south
Celastrus orbiculatus	Oriental Bittersweet	Invasive by seed
Hedera helix	English Ivy	Invasive by rhizomes
Lonicera japonica	Japanese Honeysuckle	Invasive by seed; more common east and south
Parthenocissus quinquefolia	Virginia Creeper	Native; invasive by rhizomes
Rhus radicans	Poison Ivy	Native; invasive by rhizomes
Vitis riparia	Wild Grape	Native; invasive by seed
Non-seed-producing plants		
Equisetum arvense	Common Horsetail	Native; resistant to most herbicides; Sedgehammer (halosulfuron-methyl) is fairly effective
Equisetum hyemale	Scouring Rush	Nonnative; resistant to most herbicides; Sedgehammer (halosulfuron-methyl) is fairly effective
Equisetum laevigatum	Smooth Horsetail	Native; resistant to most herbicides; Sedgehammer (halosulfuron-methyl) is fairly effective
Equisetum pratense	Meadow Horsetail	Nonnative; resistant to most herbicides; Sedgehammer (halosulfuron-methyl) is fairly effective

TABLE 12.29. INSTALLATION COST COMPARISON OF PRAIRIE VERSUS LAWN

Equipment, labor, and materials	Turfgrass cost per 1/2 acre		Prairie cost per 1/2 acre	
	Seeded turf by contractor	Seeded turf by homeowner	Seeded prairie by contractor	Seeded prairie by homeowner
Preseeding glyphosate herbiciding (1 application for turf, 3 for prairie)	$150		$450	
Preseeding herbicide only cost by homeowner @ $25 per application		$25		$75
Single-use sprayer rental by homeowner for turf @ $50 each		$50		
3 sprayer rentals by homeowner for prairie @ $50 each				$150
Tiller rental by homeowner @ $150/day		$150		$150
Pull-behind broadcast seed and fertilizer spreader rental by homeowner		$50		$50
Roller rental by homeowner for firming seed into soil		$50		$50
Starter fertilizer	$75	$75	NA	NA
Seed	$300	$300	$850	$850
Landscaper soil tillage, seeding labor, and equipment usage charges	$1,200		$1,600	
Straw mulch and contractor application labor	$750		$750	
Cost for 30 50 lb. bales of clean straw by homeowner @ $5 each		$150		$150
Total	**$2,475**	**$850**	**$3,650**	**$1,475**
Cost per square yard, installed (2420 sq. yd. = 1/2 acre)	**$1.02 per sq. yd.**	**$0.35 per sq. yd.**	**$1.51 per sq. yd.**	**$0.61 per sq. yd.**
			47.47% more than contractor turf cost	*73.53% more than homeowner turf cost*

Notes: Cost assumptions made in this table include the following: (1) site preparation is done with glyphosate herbicide, not with repeated tillage, smothering, or seeded smother crops such as buckwheat; (2) prairie does not require stater fertilizer; (3) no topsoil or other soil amendments are required for either lawn or prairie, except for lawn starter fertilizer; (4) homeowner rents sprayer, tiller, broadcast seeder, and roller rather than purchasing them; (5) no irrigation system is included for lawn in these estimates. Irrigation installation can increase the cost for the lawn by 50% to 100%; (6) no costs for landscape design is included in these estimates; (7) costs will vary by region based on local labor and general business overhead costs; and (8) estimates are in 2020 dollars.

TABLE 12.30. MAINTENANCE COST COMPARISON OF PRAIRIE VERSUS LAWN

Contractor maintained per 1/2 acre	Contractor year 1	Contractor year 2	Contractor year 3	Contractor year 4
Lawn maintenance				
Install	$2,475			
Four-step fertilization program	$600	$618	$637	$656
Aerating and overseeding		$300		$318
Contractor mowing (24/yr. @ $80 each)	$1,920	$1,978	$2,037	$2,098
Homeowner mowing (24/yr.; see below)				
Municipal water for irrigation	$1,200	$1,236	$1,273	$1,311
Total annual costs	$6,195	$4,132	$3,947	$4,383
Cumulative costs, year over year	$6,195	$10,327	$14,273	$18,656
Prairie meadow maintenance				
Using burning management				
Installation using glyphosate prep.	$3,650			
Municipal water, first 2 months	$250			
First-year mowings, 4 mowings	$600			
Second-year mowings, 3 mowings		$450		
Contractor burning every other year			$1,500	
Homeowner burning every other year				
Invasive plant control			$500	$515
Total annual costs	$4,500	$450	$2,000	$515
Cumulative costs, year over year	$4,500	$4,950	$6,950	$7,465
Cumulative savings, prairie over lawn	$1,695	$5,377	$7,323	$11,191
Percent savings, prairie over lawn	27%	52%	51%	60%
Using mowing management				
Installation using glyphosate prep.	$3,650			
Municipal water, first 2 months	$250			
First-year mowings, 4 mowings	$600			
Second-year mowings, 3 mowings		$450		
Annual fall or spring mowing	$200	$206	$212	$219
Invasive plant control			$500	$515
Total annual costs	$4,700	$656	$712	$734
Cumulative costs, year over year	$4,700	$5,356	$6,068	$6,802
Cumulative savings, prairie over lawn	$1,495	$4,971	$8,205	$11,855
Percent savings, prairie over lawn	24%	48%	57%	64%

Contractor year 5	Contractor year 6	Contractor year 7	Contractor year 8	Contractor year 9	Contractor year 10
$675	$696	$716	$738	$760	$783
	$338		$358		$380
$2,161	$2,226	$2,293	$2,361	$2,432	$2,505
$1,351	$1,391	$1,433	$1,476	$1,520	$1,566
$4,187	$4,650	$4,442	$4,933	$4,712	$5,234
$22,843	$27,493	$31,935	$36,869	$41,581	**$46,815**
$1,591		$1,688		$1,791	
$530	$546	$563	$580	$597	$615
$2,122	$546	$2,251	$580	$2,388	$615
$9,587	$10,133	$12,384	$12,964	$15,352	**$15,967**
$13,256	$17,360	$19,551	$23,905	$26,229	**$30,848**
58%	63%	61%	65%	63%	66%
				Over $30,000 saved over 10 years	
$225	$232	$239	$246	$253	$261
$530	$546	$563	$580	$597	$615
$756	$778	$802	$826	$850	$876
$7,557	$8,335	$9,137	$9,963	$10,813	**$11,689**
$15,286	$19,158	$22,798	$26,906	$30,768	**$35,126**
67%	70%	71%	73%	74%	75%

(continued)

TABLE 12.30. *(continued)*

Homeowner maintained per 1/2 acre	Homeowner year 1	Homeowner year 2	Homeowner year 3	Homeowner year 4
Lawn maintenance				
Install	$850			
Four-step fertilization program	$200	$206	$212	$219
Aerating and overseeding		$200		$212
Contractor mowing (24/yr. @ $80 each)				
Homeowner mowing (24/yr.)	$520	$536	$552	$568
Municipal water for irrigation	$1,200	$1,236	$1,273	$1,311
Total annual costs	$2,770	$2,178	$2,037	$2,310
Cumulative costs, year over year	$2,770	$4,948	$6,985	$9,295
Prairie meadow maintenance				
Using burning management				
Installation using glyphosate prep.	$1,475			
Municipal water, first 2 months	$250			
First-year mowings, 4 mowings	$300			
Second-year mowings, 3 mowings		$225		
Contractor burning every other year				
Homeowner burning every other year			$250	
Invasive plant control			$200	$206
Total annual costs	$2,025	$225	$450	$206
Cumulative costs, year over year	$2,025	$2,250	$2,700	$2,906
Cumulative savings, prairie over lawn	$745	$2,698	$4,285	$6,389
Percent savings, prairie over lawn	27%	55%	61%	69%
Using mowing management				
Installation using glyphosate prep.	$1,475			
Municipal water, first 2 months	$250			
First-year mowings, 4 mowings	$300			
Second-year mowings, 3 mowings		$225		
Annual fall or spring mowing	$100	$103	$106	$109
Invasive plant control			$200	$206
Total annual costs	$2,125	$328	$306	$315
Cumulative costs, year over year	$2,125	$2,453	$2,759	$3,074
Cumulative savings, prairie over lawn	$645	$2,495	$4,225	$6,220
Percent savings, prairie over lawn	23%	50%	60%	67%

Notes: Cost assumptions made in this table include the following: (1) Costs will vary around the country due to contractor availability, rates, materials, and travel time to site. (2) Mowing lawn weekly for 24 weeks per year. (3) Contractor lawn mowing charges based on $80 per mowing in year 1 (2020). (4) Homeowner lawn mowing cost based on $400 per year for ten year amortized mower cost and annual maintenance, plus $5 per week for gasoline in year 1 (2020). (5) Homeowner rents lawn aerator. (6) Contractor prairie mowing at $150 per mowing. (7) Homeowner prairie mowing at $75 per mowing. (8) Lawn water usage based on watering 18 times over the growing season at 1 inch per, using water rate of $6.21/ccf (1 ccf = 100 cubic feet). (9) Prairie meadow water

Homeowner year 5	Homeowner year 6	Homeowner year 7	Homeowner year 8	Homeowner year 9	Homeowner year 10
$225	$232	$239	$246	$253	$261
	$225		$239		$253
$585	$603	$621	$640	$659	$678
$1,351	$1,391	$1,433	$1,476	$1,520	$1,566
$2,161	$2,451	$2,293	$2,600	$2,432	$2,759
$11,456	$13,907	$16,199	$18,799	$21,232	$23,990
$250		$250		$250	
$212	$219	$225	$232	$239	$246
$462	$219	$475	$232	$489	$246
$3,368	$3,587	$4,062	$4,294	$4,782	$5,028
$8,088	$10,320	$12,137	$14,506	$16,449	$18,962
71%	74%	75%	77%	77%	79%
				Nearly $19,000 saved over 10 years	
$113	$116	$119	$123	$127	$130
$212	$219	$225	$232	$239	$246
$325	$334	$345	$355	$365	$376
$3,399	$3,734	$4,078	$4,433	$4,798	$5,175
$8,057	$10,173	$12,121	$14,366	$16,433	$18,815
70%	73%	75%	76%	77%	78%
				Over $18,000 saved over 10 years	

usage based on daily morning watering for 15–20 minutes in the first 2 months for a spring seeding. (10) Watering may not even be necessary with fall-dormant seedings or late-winter frost seedings. No watering is required after the first growing season. (11) Burning management for prairie meadow in areas where permitted. Assumes homeowner hires professionals. (12) Burning cost for homeowner is based on initial cost of $1000 for backpack water sprayer, drip torch, flappers, and Nomex fire-retardant suit (four burns in ten years). (13) Annual cost increases of 3% per year is assumed for all variable costs such as materials and labor.

Notes

Chapter 2

1. J. T. Curtis and Mike Partch, "Effects of Fire on the Competition between Bluegrass and Certain Prairie Plants," *American Midland Naturalist* 39 (1948): 437–43.

Chapter 4

1. R. A. Relyea, "The Impact of Insecticides and Herbicides on the Biodiversity and Productivity of Aquatic Communities," *Ecological Applications* 15, no. 2 (2005): 618–27.

Chapter 7

1. Neil Diboll, "Mowing as an Alternative to Spring Burning for Control of Cool-Season Exotic Grasses in Prairie Grass Plantings," in *Proceedings of the Ninth North American Prairie Conference, Held July 29 to August 1, 1984, Moorhead, Minnesota* (Fargo: Tri-College University Center for Environmental Studies, North Dakota State University, 1986), 204–9.

Chapter 9

1. For more information, see G. Bryant, *Propagation Handbook* (Mechanicsburg, PA: Stackpole Books, 1995); Ken Druse, *Making More Plants: The Science, Art and Joy of Propagation* (New York: Stewart, Tabori, and Chang, 2012); H. T. Hartmann, D. E. Kester, F. T. Davies, and R. L. Geneve, *Plant Propagation, Principles and Practices*, 6th ed. (Upper Saddle River, NJ: Prentice Hall, 1996) P. D. A. McMillan Browse, *Plant Propagation* (New York: Simon and Schuster, 1978); and A. Toogood, *Plant Propagation Made Easy* (Portland, OR: Timber Press, 1993).

Chapter 10

1. J. Owen, *Garden Life* (London: Chatto and Windus, 1983).

Index

Note: the letter *f* following a page number denotes a figure, and the letter *t* denotes a table.

White Blue-Eyed Grass, 52–53, 55, 57
White False Indigo, 88, 198–99, 200–201, 367t, 420
White Heath Aster, 162–63. *See also* Heath Aster
White Meadowsweet, 274–75. *See also* Meadowsweet
White Prairie Clover, 204–5, 206–7, 367t
White Turtlehead, 111, 235, 241, 278–79
Wholeleaf Rosinweed, 152–53. *See also* Rosinweed
Wild Crocus, 260–61. *See also* Pasque Flower
Wild Foxglove, 280–81
Wild Iris, 50–51
wildlife: attracted, 2, 488–95t; in ecosystem/food web, 2. *See also specific wildlife*
Wild Lupine, 214–15, 405
Wild Petunia, 68–69, 405
Wild Quinine, 98, 138–39, 157, 405
Wildryes, 312–15, 321, 342, 367t, 387, 392t, 393t
Wild Senna, 216–17, 218
Wilson, E. O., on insects, 410
Windflower, 260–61. *See also* Pasque Flower
Winter Wheat straw: mulch, in prairie gardens, 31, 31f; mulch, in prairie meadows, 355f; and propagating seeds, as protection over winter, 400
wolverines, in ecosystem, 434, 438
wolves, in ecosystem, 438–39, 438f
woodchucks (groundhogs), in ecosystem, 433
woodlands. *See* forests and woodlands
woody plants: burning, 372; for insects and birds, in ecosystem, 372, 430–31; mowing, 372, 380; in prairie meadows, controlling, 367–69; in prairie meadows, invasive seeds dropped by birds, 372; in prairie origins, 6

Xerothermic period, 9–10

Yanny, Michael, v–vi
Yarrow flowers, 419f
Yellow Coneflower, 15, 37f, 140–41, 147, 149

Zizia aptera (Heartleaf Golden Alexanders, Meadow Zizia), 72–73, 74–75
Zizia aurea (Golden Alexanders, Golden Zizia), 32f, 73, 74–75, 385